T0215890

Lecture Notes in Computer Science　9589

Commenced Publication in 1973
Founding and Former Series Editors:
Gerhard Goos, Juris Hartmanis, and Jan van Leeuwen

More information about this series at http://www.springer.com/series/7410

Dongdai Lin · XiaoFeng Wang
Moti Yung (Eds.)

Information Security and Cryptology

11th International Conference, Inscrypt 2015
Beijing, China, November 1–3, 2015
Revised Selected Papers

 Springer

Editors
Dongdai Lin
Chinese Academy of Sciences
SKLOIS Institute of Information
  Engineering
Beijing
China

Moti Yung
Computer Science Department
Columbia University
New York, NY
USA

XiaoFeng Wang
School of Informatics and Computing
Indiana University at Bloomington
Bloomington, IN
USA

ISSN 0302-9743   ISSN 1611-3349 (electronic)
Lecture Notes in Computer Science
ISBN 978-3-319-38897-7   ISBN 978-3-319-38898-4 (eBook)
DOI 10.1007/978-3-319-38898-4

Library of Congress Control Number: 2016938674

LNCS Sublibrary: SL4 – Security and Cryptology

Printed on acid-free paper

This Springer imprint is published by Springer Nature
The registered company is Springer International Publishing AG Switzerland

Preface

This volume contains the papers presented at Inscrypt 2015: The 11th China International Conference on Information Security and Cryptology, held during November 1–3, 2015, in Beijing, China. Inscrypt is a well-recognized international forum for security researchers and cryptographers to exchange ideas and present their work, and is held every year in China.

The conference received 79 submissions. Each submission was reviewed by two to four Program Committee members. The Program Committee, after some deliberation, decided to accept 27 papers. The overall acceptance rate is, therefore, 34.17 %.

Inscrypt 2015 was held in cooperation with the International Association of Cryptologic Research (IACR), and was co-organized by the State Key Laboratory of Information Security (SKLOIS) of the Chinese Academy of Sciences (CAS), and the Chinese Association for Cryptologic Research (CACR). We note that the conference could not have been a success without the support of these organizations, and we sincerely thank them for their continued assistance and help.

We would also like to thank the authors who submitted their papers to Inscrypt 2015, and the conference attendees for their interest and support. We thank the Organizing Committee for their time and efforts dedicated to arranging the conference. This allowed us to focus on selecting papers and on dealing with the scientific program. We thank the Program Committee members and the external reviewers for their hard work in reviewing the submissions; the conference would not have been possible without their expert reviews. Finally, we thank the EasyChair system and its operators for making the entire process of the conference convenient.

November 2015

Dongdai Lin
XiaoFeng Wang
Moti Yung

Inscrypt 2015

11th China International Conference on Information Security and Cryptology

Beijing, China
November 1–3, 2015

Sponsored and organized by

State Key Laboratory of Information Security
(Chinese Academy of Sciences)
Chinese Association for Cryptologic Research

in cooperation with

International Association for Cryptologic Research

Conference Chair

Dan Meng   Institute of Information Engineering, CAS, China
Jianying Zhou   Institute for Infocomm Research, Singapore

Program Co-chairs

Dongdai Lin   SKLOIS, Institute of Information Engineering,
                         CAS, China
XiaoFeng Wang   Indiana University at Bloomington, USA
Moti Yung   Google Inc. and Columbia University, USA

Local Organizing Committee Co-chairs

Chuankun Wu   SKLOIS, Institute of Information Engineering,
                         CAS, China
Yanping Yu   Chinese Association for Cryptologic Research, China

Publicity Chair

Kai Chen   SKLOIS, Institute of Information Engineering,
                         CAS, China

Technical Program Committee

Atsuko Miyaji	Japan Advanced Institute of Science and Technology, Japan
Bertram Poettering	Ruhr University Bochum, Germany
Chun-I Fan	National Sun Yat-sen University, Taiwan
Cliff Zou	University of Central Florida, USA
Cong Wang	City University of Hong Kong, SAR China
Cunsheng Ding	Hong Kong University of Science and Technology, SAR China
Danfeng Yao	Virginia Tech, USA
Dawu Gu	Shanghai Jiao Tong University, China
Debin Gao	Singapore Management University, Singapore
Di Ma	University of Michigan-Dearborn, USA
Elisa Bertino	Purdue University, USA
Erman Ayday	Bilkent University, Turkey
Fangguo Zhang	Sun Yat-sen University, China
Feng Hao	Newcastle University, UK
Florian Mendel	Graz University of Technology, Austria
Giovanni Russello	The University of Auckland, New Zealand
Giuseppe Persiano	Università di Salerno, Italy
Huaqun Guo	Institute for Infocomm Research, Singapore
Ioana Boureanu	Akamai Technologies Limited, USA
Jian Guo	Nanyang Technological University, Singapore
Jian Weng	Jinan University, China
Jintai Ding	University of Cincinnati, USA
Josef Pieprzyk	Queensland University of Technology, Australia
Kai Chen	Institute of Information Engineering, CAS, China
Kefei Chen	Hangzhou Normal University, China
Kehuan Zhang	Chinese University of Hong Kong, SAR China
Laszlo Csirmaz	Central European University, Hungary
Lei Hu	Institute of Information Engineering, CAS, China
Liqun Chen	Hewlett-Packard Laboratories, UK
Longjiang Qu	National University of Defense Technology, China
Maria Naya-Plasencia	Inria, France
Matt Henricksen	Institute For Infocomm Research, Singapore
Mehmet Sabir Kiraz	Tubitak Bilgem, Turkey
Meiqin Wang	Shandong University, China
Min Yang	Fudan University, China
Nicolas Courtois	University College London, UK
Ninghui Li	Purdue University, USA
Peng Liu	The Pennsylvania State University, USA
Seungwon Shin	KAIST, Korea
Shaohua Tang	South China University of Technology, China
Sherman S.M. Chow	Chinese University of Hong Kong, SAR China
Shouhuai Xu	University of Texas at San Antonio, USA

Contents

Web and Application Security

Cloud Security

Key Management and Public Key Encryption

Zero Knowledge and Secure Computations

Software and Mobile Security

Hash Function

Biclique Cryptanalysis of Full Round AES-128 Based Hashing Modes

Donghoon Chang, Mohona Ghosh$^{(\boxtimes)}$, and Somitra Kumar Sanadhya

Indraprastha Institute of Information Technology, Delhi (IIIT-D), New Delhi, India
{donghoon,mohonag,somitra}@iiitd.ac.in

Abstract. In this work, we revisit the security analysis of hashing modes instantiated with AES-128. We use biclique cryptanalysis as the basis for our evaluation. In Asiacrypt'11, Bogdanov et al. had proposed biclique technique for key recovery attacks on full AES-128. Further, they had shown application of this technique to find preimage for compression function instantiated with AES-128 with a complexity of $2^{125.56}$. However, this preimage attack on compression function cannot be directly converted to preimage attack on hash function. This is due to the fact that the initialization vector (IV) is a publically known constant in the hash function settings and the attacker is not allowed to change it, whereas the compression function attack using bicliques introduced differences in the chaining variable. We extend the application of biclique technique to the domain of hash functions and demonstrate second preimage attack on all 12 PGV modes.

The complexities of finding second preimages in our analysis differ based on the PGV construction chosen - the lowest being $2^{126.3}$ and the highest requiring $2^{126.6}$ compression function calls. We implement C programs to find the best biclique trails (that guarantee the lowest time complexity possible) and calculate the above mentioned values accordingly. Our security analysis requires only 2 message blocks and works on full 10 rounds of AES-128 for all 12 PGV modes. This improves upon the previous best result on AES-128 based hash functions by Sasaki at FSE'11 where the maximum number of rounds attacked is 7. Though our results do not significantly decrease the attack complexity factor as compared to brute force but they highlight the actual security margin provided by these constructions against second preimage attack.

Keywords: AES · Block ciphers · Hash functions · Cryptanalysis · Biclique · Second preimage attack

1 Introduction

Block ciphers have been favored as cryptographic primitives for constructing hash functions for a long time. In [17], Preneel et al. proposed 64 basic ways to construct a n-bit compression function from a n-bit block cipher (under a n-bit key). Black et al. [5] analyzed the security of such constructions and showed

© Springer International Publishing Switzerland 2016
D. Lin et al. (Eds.): Inscrypt 2015, LNCS 9589, pp. 3–21, 2016.
DOI: 10.1007/978-3-319-38898-4_1

12 of them to be provably secure. These modes are commonly termed as PGV hash modes. The three most popularly used modes are Davies-Meyer (DM), Matyas-Meyer-Oseas (MMO) and Miyaguchi-Preneel (MP) modes.

AES (Advanced Encryption Standard), standardized by the US NIST in October 2000 and widely accepted thereafter has been considered a suitable candidate for block cipher based hash functions in the cryptographic community. ISO standardized Whirlpool [3] is a popular example of the same. Infact, in the recently concluded SHA-3 competition also, several AES based hash functions were submitted, e.g., LANE [11], ECHO [4], Grøstl [9] etc. A significant progress has been made in the field of block cipher based hash function security. Spearheaded by rebound attacks alongwith other cryptanalytic techniques, several AES as well as other block cipher based dedicated hash functions have been reviewed and cryptanalyzed [12,14–16,18,19,21]. But all of the analysis that has been done has been performed on round-reduced versions of block ciphers. Specifically, if we refer to the previous best result on AES-128 based hash modes performed by Sasaki [18], the maximum number of rounds attacked is 7.

The reason behind this restriction was the fact that AES-128 itself was resistant to full 10 rounds attack for a considerable period of time since its advent. Until few years ago, there was no single key model attack known which could break full AES-128 better than brute force. In Asiacrypt'11, Bogdanov et al. [7] proposed a novel idea called biclique attack which allowed an attacker to recover the AES secret key 3–5 times faster than exhaustive search. Subsequently, this technique was applied to break many other block ciphers such as PRESENT [1], ARIA [22], HIGHT [10] etc. As block cipher and block cipher based hash function security are inter-related, it is imperative to analyse the hash function security against biclique technique.

Biclique cryptanalysis is a variant of meet-in-the-middle attack, first introduced by Khovratovich et al. in [13] for preimage attacks on hash functions Skein and SHA-2. The concept was taken over by Bogdanov et al. to successfully cryptanalyze full rounds of all AES variants. The biclique attack results on AES in [7] were further improved in [6,20]. Bogdanov et al. in [7] also showed conversion of biclique key recovery attack on AES-128 to the corresponding preimage attack on AES-128 instantiated compression function. The current best complexity of this attack as reported in [6] is $2^{125.56}$.

1.1 Motivation

The above biclique based preimage attack on AES-128 instantiated compression function cannot be converted to preimage attack on the corresponding hash function (and hence second preimage attack as discussed in Sect. 5 later). This is due to the fact that in the preimage attack on compression function shown in [6,7], the attacker needs to modify the chaining variable (CV) value and the message input to obtain the desired preimage. However, in hash function settings, the initialization vector (IV) is a publically known constant which cannot be altered by the attacker. Hence, the biclique trails used in the preimage attack on

AES-128 based compression function in [6, 7] cannot be adopted to find preimage for the corresponding AES-128 based hash function. This can be explained as discussed below.

Let us consider Matyas-Meyer-Oseas (MMO) mode and Davies-Meyer (DM) mode based compression functions as shown in Fig. 1(a) and (b). In case of MMO mode, the chaining variable acts as the key input to the underlying block cipher AES (as shown in Fig. 1(a)). If the chaining variable is used as the IV (in hash function settings) then it is fixed and cannot be modified. This means that the value of the key input to the block cipher should not change. However, the type of biclique trails used in [7] (as shown in Fig. 2) for compression function introduce a change both in the key input as well as all the intermediate states including the plaintext input ensuring that the final chaining variable so obtained after the attack will not be the desired IV. Hence, the kind of biclique trails we are interested in should only affect the intermediate states (an example of which is given in Fig. 3) and not the key input.

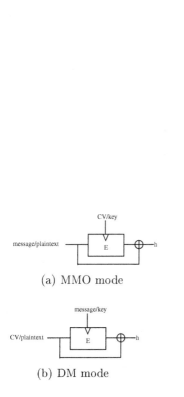

(a) MMO mode

(b) DM mode

Fig. 1. Compression function in MMO and DM mode respectively.

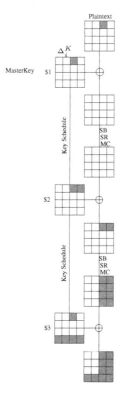

Fig. 2. An example of the trail used in [7] for preimage attack on AES-128 instantiated compression function.

Similarly, in the DM mode, the chaining variable acts as the plaintext input to the underlying block cipher (as shown in Fig. 1(b)). Therefore, if the chaining variable under consideration is the *IV* then the chosen biclique trails should not inject any difference in the plaintext input of the block cipher (an example of the same is shown in Fig. 4). Again, the biclique trails adapted for preimage attack on AES-128 instantiated compression function do not satisfy this condition (as seen in Fig. 2).

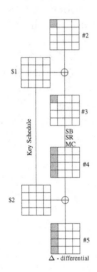

Fig. 3. An example of the desired trails that will work for attacking MMO based hash function. It is to be noted only the plaintext input and subsequent intermediate states are affected in the trail considered whereas the key input is a fixed constant.

Fig. 4. An example of the desired trail that will work for attacking DM based hash function. It is to be noted here that the plaintext input is not affected by the differential trail so chosen and is a fixed constant.

The examples discussed above warrant searching of new biclique trails which can be used to launch second preimage attack on AES-128 based hash functions. Moreover, searching these trails manually may not give the best results as demonstrated in [2,6]. Hence, automated search process is required. In this work, we implemented our restrictions in C programs to enumerate the best biclique trails which guarantee the lowest possible attack complexities. We then apply biclique technique to evaluate the security of AES-128 based hash functions against second preimage attack.

1.2   Our Contributions

The contributions of this paper are as follows:

– We re-evaluate the offered security of full 10 rounds AES-128 based hash functions against second preimage attack. The previous best result could only work on 7 rounds.

- Our analysis works on all 12 PGV modes of the hash function constructions.
- The complexities of the biclique based analysis differ depending upon the PGV construction chosen. For MP and MMO mode it is $2^{126.3}$ whereas for DM mode it is $2^{126.67}$.
- We propose new biclique trails to achieve the above results.
- All the trails have been obtained by implementing C programs which ensure that they yield the best attacks (lowest possible time complexity).

The results of our security evaluation against second preimage attack on all 12 PGV based modes are given in Table 1.

Table 1. Summary of the results obtained. In this table, we assume hash function to be instantiated with block cipher E, h is the chaining variable, m is the message input and $h \oplus m = w$.

S.No	Hash Function Modes	Second Preimage Complexity	Succ. Prob	Brute Force Complexity	Succ. Prob
1	$E_h(m) \oplus m$ - MMO	$2^{126.3}$	0.632	2^{128}	0.632
2	$E_h(m) \oplus w$ - MP	$2^{126.3}$	0.632	2^{128}	0.632
3	$E_m(h) \oplus h$ - DM	$2^{126.6}$	0.632	2^{128}	0.632
4	$E_h(w) \oplus w$ - similar to MMO	$2^{126.3}$	0.632	2^{128}	0.632
5	$E_h(w) \oplus m$ - similar to MMO	$2^{126.3}$	0.632	2^{128}	0.632
6	$E_m(h) \oplus w$ - similar to DM	$2^{126.6}$	0.632	2^{128}	0.632
7	$E_m(w) \oplus h$ - similar to DM	$2^{126.6}$	0.632	2^{128}	0.632
8	$E_m(w) \oplus w$ - similar to DM	$2^{126.6}$	0.632	2^{128}	0.632
9	$E_w(h) \oplus h$ - similar to DM	$2^{126.6}$	0.632	2^{128}	0.632
10	$E_w(h) \oplus m$ - similar to DM	$2^{126.6}$	0.632	2^{128}	0.632
11	$E_w(m) \oplus h$ - similar to MP	$2^{126.3}$	0.632	2^{128}	0.632
12	$E_w(m) \oplus m$ - similar to MMO	$2^{126.3}$	0.632	2^{128}	0.632

2   Preliminaries

In this section we give a brief overview of the key concepts used in our cryptanalysis technique to facilitate better understanding.

2.1   AES-128

AES-128 is a block cipher with 128-bit internal state and 128-bit key K. The internal state and the key is represented by a 4×4 matrix. The plaintext is xor'ed with the key, and then undergoes a sequence of 10 rounds. Each round consists of four transformations: nonlinear bytewise SubBytes, the byte permutation ShiftRows, linear transformation MixColumns, and the addition with a subkey AddRoundKey. MixColumns is omitted in the last round.

For the sake of clarity, we will follow the same notation used for description of AES-128 as used in [7]. We address two internal states in each round as follows: #1 is the state before SubBytes in round 1, #2 is the state after MixColumns in round 1, #3 is the state before SubBytes in round 2, ..., #19 is the state before SubBytes in round 10, #20 is the state after ShiftRows in round 10. The key K is expanded to a sequence of keys $K^0, K^1, K^2, \ldots, K^{10}$, which form a 4×44 byte array. Then the 128-bit subkeys $0, $1, $2, \ldots, $10 come out of the sliding window with a 4-column step. We refer the reader to [8] for a detailed description of AES.

2.2 Biclique Key Recovery Attack

In this section, we briefly discuss the independent biclique key recovery attack for AES-128. For a more detailed description of bicliques, one can refer to [7]. In this attack, the entire key space of AES-128 is first divided into non-overlapping group of keys. Then, a subcipher f that maps an internal state S to a ciphertext C under a key K, i.e. $f_K(S) = C$ is chosen. Suppose f connects 2^d intermediate states $\{S_j\}$ to 2^d ciphertexts $\{C_i\}$ with 2^{2d} keys $\{K[i,j]\}$. The 3-tuple of sets $[\{S_j\}, \{C_i\}, \{K[i,j]\}]$ is called a d-dimensional biclique, if: $\forall i, j \in \{0, \ldots, 2^d - 1\} : C_i = f_{K[i,j]}(S_j)$.

Each key in a group can be represented relative to the base key of the group, i.e., $K[0,0]$ and two key differences Δ_i^k and ∇_j^k such that: $K[i,j] = K[0,0] \oplus \Delta_i^k \oplus \nabla_j^k$. For each group we choose a base computation i.e., $S_0 \xrightarrow[f]{K[0,0]} C_0$. Then C_i and S_j are obtained using 2^d forward differentials Δ_i, i.e., $S_0 \xrightarrow[f]{K[0,0] \oplus \Delta_i^k} C_i$ and 2^d backward differentials ∇_j, i.e., $S_j \xleftarrow[f^{-1}]{K[0,0] \oplus \nabla_j^k} C_0$. If the above two differentials do not share active nonlinear components for all i and j, then the following relation: $S_0 \oplus \nabla_j \xrightarrow[f]{K[0,0] \oplus \Delta_i^k \oplus \nabla_j^k} C_0 \oplus \Delta_i$ is satisfied [7]:

Once a biclique is constructed for an arbitrary part of the cipher, meet-in-the middle (MITM) attack is used for the remaining part to recover the key. During the MITM phase, a partial intermediate state is chosen as the matching state v. The adversary then precomputes and stores in memory 2^{d+1} times full computations upto a matching state v: $\forall i, P_i \xrightarrow{K[i,0]} \vec{v}$ and $\forall j, \overleftarrow{v} \xleftarrow{K[0,j]} S_j$.

Here, plaintext P_i is obtained from ciphertexts C_i through the decryption oracle[1]. If a key in a group satisfies the following relation: $P_i \xrightarrow[h]{K[i,j]} \vec{v} = \overleftarrow{v} \xleftarrow[g^{-1}]{K[i,j]} S_j$, then the adversary proposes a key candidate. If a right key is not found in the chosen group then another group is chosen and the whole process is repeated. The full complexity of independent biclique attacks is calculated as:

$$C_{full} = 2^{k-2d}(C_{biclique} + C_{precompute} + C_{recompute} + C_{falsepos}),$$

---

[1] Under hash function settings decryption oracle is replaced by feed-forward operation.

where, $C_{precompute}$ is the cost complexity for calculating v for 2^{d+1}, $C_{recompute}$ is the cost complexity of recomputing v for 2^{2d} times and $C_{falsepos}$ is the complexity to eliminate false positives. As mentioned in [7], the full key recovery complexity is dominated by $2^{k-2d} \times C_{recomp}{}^2$.

3   Notations

To facilitate better understanding, we use the following notations in the rest of the paper.

CV	: Chaining Variable
IV	: Initialization Vector
(CV, message)	: Input tuple to hash function/compression function
(key, plaintext)	: Input tuple to underlying block cipher
n	: Input message/key size (in bits)
A$_b$	: Base State
m$_b$	: Base Plaintext
K$_b$	: Base Key
K[i, j]	: Keys generated by Δ_i and ∇_j modifications
M[i, j]	: Messages generated by Δ_i and ∇_j modifications
Nbr	: Number of AES rounds called
E$_{enc/dec}$	: One Round of AES encryption/decryption
E(x, y)	: Full AES encryption under y-bit key and x-bit message
E^{-1}(x, y)	: Full AES decryption under y-bit key and x-bit message

4   Biclique Based Preimage Attack on AES-128 Instantiated Compression Function

In this section, we examine how biclique technique discussed in Sect. 2.2 can be applied to find preimage for block cipher based compression function. This preimage attack on compression function will then be used to evaluate second preimage resistance of AES-128 based hash functions under different PGV modes as discussed in Sect. 5.

Let us consider an AES-128 based compression function (as shown in Fig. 5). To find the preimage for h, the attacker needs to find a valid $(CV, message)$ pair which generates h. In terms of the underlying block cipher E which is instantiated with AES-128, this problem translates to finding a valid (plaintext, key) pair where both the key and the plaintext are of 128-bits size. To guarantee the existence of a preimage for h (with probability 0.632), the attacker needs to test 2^{128} distinct (key, plaintext) pairs.

---

[2] C_{recomp} in turn is measured as: 2^{128} (#S-boxes recomputed in MITM phase/#Total S-boxes required in one full AES encryption) $\implies 2^{128}$ (#S-boxes recomputed in MITM phase/200).

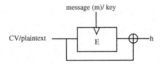

Fig. 5. AES-128 instantiated compression function in DM mode.

When biclique methodology is applied on AES-128 to recover the secret key [7], full key space, i.e., 2^{128} keys are divided into 2^{112} groups of 2^{16} size each and tested[3]. These 2^{112} groups are generated from 2^{112} base key values where each base value defines one group. However, the same biclique approach when extended to hash functions warrants the need of testing 2^{128} (key, plaintext) pairs. These 2^{128} (key, plaintext) pairs will be generated from 2^{112} (key, plaintext) base states. Hence, under hash function settings, alongwith the *base key* we introduce the term *"base message"*. Let K_b denote the base key value and A_b denote the base message value. If we apply the original biclique approach [7] on compression function, then 2^{128} (key, plaintext) pairs are generated from a combination of $2^{112}(K_b, A_b)$ as shown in (Fig. 6). Here, a single A_b is chosen and repeated across all the groups whereas 2^{112} different $K_b's$ are used. The biclique algorithm for the attack is shown in Fig. 7. In Algorithm 1, the specific (i,j) tuple for which a match is found gives us the corresponding $K[i,j]$ and $M[i,j]$ as the desired inputs for compression function. The complexity of this attack when applied for searching preimages in AES-128 instantiated compression function is $2^{125.56}$ [6].

$(K_b^{(1)}, A_b) \longrightarrow 2^{16}$ (key, message) pairs

$(K_b^{(2)}, A_b) \longrightarrow 2^{16}$ (key, message) pairs

$(K_b^{(3)}, A_b) \longrightarrow 2^{16}$ (key, message) pairs

$(K_b^{(2^{112})}, A_b) \longrightarrow 2^{16}$ (key, message) pairs

---

Algorithm 1 :

Fix a base state A_b
for *each* 2^{112} *base keys* $(K_b's)$ *and the fixed chosen* A_b **do**

    Generate 2^{16} (Δ_i^k, ∇_j^k) combinations
    Generate corresponding $2^{16} K[i,j]$
    Construct a biclique structure using these $2^{16} K[i,j]$

    for *each* $2^{16} K[i,j]$ **do**

        1. Generate $M[i,j]$ (where $M[i,j]$ = Nbr $E_{enc/dec}(K[i,j], A_b)$)
        2. Perform meet-in-the-middle attack in the rest of the rounds

Fig. 6. Generation of groups in original attack [7]

Fig. 7. Steps of the original biclique attack in [7] using the base key K_b and the base message A_b.

---

[3] Here, bicliques of dimension $d = 8$ are constructed. In our attacks, we also construct bicliques of dimension 8.

In the procedure described above, it can be seen that the attacker generates a chaining value $(M[i,j])$ of her own along with the preimage $(K[i,j])$. However, as already discussed, the IV value is a public constant in the hash function setting and cannot be altered by the attacker. In the subsequent section, we show how to utilize variants of the above framework for launching second preimage attack on AES-128 based hash functions in different PGV modes with IV being fixed.

5   Second Preimage Attack on Hash Functions

In this section, we examine the feasability of extending the biclique cryptanalysis technique for second preimage attack on AES-128 instantiated hash functions for all 12 PGV modes.

5.1   PGV Construction 1 - Matyas-Meyer-Oseas (MMO) Mode: $E_h(\mathbf{m}) \oplus \mathbf{m}$

Consider MMO based hash function as shown in Fig. 8. Here, the (chaining variable, message block) tuple act as the (key, plaintext) inputs respectively to block cipher E. In this case, the attacker is given m $= (m_0 \,||\, m_1 \,||\, pad)$ and its corresponding hash value h_2. Her aim is to find another different message, m' that will produce the same h_2. To achieve so, the attacker can consider m' as - $(m_0' \,||\, m_1 \,||\, pad)$ where the second half of $m' = m$ while for the first half, the attacker has to carry a biclique attack. For the first half, i.e., $h_1 := E_{IV}(m_0')$, the attacker knows h_1 and IV. Her aim is now to find a preimage m_0' which produces h_1 under the given IV. The attack steps are as follows:

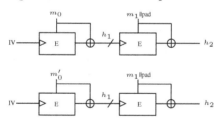

Fig. 8. Second preimage attack on MMO based hash function

1. The attacker fixes IV as the key input to the block cipher E & chooses a 128-bit base message A_b.
2. *Choice of biclique structure.* Here, the key input to the block cipher (i.e., IV) is fixed. The attacker has to choose a biclique structure such that the Δ_i and ∇_j trails only modify the message states and not the key states (since IV cannot change) plus the biclique attack should have lowest search complexity. All the existing biclique trails in literature allow modification in the keys states as well, therefore, we construct new biclique trails to suit our needs.

3. We represent the Δ and ∇ trails as Δ_i^m and ∇_j^m respectively. The biclique structure satisfying the above requirements is as shown in Fig. 9(a).

4. For the above biclique, she divides the 128-bit message space into 2^{112} groups each having 2^{16} messages with respect to intermediate state #3 as shown in Fig. 9(a). The base messages are all 16-byte values with two bytes (i.e., bytes 0 and 4) fixed to 0 whereas the remaining 14-bytes taking all possible values (shown in Fig. 10). The messages in each group $(M[i,j])$ are enumerated with respect to the base message by applying difference as shown in Fig. 11. The proof for the claim that this base message (with the corresponding Δ_i and ∇_j differences) uniquely divides the message space into non-overlapping groups is given in Appendix A.1.

5. The biclique covers 1.5 rounds (round 2 and round 3 upto *Shift Rows* operation). Δ_i^m trail activates byte 0 whereas ∇_j^m trail activates bytes 3,4,9 and 14 of #3 state.

6. Meet-in-the-middle attack is performed on the rest 8.5 rounds. In the MITM phase, partial matching is done in byte 12 of state #13. In the backward direction, Δ_i^m trail activates 4 bytes in the plaintext i.e., byte 0, 5, 10 and 15

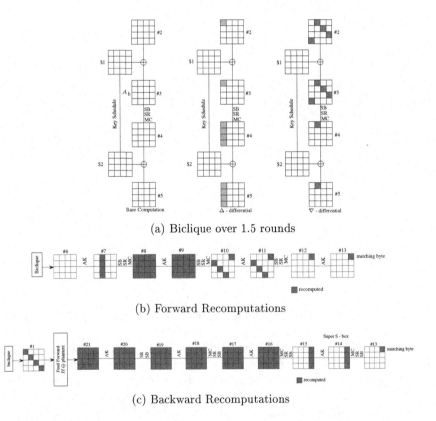

(a) Biclique over 1.5 rounds

(b) Forward Recomputations

(c) Backward Recomputations

Fig. 9. Biclique structure for MMO mode when key/IV is known

whereas ∇_j^m activates all bytes. As such, during the recomputation phase, the 4 bytes of plaintext affected by both Δ_i^m and ∇_j^m trails need to be recomputed. Similar explanation can be provided for other bytes shown to be recomputed in Fig. 9(b) and (c). In the forward propagation (starting from round 4), $4+16+4 = 24$ S-boxes and in the backward propagation (starting from round 1), $4+16+16+4+1= 41$ S-boxes are recomputed. Thus, a total of 65 S-boxes are involved in the recomputation process. One full AES encryption requires 200 S-box computations. As each group has 2^{16} messages, $C_{recomp} = 2^{16} \times \frac{65}{200}$ $= 2^{14.3}$. Hence, $C_{full} = 2^{112} \times 2^{14.3} = 2^{126.3}$.

7. For the specific (i, j) value which produces a match in the middle, the corresponding $M[i, j]$ i.e., xoring of #3 states in base computation, Δ_i and ∇_j trails (in Fig. 9(a)) yields the plaintext m'_0 for the block cipher E. The biclique algorithm, i.e., Algorithm 2 is as shown in Fig. 12.

0	0		

Fig. 10. Base message

i	j_1		
		j_2	
			j_3
j_4			

Fig. 11. Δ_i and ∇_j differences

Thus with a time complexity of $2^{126.3}$, the attacker is able to find a (IV, m'_0) pair which produces hash value h_1 and $m' = (m'_0 \parallel m_1 \parallel pad)$ forms a valid second preimage.

PGV Construction 2 - Miyaguchi-Preneel Mode (MP) Mode: $E_h(m) \oplus m \oplus h$ - The MP mode is an extended version of MMO mode. The only difference between the two constructions is the fact that output of block cipher is xor'ed both with the plaintext input as well the chaining variable input. However, this does not demand any extra attack requirements and the second preimage attack on MP mode is exactly the same as that described on MMO mode.

5.2 PGV Construction 3 Davies-Meyer (DM) Mode: $E_m(h) \oplus h$

In the DM based hash function (as shown in Fig. 13), the (chaining variable, message block) tuple act as the (plaintext, key) inputs respectively to block cipher E. We again inspect a similar scenario as described in Sect. 5.1, i.e., for a message m = $(m_0 \parallel m_1 \parallel pad)$, the attacker is given its corresponding hash value h_2. Her aim is to find another different message m' that will produce the same h_2. Consider the hash function as concatenation of two compression functions - $E_{m_0}(IV)$ and $E_{m_1 \parallel pad}(h_1)$. To get a valid second preimage, the attacker chooses m' as - $(m'_0 \parallel m_1 \parallel pad)$ i.e., she focuses on the first compression

---

Algorithm 2:

Fix a base state A_b
for *each* 2^{112} *base messages* $(A_b's)$ *and a single base key i.e.,* IV
do

  Generate 2^{16} (Δ_i^m, ∇_j^m) combinations
  Generate corresponding $2^{16} M[i,j]$
  Construct a biclique structure using these $2^{16} M[i,j]$
  for *each* $2^{16} M[i,j]$ **do**

    1. Perform meet-in-the-middle attack in the rest of the rounds

---

Fig. 12. Steps of the new biclique attack when key input to the underlying block cipher is fixed and cannot be modified by the attacker for MMO mode.

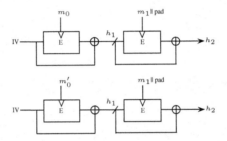

Fig. 13. Second preimage attack on DM based hash function

function and her aim is to find m_0' such that $E_{m_0'}(IV) = h_1$ when IV and h_1 are known to the attacker. The attack steps are as follows:

1. The attacker fixes the IV as the plaintext input to the block cipher.
2. **Choice of biclique structure.** Under the given attack scenario, since the message input, i.e., IV is fixed, the attacker has to choose a biclique structure such that the Δ_i and ∇_j trails do not modify the plaintext state and the biclique attack has lowest search complexity. The biclique structure satisfying the above requirements is given in Fig. 14(a).
3. For the above biclique, she divides the 128-bit key space into 2^{112} groups, each having 2^{16} keys with respect to subkey $0 i.e., the master key as the base key as shown in Fig. 14(a). The base keys are all 16-byte values with two bytes (i.e., bytes 0 and 1) fixed to 0 whereas the remaining 14-bytes taking all possible values (shown in Fig. 15). The keys in each group $(K[i,j])$ are enumerated with respect to the base key by applying difference as shown in Fig. 16. It can be easily verified that this base key uniquely divides the key space into non-overlapping groups.
4. The biclique covers the first round. Δ_i trail activates byte 0 of $0 subkey whereas ∇_j trail activates byte 1 of $ 0 subkey.

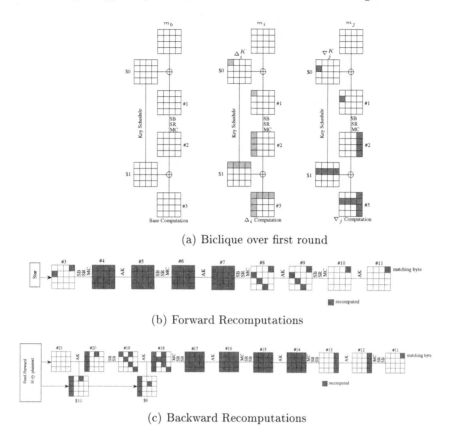

(a) Biclique over first round

(b) Forward Recomputations

(c) Backward Recomputations

Fig. 14. Biclique structure for DM mode when IV/message input is known to the attacker

5. The attacker then performs meet-in-the-middle attack on the rest of the 9 rounds. In the MITM phase, partial matching is done by byte 12 of state #11. In the forward propagation (starting from round 2), $2+16+16+4 = 38$ S-boxes and in the backward propagation (starting from round 10), $5+16+16+4+1 = 42$ S-boxes need to be recomputed (as shown in Fig. 14(b) and (c)). 2 S-box recomputations in the key schedule are also required. Thus a total of 82 S-boxes are involved in recomputation process. One full AES encryption requires 200 S-box computations. As each group has 2^{16} keys, $C_{recomp} = 2^{16} \times \frac{82}{200} = 2^{14.6}$. Hence, $C_{full} = 2^{112} \times 2^{14.6} = 2^{126.6}$.

6. For the specific (i, j) value which produces a match in the middle, the corresponding $K[i, j]$ forms the key (m_0) for the block cipher E. The biclique algorithm, i.e., Algorithm 3 is given in Fig. 17.

Thus with a time complexity of $2^{126.6}$, the attacker is able to find a (IV, m_0') pair which produces hash value h_1 and $m' = (m_0' \parallel m_1 \parallel pad)$ forms a valid second preimage. The attack procedure on two block message for other

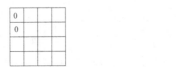

Fig. 15. Base message   **Fig. 16.** Δ_i and ∇_j differences

Algorithm 3 :

for *each* 2^{112} *base keys* $(K'_b s)$ *and the fixed chosen IV* **do**
    Generate 2^{16} (Δ_i^k, ∇_j^k) combinations
    Generate the corresponding $2^{16} K[i, j]$
    Construct a biclique structure using these $2^{16} K[i, j]$
    for *each* $2^{16} K[i, j]$ **do**

        1. Perform meet-in-the-middle attack in the rest of the rounds

---

Fig. 17. Steps of the new biclique attack when message input is fixed and known to the attacker under DM mode

constructions is similar to those discussed in Sects. 5.1 and 5.2. Their results are given in Table 1.

6   Second Preimage Attack on Hash Functions Extended to Messages with Message Length ≥ 3

The second preimage attack discussed in above sections can be extended to messages of any length >2 with same complexity as obtained for 2-block messages. To demonstrate the same, consider a MMO-based hash function with 3-block message as shown in Fig. 18. In this case, the attacker is given a message $m = (m_0 \parallel m_1 \parallel m_2 \parallel pad)$ and its corresponding hash value h_3. Her aim is to find another message m', such that $H(m') = H(m)$. The attacker knows IV and the compression function E. She will choose any m_0 of her own choice, e.g., let $m_0 = 0$, and then calculate $h_1 = E_{IV}(0)$. Once she knows h_1, the setting is reduced to the case discussed in Sect. 5.1, i.e., h_1 and h_2 are known to the attacker and her aim is to find m'_1 such that $m' = (0 \parallel m'_1 \parallel m_2 \parallel pad)$ forms a valid second preimage. This can be found with a complexity of $2^{126.3}$ which is same as that shown for a 2-block message. Similarly, the attack can be applied on other long messages for all other PGV modes.

7   Conclusions

In this paper, we evaluate the security of AES-128 based hash modes against second preimage attack. Specifically, we examine the applicability of biclique

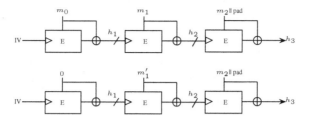

Fig. 18. MMO base hash function with $|m| = 3$

attack on all 12 PGV modes when instantiated with AES-128 and show that best biclique attack for finding preimages in AES-128 instantiated compression function does not translate to best attack for second preimage search under AES-128 based hash function settings. A natural research extension to this work would be to apply the ideas discussed in this paper to hash functions instantiated with other block ciphers. Another research direction can be to extend the methodology to carry out collision attacks on hash functions.

A   Proofs

In this section, we will prove how the base structure which we chose for bicliques in Sect. 5.1 produce non-overlapping keys/messages within a same group and between groups.

A.1   Biclique Structure When *IV* Is Known and Acts as the Message Input to Block Cipher E

For the base message (shown in Fig. 10) that is used for the biclique structure in Fig. 9(a), our aim is to prove that when Δ_i and ∇_j differences are injected in this base message (as shown in Fig. 19), we are able to partition the message space into 2^{112} groups with 2^{16} messages in each and the inter and intra group messages generated are non-overlapping. The ∇_{j1}, ∇_{j2}, ∇_{j3} and ∇_{j4} are differences produced from ∇_j as shown in Fig. 20.

Here, $b_{i,j}$ and $c_{i,j}$ $(0 \leq i,j \leq 3)$ represent the base values of corresponding bytes in the intermediate states #B and #C respectively as shown in Fig. 21. #B and #C are #3 and #4 states in Fig. 9(a).

Aim: Given any two base messages B, B', any two Δ_i differences i, i', any two ∇_j differences j, j' $(0 \leq i,j \leq 2^8)$, we want to prove that B[i,j] $\neq$ B[i',j'] i.e., messages generated are non-overlapping. We will prove this statement case-by-case. Cases (1–4) cover inter group messages whereas Cases (5–7) cover within group messages. For all the proofs discussed below, we will refer to Figs. 22, 23 and 24 for better understanding.

Case 1. Given $B \neq B'$, $i = i', j = j'$, $b_{00}=b_{10}=b'_{00}=b'_{10}=0$, to show: $B[i,j] \neq B'[i',j']$

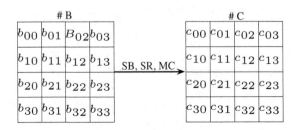

Fig. 19. Δ_i and ∇_j differences in base message

Fig. 20. Relation between $\nabla j, \nabla j_1, \nabla j_2, \nabla j_3, \nabla j_4$

Fig. 21. Relation between #B and #C states

Proof: We will prove this setting by 'proof by contraposition', i.e., if $B[i,j] = B'[i',j']$, $i=i', j=j'$, $b_{00}=b_{10}=b'_{00}=b'_{10}=0$, $\implies B=B'$

In Fig. 24, if $B[i,j] = B'[i',j'] \implies C[i,j] = C'[i',j'] \implies c_{0,2} = c'_{0,2}$, $c_{0,3} = c'_{0,3}$, $c_{1,1} = c'_{1,1}$, $c_{1,2} = c'_{1,2}$, $c_{1,3} = c'_{1,3}$, $c_{2,1} = c'_{2,1}$, $c_{2,2} = c'_{2,2}$, $c_{2,3} = c'_{2,3}$, $c_{3,1} = c'_{3,1}$, $c_{3,2} = c'_{3,2}$ and $c_{3,3} = c'_{3,3}$. Since $C[i,j] = C'[i',j'] \implies c_{0,1} \oplus j = c'_{0,1} \oplus j'$. As $j = j' \implies c_{0,1} = c'_{0,1}$. Hence, **12** bytes in state C and corresponding bytes in state C' share equal values. This relation automatically transcends to related byte positions in B and B' after application of *InvMixColumns*, *InvShiftRows* and *InvSubBytes* (as shown in Fig. 22), i.e., $b_{0,1} = b'_{0,1}$, $b_{0,2} = b'_{0,2}$, $b_{0,3} = b'_{0,3}$, $b_{1,0} = b'_{1,0}$, $b_{1,2} = b'_{1,2}$, $b_{1,3} = b'_{1,3}$, $b_{2,0} = b'_{2,0}$, $b_{2,1} = b'_{2,1}$, $b_{2,3} = b'_{2,3}$, $b_{3,0} = b'_{3,0}$, $b_{3,1} = b'_{3,1}$ and $b_{3,2} = b'_{3,2}$, 12 bytes in B and B' respectively also have same base values). As we have assumed $B[i,j] = B'[i',j'] \implies b_{1,1} = b'_{1,1}$, $b_{2,2} = b'_{2,2}$ and $b_{3,3} = b'_{3,3}$ as these base values are not affected by Δ_i and ∇_j differences (as seen in Fig. 24). Since in states B and B', $b_{0,0} = b'_{0,0} = 0$, hence all **16** byte positions in B and corresponding byte positions in B' share same base values. Hence $B = B'$. This proves that our initial proposition is correct.

Case 2. Given $B \neq B'$, $i = i'$, $j \neq j'$, $b_{00}=b_{01}=b'_{00}=b'_{01}=0$, to show: $B[i,j] \neq B'[i',j']$

Proof: We will prove this setting by 'proof by contradiction', i.e., let us assume if $B \neq B'$, $i = i'$, $j \neq j'$, $b_{00}=b_{10}=b'_{00}=b'_{10}=0$, $\implies B[i,j] = B'[i',j']$

In Fig. 24, if $B[i,j] = B'[i',j'] \implies C[i,j] = C'[i',j'] \implies c_{0,1} \oplus j = c'_{0,1} \oplus j'$. Since $j \neq j' \implies c_{0,1} \neq c'_{0,1}$. As a result after applying *InvMixColumns* and *InvSubBytes* on them the bytes generated i.e., $b_{0,1}$ and $b'_{0,1}$ should also satisfy the relation - $b_{0,1} \neq b'_{0,1}$. But $b_{0,1} = b'_{0,1} = 0$ (as seen in Fig. 21). Hence, a contradiction arises implying our assumed proposition is wrong. Therefore, our initial proposition is correct.

Case 3. Given $B \neq B'$, $i \neq i'$, $j = j'$, $b_{00}=b_{01}=b'_{00}=b'_{01}=0$, to show: $B[i,j] \neq B'[i',j']$

Proof: In this setting since $i \neq i'$, hence $B[i,j] \neq B'[i',j']$ always as they will always differ at zeroth byte position (Fig. 24).

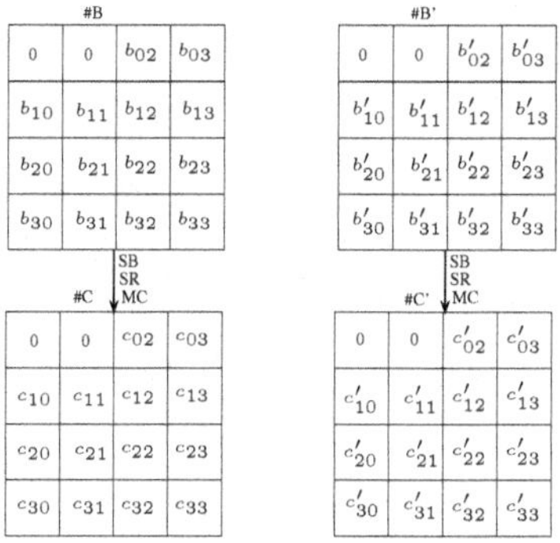

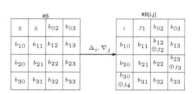

Fig. 22. Relation between base states B and C. The labels inside each box denote the base values of the corresponding byte positions

Fig. 23. Modification of state $\#B$ after applying Δ_i and ∇_j differences. Same relation exists between $\#B'$ and $\#B'[i,j]$

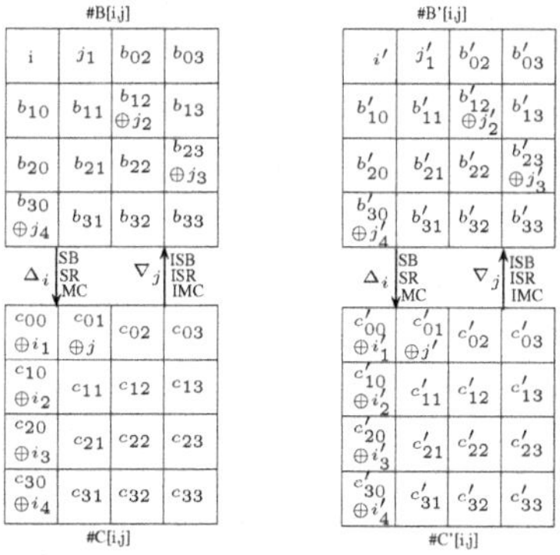

Fig. 24. Relation between states $\#B[i,j]$, $\#C[i,j]$ and $\#B'[i,j]$, $\#C'[i,j]$

Case 4. Given $B \neq B'$, $i \neq i'$, $j \neq j'$, $b_{00}=b_{01}=b'_{00}=b'_{01}=0$, to show: $B[i,j] \neq B'[i',j']$

Proof: Proof similar to as discussed in *Case 3*.

Case 5. Given $B = B'$, $i \neq i'$, $j \neq j'$, $b_{00}=b_{01}=b'_{00}=b'_{01}=0$, to show: $B[i,j] \neq B'[i',j']$

Proof: Proof similar to as discussed in *Case 3.*

Case 6. Given $B = B'$, $i \neq i'$, $j = j'$, $b_{00}=b_{01}=b'_{00}=b'_{01}=0$, to show: $B[i,j] \neq B'[i',j']$

Proof: Proof similar to as discussed in *Case 3.*

Case 7. Given $B = B'$, $i = i'$, $j \neq j'$, $b_{00}=b_{01}=b'_{00}=b'_{01}=0$, to show: $B[i,j] \neq B'[i',j']$

Proof: Since $B = B' \implies C = C' \implies c_{0,1} = c'_{0,1}$. As $j \neq j' \implies c_{0,1} \oplus j \neq c'_{0,1} \oplus j' \implies C[i,j] \neq C'[i',j']$ always as they will everytime differ at fourth byte position (Fig. 24). As a result $B[i,j] \neq B'[i',j']$ always due to bijection relation between states B and C.

Hence we proved that in all cases $M[i,j]$'s so generated are non-overlapping.

References

1. Abed, F., Forler, C., List, E., Lucks, S., Wenzel, J.: Biclique cryptanalysis of the PRESENT and LED lightweight ciphers. IACR Cryptology ePrint Archive, 2012:591 (2012)
2. Abed, F., Forler, C., List, E., Lucks, S., Wenzel, J.: A framework for automated independent-biclique cryptanalysis. In: Moriai, S. (ed.) FSE 2013. LNCS, vol. 8424, pp. 561–582. Springer, Heidelberg (2014)
3. Barreto, P.S.L.M., Rijmen, V.: Whirlpool. In: van Tilborg, H.C.A., Jajodia, S. (eds.) Encyclopedia of Cryptography and Security, 2nd edn, pp. 1384–1385. Springer US, New York (2011)
4. Benadjila, R., Billet, O., Gilbert, H., Macario-Rat, G., Peyrin, T., Robshaw, M., Seurin, Y.: SHA-3 proposal: ECHO. Submission to NIST (2008)
5. Black, J., Rogaway, P., Shrimpton, T., Stam, M.: An analysis of the blockcipher-based hash functions from PGV. J. Cryptology **23**(4), 519–545 (2010)
6. Bogdanov, A., Chang, D., Ghosh, M., Sanadhya, S.K.: Bicliques with minimal data and time complexity for AES. In: Lee, J., Kim, J. (eds.) ICISC 2014. LNCS, vol. 8949, pp. 160–174. Springer, Heidelberg (2011)
7. Bogdanov, A., Khovratovich, D., Rechberger, C.: Biclique cryptanalysis of the full AES. In: Lee, D.H., Wang, X. (eds.) ASIACRYPT 2011. LNCS, vol. 7073, pp. 344–371. Springer, Heidelberg (2011)
8. Daemen, J., Rijmen, V.: The Design of Rijndael: AES - The Advanced Encryption Standard. Information Security and Cryptography. Springer, Heidelberg (2002)
9. Gauravaram, P., Knudsen, L.R., Matusiewicz, K., Mendel, F., Rechberger, C., Schläffer, M., Thomsen, S.S.: Grøstl - a SHA-3 candidate. In: Symmetric Cryptography, Dagstuhl Seminar Proceedings, Dagstuhl, Germany (2009)
10. Hong, D., Koo, B., Kwon, D.: Biclique attack on the full HIGHT. In: Kim, H. (ed.) ICISC 2011. LNCS, vol. 7259, pp. 365–374. Springer, Heidelberg (2012)
11. Indesteege, S.: The LANE Hash Function. Submission to NIST (2008)
12. Jean, J., Naya-Plasencia, M., Schläffer, M.: Improved analysis of ECHO-256. In: Miri, A., Vaudenay, S. (eds.) SAC 2011. LNCS, vol. 7118, pp. 19–36. Springer, Heidelberg (2011)

13. Khovratovich, D., Rechberger, C., Savelieva, A.: Bicliques for preimages: attacks on skein-512 and the SHA-2 family. In: Canteaut, A. (ed.) FSE 2012. LNCS, vol. 7549, pp. 244–263. Springer, Heidelberg (2012)
14. Lamberger, M., Mendel, F., Rechberger, C., Rijmen, V., Schläffer, M.: The rebound attack and subspace distinguishers: application to whirlpool. IACR Cryptology ePrint Archive, 2010:198 (2010)
15. Matusiewicz, K., Naya-Plasencia, M., Nikolić, I., Sasaki, Y., Schläffer, M.: Rebound attack on the full LANE compression function. In: Matsui, M. (ed.) ASIACRYPT 2009. LNCS, vol. 5912, pp. 106–125. Springer, Heidelberg (2009)
16. Mendel, F., Rechberger, C., Schläffer, M., Thomsen, S.S.: Rebound attacks on the reduced Grøstl hash function. In: Pieprzyk, J. (ed.) CT-RSA 2010. LNCS, vol. 5985, pp. 350–365. Springer, Heidelberg (2010)
17. Preneel, B., Govaerts, R., Vandewalle, J.: Hash functions based on block ciphers: a synthetic approach. In: Stinson, D.R. (ed.) CRYPTO 1993. LNCS, vol. 773, pp. 368–378. Springer, Heidelberg (1994)
18. Sasaki, Y., Yasuda, K.: Known-key distinguishers on 11-round feistel and collision attacks on its hashing modes. In: Joux, A. (ed.) FSE 2011. LNCS, vol. 6733, pp. 397–415. Springer, Heidelberg (2011)
19. Schläffer, M.: Subspace distinguisher for 5/8 rounds of the ECHO-256 hash function. In: Biryukov, A., Gong, G., Stinson, D.R. (eds.) SAC 2010. LNCS, vol. 6544, pp. 369–387. Springer, Heidelberg (2011)
20. Tao, B., Wu, H.: Improving the biclique cryptanalysis of AES. In: Foo, E., Stebila, D. (eds.) ACISP 2015. LNCS, vol. 9144, pp. 39–56. Springer, Heidelberg (2015)
21. Wu, S., Feng, D., Wu, W.: Cryptanalysis of the LANE hash function. In: Jacobson Jr., M.J., Rijmen, V., Safavi-Naini, R. (eds.) SAC 2009. LNCS, vol. 5867, pp. 126–140. Springer, Heidelberg (2009)
22. Chen, S.Z., Xu, T.M.: Biclique attack of the full ARIA-256. IACR Cryptology ePrint Archive, 2012:11 (2012)

Hashing into Generalized Huff Curves

` Xiaoyang He[1,2,3], Wei Yu[1,2(✉)], and Kunpeng Wang[1,2]

[1] State Key Laboratory of Information Security,
Institute of Information Engineering, Chinese Academy of Sciences, Beijing, China
hexiaoyang@iie.ac.cn
[2] Data Assurance and Communication Security Research Center,
Chinese Academy of Sciences, Beijing, China
yuwei_1_yw@163.com
[3] University of Chinese Academy of Sciences, Beijing, China

Abstract. Huff curves are well known for efficient arithmetics to their group law. In this paper, we propose two deterministic encodings from $\mathbb{F}_q$ to generalized Huff curves. When $q \equiv 3 \pmod 4$, the first deterministic encoding based on Skalpa's equality saves three field squarings and five multiplications compared with birational equivalence composed with Ulas' encoding. It costs three multiplications less than simplified Ulas map. When $q \equiv 2 \pmod 3$, the second deterministic encoding based on calculating cube root costs one field inversion less than Yu's encoding at the price of three field multiplications and one field squaring. It costs one field inversion less than Alasha's encoding at the price of one multiplication. We estimate the density of images of these encodings with Chebotarev density theorem. Moreover, based on our deterministic encodings, we construct two hash functions from messages to generalized Huff curves indifferentiable from a random oracle.

Keywords: Elliptic curves · Generalized Huff curves · Character sum · Hash function · Random oracle

1 Introduction

Plenty of elliptic/hyperelliptic curve cryptosystems require hashing into algebraic curves. Many identity-based schemes need messages to be hashed into algebraic curves, including encryption schemes [1,2], signature schemes [3,4], signcryption schemes [5,6], and Lindell's universally-composable scheme [7]. The simple password exponential key exchange [10] and the password authenticated key exchange protocols [11] both require a hash algorithm to map the password into algebraic curves.

Boneh and Franklin [8] proposed an algorithm to map elements of $\mathbb{F}_q$ to rational points on an ordinary elliptic curve. This algorithm is probabilistic and

This work is supported in part by National Research Foundation of China under Grant No. 61502487, 61272040, and in part by National Basic Research Program of China (973) under Grant No. 2013CB338001.

D. Lin et al. (Eds.): Inscrypt 2015, LNCS 9589, pp. 22–44, 2016.
DOI: 10.1007/978-3-319-38898-4_2

fails to return a point at the probability of $1/2^k$, where k is a predetermined bound. One disadvantage of this algorithm is that its total number of running steps depends on the input $u \in \mathbb{F}_q$, hence is not constant. Thus the algorithm may be threaten by timing attacks [9], and the information of the message may leaked out. Therefore, it is significant to find algorithms hashing into curves in constant number of operations.

There exist various algorithms encoding elements of $\mathbb{F}_q$ into elliptic curves in deterministic polynomial time. When $q \equiv 3 \pmod 4$, Shallue and Woestijne proposed an algorithm [12] based on Skalba's equality [13], using a variation of Tonelli-Shanks algorithm to calculate square roots efficiently as $x^{1/2} = x^{(q+1)/4}$. Fouque and Tibouchi [14] simplified this encoding by applying brief version of Ulas' function [15]. Moreover, they generalized Shallue and Woestijne's method so as to hash into some special hyperelliptic curves. When $q \equiv 2 \pmod 3$, Icart [16] gave an algorithm based on computing cube roots efficiently as $x^{1/3} = x^{(2q-1)/3}$ in Crypto 2009. Both algorithms encode elements of $\mathbb{F}_q$ into curves in short Weierstrass form.

After initial algorithms listed above, hashing into Hessian curves [17] and Montgomery curves [18] were proposed. Alasha [19] constructed deterministic encodings into Jacobi quartic curves, Edwards curves and Huff curves. Yu constructed a hash function from plaintext to $C_{34}-$ curves by finding a cube root [20].

Huff curves, first introduced by Huff [21] in 1948, were utilized by Joye, Tibouchi and Vergnaud [22] to develop an elliptic curve model over a finite field K where $char(K) > 2$. They also presented the efficient explicit formulas for adding or doubling points on Huff curves. In 2011, Ciss and Sow [27] introduced generalized Huff curves: $ax(y^2 - c) = by(x^2 - d)$ with $abcd(a^2c - b^2d) \neq 0$, which contain classical Huff curves [22] as special cases. Wu and Feng [23] independently presented another kind of curves they also called generalized Huff curves: $x(ay^2 - 1) = y(bx^2 - 1)$, which is in fact an equivalent variation of Ciss and Sow's construction. Wu and Feng constructed arithmetic and pairing formulas on generalized Huff curves. Generalized Huff curves own an effective group law and unified addition-doubling formula, hence are resistant to side channel attacks [24]. Devigne and Joye also analyzed Huff curves over binary fields [28]: $ax(y^2 + cy + 1) = by(x^2 + cx + 1)$ with $abc(a - b) \neq 0$.

We propose two deterministic encodings directly from $\mathbb{F}_q$ to generalized Huff curves: brief Shallue-Woestijne-Ulas (SWU) encoding and cube root encoding. Based on Skalba's equality [13], brief SWU encoding costs three field squarings and five multiplications less than birational equivalence from short Weierstrass curve to generalized Huff curve composed with Ulas' original encoding [15]. It saves three squarings less than birational equivalence from short Weierstrass curve to generalized Huff curve composed with simplified Ulas map [26]. To prove our encoding's B-well-distributed property, we estimate the character sum of an arbitrary non-trivial character defined over generalized Huff curves through brief SWU encoding. We also estimate the size of image of brief SWU encoding. Based on calculating cube root of elements in $\mathbb{F}_q$, cube root encoding saves one field inversion compared with Yu's encoding function at the price of one field multiplication. It saves one field inversion compared with Alasha's encoding at the price of one field squaring and three field multiplications. We estimate the relevant character sum and the size of image of cube root encoding in similar way.

Based on brief SWU encoding and cube root encoding, we construct two hash functions efficiently mapping binary messages into generalized Huff curves, which are both indifferentiable from random oracle.

We do experiments over 192−bit prime field $\mathbb{F}_{P192}$ and 384-bit prime field $\mathbb{F}_{P384}$ recommended by NIST in the elliptic curve standard [25]. On both fields, there exist efficient algorithms to calculate the square root and cube root for each element. On $\mathbb{F}_{P192}$, our cube root encoding f_I saves 13.20 % running time compared with Alasha's encoding function f_A, 8.97 % with Yu's encoding f_Y, on $\mathbb{F}_{P384}$, f_I saves 7.51 % compared with f_A and 4.40 % with f_Y. Our brief SWU encoding f_S also runs faster than f_U, birational equivalence composed with Ulas' encoding function and f_E, birational equivalence composed with Fouque and Tibouchi's brief encoding. Experiments show that f_S saves 9.19 % compared with f_U and 7.69 % with f_E on $\mathbb{F}_{P192}$, while it saves 5.92 % compared with f_U and 5.17 % with f_E on $\mathbb{F}_{P384}$.

Organization of the Paper. In Sect. 2, we recall some basics of generalized Huff curves. In Sect. 3, we introduced brief SWU encoding, prove its B-well-distributed property by estimating the character sum of this encoding, and calculate the density of image of the encoding. In Sect. 4, we proposed the cube root encoding, also prove its B-well-distributed property and calculate the density of image of the encoding by similar methods. In Sect. 5, we construct 2 hash functions indifferentiable from random oracle. In Sect. 6, time complexity of given algorithms is analysed, and we presented the practical results. Section 7 is the conclusion of the paper.

2 Generalized Huff Curves

Suppose $\mathbb{F}_q$ is a finite field whose characteristic is greater than 2.

Definition 1 ([27]). *Generalized Huff curve can be written as:*

$$ax(y^2 - c) = by(x^2 - d),$$

where $a, b, c, d \in \mathbb{F}_q$ with $abcd(a^2c - b^2d) \neq 0$.

For generalized Huff curve E, if $c = \gamma^2, d = \delta^2$ are squares of $\mathbb{F}_q$, let $(x, y) = (\delta x', \gamma y')$, we find that E is $\mathbb{F}_q$-isomorphic to classical Huff curve $(a\delta\gamma^2)x'(y'^2 - 1) = (b\delta^2\gamma)y'(x'^2 - 1)$. If c or d is not a square of $\mathbb{F}_q$, there exists no relevant classical Huff curve which is $\mathbb{F}_q$-isomorphic to E. Therefore, generalized Huff curves contain classical Huff curves as a proper subset.

Consider the point sets on projective plane $(X : Y : Z) \in \mathbb{P}^2(\overline{\mathbb{F}_q})$, generalized Huff curve can be written as:

$$aX(Y^2 - cZ^2) = bY(X^2 - dZ^2).$$

Generalized Huff curve has 3 infinity points: $(1 : 0 : 0), (0 : 1 : 0), (a : b : 0)$. We give a picture of generalized curve $3x\left(y^2 - 1\right) = -5y\left(x^2 - 2\right)$ as shown in Fig. 1 (over $\mathbb{R}$):

According to [23], a generalized Huff curve over $\mathbb{F}_q$ contains a copy of $\mathbb{Z}/2\mathbb{Z} \times \mathbb{Z}/2\mathbb{Z}$. In fact, every elliptic curve with 3 points of order 2 is $\mathbb{F}_q$-isomorphism to a generalized Huff curve. In particular, $ax(y^2 - c) = by(x^2 - d)$ is $\mathbb{F}_q$-isomorphic to $y^2 = x(x + a^2c)(x + b^2d)$.

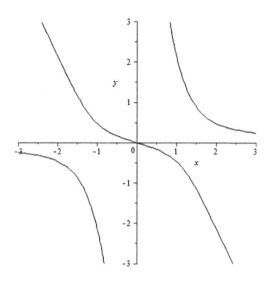

Fig. 1. Generalized Huff Curve $3x\left(y^2 - 1\right) = -5y\left(x^2 - 2\right)$

3   Brief SWU Encoding

For $q \equiv 3 \pmod 4$, Ulas presented an encoding function from $\mathbb{F}_q$ to curve $y^2 = x^n + ax^2 + bx$ [15]. We construct our deterministic encoding function f_S by generalizing his method, mapping $u \in \mathbb{F}_q$ to $(x, y) \in E(\mathbb{F}_q)$.

3.1   Algorithm

Input: a, b, c, d and $u \in \mathbb{F}_q$.
Output: A point $(x, y) \in E(\mathbb{F}_q)$.

1. If $u = 0$ then return $(0, 0)$.
2. $X(u) = \dfrac{a^2b^2cd}{a^2c + b^2d}(u^2 - 1)$.
3. Calculate $g(X(u))$ where $g(s) = s^3 + (a^2c + b^2d)s^2 + a^2b^2cds$.
4. $Y(u) = -\dfrac{a^2b^2cd}{a^2c + b^2d} \cdot \left(1 - \dfrac{1}{u^2}\right)$.

5. Calculate $g(Y(u))$.
6. If $g(X(u))$ is a quadratic residue, then $(s,t) = \left(X(u), -\sqrt{g(X(u))} \right)$,

    else $(s,t) = \left(Y(u), \sqrt{g(Y(u))} \right)$.
7. $(x,y) = \left(\dfrac{bd(s+a^2c)}{t}, \dfrac{ac(s+b^2d)}{t} \right)$.

According to [14], there exists a function $U(u) = u^3 g(Y(u))$, such that the equality

$$U(u)^2 = -g(X(u))g(Y(u)) \tag{1}$$

holds. Thus either $g(X(u))$ or $g(Y(u))$ is a quadratic residue. Choose the one which has square roots in $\mathbb{F}_q$. Note that $q \equiv 3 \pmod{4}$, we can efficiently calculate the standard square root by $\sqrt{a} = a^{(q+1)/4}$. Hence the mapping $u \mapsto (s,t)$ satisfying $t^2 = g(s)$ is constructed. Then in step 7, we transfer (s,t) to $(x,y) \in E(\mathbb{F}_q)$ by a birational equivalence. It is easy to check that this birational equivalence is one-to-one and onto when it is extended to a map between projective curves. The image of $(0,0),(-a^2c,0),(-b^2d,0)$ are infinite points $(a:b:0),(0:1:0),(1:0:0)$ respectively while the image of $(0:1:0)$ is $(0,0)$ on E. Denote the map $u \mapsto (s,t)$ by ρ, and denote the map $(s,t) \mapsto (x,y)$ by ψ, we call the composition $f_S = \psi \circ \rho$ brief SWU encoding. Therefore given $(s,t) \in Im(\rho)$, either $t = \sqrt{g(s)}$ hence s is the image of $Y(u)$ and has at most 2 preimages, or $t = -\sqrt{g(s)}$ hence s is the image of $X(u)$ and has still at most 2 preimages. Moreover, it is easy to check that ψ is one-to-one. Therefore for each finite point on $E(\mathbb{F}_q)$, and for the infinite point $(a:b:0)$, f_S has at most 2 preimages, but for the rest 2 infinite points of $E(\mathbb{F}_q)$, whose projective coordinates are $(1:0:0)$ and $(0:1:0)$, f_S has at most 4 preimages since the corresponding t vanishes.

3.2   Theoretical Analysis of Time Cost

Let S denote field squaring, M denote field multiplication, I field inversion, E_S the square root, E_C the cube root, D the determination of the square residue. Suppose $a,b,c,d \in \mathbb{F}_q$. In this paper we make the assumption that $S = M$, $I = 10M$ and $E_S = E_C = E$.

The cost of f_S can be calculated as follows:

1. Calculating u^2 costs S, multiplying $u^2 - 1$ by $\dfrac{a^2b^2cd}{a^2c+b^2d}$ costs M, and it is enough to calculate $X(u)$.
2. To compute $Y(u)$, we need to calculate the inversion of u^2 for $I + M$.
3. When s is known, computing $g(s) = s(s^2 + (a^2c + b^2d)s + a^2b^2cd) = s(s + a^2c)(s + b^2d)$ takes $2M$. To make sure that the algorithm be run in constant time, both $g(X(u))$ and $g(Y(u))$ must be calculated and it requires $4M$.

4. In general case, exact one of $g(X(u))$ and $g(Y(u))$ is a quadratic residue. We only need to check once and it takes D, then compute the square root E_S of the quadratic residue. Then values of s and t are known.
5. Finally, we calculate the inverse of t, which requires I. Then multiplying the inverse by $s + a^2c$ and $s + b^2d$ costs $2M$, then calculating x and y costs $2M$, hence it requires $I + 4M$ in this step.

Therefore, f_S requires $E_S + 2I + 10M + S + D = E + 31M + D$ in all.

3.3   B-Well-Distributed Property of Brief SWU Encoding

Definition 2 (Character Sum). *Suppose f is an encoding from $\mathbb{F}_q$ into a smooth projective elliptic curve E, and $J(\mathbb{F}_q)$ denotes the Jacobian group of E. Assume that E has an $\mathbb{F}_q - rational$ point O, by sending $P \in E(\mathbb{F}_q)$ to the $\deg 0$ divisor $(P) - (O)$, we can regard f as an encoding to $J(\mathbb{F}_q)$. Let χ be an arbitrary character of $J(\mathbb{F}_q)$. We define the character sum*

$$S_f(\chi) = \sum_{s \in \mathbb{F}_q} \chi(f(s)).$$

We say that f is B-well-distributed if for any nontrivial character χ of $J(\mathbb{F}_q)$, the inequality $|S_f(\chi)| \leq B\sqrt{q}$ holds [29].

Lemma 1 (Corollary 2, Sect. 3, [29]). *If f is a B-well-distributed encoding into a curve E, then the statistical distance between the distribution defined by $f^{\otimes s}$ on $J(\mathbb{F}_q)$ and the uniform distribution is bounded as:*

$$\sum_{D \in J(\mathbb{F}_q)} |\frac{N_s(D)}{q^s} - \frac{1}{\#J(\mathbb{F}_q)}| \leq \frac{B^s}{q^{s/2}} \sqrt{\#J(\mathbb{F}_q)},$$

where

$$f^{\otimes s}(u_1, \ldots, u_s) = f(u_1) + \ldots + f(u_s),$$

$$N_s(D) = \#\{(u_1, \ldots, u_s) \in (\mathbb{F}_q)^s | D = f(u_1) + \ldots + f(u_s)\},$$

i.e., $N_s(D)$ is the size of preimage of D under $f^{\otimes s}$. In particular, when s is greater than the genus of E, the distribution defined by $f^{\otimes s}$ on $J(\mathbb{F}_q)$ is statistically indistinguishable from the uniform distribution. Especially, in the elliptic curves' case, $g_E = 1$, let $s = g_E + 1 = 2$, the hash function construction

$$m \mapsto f^{\otimes 2}(h_1(m), h_2(m))$$

is indifferentiable from random oracle if h_1, h_2 are seen as independent random oracles into $\mathbb{F}_q$ (See [29]).

Hence, it is of great importance to estimate the character sum of an encoding into an elliptic curve, and we will study the case of generalized Huff curves.

Definition 3 (Artin Character). *Let E be a smooth projective elliptic curve, $J(\mathbb{F}_q)$ be Jacobian group of E. Let χ be a character of $J(\mathbb{F}_q)$. Its extension is a multiplicative map $\overline{\chi} : Div_{\mathbb{F}_q}(E) \to \mathbb{C}$,*

$$\overline{\chi}(n(P)) = \begin{cases} \chi(P)^n, & P \in S, \\ 0, & P \notin S. \end{cases}$$

Here P is a point on $E(\mathbb{F}_q)$, S is a finite subset of $E(\mathbb{F}_q)$, usually denotes the ramification locus of a morphism $Y \to X$. Then we call $\overline{\chi}$ an Artin character of X.

Theorem 1. *Let $h : \tilde{X} \to X$ be a nonconstant morphism of projective curves, and χ is an Artin character of X. Suppose that $h^*\chi$ is unramified and nontrivial, φ is a nonconstant rational function on $\tilde{X}$. Then*

$$\left| \sum_{P \in \tilde{X}(\mathbb{F}_q)} \chi(h(P)) \left(\frac{\varphi(P)}{q} \right) \right| \leqslant (2\tilde{g} - 2 + 2 \deg \varphi)\sqrt{q},$$

where $\left(\dfrac{\cdot}{q} \right)$ denotes Legendre symbol, and $\tilde{g}$ is the genus of $\tilde{X}$.

Proof. See Theorem 3, [29].

Theorem 2. *Let f_S be the brief SWU encoding encoding from $\mathbb{F}_q$ to generalized Huff curve E, $q \equiv 2 \pmod 3$. For any nontrivial character χ of $E(\mathbb{F}_q)$, the character sum $S_{f_S}(\chi)$ satisfies:*

$$|S_{f_S}(\chi)| \leqslant 16\sqrt{q} + 45.$$

Proof. Let $S = \{0\} \bigcup \{\text{roots of } g(X(u)) = 0\} \bigcup \{\text{roots of } g(Y(u)) = 0\}$ where $X(\cdot)$ and $Y(\cdot)$ are defined as in Sect. 3.1. For any $u \in \mathbb{F}_q \backslash S$, $X(u)$ and $Y(u)$ are both well defined and nonzero. Let $C_X = \{(u,s,t) \in \mathbb{F}_q^3 | s = X(u), t = -\sqrt{g(X(u))}\}, C_Y = \{(u,s,t) \in \mathbb{F}_q^3 | s = Y(u), t = \sqrt{g(Y(u))}\}$ be the smooth projective curves. It is trivial to see there exist one-to-one map $P_X : u \mapsto (u, s \circ \rho_X(u), t \circ \rho_X(u))$ from $\mathbb{P}^1(\mathbb{F}_q)$ to $C_X(\mathbb{F}_q)$ and $P_Y : u \mapsto (u, s \circ \rho_Y(u), t \circ \rho_Y(u))$ from $\mathbb{P}^1(\mathbb{F}_q)$ to $C_Y(\mathbb{F}_q)$. Let h_X and h_Y be the projective maps on C_X and C_Y satisfying $\rho_X(u) = h_X \circ P_X(u)$ and $\rho_Y(u) = h_Y \circ P_Y(u)$. Let $g_X = P_X^{-1}$, $g_Y = P_Y^{-1}$, $S_X = g_X^{-1}(S \bigcup \{\infty\}) = P_X(S) \bigcup P_X(\infty)$, $S_Y = g_Y^{-1}(S \bigcup \{\infty\}) = P_Y(S) \bigcup P_Y(\infty)$.

To estimate $S_{f_S}(\chi)$,

$$S_{f_S}(\chi) = \left| \sum_{u \in \mathbb{F}_q \backslash S} (f_S^* \chi)(u) + \sum_{u \in S} (f_S^* \chi)(u) \right|$$

$$\leqslant \left| \sum_{u \in \mathbb{F}_q \backslash S} (f_S^* \chi)(u) \right| + \#S,$$

we deduce as follows,

$$\left| \sum_{u \in \mathbb{F}_q \backslash S} (f_S^* \chi)(u) \right| = \left| \sum_{\substack{P \in C_Y(\mathbb{F}_q) \backslash S_Y \\ \left(\frac{t(P)}{q}\right) = 1}} (h_Y^* \psi^* \chi)(P) + \sum_{\substack{P \in C_X(\mathbb{F}_q) \backslash S_X \\ \left(\frac{t(P)}{q}\right) = -1}} (h_X^* \psi^* \chi)(P) \right|$$

$$\leqslant \# S_Y + \# S_X + \left| \sum_{\substack{P \in C_Y(\mathbb{F}_q) \\ \left(\frac{t(P)}{q}\right) = +1}} (h_Y^* \psi^* \chi)(P) \right| + \left| \sum_{\substack{P \in C_X(\mathbb{F}_q) \\ \left(\frac{t(P)}{q}\right) = -1}} (h_X^* \psi^* \chi)(P) \right| ,$$

and

$$2 \left| \sum_{\substack{P \in C_Y(\mathbb{F}_q) \\ \left(\frac{t(P)}{q}\right) = +1}} (h_Y^* \psi^* \chi)(P) \right|$$

$$= \left| \sum_{P \in C_Y(\mathbb{F}_q)} (h_Y^* \psi^* \chi)(P) + \sum_{P \in C_Y(\mathbb{F}_q)} (h_Y^* \psi^* \chi)(P) \cdot \left(\frac{t(P)}{q}\right) \right.$$

$$\left. - \sum_{\left(\frac{t(P)}{q}\right) = 0} (h_Y^* \psi^* \chi)(P) \right|$$

$$\leqslant \left| \sum_{P \in C_Y(\mathbb{F}_q)} (h_Y^* \psi^* \chi)(P) \right| + \left| \sum_{P \in C_Y(\mathbb{F}_q)} (h_Y^* \psi^* \chi)(P) \cdot \left(\frac{t(P)}{q}\right) \right|$$

$$+ \#\{\text{roots of } g(Y(u)) = 0\}.$$

From the covering $\psi \circ h_Y : C_Y \to E$, $Y(u) = s \circ \psi^{-1}(x, y)$, which implies

$$T(u) = (a^3 cy - b^3 dx)u^2 - (acx - bdy)ab = 0.$$

$$\Leftrightarrow u^2 = \frac{ab(acx - bdy)}{a^3 cy - b^3 dx}.$$

Indeed, $\psi \circ h_Y$ is ramified if and only if $T(u)$ has multiple roots, which occurs when $u = 0$ or at infinity. Hence by Riemann-Hurwitz formula,

$$2g_{C_Y} - 2 = 0 + 1 + 1 = 2.$$

Hence curve C_Y is of genus 2. Similarly, C_X is also of genus 2.
    Observe that

$$\deg t = [\mathbb{F}_q(s, t, u) : \mathbb{F}_q(t)] = [\mathbb{F}_q(s, t, u) : \mathbb{F}_q(s, t)][\mathbb{F}_q(s, t) : \mathbb{F}_q(t)] = 2 \cdot 3 = 6.$$

Further more, by Theorem 3 in [29], $\left| \sum_{P \in C_Y(\mathbb{F}_q)} (h_Y^* \psi^* \chi)(P) \right| \leqslant (2g_{C_Y} - 2)\sqrt{q} = 2\sqrt{q}$, $\left| \sum_{P \in C_Y(\mathbb{F}_q)} (h_Y^* \psi^* \chi)(P) \cdot \left(\frac{t(P)}{q} \right) \right| \leqslant (2g_{C_Y} - 2 + 2 \det t)\sqrt{q} = 14\sqrt{q}$, and $g(Y(u)) = 0$ is sextic polynomial, we can derive

$$\left| \sum_{\substack{P \in C_Y(\mathbb{F}_q) \\ \left(\frac{t(P)}{q} \right) = +1}} (h_Y^* \psi^* \chi)(P) \right| \leqslant 8\sqrt{q} + 3.$$

And

$$\left| \sum_{\substack{P \in C_X(\mathbb{F}_q) \\ \left(\frac{t(P)}{q} \right) = -1}} (h_X^* \psi^* \chi)(P) \right|$$

has the same bound.

Hence $|S_{f_S}(x)| \leqslant 16\sqrt{q} + 6 + \#S_Y + \#S_X + \#S$. Note that $g(X(u)) = 0$ and $g(Y(u)) = 0$ have common roots, we can deduce that $\#S \leqslant 1 + 6 = 7$. Thus $\#S_X \leqslant 2(\#S + 1) \leqslant 16$. By the same reason, $\#S_Y \leqslant 16$. Then $|S_{f_S}(x)| \leqslant 16\sqrt{q} + 45$. Thus f_S is well-distributed encoding using the Theorem 3 in [29]. ∎

3.4   Calculating the Density of the Image

In the case of dealing with short Weierstrass curves, Icart conjectured that the density of image $\frac{\#Im(f)}{\#E(\mathbb{F}_q)}$, is near $\frac{5}{8}$, see [16]. Fouque and Tibouchi proved this conjecture [14] using Chebotarev density theorem. Now we apply this theorem onto generalized Huff curves, and give their sizes of images of deterministic encodings.

Theorem 3 (Chebotarev, [31]). *Let K be an extension of $\mathbb{F}_q(x)$ of degree $n < \infty$ and L a Galois extension of K of degree $m < \infty$. Assume $\mathbb{F}_q$ is algebraically closed in L, and fix some subset φ of $Gal(L/K)$ stable under conjugation. Let $s = \#\varphi$ and $N(\varphi)$ the number of places v of K of degree 1, unramified in L, such that the Artin symbol $\left(\frac{L/K}{v} \right)$ (defined up to conjugation) is in φ. Then*

$$|N(\varphi) - \frac{s}{m}q| \leqslant \frac{2s}{m}((m + g_L) \cdot q^{1/2} + m(2g_K + 1) \cdot q^{1/4} + g_L + nm)$$

where g_K and g_L are genera of the function fields K and L.

Theorem 4. *Let E be the generalized Huff curve over $\mathbb{F}_q$ defined by equation $ax(y^2 - c) = by(x^2 - d), abcd(a^2c - b^2d) \neq 0$, f_S is the corresponding brief SWU encoding function. Then*

$$|\#Im(f_S) - \frac{1}{2}q| \leqslant 4q^{1/2} + 6q^{1/4} + 27.$$

Proof. K is the function field of E which is the quadratic extension of $\mathbb{F}_q(x)$, hence $d = 2$, and by the property of elliptic curve, $g_K = 1$.

$Gal(L/K) = S_2$, hence $m = \#S_2 = 2$. φ is the subset of $Gal(L/K)$ consisting a fixed point, which is just $(1)(2)$, then $s = 1$.

Let W be the preimage of the map ψ, $W(\mathbb{F}_q)$ be the corresponding rational points on W. By the property that ψ is one-to-one rational map, $\#Im(f_S) = \#Im(\psi^{-1} \circ f) = I_X + I_Y + I_0$, where $I_X = \#\{(s,t) \in W(\mathbb{F}_q) | \exists u \in \mathbb{F}_q, s = X(u), y = -\sqrt{g(X(u))} \neq 0\}$, $I_Y = \#\{(s,t) \in W(\mathbb{F}_q) | \exists u \in \mathbb{F}_q, s = Y(u), t = \sqrt{g(Y(u))} \neq 0\}$, $I_0 = \#\{(s,0) \in W(\mathbb{F}_q) | g(X(u)) = 0 \text{ or } g(Y(u)) = 0\}$. It is trivial to see that $I_0 \leqslant 3$.

Let N_X denote the number of rational points on the curve W with an s-coordinate of the form $X(u)$ and N_Y denote the number of rational points on the curve W with an s-coordinate of the form $Y(u)$, we have

$$2I_X \leqslant N_X \leqslant 2I_X + I_0 \leqslant 2I_X + 3,$$
$$2I_Y \leqslant N_Y \leqslant 2I_Y + 3.$$

Hence $I_X + I_Y \leqslant \dfrac{1}{2}(N_X + N_Y) \leqslant I_X + I_Y + 3$.

Since the place v of K of degree 1 correspond to the projective unramified points on $E(\mathbb{F}_q)$, hence $|N_X - N(\varphi)| \leqslant 12 + 3 = 15$, where 3 represents the number of infinite points, 12 represents the number of ramified points. Then we have

$$|N_X - \frac{1}{2}q| \leqslant |N_X - N(\varphi)| + |N(\varphi) - \frac{1}{2}q|$$
$$\leqslant 15 + (4q^{1/2} + 6q^{1/4} + 6) = 4q^{1/2} + 6q^{1/4} + 21.$$

Analogously, $|N_Y - \frac{1}{2}q| \leqslant 4q^{1/2} + 6q^{1/4} + 21$.

Therefore, we have

$$|\#Im(f_S) - \frac{1}{2}q| \leqslant |\#Im(f_S) - \frac{N_X + N_Y}{2}| + |\frac{N_X + N_Y}{2} - \frac{1}{2}q|$$
$$\leqslant I_0 + |I_X - \frac{N_X}{2}| + |I_Y - \frac{N_Y}{2}| + (4q^{1/2} + 6q^{1/4} + 21)$$
$$\leqslant 3 + \frac{3}{2} + \frac{3}{2} + (4q^{1/2} + 6q^{1/4} + 21)$$
$$= 4q^{1/2} + 6q^{1/4} + 27. \qquad \blacksquare$$

4   Cube Root Encoding

4.1   Algorithm

When $q \equiv 2 \pmod 3$ is a power of odd prime number, we give our deterministic construction $f_I : u \mapsto (x, y)$ in the following way:

Input: $a, b, c, d,$ and $u \in \mathbb{F}_q$.
Output: A point $(x, y) \in E(\mathbb{F}_q)$.

1. $t = u^2 - a^2 c - b^2 d$.
2. $r = \dfrac{1}{2}\left(a^2 b^2 cd - \dfrac{1}{3}t^2\right)$.
3. $s = \dfrac{ut}{3} + \sqrt[3]{ur^2 - (\dfrac{ut}{3})^3}$.
4. $(x, y) = \left(\dfrac{bd(s + a^2 cu)}{su + r}, \dfrac{ac(s + b^2 du)}{su + r}\right)$.

   In step 3, since $q \equiv 2 \pmod 3$, we can efficiently calculate the the cube root by $\sqrt[3]{a} = a^{(2q-1)/3}$.

4.2   Theoretical Analysis of Time Cost

Let M, S, I and E_C represent the same as in Sect. 3.2. The cost of encoding function f_I can be estimated as follows:

1. Computing u^2 costs S. Then t can be calculated.
2. To compute r, we need S.
3. We use $S + M$ to calculate ur^2, then use M to get ut and $S + M$ to calculate $(\dfrac{ut}{3})^2$, take E_C to calculate s.
4. Finally, to calculate the inversion of $su+r$, we need $M+I$. Calculating $\dfrac{s}{us+r}$ and $\dfrac{u}{us+r}$ cost $2M$. Calculating $\dfrac{a^2 bcdu}{su+r}, \dfrac{bds}{su+r}, \dfrac{b^2 acdu}{su+r}, \dfrac{acs}{su+r}$ cost $4M$ with pre-computations.

   Therefore, f_I requires $E_C + I + 4S + 10M = E + 24M$.

4.3   Properties of Cube Root Encodings

Lemma 2. *Suppose $P(x, y)$ is a point on generalized Huff curve E, then equation $f_I(u) = P$ has solutions satisfying $H(u; x, y) = 0$.*

When $a^4 c^2 + b^4 d^2 \neq a^2 b^2 cd$,

$$H(u; x, y) = (acx - bdy)u^4 + (2b^3 d^2 y - 2a^3 c^2 x + 4abcd(bx - ay))u^2$$
$$+ 6abcd(a^2 c - b^2 d)u + (acx - bdy)(a^4 c^2 + b^4 d^2 - a^2 b^2 cd).$$

When $a^4 c^2 + b^4 d^2 = a^2 b^2 cd$,

$$H(u; x, y) = (acx - bdy)u^3 + (2b^3 d^2 y - 2a^3 c^2 x + 4abcd(bx - ay))u$$
$$+ 6abcd(a^2 c - b^2 d). \tag{2}$$

Proof. By the algorithm in Sect. 4.1, we have

$$
\begin{cases}
(xu - bd)s = a^2bcdu - xr \\
(yu - ac)s = ab^2cdu - yr
\end{cases}
\Rightarrow
\frac{xu - bd}{yu - ac} = \frac{a^2bcdu - xr}{ab^2cdu - yr}
$$

$$
\Rightarrow (-bdy + acx)u^4 + (-2\,a^3c^2x + 4\,xab^2cd - 4\,bdya^2c + 2\,b^3d^2y)u^2
$$
$$
+ 6\,abcd(a^2c - b^2d)u + (-b^2da^2c + a^4c^2 + b^4d^2)(-bdy + acx) = 0. \tag{3}
$$

When $a^4c^2 + b^4d^2 = a^2b^2cd$, the constant coefficient of this equation is 0. Then eliminate u, we get (2).

Meanwhile, if $H(u; x, y) = 0$ and $(x, y) \in E$, we have

$$
\begin{cases}
ax(y^2 - c) = by(x^2 - d) \\
(acx - bdy)\left(b^2da^2c - \dfrac{(a^2c + b^2d - u^2)^2}{3}\right) = 2u((xu - bd)ab^2cd \\
\qquad\qquad\qquad\qquad\qquad\qquad\qquad\qquad\qquad\qquad -a^2bcdt(yu - ac))
\end{cases}
$$

which leads to

$$
(xu - bd)ab^2cd - (yu - ac)a^2bcd = (acx - bdy)(a^2b^2cd - \frac{1}{3}(a^2c + b^2d - u^2)^2)/2u,
$$

from this equation and the definition of s, r, we get

$$
\begin{cases}
x = \dfrac{bd(s + a^2cu)}{su + r} \\
y = \dfrac{ac(s + b^2du)}{su + r}.
\end{cases}
\Rightarrow (x, y) = f_I(u).
\qquad\blacksquare
$$

4.4   The Genus of Curve C

Denote F by the algebraic closure of $\mathbb{F}_q$. We consider the graph of f_I:

$$
\begin{aligned}
C &= \{(x, y, u) \in E \times \mathbb{P}^1(F)|\quad f_I(u) = (x, y)\} \\
&= \{(x, y, u) \in E \times \mathbb{P}^1(F)|\quad H(u; x, y) = 0\},
\end{aligned}
$$

which is the subscheme of $E \times \mathbb{P}^1(F)$.

Now we calculate the genus of C. In the case $a^4c^2 + b^4d^2 \neq a^2b^2cd$, the projection $g : C \to E$ is a morphism of degree 4, hence the fiber at each point of E contains 4 points. The branch points of E are points $(x, y) \in E$ where $H(u; x, y)$ has multiple roots, which means the discriminant $D = disc(H)$ vanishes at (x, y). By substituting $x^2 = -\dfrac{-axy^2 + axc - byd}{by}$ into D, it can be represented as

$$
D = -\frac{16a^3(P_1(y)x + P_2(y))}{b^5y^5} \Rightarrow x = -\frac{P_2(y)}{P_1(y)},
$$

where $P_1(y)$ is a polynomial of degree 10, $P_2(y)$ is a polynomial of degree 11. Substituting $x = -\dfrac{P_2(y)}{P_1(y)}$ into $E(x,y) = 0$, we find that y satisfies $y^{11} \cdot Q(y) = 0$, where $Q(y)$ is a polynomial of degree 12. Hence there are at most 12 branch points on E other than $(0,0)$. It is easy to check that $(x,y) = (0,0)$ is a branch point, since the multiplicity of $u = \infty$ is 3. If $H(u;x,y)$ has triple roots at (x,y), we have:

$$\begin{cases} E(x,y) = 0 \\ H(u;x,y) = 0 \\ \dfrac{d}{du}H(u;x,y) = 0 \\ \dfrac{d^2}{du^2}H(u;x,y) = 0. \end{cases} \qquad (4)$$

In general cases, when $(x,y) \neq (0,0)$, (4) has no solution, thus all 12 branch points have ramification index 2. By Riemann-Hurwitz formula, $2g_C - 2 \leqslant 4 \cdot (2 \cdot 1 - 2) + 12 \cdot (2 - 1) + 1 \cdot (3 - 1)$, we get $g_C \leqslant 8$.

In the case that $a^4 c^2 + b^4 d^2 = a^2 b^2 cd$, analogous to previous proof, we can show that g is a morphism of degree 3, D is a cubic function of u and hence has 3 different roots unless $disc(D) = 0$. By similar calculation, we find that only when y satisfies some sextic function, the point $(x,y) \in E$ is a branch point. Hence there are at most 6 branch points on E, with ramification index 2. By Riemann-Hurwitz formula, $2g_C - 2 \leqslant 3 \cdot (2 \cdot 1 - 2) + 6 + (3 - 1)$, we get $g_C \leqslant 5$.

Hence we have

Theorem 5. *If $a^4 c^2 + b^4 d^2 \neq a^2 b^2 cd$, the genus of curve C is at most 8; if $a^4 c^2 + b^4 d^2 = a^2 b^2 cd$, the genus of curve C is at most 5.*

Next, we will utilize this theorem to estimate the upper bound of the character sum for an arbitrary nontrivial character of $E(\mathbb{F}_q)$.

4.5   Estimating Character Sums on the Curve

Theorem 6. *Let f_I be the cube root encoding from $\mathbb{F}_q$ to generalized Huff curve E, $q \equiv 3 \pmod 4$. For any nontrivial character χ of $E(\mathbb{F}_q)$, the character sum $S_{f_I}(\chi)$ satisfies:*

$$|S_{f_I}(\chi)| \leqslant \begin{cases} 14\sqrt{q} + 3, & a^4 c^2 + b^4 d^2 \neq a^2 b^2 cd, \\ 8\sqrt{q} + 3, & a^4 c^2 + b^4 d^2 = a^2 b^2 cd. \end{cases} \qquad (5)$$

Proof. Let $K = \mathbb{F}_q(x,y)$ be the function field of E. Recall that a point $(x,y) \in E$ is the image of u if and only if

$$H(u;x,y) = 0.$$

Then a smooth projective curve $C = \{(x,y,u)|(x,y) \in E, H(u;x,y) = 0\}$ is introduced, whose function field is the extension $L = K[u]/(H)$. By field inclusions $\mathbb{F}_q(u) \subset L$ and $K \subset L$ we can construct birational maps $g : C \to \mathbb{P}^1(\mathbb{F}_q)$ and $h : C \to E$. Then g is a bijection and $f_I(u) = H \circ g^{-1}(u)$.

Since the genus of curve C is at most 8, by Theorem 1, we have

$$|S_{f_I}(\chi) + \sum_{P\in C(\mathbb{F}_q), u(P)=\infty} \chi \circ h(P)| = | \sum_{P\in C(\mathbb{F}_q)} \chi \circ h(P)| \leqslant (2\cdot 8 - 2)\sqrt{q} = 14\sqrt{q}.$$

For $(x,y) = (0,0)$, function $H(u; x, y) = 0$ has only one finite solution, hence there exist 3 infinite solutions of u; for other points on $\bar{E}$, it can be check that all solutions of $H(u; x, y) = 0$ are finite. Therefore $|\sum_{P\in C(\mathbb{F}_q), u(P)=\infty} \chi \circ h(P)| \leqslant 3$. Hence $|S_{f_I}(\chi)| \leqslant 14\sqrt{q} + 3$.

In the case that $a^4c^2 + b^4d^2 = a^2b^2cd$, it is proved that the genus of C is at most 5. Analogous to previous discussion, we have $|S_{f_I}(\chi)| \leqslant 8\sqrt{q} + 3$. ∎

4.6   Galois Group of Field Extension

Let $K = F(x,y)$ be the function field of generalized Huff curve E, L be the function field of C. To estimate the character sum of any character of Jacobian group of E, or to estimate the size of image of f_I, we need know the structure of $Gal(L/K)$. By [31], when L/K is a quartic extension, then $Gal(L/K) = S_4$ if and only if

1. $H(u)$ is irreducible over $F(x,y)$.
2. Let $R(u)$ be the resolvent cubic of $H(u)$, then $R(u)$ is irreducible over $F(x,y)$.
3. The discriminant of $R(u)$ is not a square in $F(x,y)$.

if L/K is a cubic extension, then $Gal(L/K) = S_3$ if and only if

1. $H(u)$ is irreducible over $F(x,y)$.
2. The discriminant of $H(u)$ is not a square in $F(x,y)$.

When L/K is a quartic extension, we have to prove 3 following lemmas:

Lemma 3. *The polynomial $H(u)$ is irreducible over $F(x,y)$.*

Proof. Substitute $x = \dfrac{bd(s + a^2c)}{t}$ and $y = \dfrac{ac(s + b^2d)}{t}$ into $H(u; x, y)$, we only need to show

$$\tilde{H}(u; s, t) = \begin{cases} u^4 - (2a^2c + 2b^2d - 6s)u^2 + 6tu + (a^4c^2 + b^4d^2 - a^2b^2cd), \\ \text{when } a^4c^2 + b^4d^2 \neq a^2b^2cd, \\ \\ u^3 + \left(-2a^2c - 6s - 2b^2d\right)u + 6v, \\ \text{when } a^4c^2 + b^4d^2 = a^2b^2cd \end{cases}$$

is irreducible over $F(s,t) = F(x,y) = K$. Let σ be the non trivial Galois automorphism in $Gal(F(s,t)/F(t))$, which maps t to $-t$, it remains to show $\tilde{H}_0(u; s, t) = \tilde{H}(u; s, t)\tilde{H}(u; s, t)^\sigma$ is irreducible over $F(t)$. Let $v = u^2$, Note that $\tilde{H}_0(u)$ can be represented as polynomial of v:

$$J_0(v) = v^4 + (-4\,ca^2 - 12\,s - 4\,db^2)v^3 + (6\,b^4d^2 + 6\,a^2b^2cd + 6\,a^4c^2 + 24\,sca^2 +$$
$$36\,s^2 + 24\,sdb^2)v^2 + (-12\,b^4d^2s - 4\,b^6d^3 - 24\,a^2b^2cds - 4\,a^6c^3 - 12\,a^4c^2s$$
$$- 36\,s^3 - 36\,db^2s^2 - 36\,ca^2s^2)t + (b^4d^2 - a^2b^2cd + a^4c^2)^2,$$
$$\text{if } a^4c^2 + b^4d^2 \neq a^2b^2cd, \tag{6}$$

or

$$J_0(v) = v^3 + (-4\,a^2c - 12\,s - 4\,b^2d)v^2 + 4\,(a^2c + 3\,s + b^2d)^2v$$
$$- 36\,s(s^2 + a^2c + b^2d + b^2da^2c), \tag{7}$$

if $a^4c^2 + b^4d^2 = a^2b^2cd$.

From (6), by Theorem 1.2.3 in [31], if $J_0(v)$ is reducible over $F(s)$, then either it can be decomposed as

$$J_0(v) = (v + A)(v^3 + Bv^2 + Cv + D)$$
$$= v^4 + (A + B)v^3 + (AB + C)v^2 + (AC + D)v + AD,$$

or it can be decomposed as

$$J_0(v) = (v^2 + Av + B)(v^2 + Cv + D)$$
$$= v^4 + (A + C)v^3 + (B + AC + D)v^2 + (BC + AD)v + BD,$$

where $A, B, C, D \in F[s]$.

In the first case, note that $AD = (b^4d^2 - a^2b^2cd + a^4c^2)^2$, A and D are both constant. Since $A + B = -4\,ca^2 - 12\,s - 4\,db^2$, B is of degree 1. Since the coefficient of v^2 is 2, degree of C is 2, which can lead to the inference that the degree of v is also 2, a contradiction to the fact it is 3.

In the second case, B and D are constants. Hence summation of the degree of A and the degree of C equals to 2, which shows that the coefficient of v is at most 2, also a contradiction.

Then we have shown that $J_0(v)$ is irreducible over $F(s)$. Let z be a root of $H_0(u)$. Then

$$[F(s, z) : F(s)] = [F(s, z) : F(s, z^2)] \cdot [F(s, z^2) : F(s)] = 4[F(s, z) : F(s, z^2)].$$

Since $\tau \in Gal(F(s, z)/F(s, z^2))$ which maps z to $-z$ is not an identity, hence $Gal(F(s, z)/F(s, z^2)) \neq \{\iota\}$, then $[F(s, z) : F(s, z^2)] \geqslant 2$. Hence $[F(s, z) : F(s)] \geqslant 8$, which shows that $H_0(u)$ is irreducible over $F(s)$.

From (7), $J_0(v)$ is cubic, then if it is reducible, it should have a root in $F(s)$, which is factor of the constant coefficient of $J_0(v)$. However, we can confirm that such root does not exist by enumerating all the possibilities. The remaining step is similar to previous case. ∎

Lemma 4. *The resolvent polynomial $R(u; x, y)$ is irreducible over $F(x, y)$.*

Proof. In the case that $H(u; x, y)$ is quartic, the resolvent cubic of $H(u)$ is

$$
\begin{aligned}
R(u; x, y) = {}& (acx - bdy)^2 u^3 + 2(acx - bdy)(-2cxb^2 ad + a^3 c^2 x + 2a^2 bcdy) \\
& - b^3 d^2 y)u^2 - 4(a^4 c^2 + b^4 d^2 - a^2 b^2 cd)(acx - bdy)^2 u - 36\, a^6 b^2 c^4 d^2 \\
& + 72\, a^4 b^4 c^3 d^3 - 36\, a^2 b^6 c^2 d^4 + 16\, b^6 d^3 a^2 c^2 x^2 + \quad 24\, b^6 d^4 a^2 cy^2 \\
& - 24\, b^4 d^2 x^2 a^4 c^3 - 24\, a^4 b^4 c^2 d^3 y^2 + 24\, a^6 b^2 c^4 dx^2 + 16\, a^6 c^3 b^2 d^2 y^2 \\
& - 8\, b^7 d^4 yacx - 8\, a^7 c^4 bdyx - 8\, b^8 d^5 y^2 - 8\, a^8 c^5 x^2
\end{aligned}
\tag{8}
$$

Similar to previous lemma, we only need to show $\tilde{R}(u; s, t)$, the transformation of $R(u; x, y)$ such that it is defined on $\psi^{-1}(E)$, is irreducible over $F(s, t)$. Represent x, y with variable s, t, we have

$$
\begin{aligned}
\tilde{R}(u; s) = {}& u^3 + (2\, ca^2 + 6\, s + 2\, db^2)u^2 + (-4\, b^4 d^2 + 4\, a^2 b^2 cd - 4\, a^4 c^2)u \\
& - 24\, a^4 c^2 s - 12\, b^2 da^2 cs - 24\, b^4 d^2 s - 8\, a^6 c^3 - 36\, s^2 a^2 c - 36\, s^2 b^2 d \\
& - 8\, b^6 d^3 - 36\, s^3
\end{aligned}
\tag{9}
$$

If $\tilde{R}(u; s)$ is reducible, it must have a degree 1 factor $u + A$, where $A \in F[s, t]$. If $A \notin F[s]$, then $(u + A)^\sigma$ is a factor of $\tilde{R}(u; s)^\sigma = \tilde{R}(u; s)$. Hence $\dfrac{\tilde{R}(u; s)}{(u + A)(u + A)^\sigma} \in F[s]$. Without loss of generality, we suppose $A \in F[s]$. Hence $\tilde{R}(u; s) = (u + A)(u^2 + Bu + C)$, $A, B, C \in F[s]$. In this case, $\tilde{R}(u; s)$ has a solution in $F[s]$ whose degree is 1, since when the value of u is a polynomial with degree $\neq 1$, $\tilde{R}(u; s)$ will be equal to a polynomial whose degree greater than 0. Suppose $A = Ps + Q$, $P, Q \in F$, then

$$
\begin{cases}
B & = 6s + 2b^2 d + 2a^2 c - (Ps + Q) \\
C & = -4b^4 d^2 + 4a^2 b^2 cd - 4a^4 c^2 - AB \\
AC & = -24a^4 c^2 s - 12b^2 da^2 cs - 24b^4 d^2 s - 8a^6 c^3 - 36s^2 a^2 c - 36s^2 b^2 d \\
& \quad -8b^6 d^3 - 36s^3.
\end{cases}
$$

Then P and Q satisfies

$$
\begin{aligned}
& P^2(P - 6)s^3 + P(3\,QP - 12\,Q - 2\,Pb^2 d - 2\,Pa^2 c)s^2 + \\
& (3\,Q^2 P - 6\,Q^2 - 4\,QPb^2 d - 4\,QPa^2 c - 4\,Pa^4 c^2 - 4\,Pb^4 d^2 + 4\,Pa^2 b^2 cd)s + \\
& Q(Q^2 - 4\,b^4 d^2 + 4\,a^2 b^2 cd - 4\,a^4 c^2 - 2\,Qb^2 d - 2\,Qa^2 c) = 0
\end{aligned}
\tag{10}
$$

where s is the variable. When $char(F) \geqslant 3$, it can be checked that solutions of P and Q do not exist. ∎

Lemma 5. *Let $D(x, y)$ be the discriminant of $R(u; x, y)$, then $D(x, y)$ is not a square in $F(x, y)$.*

Proof. Similar to previous proof, we only need to show that

$$
\tilde{D}(s, t) = D(x(s, t), y(s, t))
$$

is not a square in $F(s, t)$. After simplification,

$$\tilde{D}(s,t) = -\frac{2^7 \cdot 3^5 \cdot (abcd(a^2c - b^2d))^8}{t^8} \cdot (27\,s^6 - (-54\,a^2c - 54\,b^2d)s^5 - (-27\,a^4c^2$$
$$- 108\,a^2b^2cd - 27\,b^4d^2)s^4 + 2\,(a^2c + b^2d)(8\,b^4d^2 + 7\,a^2b^2cd + 8\,a^4c^2)s^3$$
$$+ 3\,a^2b^2cd(8\,a^4c^2 - 23\,a^2b^2cd + 8\,b^4d^2)s^2 - 24\,a^4b^4c^2d^2(a^2c + b^2d)s$$
$$- 16\,b^4d^2a^4c^2(a^4c^2 + b^4d^2 - a^2b^2cd)), \tag{11}$$

In fact, we only need to show that $\tilde{G}(s,t) = -\dfrac{t^8}{2^7 \cdot 3^5 \cdot (abcd(a^2c - b^2d))^8}\tilde{D}(s,t)$
is irreducible over $F(s,t)$.

Suppose $\tilde{G}$ is a square in $F(s,t)$, then $F(s,t) \supseteq F(s,\sqrt{\tilde{G}}) \supseteq F(s)$. Note that $[F(s,t) : F(s)] = 2$, either $F(s,\sqrt{\tilde{G}}) = F(s,t)$ or $F(s,\sqrt{\tilde{G}}) = F(s)$.

In the first case, $\tilde{G}$ is $s(s + a^2c)(s + b^2d) = t^2$ times a square in $F(s)$. But divide $\tilde{G}$ by $s(s + a^2c)(s + b^2d)$, the remainder vanishes if and only if $a^4c^2 + b^4d^2 - a^2b^2cd = 0$.

In the second case, $\tilde{G}$ is a square over $F(s)$. Suppose

$$\tilde{G}(s) = \left(\sqrt{27}s^3 + Bs^2 + Cs \pm 4a^2b^2cd\sqrt{a^2b^2cd - a^4c^2 - b^4d^2}\right)^2,$$

expand the right hand side of this equation and compare its coefficients of $s^i, i = 1$ to 5 with the left hand side, and it is checked there are no $B, C \in F$ s.t the equality holds. ∎

Remark: by similar method, we can also prove that when L/K is a cubic extension, $H(u;x,y)$ is irreducible over $F(x,y)$ and its discriminant is not a square in $F(x,y)$.

Summarize these lemmas, we directly deduce:

Theorem 7. *Let $K = \mathbb{F}_q(x,y)$ be the function field of E. The polynomial $H(u;x,y)$ introduced in (3) is irreducible over K, then when $a^4c^2 + b^4d^2 \neq a^2b^2cd$, its Galois group is S_4; when $a^4c^2 + b^4d^2 = a^2b^2cd$, its Galois group is S_3.*

In Sect. 5.2, we will use this theorem to construct a hash function indifferentiable from random oracle.

4.7   Calculating the Density

Similar to Sect. 3.4, we apply Chebotarev density theorem to estimate the size of image of f_I.

Theorem 8. *Let E be the generalized Huff curve over $\mathbb{F}_q$ defined by equation $ax(y^2 - c) = by(x^2 - d), abcd(a^2c - b^2d) \neq 0$, f_I is the corresponding hash function defined in Sect. 4.1. Then if $a^4c^2 + b^4d^2 \neq a^2b^2cd$, we have*

$$|\#Im(f_I) - \frac{5}{8}q| \leq \frac{5}{4}(31q^{1/2} + 72q^{1/4} + 67),$$

and if $a^4c^2 + b^4d^2 = a^2b^2cd$, we have

$$|\#Im(f_I) - \frac{2}{3}q| \leqslant \frac{4}{3}(10q^{1/2} + 18q^{1/4} + 30).$$

Proof. K is the function field of E which is the quadratic extension of $\mathbb{F}_q(x)$, hence $d = 2$, and by the property of elliptic curve, $g_K = 1$.

In the case that $a^4c^2 + b^4d^2 \neq a^2b^2cd$, $Gal(L/K) = S_4$, hence $m = \#S_4 = 24$. φ is the subset of $Gal(L/K)$ consisting at least 1 fixed point, which are conjugates of $(1)(2)(3)(4), (12)(3)(4)$ and $(123)(4)$, then $s = 1+6+8 = 15$. Since the place v of K of degree 1 correspond to the projective unramified points on $E(\mathbb{F}_q)$, hence $|\#Im(f_I) - N(\varphi)| \leqslant 12 + 3 = 15$, where 3 represents the number of infinite points, 12 represents the number of ramified points. Then we have

$$|\#Im(f_I) - \frac{5}{8}q| \leqslant |\#Im(f_I) - N(\varphi)| + |N(\varphi) - \frac{5}{8}q|$$

$$\leqslant 15 + \frac{5}{4}(31q^{1/2} + 72q^{1/4} + 55)$$

$$= \frac{5}{4}(31q^{1/2} + 72q^{1/4} + 67).$$

In the case that $a^4c^2 + b^4d^2 = a^2b^2cd$, $Gal(L/K) = S_3$, hence $m = \#S_3 = 6$. The corresponding s has the value of 4. $|\#Im(f_I) - N(\varphi)| \leqslant 6 + 3 = 9$, where 3 represents the number of infinite points, 6 represents the number of ramified points. Hence

$$|\#Im(f_I) - \frac{2}{3}q| \leqslant |\#Im(f_I) - N(\varphi)| + |N(\varphi) - \frac{2}{3}q|$$

$$\leqslant 9 + \frac{2}{3}(10q^{1/2} + 18q^{1/4} + 16)$$

$$= \frac{2}{3}(10q^{1/2} + 18q^{1/4} + 30). \qquad \blacksquare$$

5   Construction of Hash Function Indifferentiable from Random Oracle

Let h be a classical hash function from messages to finite field $\mathbb{F}_q$, we can show that both $f_S \circ h$ and $f_I \circ h$ are one-way and collision-resistance according to the fact that each point on E has finite preimage through f_S and f_I [16]. Hence $f_S \circ h$ and $f_I \circ h$ are both hash functions mapping messages to $E(\mathbb{F}_q)$. However, since f_S and f_I are not surjective, $f_S \circ h$ and $f_I \circ h$ are easy to be distinguished from a random oracle even when h is modeled as a random oracle to $\mathbb{F}_q$ [33]. Therefore, we introduce 2 new constructions of hash functions which are indifferentiable from a random oracle.

5.1   First Construction

Suppose $f : \mathbb{S} \to \mathbb{G}$ is a weak encoding [26] to a cyclic group $\mathbb{G}$, where $\mathbb{S}$ denotes prime field $\mathbb{F}_q$, $\mathbb{G}$ denotes $E(\mathbb{F}_q)$ which is of order N with generator G, $+$ denotes elliptic curve addition. According to the proof of random oracle, we can construct a hash function $H_R : \{0,1\}^* \to \mathbb{G}$:

$$H_R(m) = f(h_1(m)) + h_2(m)G,$$

where $h_1 : \{0,1\}^* \to \mathbb{F}_q$ and $h_2 : \{0,1\}^* \to \mathbb{Z}/N\mathbb{Z}$ are both classical hash functions. $H_R(m)$ is indifferentiable from a random oracle in the random oracle model for h_1 and h_2.

We only need to show f_S, f_I are both weak encodings to prove that $H_S(m) = f_S(h_1(m)) + h_2(m)G$ and $H_I(m) = f_I(h_1(m)) + h_2(m)G$ are indifferentiable from a random oracle in the random oracle model for h_1 and h_2. By the definition of weak encoding [26], f_S is a $\dfrac{2N}{q}$-weak encoding and f_I is a $\dfrac{4N}{q}$-weak encoding, both $\dfrac{2N}{q}$ and $\dfrac{4N}{q}$ are polynomial functions of the security parameter.

5.2   Second Construction

Another construction is as follows:

$$\begin{cases} H_{S'} = f_S(h_1(m)) + f_S(h_2(m)) \\ H_{I'} = f_I(h_1(m)) + f_I(h_2(m)). \end{cases}$$

We have proved that f_S, f_I are both well distributed encodings in Sects. 3.3 and 4.5. According to corollary 2 of [29], $H_{I'}$ and $H_{S'}$ are both indifferentiable from a random oracle, where h_1 and h_2 are regarded as independant random oracles with values in $\mathbb{F}_q$.

6   Time Comparison

When $q \equiv 3 \pmod 4$, the key step of an encoding function is calculating square root for given element of $\mathbb{F}_q$. For convenience to make comparisons, we first introduce a birational map between generalized Huff curve E and short Weierstrass curve

$$E_W : t^2 = s^3 + \frac{a^2b^2cd - a^4c^2 - b^4d^2}{3}s + \frac{1}{27}(2\,a^6c^3 - 3\,a^4c^2b^2d - 3\,a^2cb^4d^2 + 2\,b^6d^3),$$

via maps

$$\vartheta : E \to E_W :$$

$$(x,y) \mapsto (s,t) = \left(\frac{1}{3}\frac{2\,a^2bcdy - 2\,ab^2cdx + xa^3c^2 - b^3d^2y}{axc - byd},\ \frac{bdac\left(a^2c - b^2d\right)}{axc - byd} \right),$$

$$\varsigma : E_W \to E :$$

$$(s,t) \mapsto (x,y) = \left(\frac{bd\left(s + \frac{2}{3}a^2c - \frac{1}{3}b^2d\right)}{t},\ \frac{ac\left(s + \frac{2}{3}b^2d - \frac{1}{3}a^2c\right)}{t} \right). \quad (12)$$

Table 1. Theoretic time cost of different deterministic encodings

Encoding	Cost	Converted cost
f_S	$E_S + 2I + D + S + 10M$	$E + D + 31M$
f_U	$E_S + 2I + D + 4S + 15M$	$E + D + 39M$
f_E	$E_S + 2I + D + 4S + 10M$	$E + D + 34M$
f_I	$E_C + I + 4S + 10M$	$E + 24M$
f_Y	$E_C + 2I + 3S + 7M$	$E + 30M$
f_A	$E_C + 2I + 4S + 9M$	$E + 33M$

Table 2. NIST primes

Prime	Value	Residue (mod 3)	Residue (mod 4)
$P192$	$2^{192} - 2^{64} - 1$	2	3
$P384$	$2^{384} - 2^{128} - 2^{96} + 2^{32} - 1$	2	3

Table 3. Time cost (ms) of different square root methods on NIST

Prime	$P192$	$P384$
f_S	0.053	0.235
f_E	0.057	0.248
f_U	0.058	0.250

Therefore, we compare our encoding f_S with 2 encodings: birational equivalence ς in (12) composed with Ulas' encoding function [15], denoted by f_U; ς composed with simplified Ulas map given by Eric Brier et al., denoted by f_E.

When $q \equiv 2 \pmod 3$, the essential of an encoding function is calculating the cube root for elements of $\mathbb{F}_q$. We compare our encoding f_I with Alasha's work [19] denoted by f_A and Yu's encoding function [32] denoted by f_Y. In comparison with f_A, we let $c = \dfrac{1}{a}, d = \dfrac{1}{b}$ since Alasha only treats this special case; in comparison with f_Y, we let $c = d = 1$, since Yu's work can only be applied on classical Huff curves.

We have shown that f_S costs $E+D+31M$, f_I costs $E+24M$. For comparison, f_U costs $(E_S+I+4S+11M+D)+(I+4M) = E+D+39M$ by Theorem 2.3(2), [15] and the map ς in (12), while f_E costs $(E_S+I+4S+6M+D)+(I+4M) = E + D + 34M$ by [14]. Yu's encoding f_Y costs $E_C + 2I + 3S + 7M = E + 30M$, Alasha's encoding f_A costs $E_C + 9M + 4S + 2I = E + 33M$ (Table 1).

We do experiments on prime field $\mathbb{F}_{P192}$ and $\mathbb{F}_{P384}$ (see Table 2). General Multiprecision PYthon project (GMPY2) [34], which supports the GNU Multiple Precision Arithmetic Library (GMP) [35] is used for big number arithmetic. The experiments are operated on an Intel(R) Core(TM) i5-4570, 3.20 GHz processor. We ran f_S, f_U, f_E, f_I, f_Y and f_A $1,000,000$ times each, where u is randomly chosen on $\mathbb{F}_{P192}$ and $\mathbb{F}_{P384}$.

Table 4. Time cost (ms) comparison between f_I and f_A

Prime	$P192$	$P384$
f_I	0.053	0.233
f_A	0.061	0.252

Table 5. Time cost (ms) comparison between f_I and f_Y

Prime	$P192$	$P384$
f_I	0.052	0.233
f_Y	0.058	0.244

From the average running times listed in Table 3, f_S is the fastest among encodings which need calculate square roots. On $\mathbb{F}_{P192}$, it saves 9.19 % running time compared with f_U, 7.69 % running time compared with f_E. On $\mathbb{F}_{P384}$, f_S saves 5.92 % running time compared with f_U and 5.17 % running time compared with f_E. f_I is also the fastest among encodings which need to calculate cube roots. On $\mathbb{F}_{P192}$, it saves 13.20 % of running time compared with f_A and 8.97 % compared with f_Y. On $\mathbb{F}_{P384}$, the relevant percentages are 7.51 % and 4.40 % (see Tables 4 and 5).

7   Conclusion

We provide two constructions of deterministic encoding into generalized Huff curves over finite fields, namely, brief SWU encoding and cube root encoding. We do theoretical analysis and practical implementations to show that when $q \equiv 3 \pmod 4$, SWU encoding is the most efficient among existed methods mapping $\mathbb{F}_q$ into generalized Huff curve E, while cube root encoding is the most efficient one when $q \equiv 2 \pmod 3$. For any nontrivial character χ of $E(\mathbb{F}_q)$, we estimate the upper bound of the character sums of both encodings. As a corollary, hash functions indifferentiable from random oracle are constructed. We also estimate image sizes of our encodings by applying Chebotarev density theorem.

References

1. Baek, J., Zheng, Y.: Identity-based threshold decryption. In: Bao, F., Deng, R., Zhou, J. (eds.) PKC 2004. LNCS, vol. 2947, pp. 262–276. Springer, Heidelberg (2004)
2. Horwitz, J., Lynn, B.: Toward hierarchical identity-based encryption. In: Knudsen, L.R. (ed.) EUROCRYPT 2002. LNCS, vol. 2332, pp. 466–481. Springer, Heidelberg (2002)

3. Boneh, D., Gentry, C., Lynn, B., Shacham, H.: Aggregate and verifiably encrypted signatures from bilinear maps. In: Biham, E. (ed.) EUROCRYPT 2003. LNCS, vol. 2656, pp. 416–432. Springer, Heidelberg (2003)
4. Zhang, F., Kim, K.: ID-based blind signature and ring signature from pairings. In: Zheng, Y. (ed.) ASIACRYPT 2002. LNCS, vol. 2501, pp. 533–547. Springer, Heidelberg (2002)
5. Boyen, X.: Multipurpose identity-based signcryption. In: Boneh, D. (ed.) CRYPTO 2003. LNCS, vol. 2729, pp. 383–399. Springer, Heidelberg (2003)
6. Libert, B., Quisquater, J.-J.: Efficient signcryption with key privacy from gap Diffie-Hellman groups. In: Bao, F., Deng, R., Zhou, J. (eds.) PKC 2004. LNCS, vol. 2947, pp. 187–200. Springer, Heidelberg (2004)
7. Lindell, Y.: Highly-efficient universally-composable commitments based on the DDH assumption. In: Paterson, K.G. (ed.) EUROCRYPT 2011. LNCS, vol. 6632, pp. 446–466. Springer, Heidelberg (2011)
8. Boneh, D., Franklin, M.: Identity-based encryption from the weil pairing. In: Kilian, J. (ed.) CRYPTO 2001. LNCS, vol. 2139, pp. 213–229. Springer, Heidelberg (2001)
9. Boyd, C., Montague, P., Nguyen, K.: Elliptic curve based password authenticated key exchange protocols. In: Varadharajan, V., Mu, Y. (eds.) ACISP 2001. LNCS, vol. 2119, pp. 487–501. Springer, Heidelberg (2001)
10. Jablon, D.P.: Strong password-only authenticated key exchange. SIGCOMM Comput. Commun. Rev. **26**(5), 5–26 (1996)
11. Boyko, V., MacKenzie, P.D., Patel, S.: Provably secure password-authenticated key exchange using Diffie-Hellman. In: Preneel, B. (ed.) EUROCRYPT 2000. LNCS, vol. 1807, pp. 156–171. Springer, Heidelberg (2000)
12. Shallue, A., van de Woestijne, C.E.: Construction of rational points on elliptic curves over finite fields. In: Hess, F., Pauli, S., Pohst, M. (eds.) ANTS 2006. LNCS, vol. 4076, pp. 510–524. Springer, Heidelberg (2006)
13. Skalba, M.: Points on elliptic curves over finite fields. Acta Arith. **117**, 293–301 (2005)
14. Fouque, P.-A., Tibouchi, M.: Estimating the size of the image of deterministic hash functions to elliptic curves. In: Abdalla, M., Barreto, P.S.L.M. (eds.) LATINCRYPT 2010. LNCS, vol. 6212, pp. 81–91. Springer, Heidelberg (2010)
15. Ulas, M.: Rational points on certain hyperelliptic curves over finite fields. Bull. Polish Acad. Sci. Math. **55**, 97–104 (2007)
16. Icart, T.: How to hash into elliptic curves. In: Halevi, S. (ed.) CRYPTO 2009. LNCS, vol. 5677, pp. 303–316. Springer, Heidelberg (2009)
17. Farashahi, R.R.: Hashing into hessian curves. In: Nitaj, A., Pointcheval, D. (eds.) AFRICACRYPT 2011. LNCS, vol. 6737, pp. 278–289. Springer, Heidelberg (2011)
18. Yu, W., Wang, K., Li, B., Tian, S.: About hash into montgomery form elliptic curves. In: Deng, R.H., Feng, T. (eds.) ISPEC 2013. LNCS, vol. 7863, pp. 147–159. Springer, Heidelberg (2013)
19. Alasha, T.: Constant-time encoding points on elliptic curve of diffierent forms over finite fields (2012). http://iml.univ-mrs.fr/editions/preprint2012/files/tammam_alasha-IML_paper_2012.pdf
20. Yu, W., Wang, K., Li, B., Tian, S.: Construct hash function from plaintext to C_{34} curves. Chin. J. Comput. **35**(9), 1868–1873 (2012)
21. Huff, G.B.: Diophantine problems in geometry and elliptic ternary forms. Duke Math. J. **15**(2), 443–453 (1948)
22. Joye, M., Tibouchi, M., Vergnaud, D.: Huff's model for elliptic curves. In: Hanrot, G., Morain, F., Thomé, E. (eds.) ANTS-IX. LNCS, vol. 6197, pp. 234–250. Springer, Heidelberg (2010)

23. Wu, H., Feng, R.: Elliptic curves in Huff model. Wuhan Univ. J. Nat. Sci. **17**(6), 473–480 (2011)

24. Elmegaard-Fessel, L.: Efficient Scalar Multiplication and Security against Power Analysis in Cryptosystems based on the NIST Elliptic Curves Over Prime Fields. Eprint, 2006/313. http://eprint.iacr.org/2006/313

25. Standards for Efficient Cryptography: Elliptic Curve Cryptography Ver. 5 (1999). http://www.secg.org/drafts.html

26. Brier, E., Coron, J.-S., Icart, T., Madore, D., Randriam, H., Tibouchi, M.: Efficient indifferentiable hashing into ordinary elliptic curves. In: Rabin, T. (ed.) CRYPTO 2010. LNCS, vol. 6223, pp. 237–254. Springer, Heidelberg (2010)

27. Ciss, A.A., Sow, D.: On a new generalization of Huff curves. Cryptology ePrint Archive: Report 2011/580 (2011). http://eprint.iacr.org/2011/580.pdf

28. Devigne, J., Joye, M.: Binary Huff curves. In: Kiayias, A. (ed.) CT-RSA 2011. LNCS, vol. 6558, pp. 340–355. Springer, Heidelberg (2011)

29. Farashahi, R.R., Fouque, P.-A., Shparlinski, I.E., Tibouchi, M., Voloch, J.F.: Indifferentiable deterministic hashing to elliptic and hyperelliptic curves. Math. Comp. **82**, 491–512 (2013)

30. Farashahi, R.R., Shparlinski, I.E., Voloch, J.F.: On hashing into elliptic curves. J. Math. Cryptol. **3**(4), 353–360 (2009)

31. Roman, S.: Field Theory. Graduate Texts in Mathematics, vol. 158, 2nd edn. Springer, New York (2011)

32. Wei, Y., Wang, K., Li, B.: Constructing hash function from plaintext to Huff curves. J. Univ. Sci. Tech. China (10), 835–838 (2014)

33. Tibouchi, M.: Impossibility of surjective icart-like encodings. In: Chow, S.S.M., Liu, J.K., Hui, L.C.K., Yiu, S.M. (eds.) ProvSec 2014. LNCS, vol. 8782, pp. 29–39. Springer, Heidelberg (2014)

34. GMPY2, General Multiprecision Python (Version 2.2.0.1). https://gmpy2.readthedocs.org

35. GMP: GNU Multiple Precision Arithmetic Library. https://gmplib.org/

Signature Schemes

Cubic Unbalance Oil and Vinegar Signature Scheme

Xuyun Nie[1,2,3]($\boxtimes$), Bo Liu[1], Hu Xiong[1], and Gang Lu[1]

[1] School of Information and Software Engineering,
University of Electronic Science and Technology of China, Chengdu 610054, China
xynie@uestc.edu.cn, boblv587@gmail.com, xionghuxiexie@163.com
[2] State Key Laboratory of Information Security,
Institute of Information Engineering, Chinese Academy of Sciences,
Beijing 100093, China
[3] Network and Data Security Key Laboratory of Sichuan Province, Chengdu, China

Abstract. As one fundamental primitive of multivariate public key cryptosystem, Unbalance Oil-Vinegar (UOV) signature scheme can offer the function of digital signature with the resistance to the quantum algorithm attack. By considering the large size of public key and signature length, we propose a cubic UOV scheme by adopting the stepwise iteration method in this paper. Comparing to the existing work such as the original UOV and its improvements, the proposed scheme enjoys shorter signature size and faster signing operation under the same security level at the cost of larger public key size. This feature is especially desirable in the environments where the computation resource of signer is limited and the communication overhead matters.

Keywords: Unbalance oil and vinegar signature scheme · Multivariate cryptosystem · Cubic polynomial

1 Introduction

The development of quantum computers will pose a threat to the safety of the traditional public key cryptosystems based on number theoretic hard problems. Multivariate Public Key Cryptography (MPKC) arises at this historic moment which can be seen as a candidate to resist quantum algorithm attack. Its security is based on the hardness of solving Multivariate Quadratic (MQ) polynomials equation system, which is NP hard problem in the worst case. Compared with traditional public key cryptosystems, MPKCs are very fast.

Patarin [1] proposed a multivariate public key signature scheme, named Oil and Vinegar (OV) scheme. The key idea of OV is to construct several so-called OV polynomials, in which the polynomials would be one degree polynomials on the oil variables given the values of the vinegar variables. In OV scheme, the number of oil variables is equal to the number of vinegar variables. Kipnis and Shamir [2] found that there is no quadratic terms on the oil variables which

© Springer International Publishing Switzerland 2016
D. Lin et al. (Eds.): Inscrypt 2015, LNCS 9589, pp. 47–56, 2016.
DOI: 10.1007/978-3-319-38898-4_3

made it possible to separate the oil and vinegar variables. Then, Kipnis et al. [3] proposed the unbalance oil and vinegar signature scheme (UOV) in which the number of vinegar variables is greater than the oil's. Unfortunately, the Oil-Vinegar separation attack can also work on it. In order to resist this attack, the number of the vinegar variables should be 2 times over the oils. Hence the public key and the length of signature were increased.

Since then, many people followed their work on how to break the UOV scheme and how to improve the efficiency. Ding et al. [4] proposed a multi-layered UOV scheme, named Rainbow. This scheme greatly improved efficiency of UOV, but it is also facing the threat of separating the oil and vinegar space [5]. Petzoldt et al. [6–9] used the linear recurring sequences (LRS) and the cyclic matrix methods to optimize Rainbow and UOV's key generation. This method makes the size of public key reduced by 86 % and 62 % respectively, while it also accelerate the speed of signature verification. But the length of signatures in them are the same as in the original UOV and Rainbow. Although Petzoldt et al. reduced the public key size of the UOV, the length of signature is still 3 times over the length of the message or its hash to be signed. Furthermore, in order to against the hybrid approach [10] for solving multivariate systems over finite fields, the recommended parameters of UOV and UOVLRS are set to be $q \geq 2^8$, $o \geq 26$, and $v \geq 52$.

In order to resist the oil-vinegar separating attack and to shorten the length of signature, we combine the UOV skill with stepwise iteration method to reconstruct the central map in UOV scheme. Firstly, we randomly choose an OV polynomial f_1 and a set of one degree polynomials $f_2, \ldots, f_o$ on the oil and vinegar variables. Then we construct some cubic polynomials and quadratic polynomials by multiplying $f_1, f_2, \ldots, f_o$ and plus some random quadratic polynomials on the vinegar variables. In signature generation, we can use stepwise iteration method to inverse the central map. The new scheme has enough cross-items on the oil variables. It can resist the oil and vinegar separation attack and the number of vinegar variables need not be bigger than the number of the oils. According to our analysis, the cubic UOV scheme has smaller public key size, shorter signature length, and higher efficiency than the original scheme under the same secure level. But our scheme has bigger public key size and lower speed of verification than some improvements of UOV, such as UOVLSR2, Rainbow, cyclicRainbow etc.

The paper is organized as follows. In Sect. 2, we describe the basic idea of original unbalance oil and vinegar scheme and its cryptanalysis. Then we present our improved unbalance UOV scheme in Sect. 3. In Sect. 4, we give cryptanalysis of our scheme. We suggest the parameters in practice and present the efficiency comparison to the original UOV scheme and its improvements in Sect. 5. Finally, we conclude the paper.

2   The Original UOV Signature Scheme

In this section, we will give brief description of the OV and UOV scheme and the security of these schemes.

2.1   OV and UOV Signature Scheme

We use the same notations as in [3]. Let $\mathbb{K} = F_q$ be a finite field with q elements. Let o and v be two positive integers and set $n = o+v$. Let $y = (y_1, \ldots, y_o)$ be the message (or its hash) to be signed, where $y_i \in \mathbb{K}$. The signature x is represented as an element of $\mathbb{K}^n$, denoted by $x = (x_1, \ldots, x_n)$. Let $u_1, \ldots, u_o$ be oil variables and $\hat{u}_1, \cdots, \hat{u}_v$ are vinegar variables, denoted $u = (u_1, \ldots, u_o, \hat{u}_1, \cdots, \hat{u}_v)$. The central map F of UOV consists of polynomials of the form:

$$y_k = f_k(u_1, \ldots, u_o, \hat{u}_1, \cdots, \hat{u}_v)$$
$$= \sum_{i=1}^{o}\sum_{j=1}^{v} a_{ij} u_i \hat{u}_j + \sum_{i=1}^{v}\sum_{j=1}^{v} b_{ij} \hat{u}_i \hat{u}_j + \sum_{i=1}^{o} c_i u_i + \sum_{j=1}^{v} d_j \hat{u}_j + e$$

To invert F, one can randomly chooses the values of vinegar variables $\hat{u}_1, \cdots, \hat{u}_v$ and substitutes them into F. Then one can gets a system of linear equations in the oil variables $u_1, \ldots, u_o$. Solving this system can get the values of the oil variables. If there is no solution of this system, one has to choose other values of the vinegar variables.

To hide the structure of F the central map in the public key, one should randomly choose an invertible affine map $S : \mathbb{K}^n \to \mathbb{K}^n$: $(u_1, \ldots, u_o, \hat{u}_1, \ldots, \hat{u}_v) \mapsto S(x_1, \ldots, x_n)$.

So, the public key is the map $P = F \circ S$ and the private keys consist of F and S. The more details of signature and verification process can be seen in [3].

In Patarins original paper [1], it was suggested to choose $o = v$ (Balanced Oil and Vinegar (OV)). After this scheme was broken by Kipnis and Shamir [2], it was recommended to choose $v > o$ (Unbalanced Oil and Vinegar (UOV) [3]).

2.2   Security of UOV

In the expressions of the central map of both OV and UOV, there are no quadratic terms in the oil variables. Due to this fact, the Kipnis-Shamir attack can work on both OV and UOV. In order to resist this attack, one should increase the rate of vinegar variables in the central map. In [3], it was recommended that $v = 2o$. Furthermore, for $o \geq 26$ equations and $v = 2o$ Vinegar variables, the UOV scheme over $GF(2^8)$ seems to be secure against Hybrid approach attack [10].

3   Cubic UOV

We use the same notations as in Sect. 2.1

3.1   Basic Idea of Improvement

In order to be immune to the Kipnis-Shamir attack, we want to reconstruct the central map of UOV using polynomial products which can introduce many quadratic terms in the oil variables in the central map. Randomly choose one oil-vinegar polynomial $f_1(u_1, \ldots, u_o, \hat{u}_1, \ldots, \hat{u}_v)$ and $o - 1$ linear polynomials $f_i(u_1, \ldots, u_o, \hat{u}_1, \ldots, \hat{u}_v)$, $2 \leq i \leq o$ in $\mathbb{K}$, we get a map $\bar{F} : K^n \to K^o$, as follows, where all the coefficients are belong to $\mathbb{K}$.

$$
\begin{cases}
z_1 = f_1(u_1, \ldots, u_o, \hat{u}_1, \ldots, \hat{u}_v) \\
\quad = \sum_{i=1}^{o} \sum_{j=1}^{v} a_{1ij} u_i \hat{u}_j + \sum_{i=1}^{v} \sum_{j=1}^{v} b_{1ij} \hat{u}_i \hat{u}_j + \sum_{i=1}^{o} c_{1i} u_i + \sum_{j=1}^{v} d_{1j} \hat{u}_j + e_1 \\
z_2 = f_2(u_1, \ldots, u_o, \hat{u}_1, \ldots, \hat{u}_v) = \sum_{i=1}^{o} a_{2i} u_i + \sum_{j=1}^{v} b_{2j} \hat{u}_j + c_2 \\
z_3 = f_3(u_1, \ldots, u_o, \hat{u}_1, \ldots, \hat{u}_v) = \sum_{i=1}^{o} a_{3i} u_i + \sum_{j=1}^{v} b_{3j} \hat{u}_j + c_3 \\
\quad \vdots \\
z_o = f_o(u_1, \ldots, u_o, \hat{u}_1, \ldots, \hat{u}_v) = \sum_{i=1}^{o} a_{ni} u_i + \sum_{j=1}^{v} b_{nj} \hat{u}_j + c_n
\end{cases}
\tag{1}
$$

Then we construct a map $\hat{F} : K^n \mapsto K^o$ like follows:

$$
\begin{cases}
y_1 = r_1(z_1 + z_1 z_2) + g_1 \\
y_2 = r_2 z_1 z_2 + g_2 \\
y_3 = r_3(z_1 + z_2) z_3 + g_3 \\
\quad \vdots \\
y_o = r_o(z_{o-2} + z_{o-1}) z_o + g_o
\end{cases}
\tag{2}
$$

where $r_i \neq 0 (1 \leq i \leq o)$ are the elements randomly chosen from $\mathbb{K}$. $g_i (1 \leq i \leq 3)$ are random cubic polynomials in the vinegar variables $\hat{u}_1, \cdots, \hat{u}_v$ while the others are quadratic. $y = (y_1 \ldots, y_o)$ be the message (or its hash) to be signed.

Let $F = \hat{F} \circ \bar{F}$. Given the values of $y' = (y'_1, \ldots, y'_o)$, the inverse of F can be derived as follows. Randomly choosing the values of the vinegar variables $\hat{u}_1 = \hat{u}'_1, \cdots, \hat{u}_v = \hat{u}'_v$ and substituting them into the system (2), we get a system in the unknowns z_i $(1 \leq i \leq o)$. From the first two equations in this system, we can obtain the value of $z_1 = z'_1$. And then we can calculate the values of $z_i = z'_i (2 \leq i \leq o)$ step by step. Substituting $z'_i (1 \leq i \leq o)$ and the values of the vinegar variables $\hat{u}_1 = \hat{u}'_1, \cdots, \hat{u}_v = \hat{u}'_v$ into the system (1), we get a linear equations system in the oil variables $u_1, \ldots, u_o$. Solving this system, we obtain the values of the oil variables, denoted by $u'_1, \ldots, u'_o$. So, the vector $(u'_1, \ldots, u'_o, \hat{u}'_1, \ldots, \hat{u}'_v)$ is the inverse of F corresponding to the $y' = (y'_1, \ldots, y'_o)$.

Remark 1. Note that, if $z_i = 0$ or $z_{i+1} = 0$ $(1 \leq i \leq o)$, we can not find the inverse of F. In this case, we should reselect the values of vinegar variables.

Remark 2. In formula (2), there are three cubic polynomials y_1, y_2, y_3 and $o - 3$ quadratic polynomials.

3.2   Proposed Scheme

We use the map $F = \hat{F} \circ \bar{F}$ as the central map in UOV. To hide the structure of the central map, we should also randomly choose an invertible affine map $S : \mathbb{K}^n \to \mathbb{K}^n$:

$(u_1, \ldots, u_o, \hat{u}_1, \ldots, \hat{u}_v) = S(x_1, \ldots, x_n).$

Let $P = F \circ S$. We can now fully describe the cubic UOV scheme.

Public Key. The public key consists of the following items.

(1) The field $\mathbb{K}$, including the additive and multiplicative structure.
(2) The map $P = F \circ S$ or equivalently, its components: $p_1, p_2, \ldots, p_o \in \mathbb{K}[x_1, x_2, \ldots, x_n].$

Private Information. The private key consists of the following items.

(1) The invertible affine transformation $S : \mathbb{K}^n \to \mathbb{K}^n$.
(2) The map $\hat{F}$ and $\bar{F}$.

Signature Generation. Let $y' = (y'_1, \ldots, y'_o)$ be the document or its hash to be signed. First the signer computes

$$(z'_1, \ldots, z'_o) = \hat{F}^{-1}(y'_1, \ldots, y'_o),$$

for some random choice of $(\hat{u}'_1, \ldots, \hat{u}'_v) \in K^v$. And then the signer computes

$$(u'_1, \ldots, u'_o) = \bar{F}^{-1}(z'_1, \ldots, z'_o, \hat{u}'_1, \ldots, \hat{u}'_v).$$

At last, the signer computes the signature of $y' = (y'_1, \ldots, y'_o)$ as

$$(x'_1, \ldots, x'_n) = S^{-1}(u'_1, \ldots, u'_o, \hat{u}'_1, \ldots, \hat{u}'_v).$$

Signature Verification. To verify $(x'_1, \ldots, x'_n)$ is indeed a valid signature of the message $y' = (y'_1, \ldots, y'_o)$, the recipient determines whether or not

$$P(x'_1, \ldots, x'_n) = (y'_1, \ldots, y'_o).$$

Remark 3. Due to the special structure of our central map, the public key of our scheme contains three cubic polynomials and $o - 3$ quadratic polynomials.

4   Security of Cubic UOV

In this section, we will study the known attacks against UOV signature scheme and their effect on our scheme.

4.1   Linearized Equations Attack

Linearization equations attack is an important tool against MPKC [11,12]. In general, one should consider the first order linearization Eq. (3) and second order linearization Eq. (4) of form as follows.

$$\sum_{i=1,j=1}^{n,o} a_{ij}x_iy_j + \sum_{i=1}^{n} b_ix_i + \sum_{j=1}^{o} c_jy_j + d = 0. \tag{3}$$

$$\sum a_{ijk}x_iy_jy_k + \sum b_{ij}x_iy_j + \sum c_ix_i + \sum d_{jk}y_jy_k + \sum e_jy_j + f = 0. \tag{4}$$

If one can find some linearization equations, he/she may forge a valid signature with the help of linearization equations.

Due to the $g_i(1 \leq i \leq o)$ are randomly chosen cubic polynomials or quadratic polynomials, we can not deduce linearization equations like Eqs. (3) and (4) in theoretical analysis. Hence, we did many computer experiments to check them. To find a linearization equation is to find coefficients in Eqs. (3) or (4). To do this, we first calculate the number N of coefficients in Eqs. (3) or (4). Then, randomly generate sufficiently many (greater than N) message/signature pairs and substitute them into Eqs. (3) or (4), we get a linear equation system of unknown coefficients. In all of our experiments, the linear equation systems have no solution. So, the linearization equations attack can not work on our scheme.

4.2   Rank Attack

There are two different types of rank attack. The first one is called MinRank attack, the other is called HighRank attack. We will consider these two attacks against the Cubic UOV.

In the MinRank attack, one wants to recover the private key of MPKCs whose quadratic form associated to the homogeneous quadratic parts is of low rank. In this attack, one tries to find linear combinations $H = \sum_{i=1}^{m} a_iH_i$, where all H_i have low rank. According to [13], the complexity of MinRank attack is $q^{ts}m^3$, where s is the minimum rank, $t = \lceil \frac{m}{n} \rceil$. In our scheme, $s \approx o$, $t = 1$, $q = 2^8$, when $o \geq 10$, the complexity is at least $O(2^{80})$.

In HighRank attack, one tries to find the variables appearing the lowest number of times in the central map. These are the oil variables in our scheme. According to [13], the complexity of HighRank is $q^w(wn^2 + n^3/6)$, where w is the minimal number of appearance in central map for any plaintext variables. In our scheme, $w \approx o$. When $o \geq 10$, the complexity of this attack is greater than $O(2^{80})$.

So when we choose $q = 2^8$ and $o \geq 10$, our scheme would immune to Rank attacks.

4.3   Oil-Vinegar Separation Attack

The key point of Oil-vinegar Separation attack is that the quadratic part of each oil-vinegar polynomial can be represented as a quadratic form with a corresponding $n \times n$ matrix of the form:

$$\begin{pmatrix} 0 & A \\ B & C \end{pmatrix}.$$

The up left $o \times o$ zero submatrix is due to the fact that there are no quadratic terms in the oil variables. This resulted in separating the oil variables to the vinegar variables. Consequently, the adversary can create an equivalent private key and therefore can forge valid signatures for arbitrary messages. The more details of this attack can be see in [2,3].

In our scheme, the quadratic part of each polynomial in the central map can be represented as a quadratic form with a corresponding $n \times n$ matrix of the form:

$$\begin{pmatrix} D & A \\ B & C \end{pmatrix}.$$

The up left $o \times o$ submatrix D is not a zero matrix due to the fact that there are many quadratic terms in the oil variables.

Hence, the Oil-vinegar Separation attack cannot work on our scheme.

4.4   Direct Attacks

A direct attack on a MPKC signature scheme is that solve the system $P(x) = y'$ to forge a signature for a given message y' by Gröbner Basis method and its variants such as F_4 and F_5 [14]. According to the paper [15], the complexity of F_5 is upper bounded by

$$O\left(\left(\frac{n_x + n_y + \min(n_x + 1, n_y + 1)}{\min(n_x + 1, n_y + 1)} \right)^{\omega} \right),$$

where n_x is the number of plaintext variables, n_y is the number of ciphertext variables and $2 \leq \omega \leq 3$ is the linear algebra constant. Let $o = 30$, $v = 10$, $q = 2^8$, then the complexity of the direct attack on our scheme by F_5 is greater than 2^{87}.

Hybrid approach [10], like FXL [16], mixes exhaustive search with Gröbner bases techniques. Instead of computing one single Gröbner basis of the whole system, this approach compute the Gröbner bases of $\#\mathbb{K}^r$ subsystems which obtained by fixing r variables. The complexity of Hybrid is:

$$o\left(\min_{0 \leq r \leq m} \left((\#\mathbb{K})^r \cdot \left(m \cdot \left(\frac{m - r - 1 + d_{reg}(m - r, m, d)}{d_{reg}(m - r, m, d)} \right) \right)^{\omega} \right) \right).$$

Let $o = 30$, $v = 10$, $q = 2^8$ in our scheme. The best tradeoff for our scheme is to fix 7 variables. Then the complexity of the Hybrid approach on our scheme is greater than 2^{83}.

5   Comparison and Efficiency

In Table 1, we compare the cubic UOV, the original UOV, UOVLSR2, and Rainbow in the key size and the efficiency under the same secure level. The results of UOV, UOVLSR2, and Rainbow are all come from the reference [17]. According the Table 1, our scheme has shorter signature than the original UOV, Rainbow, UOVLRS2, cyclicUOV, and cyclicRainbow under the same secure level, though the public key size is greater than those improvements of UOV.

And in Table 2, we compare the running time about the each process of the original UOV, Rainbow, UOVLRS2, cyclicUOV, cyclicRainbow and CUOV under the same secure level. According Table 2, the speeds of key generation and signature in our scheme are faster than the original UOV, Rainbow, UOVLRS2, cyclicUOV. But, the verification is slower than the improvements of UOV. All of

Table 1. Sizes comparison

Scheme	Hash length (bit)	Signature length (bit)	Public key size (KB)
UOV(2^8,28,56)	224	672	99.9
UOVLRS2(2^8,28,56)	224	672	13.5
cyclicUOV(2^8,28,56)	224	672	16.5
Rainbow(2^8,17,13,13)	208	344	25.1
cyclicRainbow(2^8,17,13,13)	208	344	10.4
CUOV(2^8, 28, 8)	224	**272**	33
UOV(2^8,30,60)	240	720	122.6
UOVLRS2(2^8,30,60)	240	720	16.4
cyclicUOV(2^8,30,60)	240	720	20.0
Rainbow(2^8,20,18,9)	216	376	31.0
cyclicRainbow(2^8,20,18,9)	216	376	12.8
CUOV(2^8, 30, 10)	240	**320**	47.5

Table 2. Efficiency comparison

Scheme	Key generation (s)	Signature generation (ms)	Signature verify (ms)
UOV(2^8,24,48)	53.046	56.01	25.05
UOVLRS2(2^8,28,56)	37.152	4.521	0.20
cyclicUOV(2^8,28,56)	37.152	4.521	0.23
Rainbow(2^8,17,13,13)	4.923	4.163	0.29
cyclicRainbow(2^8,17,13,13)	2.377	2.01	0.14
CUOV(2^8, 28, 8)	**0.531**	**3.56**	8.28

our experiments were performed on a normal computer, with Intel Core i5-3470 CPU, 3.2 GHz, 4 GB RAM by Magma.

6 Conclusion

In this paper, we proposed a cubic UOV signature scheme by combining UOV and stepwise iteration method. Our scheme can avoid oil-vinegar separation attack. And our scheme can resist Gröbner basis attack and Hybrid approach attack for carefully choosing parameters, for example, $o = 30$, $v = 10$, $q = 2^8$. Moreover, our scheme has lower public key size, shorter signature and faster than the original UOV under the same secure level.

Acknowledgements. We want to thank the anonymous reviewers for their comments which helped to improve the paper. This work is supported by the National Key Basic Research Program of China (2013CB834203), the National Natural Science Foundation of China (No. 61370026, 61472064), The science and technology foundation of Sichuan Province (No. 2014GZ0109). The authors would like to thank Prof. Lei Hu for his helpful comments.

References

1. Patarin, J.: The oil and vinegar signature scheme. In: Presented at the Dagstuhl Workshop on Cryptography, September 1997
2. Kipnis, A., Shamir, A.: Cryptanalysis of the oil and vinegar signature scheme. In: Krawczyk, H. (ed.) Advances in Cryptology – CRYPTO 1998. LNCS, vol. 1462, pp. 257–266. Springer, Heidelberg (1998)
3. Kipnis, A., Patarin, J., Goubin, L.: Unbalanced oil and vinegar signature schemes. In: Stern, J. (ed.) EUROCRYPT 1999. LNCS, vol. 1592, pp. 206–222. Springer, Heidelberg (1999)
4. Ding, J., Schmidt, D.: Rainbow, a new multivariable polynomial signature scheme. In: Ioannidis, J., Keromytis, A.D., Yung, M. (eds.) ACNS 2005. LNCS, vol. 3531, pp. 164–175. Springer, Heidelberg (2005)
5. Ding, J., Yang, B.-Y., Chen, C.-H.O., Chen, M.-S., Cheng, C.-M.: New differential-algebraic attacks and reparametrization of rainbow. In: Bellovin, S.M., Gennaro, R., Keromytis, A.D., Yung, M. (eds.) ACNS 2008. LNCS, vol. 5037, pp. 242–257. Springer, Heidelberg (2008)
6. Petzoldt, A., Bulygin, S., Buchmann, J.: CyclicRainbow – a multivariate signature scheme with a partially cyclic public key. In: Gong, G., Gupta, K.C. (eds.) INDOCRYPT 2010. LNCS, vol. 6498, pp. 33–48. Springer, Heidelberg (2010)
7. Petzoldt, A., Bulygin, S., Buchmann, J.: Linear recurring sequences for the UOV key generation. In: Catalano, D., Fazio, N., Gennaro, R., Nicolosi, A. (eds.) PKC 2011. LNCS, vol. 6571, pp. 335–350. Springer, Heidelberg (2011)
8. Petzoldt, A., Bulygin, S.: Linear recurring sequences for the UOV key generation revisited. In: Kwon, T., Lee, M.-K., Kwon, D. (eds.) ICISC 2012. LNCS, vol. 7839, pp. 441–455. Springer, Heidelberg (2013)
9. Petzoldt, A., Bulygin, S., Buchmann, J.: Fast verification for improved versions of the UOV and rainbow signature schemes. In: Gaborit, P. (ed.) PQCrypto 2013. LNCS, vol. 7932, pp. 188–202. Springer, Heidelberg (2013)

10. Bettale, L., Faugere, J.C., Perret, L.: Hybrid approach for solving multivariate systems over finite fields. J. Math. Cryptology **3**(3), 177–197 (2010)
11. Patarin, J.: Cryptanalysis of the Matsumoto and Imai public key scheme of euro-crypt'88. In: Coppersmith, D. (ed.) CRYPTO 1995. LNCS, vol. 963, pp. 248–261. Springer, Heidelberg (1995)
12. Nie, X., Petzoldt, A., Buchmann, J.: Linearization equation attack on 2-layer non-linear piece in hand method. IEICE Trans. **97−A**(9), 1952–1961 (2014)
13. Yang, B.-Y., Chen, J.-M.: Building secure tame-like multivariate public-key cryp-tosystems: the new TTS. In: Boyd, C., González Nieto, J.M. (eds.) ACISP 2005. LNCS, vol. 3574, pp. 518–531. Springer, Heidelberg (2005)
14. Faugre, J.C.: A new efficient algorithm for computing Gröbner bases without reduc-tion to zero (F5). In: Proceedings of the International Symposium on Symbolic and Algebraic Computation in ISSAC 2002, pp. 75–83 (2002)
15. Faugre, J.C., El Din, M.S., Spaenlehauer, P.J.: Gröbner bases of bihomogeneous ideals generated by polynomials of bidegree (1, 1): algorithms and complexity. J. Symbolic Comput. **46**(4), 406–437 (2010)
16. Courtois, N.T., Klimov, A.B., Patarin, J., Shamir, A.: Efficient algorithms for solving overdefined systems of multivariate polynomial equations. In: Preneel, B. (ed.) EUROCRYPT 2000. LNCS, vol. 1807, pp. 392–407. Springer, Heidelberg (2000)
17. Petzoldt, A.: Selecting and reducing key sizes for multivariate cryptography, the thesis of Ph.D. Technischen Universitat Darmstadt (2013)

Two Approaches to Build UOV Variants with Shorter Private Key and Faster Signature Generation

Yang Tan and Shaohua Tang[(✉)]

School of Computer Science and Engineering, South China University of Technology,
Guangzhou, Guangdong, China
csshtang@scut.edu.cn, shtang@IEEE.org

Abstract. UOV is one of the earliest signature schemes in Multivariate Public Key Cryptography (MPKC). It also poses a strong security and none of the existing attacks can cause severe security threats to it. However, it suffers from a large key size. In this paper, we will propose two approaches to build variants of UOV with shorter private key size and faster signature generating process.

Keywords: Multivariate public cryptography · UOV · Rainbow

1 Introduction

In post-quantum era, with the emergence of the powerful quantum computers, public key algorithms based on traditional number theory will be extremely vulnerable. Popular algorithms such as RSA, ECC, Elgamal will be broken in a polynomial time according to Shor's algorithm [17,18]. Thereby, finding an alternative of these algorithms is very urgent. Multivariate public key cryptography (MPKC) is one of the most promising candidates in post-quantum cryptography. Other important branches include: (1) Lattice-based cryptography; (2) Code-based Cryptography; (3) Hash-based Cryptography and so on.

A MPKC scheme is usually built as $P = S \circ F \circ T$ in which S and T are invertible linear affine transformations used to cover the structure of central map F and make P look random. F is a special set of quadratic multivariate polynomials which could be inverted efficiently. To sign a message M, the signer could compute $X = S^{-1}(M)$, $Y = F^{-1}(X)$, $Z = T^{-1}(Y)$ in order and output Z as the signature. To verify the correctness of this signature, signature receiver could check if $M = P(Z)$. If it matches, accept. Otherwise, reject. Because of the property that they are normally efficient in computing, MPKC schemes are appropriate for applications on portable devices such as smart card, RFID. Its security relies on a hard problem that solving a random system of multivariate quadratic equations over a finite field is NP-hard. Current research indicates that a quantum computer couldn't solve this kind of problem in a polynomial time.

© Springer International Publishing Switzerland 2016
D. Lin et al. (Eds.): Inscrypt 2015, LNCS 9589, pp. 57–74, 2016.
DOI: 10.1007/978-3-319-38898-4_4

Since the first MPKC scheme: MI [8] was proposed in 1988, this area has undergone a rapid development in last two or three decades. A lot of encryption and signature schemes have been proposed, e.g., HFE [11], TTS [20], etc. Among them, UOV is one of the signature schemes with the most strong security. None of the existing attack poses real security threats to it while a lot of other MPKC schemes are vulnerable to attacks like MinRank [20], High Rank attack [5,20], Direct attack, Differential attack [5], Rainbow Band Separation attack [5,19] and so on. Consequently, UOV is an ideal choice for a signature scheme when security is the top priority.

However, like a regular MPKC scheme, it also suffers from overlarge key size (includes public key and private key size). This flaw can restrict its applications on some devices with limit storage. Thereby, to come up with a secure signature scheme with compact key size is one of the main goals in the research of MPKC. A lot of effort has been made in this area in the recent years. For example, TTS could be viewed as a special case of Rainbow with sparse private key. In [12–14], the authors proposed to insert some special sequences into the generation of public key to save some memory. This method's effect is quite obvious. The public key size is reduced by a factor up to more than 7 for UOV. This method could also speed up the verification process according to the conclusion made in [16]. In [21–23], the authors were enlightened and proposed two ways to reduce the private key size of Rainbow and improve the efficiency of signature generation in the meanwhile. In [22], the author combined those two methods to further reduce the private key size and improve the efficiency.

However, research of reducing the private key size of UOV hasn't been made yet. In this paper, inspired by the previous research, we will propose two variants of UOV with shorter private key size and higher signature generation efficiency.

The structure of this paper is as follows: First of all, we'd like to introduce the general UOV [4] and Rainbow [4] signature schemes. Then we introduce the existing methods that could be used to reduce the public key or private key size. Next, inspired by the existing methods, we would like to propose two variants of UOV with shorter private key size and faster signature generation. At fourth, we'd like to make a security analysis of our schemes by applying existing known attacks to them. During the security analysis, we also slightly modify the existing Kipis-Shamir attack [2] against UOV based on even characteristic field since the original one doesn't work. Fifth, we make an overall comparison with the original UOV concerning to two widely accepted security levels: 2^{80} and 2^{100}. Finally, we make a conclusion.

2   An Introduction of the Regular UOV and Rainbow

UOV and Rainbow are two well-known MPKC signature schemes. Both of them are based on a small field and Rainbow could be regarded as a multi-layer extension of UOV. In this section, we will introduce those two signature schemes.

2.1   UOV

To figure out what UOV is, first of all, we'd like to introduce the concept of Oil-Vinegar polynomial with the following form:

$$f = \sum_{i=1}^{o}\sum_{j=1}^{v} a_{ij}x_i x'_j + \sum_{i=1}^{v}\sum_{j=1}^{v} b_{ij}x'_i x'_j$$
$$+ \sum_{i=1}^{o} c_i x_i + \sum_{j=1}^{v} d_j x'_j + e \qquad (1)$$

Variables are divided into two kinds in the above polynomial: Oil variables (x_i) and Vinegar variables (x'_j). The number of Oil variables is o and the number of the Vinegar variables is v. Central map F can be composed of o Oil-Vinegar polynomials. The invertibility of the central map comes from the fact that once random values are assigned to the vinegar variables set, it becomes a set of linear equations of Oil variables and can be efficiently solved by Gaussian Elimination.

Once the central map F is determined, the public key can be calculated as:

$$P = F \circ T \qquad (2)$$

in which T is a linear affine transformation. There's no need to composite a linear affine transformation on the left side of the central map F since it will not affect its security (The central map polynomials will still be the Oil-Vinegar form after the composition of the linear affine transformation on the left).

Define $d = v - o$, when $d = 0$, it's called balanced Oil-Vinegar scheme (OV for short) while when $d > 0$, it's known as Unbalanced Oil-Vinegar scheme (UOV) [9]. The balanced Oil-Vinegar scheme can be easily broken by the Kipnis-Shamir attack [10]. The extended Kipnis-Shamir attack could also be used to attack UOV [9]. The complexity of this attack can be determined by: $q^{v-o-1}o^4$. Thereby, the designer could adjust corresponding parameters to meet the required security level.

2.2   Rainbow

Rainbow is a multi-layer extension of UOV. Each layer is an independent UOV scheme. All the variables of the previous layer could be viewed as the vinegar variables of the next layer. More specifically, the relations of the variables of different layers could be denoted as:

$$[x_1, ..., x_{v_1}]\{x_{v_1+1}, ..., x_{v_2}\}$$
$$[x_1, ..., x_{v_1}, x_{v_1+1}, ..., x_{v_2}]\{x_{v_2+1}, ..., x_{v_3}\}$$
$$[x_1, ..., x_{v_1}, x_{v_1+1}, ..., x_{v_2}, x_{v_2+1}, ..., x_{v_3}]\{x_{v_3+1}, ..., x_{v_4}\}$$
$$\vdots$$
$$[x_1, ..., ..., ..., ..., ..., ..., ..., ..., ..., ..., x_{v_u}]\{x_{v_u+1}, ..., x_n\}$$

in which $[x_1, ..., x_{v_i}]$ represents ith layers' Vinegar variables and $\{x_{v_i+1}, ..., x_{v_{i+1}}\}$ represents ith layers' Oil variables. Also, in those layers, v_i represents the number of the Vinegar variables and o_i represents the number of Oil variables of ith layer and we have $v_{i+1} = v_i + o_i$ apparently. Each layer has o_i Oil-Vinegar polynomials and $m = o_1 + o_2 +, ..., +o_u$ polynomials in total. The number of variables in total is $v_1 + o_1 + o_2, ..., +o_l = v_{u+1} = n$. The structure of a u-layer Rainbow is denoted as:

$$(v_1, o_1, ..., o_{u-1}, o_u).$$

The public key of Rainbow is built as:

$$P = L_1 \circ F \circ L_2 \tag{3}$$

Unlike UOV, to build the public key of Rainbow, a bijective linear transformation L_1 must be composited to cover the structure difference of different layers.

In the signing process, to invert the central map F of Rainbow, the signer needs to assign a random set of values to the Vinegar variables of the first layer and solve the Oil variables of the first layer. Next, the signer substitutes all the variables of the first layer to the second layer as the Vinegar variables of this layer and solve this layer's Oil variables. The signer repeat this process till all layer's variables are solved and outputted as the solution. The rest part of the signing process is the same as a regular MPKC scheme.

3 Existing Methods to Reduce the Public Key and Private Key Size

In this section, we will describe some existing methods that could be used to reduce the public key or private key size of UOV and Rainbow.

3.1 Methods to Reduce the Public Key Size

In [13–15], the authors proposed two methods to reduce the public key size of UOV and Rainbow by using two kinds of sequence: Cyclic Sequence and Linear Recurring Sequence and insert them to the public key polynomials' corresponding matrices.

In those two methods, the public polynomials' coefficients of quadratic terms are no longer denoted as m traditional $n \times n$ matrices (Symmetric or Upper-triangular). They are denoted as a single $\frac{(n+1) \cdot (n+2)}{2} \times m$ maucaley matrix M_P. Each row corresponds to one public key polynomials' coefficients of quadratic terms. The authors further divide M_P into tow parts:

$$M_P = (B/C) \tag{4}$$

in which B denotes the coefficients of the Vinegar-Vinegar quadratic cross-terms and Oil-Vinegar quadratic cross-terms, and C denotes Oil-Oil cross-terms. The corresponding central map can be denoted as a matrix:

$$M_F = (Q/0) \tag{5}$$

Since P is generated by F and T, M_P and M_F also have the relation:

$$M_P = M_F \cdot A \tag{6}$$

where A can be computed by coefficients of T.

The essential part of building the public key with reduced key size is to generate B in M_P using a particular sequence. Once B is assigned values, M_F can be computed by the relations revealed in (6). As long as M_F is known, the rest of $M_P{:}C$ can be further computed. As to the sequences to generate B, Cyclic Sequence and Linear Recurring sequence are involved. In the case of Cyclic Sequence, matrix B can be represented as:

$$\begin{bmatrix} b_1 & b_2 & \cdots & b_D \\ b_D & b_1 & \cdots & b_{D-1} \\ \vdots & \vdots & \ddots & \vdots \\ b_{D-m+2} & b_{D-m+3} & \cdots & b_{D-m+1} \end{bmatrix}$$

From the above form, we can see that only the first row of B is generated by random, the ith row of B is generated by cyclic right shifting $i-1$ position of the first row. This method is later extended in [23] to build the central map of Rainbow. In this paper, the authors used a rotation sequence of matrices rather than a Cyclic Sequence of rows in a single matrix.

The other sequence that could be used to reduce the public key size is called Linear Recurring Sequence. The definition of this sequence is given as:

Definition: Given a positive constant number L, and L random elements in $GF(q)$: $\alpha_1, \alpha_2, ..., \alpha_L$. Given the initial values: $\{s_1, s_2, ..., s_L\}$, the Linear Recurring Sequence (LRS) is a sequence $\{s_1, s_2, ...\}$ generated by: $s_j = \alpha_1 s_{j-1} + \alpha_2 s_{j-2} + ... + \alpha_L s_{j-L} (j > L)$.

L is defined as the length of this sequence. Apparently, this linear recurring sequence can be also used to generate matrix B to reduce the public key size. Elements of B in (4) can be computed as: $b_{ij} = s_{D(i-1)+j} (i = 1, ..., m, j = 1, ..., D)$. Variables need to store are: $\alpha_1, \alpha_2, ..., \alpha_L$ and $\{s_1, s_2, ..., s_L\}$.

3.2   Method to Reduce the Private Key Size of Rainbow

In [21], authors proposed another way to reduce the private key size of Rainbow: Matrix-based Rainbow. Assume the rainbow has l layers and with structure: $(v_1, o_1, ..., o_l)$. Solving each layer's variables actually ends up with solving a system linear equations with the form:

$$L \cdot X = V \tag{7}$$

in which L is a $o_i \times o_i$ coefficient matrix generated after assigning values to vinegar variables of i th layer, and X is the vector of length o_i composed of the unknown oil variables. V is the constant vector with length o_i. To save storage of the private key, the above equation can be further divided into d_i parts. Assume $o_i = d_i \times o_i'$, $V = (V_1, V_2, \cdots, V_{d_i})$, $X = (X_1, X_2, \cdots, X_{d_i})$, we will have:

$$L_k \cdot X_k = V_k, \ k = 1, ..., d_i; \tag{8}$$

To solve Eq. (7), the signer can solve (8) separately to improve the efficiency. To achieve this goal, L should have the following form:

$$L = \begin{pmatrix} A & 0 & \cdots & 0 \\ 0 & A & \cdots & 0 \\ \vdots & \vdots & \ddots & \vdots \\ 0 & 0 & \cdots & A \end{pmatrix} \tag{9}$$

in which A is a $o_i' \times o_i'$ matrix. To solve Eq. (7) by regular Gaussian Elimination method, the complexity will be around $O(o_i^3)$. However, in this new method, it will drop to $o_i'^3$ (equation in (8) can be solved simultaneously, since L_k is identical corresponding to A in L). As to how to construct a L in the above form, the constructor needs to choose coefficients of each layer's Oil-Vinegar cross-terms and Oil linear terms accordingly. Details won't be described here, interested readers could refer to [21].

In [22], the author combined those two methods: Matrix-based Rainbow and NT-Rainbow to build a more compact and efficient rainbow.

4 Our Construction

In our construction, inspired by the previous works, we'd like to extend the rotating matrix sequence and Matrix-based Rainbow method to UOV. Also, we'd like to bring in the Linear Recurring Sequence method to build another variant of UOV with shorter key size and faster signature generation.

4.1 UOV Variant 1

First of all, we determine the values of the coefficients of Vinegar-Vinegar cross-terms. Apparently, these coefficients can also be denoted as a rotating sequence of matrix. Suppose an UOV is based on $GF(q)$, the number of Vinegar-variables is: v and the number of Oil-variables is: o. We have $n = v + o$, $m = o$. The coefficients of Vinegar-Vinegar cross-terms are denoted by the following matrix sequence.

$$B^1 = \begin{bmatrix} b_{1,1} & b_{1,2} & \cdots & b_{1,v} \\ b_{2,1} & b_{2,2} & \cdots & b_{2,v} \\ \vdots & \vdots & \ddots & \vdots \\ b_{v,1} & b_{v,2} & \cdots & b_{v,v} \end{bmatrix}, B^2 = \begin{bmatrix} b_{v,1} & b_{v,2} & \cdots & b_{v,v} \\ b_{1,1} & b_{1,2} & \cdots & b_{1,v} \\ \vdots & \vdots & \ddots & \vdots \\ b_{v-1,1} & b_{v-1,2} & \cdots & b_{v-1,v} \end{bmatrix},$$

$$..., B^l = \begin{bmatrix} b_{v-l+2,1} & b_{v-l+2,2} & \cdots & b_{v-l+2,v} \\ b_{v-l+3,1} & b_{v-l+3,2} & \cdots & b_{v-l+3,v} \\ \vdots & \vdots & \ddots & \vdots \\ b_{v-l+1,1} & b_{v-l+1,2} & \cdots & b_{v-l+3,v} \end{bmatrix}$$

in which $1 \leq l \leq o$. The lth matrix is generated by down rotating $l - 1$ rows of the first matrix. All the indexes are values modulus v.

Also, the method used to build Matrix-based Rainbow can also be applied to our UOV construction. First of all, we need to generate the corresponding coefficients to make sure the linear equations to solve during the signing process is in the form of (8) and L be the form of (9).

First of all, assume $o = d \times o'$. If central map's polynomials are denoted as:

$$g^{(v+l)}(x) = x^T A^{(v+l)} x + B^{(v+l)} x + C^{(v+l)}, \ x = (x_1, x_2, ..., x_n)^T, l = 1, ..., o \quad (10)$$

in which $A^{(v+l)}$ is a $n \times n$ matrix, $B^{(v+l)}$ is a vector with length n and $C^{(v+l)}$ is a constant. $A^{(v+l)}$ can be further denoted as:

$$A^{(v+l)} = \begin{pmatrix} A_0^{(v+l)} & A_1^{(v+l)} \\ 0 & 0 \end{pmatrix}, \ l = 1, ..., o \quad (11)$$

where $A_0^{(v+l)} = c(i, j)$ is a $v \times v$ matrix corresponding to the coefficients of Vinegar-Vinegar cross-terms and can be denoted by B^l. If it's in a upper-triangular form, then:

$$c(i, j) = \begin{cases} b_{i-l+1,j} + b_{j-l+1,i} & (i < j) \\ b_{i-l+1,j} & (i = j) \\ 0 & (else) \end{cases} \quad (12)$$

On the other hand, $A_0^{(v+l)}$ can also be represented by B^l directly.

Also, to extend the Matrix-based Rainbow to UOV construction, the $v \times o$ matrix $A_1^{(v+l)}$ is the most crucial part to this construction. It determines L's form. $A_1^{(v+l)}$ should have the shape:

$$A_1^{(v+io'+j)} = (\overbrace{0, ..., 0}^{io'}, a_j, \overbrace{0, ..., 0}^{(d-i-1)o'}), (0 \leq i \leq d, 0 \leq j \leq o') \quad (13)$$

in which 0 represents a zero v-dimensional vector and a_j represents a random $v \times o'$ sub-matrix.

Next, we determine the values of $B^{(v+l)}$ and it can be further divided into:

$$B^{(v+l)} = (B_0^{(v+l)}, B_1^{(v+l)}), \ l = 1, ..., o \quad (14)$$

in which $B_0^{(v+l)}$ is a random vector in length v and $B_1^{(v+l)}$ is a vector with length o of the form:

$$B_1^{(v+ho'+j)} = (\overbrace{0, ..., 0}^{ho'}, b_j, \overbrace{0, ..., 0}^{(d-h-1)o'}), (0 \leq h < d, 0 < j < o') \quad (15)$$

in which 0 represents 0 and b_j is a vector of length o' . Together, b_j and a_j can determine a row of A in Eq. (9).

At last, $C^{(v+l)}$ can be a random constant.

This construction is similar to the case of building a sparse Rainbow with higher efficiency in signing [22]. It could be viewed as an extension to UOV.

4.2   UOV Variant 2

Moreover, since the linear recurring sequence can also be used to build a MPKC scheme with shorter public key, we'd like to extend this method to UOV.

Normally, a regular linear recurring sequence is hard to explore a property of improving signing efficiency. In [12], authors proposed a special linear recurring sequence which could be used to reduce the public key size and enhance verifying efficiency in the meanwhile. Instead of using one linear recurring sequence, the authors used o different linear recurring sequences with length 1. All the initial values for these o sequences are 1 for simplicity.

Inspired by that, we would also like to explore a special way to utilize linear recurring sequence to reduce private key size and enhance signing efficiency. In our new construction, the linear recurring sequence is used to generate the coefficients of Vinegar-Vinegar cross-terms. We use a upper-triangular matrix to represent the corresponding cross-terms' coefficients. Under this circumstance, $\frac{v \cdot (v+1)}{2}$ linear recurring sequences of length 1 are needed. The first matrix is composed of initial values of these sequences and they are totally random instead of being 1. The rest matrices are generated by the elements of these linear recurring sequences. They should be in the form of:

$$B^1 = \begin{bmatrix} b_{11} & b_{12} & \dots & b_{1v} \\ 0 & b_{22} & \dots & b_{2v} \\ \vdots & \vdots & \ddots & \vdots \\ 0 & 0 & \dots & b_{vv} \end{bmatrix}, B^2 = \begin{bmatrix} b_{11} \cdot \alpha & b_{12} \cdot \alpha & \dots & b_{1v} \cdot \alpha \\ 0 & b_{22} \cdot \alpha & \dots & b_{2v} \cdot \alpha \\ \vdots & \vdots & \ddots & \vdots \\ 0 & 0 & \dots & b_{vv} \cdot \alpha \end{bmatrix},$$

$$\dots, B^l = \begin{bmatrix} b_{11} \cdot \alpha^{l-1} & b_{12} \cdot \alpha^{l-1} & \dots & b_{1v} \cdot \alpha^{l-1} \\ 0 & b_{22} \cdot \alpha^{l-1} & \dots & b_{2v} \cdot \alpha^{l-1} \\ \vdots & \vdots & \ddots & \vdots \\ 0 & 0 & \dots & b_{vv} \cdot \alpha^{l-1} \end{bmatrix}$$

Apparently, this matrix sequence can be directly substituted into $A_0^{(v+l)}$.

As to how to determine the values of $A_1^{(v+l)}$, $B^{(v+l)}$ and $C^{(v+l)}$, we do the same as UOV variant 1.

4.3   Parameters Summarization

Based on the previous description of how to construct our UOV variants, we list the parameters needed to build the central map of them:

(1) a_j: a $v \times o'$ sub-matrix corresponding to the non-zero coefficients of cross-terms between Vinegar variables and Oil variables, $j = 1, \dots, o'$.
(2) B^1: the initial matrix of a matrix sequence corresponding to the coefficients of cross-terms between Vinegar variables and Vinegar variables. The matrix sequence can be generated by a rotation sequence of matrix or some linear recurring sequences.
(3) α: If coefficients of Vinegar-Vinegar cross-terms are generated by linear recurring sequences of length 1, this element is needed.

(4) b_j: a o' dimensional vector corresponding to non-zero coefficients of linear terms of Oil variables, $j = 1, ..., o'$.

(5) $B_0^{(v+l)}$: vector in length v corresponding to coefficients of linear terms of Vinegar variables, $l = 1, ..., o$.

(6) $C^{(v+l)}$: the constant part in central map, $l = 1, ..., o$.

4.4 Private Key Size

After giving the parameters needed to generate the central map, we can calculate the required private key size to build these tow UOV variants.

First of all, we calculate the required storage size for central map. This can be calculated by simply add the size of (1)–(6) in previous section which is:

Variant 1 (Rotation Sequence):

$$v \times o' \times o' + v \times v + o' \times o' + v \times o + o = (v + 1)o'^2 + v \times n + o$$

Variant 2 (Linear Recurring Sequence):

$$v \times o' \times o' + v \times (v+1)/2 + 1 + o' \times o' + v \times o + o = (v+1)o'^2 + v \times (n+o+1)/2 + o + 1$$

Secondly, we take the size of linear affine transformation into account which is: $n \cdot (n + 1)$.

The total storage size needed for private key is:

(1) UOV variant 1: $(v + 1) \cdot o'^2 + v \cdot n + o + n \cdot (n + 1)$
(2) UOV variant 2: $(v + 1) \cdot o'^2 + v \cdot (n + 1) + o + n \cdot (n + 1)$

On the other hand, the private key size of a regular UOV is:

$$o(v \cdot o + \frac{v(v + 1)}{2} + n + 1) + n \cdot (n + 1)$$

Values of these equations are measured by the size of a finite field element.

In this section, we only give the formulas to calculate corresponding private key size. In the following section, specific storage size will be given after the parameters of UOV and UOV variants are determined.

4.5 Signing Process of Our UOV Variants and Their Efficiency

In this section, we are going to describe the signing process of our tow variants and their efficiency.

Assume the document to sign is M. First of all, the signer needs to invert the central map $S' = F^{-1}(M)$. Secondly, invert affine linear transformation T by calculating $S = T^{-1}(S')$. For our variants, this part is the same as the original UOV. Each element in T involves an addition operation and multiplication operation. The complexity of this process is $O(n^2)$.

As to inverting the central map, two parts are involved:

(1) Calculate the constant parts: V of the linear equation system in (7).
(2) Invert this linear equation system: (7).

These two parts are calculated after assigning random values to Vinegar variables. Assume the random values set assigned to Vinegar variables is: b.

Calculation of V. Specifically, V is the sum of following three parts:

(1) After assigning random values to Vinegar variables, cross-terms between Vinegar variables $x_i x_j (i, j = 1, ..., v)$ become constants;
(2) After assigning random values to Vinegar variables, linear terms of Vinegar variables $x_i (i = 1, ..., v)$ also become constants;
(3) The constant parts in central map: $C^{(v+l)}, l = 1, ..., o$.

Firstly, we talk about the calculation of part (1):

UOV Variant 1: To calculate constant part (1), we substitute a v dimensional vector b into Vinegar variables and calculate $bB^l b^T$ for each central equation $(l = 1, ..., o)$. We further denote $B^l b^T$ as $b'^{(l)T}$ of length v. Since B^l is generated by cyclicly down rotating $l - 1$ rows of B^1, $b'^{(l)T}$ could be generated by cyclicly down rotating $l - 1$ positions of $b'^{(1)T}$ accordingly. Anyway, the signer only needs to compute $b'^{(1)T}$, the rest of $b'^{(l)T}$ are generated by its down rotating sequence. At last, compute $b \cdot b'^{(l)T}$ as the constant produced by Vinegar-Vinegar cross-terms.

UOV Variant 2: For this variant, it also substitutes a v dimensional vector b into Vinegar variables and calculates $bB^l b^T$ for each central equation $(l = 1, ..., o)$. Since $bB^l b^T = \sum_{i=1}^{v}\sum_{j=1}^{v} b_i b_j B_{ij}^l$ and $B_{ij}^l = B_{ij}^1 \cdot \alpha^{l-1}$, we have

$$bB^l b^T = \sum_{i=1}^{v}\sum_{j=1}^{v} b_i b_j B_{ij}^l = \sum_{i=1}^{v}\sum_{j=1}^{v} b_i b_j B_{ij}^1 \cdot \alpha^{l-1} = bB^1 b^T \cdot \alpha^{l-1}.$$ Thereby, the

signer only needs to compute $bB^1 b^T$ and $bB^l b^T$ can be computed by simply multiply α^{l-1}.

The complexities of calculating part (1) are $O(v^2 + ov) = O(v^2)$ and $O(v^2 + o - 1) = O(v^2)$ for Variant 1 and 2 respectively. Corresponding complexity of a Regular UOV is $O(o \cdot v^2)$.

On the other hand, the processes of calculating part (2) and part (3) are the same as a regular UOV and their complexities are negligible to computing part (1).

Solving $L \cdot X = V$. In the signing process of a regular UOV, after assigning random values to Vinegar variables, the remaining Oil variables can be solved by a set of linear Eq. (7). Normally, L looks random. However, in our construction, L will be in the form of Eq. (9) in which A can be computed as:

$$A = \begin{pmatrix} b \cdot a_1 \\ \vdots \\ b \cdot a_{o'} \end{pmatrix} + \begin{pmatrix} b_1 \\ \vdots \\ b_{o'} \end{pmatrix} \qquad (16)$$

Furthermore, V and X are divided into d parts $V = (V_1, V_2, \cdots, V_d)$, $X = (X_1, X_2, \cdots, X_d)$. Consequently, the equations need to solve can be transformed into solving a set of $A \cdot X_i = V_i$ independently. This can be done by a typical Gaussian Elimination in $O(o'^3)$ operations. One thing to note is that these equations can be solved simultaneously since all the coefficients matrices are identical: A. Thereby, in the solving process, all equations proceed the same row transformations of A or we can directly calculate $X_i = V_i \times A^{-1}$.

The complexity of this process is $O(o'^3)$ for our UOV variants while for a general UOV, it's $O(o^3) = O(d^3 \cdot o'^3)$.

4.6   General Description of Our Schemes

In this section, we are going to give a general description of our scheme.

Key Generation.

(1) Private Key: According to the required security level, choose the appropriate set of parameters including finite field $k = GF(q)$, number of Vinegar variables v, number of Oil variables: o and o', d. Generate the quintuple of parameters: $(a_j, B^1, b_r, B_0^{(v+l)}, C^{(v+l)})$ for Variant 1 or six-tuple of parameters $(a_j, B^1, \alpha, b_r, B_0^{(v+l)}, C^{(v+l)})$ for Variant 2 to construct central map F. Moreover, generate the invertible affine transformation: $T : k^n \rightarrow k^n$.
(2) Public Key: Generated by $P = F \circ T : k^n \rightarrow k^m$.

Sign. Input the document to sign: $M \in k^m$. First, invert the central map: $S' = F^{-1}(M)$. This can be done by the process given in Sect. 4.5. Next, invert the linear affine transformation: $S = T^{-1}(S')$. Output $S \in k^n$ as the signature.

Verify. The signer sends document-signature pair to a receiver: (M, S). Receiver verifies the correctness of the signature by check if $P(S) = M$. If it matches, the signature is legitimate. Otherwise, reject it.

Parameters Representation of Our Scheme. Normally, a UOV's parameters can be denoted as: (k, v, o). We have two variants and $o = o' \cdot d$. Thereby, our schemes parameters can be denoted as: $V1\,(GF(k), v, o' \times d)$, $V2(GF(k), v, o' \times d)$ for Variant 1 and 2 respectively.

5   Security of Our UOV Variants

In this section, we are going to make a security analysis of our UOV variants by applying known existing attacks to them and make a comparison with regular UOV.

5.1   Direct Attack and UOV Reconciliation Attack

Direct attack treat UOV public key as a set of quadratic equations and solve them directly. Known efficient algorithms include Grobner bases attack F4 [6], F5 [7] and XL algorithm [3].

UOV reconciliation attack [5] could viewed as an improved version of brute force attack. It tries to find a sequence of basis that could transform the public key of UOV to the central Oil-Vinegar form. However, the main part of this attack is still direct attack. It's complexity could be transformed into directly solving a quadratic system of $m = o$ equations in v variables.

For a regular UOV, since $v > o$, directly solving public key of UOV or using reconciliation attack could all be transferred to directly solving an under-defined system (number of variables is greater than the number of equations). Before applying direct attack to an under-defined system, one should assign random values to variables to make the whole system a generic one or over-defined one [1]. Consequently, reconciliation attack against UOV is as difficult as a direct attack against it since both of them end up with directly solving a generic or over-defined system of quadratic equations with the same number of equations: o.

On the other hand, because of the linear affine transformation T, despite the difference of our construction of central map from a regular UOV, our variants' public key also look totally random. Thereby, we expect our UOV variants have the same security level against direct attack as a regular UOV.

To verify our conclusions, we are able to write magma programs about our UOV variants and regular UOV against direct attack on a workstation: Dell Precision T5610. We choose three small-scale groups of parameters for each scheme. For each scheme, we test for 100 times and record their average attacking time. The results are listed Table 1.

Table 1. Comparisons between UOV variants and regular UOV against direct attack

Schemes	Regular UOV	UOV variant 1	UOV variant 2
Group 1	$(GF(2^2),4,8)$	$(GF(2^2),2 \times 2,8)$	$(GF(2^2),2 \times 2,8)$
	0.539 s	0.496 s	0.514 s
Group 2	$(GF(5),4,8)$	$(GF(5),2 \times 2,8)$	$(GF(5),2 \times 2,8)$
	3.221 s	3.051 s	3.435 s
Group 3	$(GF(7),3,6)$	$(GF(7),3 \times 1,6)$	$(GF(7),3 \times 1,6)$
	0.728 s	0.752 s	0.732 s

From this table, we can clearly see that our UOV variants' performances against direct attack are almost the same as regular UOV's.

5.2 Kipnis-Shamir Attack

This attack was proposed in [10] by Kipnis and Shamir to attack balanced Oil-Vinegar scheme and later extended to evaluate the security of UOV scheme [9]. At first, this attack could only work on UOV based on odd characteristic field. The complexity of this attack is $q^{d-1} \times m^4$ [9]. In [2], the authors extend this attack on UOV with even characteristic field by making a small modification. The complexity of the modified attack against UOV with even characteristic field is $q^{d+1} \times m^4$

The essence of Kipnis-Shamir attack is to use public key polynomials' corresponding matrices to find the desired the hidden Oil space which could be used to construct an equivalent private key of linear affine transformation: T.

Here, we give a brief description of the process of Kipnis-Shamir attack in [2]:

(1) Produce the corresponding symmetric matrices for the homogeneous quadratic parts of public key's polynomials: $W_1, W_2, ..., W_m$. If the scheme is based on even characteristic, the entry $(1,1)$ of each matrix is set to 1.
(2) Randomly choose two linear combination of $W_1, W_2, ..., W_m$ and still denote them as W_1 and W_2 in which W_1, W_2 is invertible. Calculate $W_{12} = W_1 \times W_2^{-1}$.
(3) Compute the characteristic polynomial of W_{12} and find its linear factor of multiplicity 1. Denote such factor as $h(x)$. Computer $h(W_{12})$ and its corresponding kernel.
(4) For each vector o in the kernel of Step 3, use $oW_i o = 0, (1 \le i \le m)$ to test if o belongs to the hidden oil space. Choose linear dependent vectors among them and append them to set T.
(5) If T contains only one vector or nothing, go back to step 2.
(6) If necessary, find more vectors in T: $o_3, o_4, ...$ Calculate $K_{o_1} \cap \cdots \cap K_{o_t}$ to find out the hidden Oil space in which K_{o_t} is a space from which the vectors x satisfy that $o_t W_i x = 0, (1 \le i \le m)$.
(7) Extract a basis of hidden Oil space and extend it to a basis of k^n and use it to transform the public key polynomials to basic Oil-Vinegar polynomials form.

For further explanation of this process, readers could refer to [2].

According to these steps, we are able to write a magma programs to test on some small scale UOV schemes. However, we found out that step (4) isn't enough to test if o is in the hidden Oil space. In our experiment, it is highly possible that vectors we found satisfy the conditions in Step (4) but don't belong to the hidden Oil space.

Consequently, we made a few changes to the Kipnis-Shamir attack in [2]: If we found a new possible vector o_t in hidden Oil space in step (4), we use $o_t W_i o_j = 0, (1 \le i \le m, 1 \le j \le t-1)$ to test if the new vector truly belongs to hidden Oil space and this new condition is strongly enough to find the desired vectors. If the new vector doesn't satisfy this condition, discard it and go back to step (2) to find a new one. Also, it is possible that after enough tries, we still can't find the desired new o_t. The reason is that the previous vectors $o_1, ..., o_{t-1}$

are also not in the hidden Oil space. Then we should discard them all and do it
from the scratch.

In our new modified Kipnis-Shamir attack, we are able to run some tests
about this attack on some small-scale regular UOV schemes. Each scheme we
test for 50 times and record its average attacking time (Table 2).

Table 2. Kipnis-Shamir attack against regular UOV

Schemes	$(GF(3),3,6)$	$(GF(3),4,8)$	$(GF(2^2),3,6)$
Attacking time	12.556 s	1155.568 s	16.861 s

However, when we try to apply our modified Kipnis-Shamir attack on our
new UOV variants, we found out that the vectors in the hidden Oil space can't be
found. By further analyzing the experiments' results, the failure of this attack
is caused by the reason that invertible matrices W_1, W_2 can't be found. As a
matter of fact, all the corresponding symmetric matrices of public key's polynomials: $W_1, W_2, ..., W_m$ are not invertible. Ergo, W_{12} can't calculated under this
circumstance. As is stated in [2,9], it's a necessary condition that public key's
corresponding matrices being invertible and symmetric for Kipnis-Shamir attack
to work. Consequently, this attack is futile against our new UOV variants.

Next, we are going to illustrate why all the corresponding symmetric matrices
of public key's polynomials: $W_1, W_2, ..., W_m$ are non-invertible:

In the construction of our UOV variants, the private key polynomials' coefficients are represented by asymmetric matrix. To represent the private key
polynomials' coefficients in a symmetric matrix form, first of all, what we need
to do is transform the original asymmetric matrix into quadratic polynomials:
$\sum_{i=1}^{n} \sum_{j=i}^{n} a_{ij}x_ix_j$. The corresponding matrix's coefficients is computed as:

$$\begin{cases} c_{i,j} = a_{ij}, & i = j \\ c_{i,j} = a_{ij}/2, & i \neq j \end{cases}$$

We further notice the original matrix of private key polynomials of our UOV
variants 1 is in the form of:

$$\begin{pmatrix} * & * & \cdots & * & 0 & \cdots & 0 & * & \cdots & * & 0 & \cdots & 0 \\ * & * & \cdots & * & 0 & \cdots & 0 & * & \cdots & * & 0 & \cdots & 0 \\ \vdots & \vdots & \ddots & \vdots & \vdots & & \vdots & \vdots & \ddots & \vdots & \vdots & & \vdots \\ * & * & \cdots & * & 0 & \cdots & 0 & * & \cdots & * & 0 & \cdots & 0 \\ 0 & 0 & \cdots & 0 & 0 & \cdots & 0 & 0 & \cdots & 0 & 0 & \cdots & 0 \\ \vdots & \vdots & \ddots & \vdots & \vdots & & \vdots & \vdots & \ddots & \vdots & \vdots & & \vdots \\ 0 & 0 & \cdots & 0 & 0 & \cdots & 0 & 0 & \cdots & 0 & 0 & \cdots & 0 \end{pmatrix}$$

and for UOV variant 2, it's in the form of:

$$\begin{pmatrix} * & * & \cdots & * & 0 & \cdots & 0 & * & \cdots & * & 0 & \cdots & 0 \\ 0 & * & \cdots & * & 0 & \cdots & 0 & * & \cdots & * & 0 & \cdots & 0 \\ \vdots & \vdots & \ddots & \vdots & \vdots & \ddots & \vdots & \vdots & \ddots & \vdots & \vdots & \ddots & \vdots \\ 0 & 0 & \cdots & * & 0 & \cdots & 0 & * & \cdots & * & 0 & \cdots & 0 \\ 0 & 0 & \cdots & 0 & 0 & \cdots & 0 & 0 & \cdots & 0 & 0 & \cdots & 0 \\ \vdots & \vdots & \ddots & \vdots & \vdots & \ddots & \vdots & \vdots & \ddots & \vdots & \vdots & \ddots & \vdots \\ 0 & 0 & \cdots & 0 & 0 & \cdots & 0 & 0 & \cdots & 0 & 0 & \cdots & 0 \end{pmatrix}$$

in which * denotes non-zero elements. The left-upper non-zero block represents the coefficients of Vinegar-Vinegar cross-terms and non-zero columns are the coefficients of Oil-Vinegar cross-terms. Both these matrices are transformed to symmetric form which should be:

$$\begin{pmatrix} * & * & \cdots & * & 0 & \cdots & 0 & * & \cdots & * & 0 & \cdots & 0 \\ * & * & \cdots & * & 0 & \cdots & 0 & * & \cdots & * & 0 & \cdots & 0 \\ \vdots & \vdots & \ddots & \vdots & \vdots & \ddots & \vdots & \vdots & \ddots & \vdots & \vdots & \ddots & \vdots \\ * & * & \cdots & * & 0 & \cdots & 0 & * & \cdots & * & 0 & \cdots & 0 \\ 0 & 0 & \cdots & 0 & 0 & \cdots & 0 & 0 & \cdots & 0 & 0 & \cdots & 0 \\ \vdots & \vdots & \ddots & \vdots & \vdots & \ddots & \vdots & \vdots & \ddots & \vdots & \vdots & \ddots & \vdots \\ 0 & 0 & \cdots & 0 & 0 & \cdots & 0 & 0 & \cdots & 0 & 0 & \cdots & 0 \\ * & * & \cdots & * & 0 & \cdots & 0 & 0 & \cdots & 0 & 0 & \cdots & 0 \\ \vdots & \vdots & \ddots & \vdots & \vdots & \ddots & \vdots & \vdots & \ddots & \vdots & \vdots & \ddots & \vdots \\ * & * & \cdots & * & 0 & \cdots & 0 & 0 & \cdots & 0 & 0 & \cdots & 0 \\ 0 & 0 & \cdots & 0 & 0 & \cdots & 0 & 0 & \cdots & 0 & 0 & \cdots & 0 \\ \vdots & \vdots & \ddots & \vdots & \vdots & \ddots & \vdots & \vdots & \ddots & \vdots & \vdots & \ddots & \vdots \\ 0 & 0 & \cdots & 0 & 0 & \cdots & 0 & 0 & \cdots & 0 & 0 & \cdots & 0 \end{pmatrix}$$

We can see that this symmetric matrix have zero rows. According to the description of our UOV variants, there will be $(d-1) \times o'$ zero-rows and $v + o'$ non-zero rows. Assume this matrix is W, apparently, we have $Rank(W) \leq v + o' < n$. Thereby, this symmetric matrix is not invertible matrix since it's not a full-rank matrix. Hence, all the corresponding symmetric matrices of public key's polynomials: $W_1, W_2, ..., W_m$ are non-invertible. Consequently, W_{12} can't be calculated. Under this circumstance, Kipnis-Shamir attack is not applicable to our UOV variants.

6   An Overall Comparison with Original UOV

In this section we are going to give an overall comparison between regular UOV and our UOV variants under the same security level requirements.

First of all, we choose the security level requirements that our UOV variants and the regular UOV should satisfy. Currently, the most prevailing ones are 2^{80} and 2^{100}. According to the conclusions made in [1], the sets of parameters of regular UOV $(GF(2^8), v = 56, o = 28)$ and $(GF(2^8), v = 72, o = 36)$ can achieve security level 2^{80} and 2^{100} respectively. Accordingly, corresponding sets of parameters of our variants can be given as UOV Variant 1: $(GF(2^8), v = 56, o = 4 \times 7)$ and $(GF(2^8), v = 72, o = 4 \times 9)$, UOV Variant 2: $(GF(2^8), v = 56, o = 4 \times 7)$ and $(GF(2^8), v = 72, o = 4 \times 9)$.

After picking the appropriate parameters, in our comparisons, we will record the scheme generating time, signature generating time, verifying time, private key size, public key size and signature length. All those records are produced by calculating the average values of 100 trials. We run these tests on a Dell Precision T5610 with Magma programs. The results are listed in Tables 3 and 4.

Table 3. Overall Comparison between regular UOV and our UOV variants under security level requirement 2^{80}

Security level 2^{80}	Regular UOV	UOV variant 1	UOV variant 2
Key generating time	9.496 s	6.615 s	6.422 s
Signature generating time	0.443 s	0.007 s	0005 s
Signature verifying time	0.015 s	0.015 s	0.014 s
Public key size	99.941 KB	99.941 KB	99.941 KB
Private key size	95.813 KB	14.321 KB	12.818 KB
Signature length	84 B	84 B	84 B

Table 4. Overall Comparison between regular UOV and our UOV variants under security level requirement 2^{100}

Security level 2^{100}	Regular UOV	UOV variant 1	UOV variant 2
Key generating time	29.533 s	23.547 s	25.704 s
Signature generating time	1.145 s	0.045 s	0.009 s
Signature verifying time	0.025 s	0.022 s	0.027 s
Public key size	206.930 KB	206.930 KB	206.930 KB
Private key size	198.844 KB	24.899 KB	22.404 KB
Signature length	108 B	108 B	108 B

From these two tables, we can see that our UOV variants do have obvious advantages over Private Key size and Signature Generating time. They verify our intentions to build the UOV variants.

7   Conclusions

In this paper, inspired by the existing methods to build UOV and Rainbow with shorter public key size or Rainbow with shorter private key size, we introduced two UOV variants which have shorter private key size and higher efficiency in signature generation. Then we made a security analysis of UOV variants by applying existing known attacks which could be used against UOV to our UOV variants. During the security analysis, we also made a small change to the existing Kipis-Shamir attack [2] against UOV based on even characteristic field since the original one didn't work. At last, we made an overall comparison between regular UOV and our UOV variants.

Acknowledgements. This work was supported by the National Natural Science Foundation of China [U1135004, 61170080], the 973 Program [2014CB360501], Guangdong Provincial Natural Science Foundation [2014A030308006], and Guangdong Province Universities and Colleges Pearl River Scholar Funded Scheme (2011).

References

1. Bettale, L., Faugère, J.C., Perret, L.: Hybrid approach for solving multivariate systems over finite fields. J. Math. Crypt. **3**(3), 177–197 (2009)
2. Cao, W., Hu, L., Ding, J., Yin, Z.: Kipnis-Shamir attack on unbalanced oil-vinegar scheme. In: Bao, F., Weng, J. (eds.) ISPEC 2011. LNCS, vol. 6672, pp. 168–180. Springer, Heidelberg (2011)
3. Courtois, N., Klimov, A., Patarin, J., Shamir, A.: Efficient algorithms for solving overdefined systems of multivariate polynomial equations. In: Preneel, B. (ed.) EUROCRYPT 2000. LNCS, vol. 1807, pp. 392–407. Springer, Heidelberg (2000)
4. Ding, J., Schmidt, D.: Rainbow, a new multivariable polynomial signature scheme. In: Ioannidis, J., Keromytis, A.D., Yung, M. (eds.) ACNS 2005. LNCS, vol. 3531, pp. 164–175. Springer, Heidelberg (2005)
5. Ding, J., Yang, B.-Y., Chen, C.-H.O., Chen, M.-S., Cheng, C.-M.: New differential-algebraic attacks and reparametrization of rainbow. In: Bellovin, S.M., Gennaro, R., Keromytis, A.D., Yung, M. (eds.) ACNS 2008. LNCS, vol. 5037, pp. 242–257. Springer, Heidelberg (2008)
6. Faugere, J.: A new efficient algorithm for computing Gröbner bases (F4). J. Pure Appl. Algebra **139**(1–3), 61–88 (1999)
7. Faugere, J.: A new efficient algorithm for computing Gröbner bases without reduction to zero F5. In: International Symposium on Symbolic and Algebraic Computation Symposium-ISSAC 2002 (2002)
8. Imai, H., Matsumoto, T.: Algebraic methods for constructing asymmetric cryptosystems. In: Algebraic Algorithms and Error-Correcting Codes, pp. 108–119 (1986)
9. Kipnis, A., Patarin, J., Goubin, L.: Unbalanced oil and vinegar signature schemes. In: Stern, J. (ed.) EUROCRYPT 1999. LNCS, vol. 1592, pp. 206–222. Springer, Heidelberg (1999)
10. Kipnis, A., Shamir, A.: Cryptanalysis of the oil and vinegar signature scheme. In: Krawczyk, H. (ed.) CRYPTO 1998. LNCS, vol. 1462, pp. 257–266. Springer, Heidelberg (1998)

11. Patarin, J.: Hidden Fields Equations (HFE) and Isomorphisms of Polynomials (IP): two new families of asymmetric algorithms. In: Maurer, U.M. (ed.) EUROCRYPT 1996. LNCS, vol. 1070, pp. 33–48. Springer, Heidelberg (1996)
12. Petzoldt, A., Bulygin, S.: Linear recurring sequences for the UOV key generation revisited. In: Kwon, T., Lee, M.-K., Kwon, D. (eds.) ICISC 2012. LNCS, vol. 7839, pp. 441–455. Springer, Heidelberg (2013)
13. Petzoldt, A., Bulygin, S., Buchmann, J.: CyclicRainbow – a multivariate signature scheme with a partially cyclic public key. In: Gong, G., Gupta, K.C. (eds.) INDOCRYPT 2010. LNCS, vol. 6498, pp. 33–48. Springer, Heidelberg (2010)
14. Petzoldt, A., Bulygin, S., Buchmann, J.: A multivariate signature scheme with a partially cyclic public key. In: Proceedings of SCC 2010. Citeseer (2010)
15. Petzoldt, A., Bulygin, S., Buchmann, J.: Linear recurring sequences for the UOV key generation. In: Catalano, D., Fazio, N., Gennaro, R., Nicolosi, A. (eds.) PKC 2011. LNCS, vol. 6571, pp. 335–350. Springer, Heidelberg (2011)
16. Petzoldt, A., Bulygin, S., Buchmann, J.: Fast verification for improved versions of the UO and rainbow signature schemes. In: Gaborit, P. (ed.) PQCrypto 2013. LNCS, vol. 7932, pp. 188–202. Springer, Heidelberg (2013)
17. Shor, P.: Algorithms for quantum computation: discrete logarithms and factoring. In: 1994 Proceedings of 35th Annual Symposium on Foundations of Computer Science, pp. 124–134. IEEE (1994)
18. Shor, P.: Polynomial-time algorithms for prime factorization and discrete logarithms on a quantum computer. SIAM J. Comput. 26, 1484–1509 (1996)
19. Thomae, E.: A generalization of the rainbow band separation attack and its applications to multivariate schemes. IACR Cryptology ePrint Archive 2012, 223 (2012)
20. Yang, B.-Y., Chen, J.-M.: Building secure tame-like multivariate public-key cryptosystems: the new TTS. In: Boyd, C., González Nieto, J.M. (eds.) ACISP 2005. LNCS, vol. 3574, pp. 518–531. Springer, Heidelberg (2005)
21. Yasuda, T., Ding, J., Takagi, T., Sakurai, K.: A variant of rainbow with shorter secret key and faster signature generation. In: Proceedings of the First ACM Workshop on Asia Public-key Cryptography, pp. 57–62. ACM (2013)
22. Yasuda, T., Takagi, T., Sakurai, K.: Efficient variant of rainbow using sparse secret keys. J. Wirel. Mobile Netw. Ubiquitous Comput. Dependable Appl. (JoWUA) 5(3), 3–13 (2014)
23. Yasuda, T., Takagi, T., Sakurai, K.: Efficient variant of rainbow without triangular matrix representation. In: Linawati, Mahendra, M.S., Neuhold, E.J., Tjoa, A.M., You, I. (eds.) ICT-EurAsia 2014. LNCS, vol. 8407, pp. 532–541. Springer, Heidelberg (2014)

A Secure Variant of Yasuda, Takagi and Sakurai's Signature Scheme

Wenbin Zhang$^{(\boxtimes)}$ and Chik How Tan

Temasek Laboratories, National University of Singapore,
Singapore, Singapore
{tslzw,tsltch}@nus.edu.sg

Abstract. Yasuda, Takagi and Sakurai proposed a new signature scheme in PQCrypto 2013 using quadratic forms over finited fields of odd characteristic. Later on two independent attacks were proposed by Hashimoto in PQCrypto 2014 and by Zhang and Tan in ICISC 2014 to break their scheme. The purpose of this paper is to fix the security problem of Yasuda, Takagi and Sakurai's scheme. We achieve this purpose by mixing their scheme with a special type HFEv polynomials to produce a new scheme, YTS-HFEv. We analyze its security and propose a practical parameter set with public key size about 57 KB and security level 2^{80}.

Keywords: Post-quantum cryptography · Multivariate public key cryptosystem · Digital signature · HFEv

1 Introduction

Since the threat of quantum computer to public key cryptography [Sho97], it has been active to search possible alternatives to current widely used RSA. One of such directions is multivariate public key cryptosystems (MPKC) [DGS06] whose trapdoor one-way functions are of the form of multivariate polynomials over finite fields. The security of MPKC relies on the problem of solving a general set of multivariate polynomial equations over finite fields which is proved to be NP-hard [GJ79]. Current main trapdoor one-way functions of MPKC are usually represented by quadratic polynomials and of course cannot be a random set of polynomials. They are usually designed [DGS06] by composing a polynomial map $F : \mathbb{F}^n \to \mathbb{F}^m$ with two affine maps $\bar{F} = L \circ F \circ R : \mathbb{F}^n \to \mathbb{F}^m$. The public key is $\bar{F}$ while the secret key usually consists of L, R, F. It should be efficient to invert the central map F but infeasible to invert $\bar{F}$ unless one knows L, R, F.

Many such trapdoor one-way functions have been proposed since 1980's for encryption and signature schemes. According to Wolf and Preneel's taxonomy of MPKC [WP05], they may be categorized as basic trapdoors and modifiers. Namely they are constructed from those basic trapdoors by applying some modification methods. In Wolf and Preneel's taxonomy of many of them proposed until 2005, there are various modification methods, but only four basic trapdoors: the Matsumoto-Imai scheme [MI88], hidden field equation (HFE) [Pat96], the unbalanced oil and vinegar schemes [KPG99], the stepwise triangular systems

© Springer International Publishing Switzerland 2016
D. Lin et al. (Eds.): Inscrypt 2015, LNCS 9589, pp. 75–89, 2016.
DOI: 10.1007/978-3-319-38898-4_5

[WBP05, WBP06]. These basic trapdoors have been used to produce various new schemes with many modifiers. Since then, there have been a few new trapdoors proposed, for example, MFE [WYHL06], ℓIC [DWY07], Square [Clo09], ABC [TDTD13], ZHFE [PBD14] and so on. However, most of schemes so far have been broken and it seems very challenging to construct new secure schemes.

Recently Yasuda, Takagi and Sakurai [YTS13] proposed a new and interesting signature scheme using quadratic forms over finite fields of odd characteristic, suitable for signature schemes. The mathematical foundation of their construction is the classification of quadratic forms over finite fields of odd characteristic. Their scheme is different from all others and is regarded as a new basic trapdoor. However, it is then soon be broken by two independent attacks [Has14] and [ZT15a, ZT15b], and there seems no obvious secure variant of it. So it becomes an open problem how to repair Yasuda, Takagi and Sakurai's (YTS' for short) scheme to make it secure.

Notice that the mathematical foundation of YTS' scheme is the classification of quadratic forms over finite fields of odd characteristic, and when the characteristic of the base field is two, there is also a similar but more complicated classification of quadratic forms. Thus it is curious that if this case could provide an analogous scheme and especially if it is secure or not. In this paper we propose such an analogous scheme over finite fields of characteristic two. However we find that it is neither secure. The attack [ZT15a, ZT15b] is still applicable.

Although the analogous scheme is not secure, we are then motivated by a most recent paper [ZT15c] which proposed an idea of using a special type of HFEv polynomials to enhance the security of signature schemes. By applying this idea to YTS' scheme and our analog, we then construct a new variant of YTS' scheme, YTS-HFEv. This new scheme is also an HFEv scheme but different from current known HFEv schemes. We show that this new scheme can resist attacks to YTS' scheme and other major attacks. We also propose a practical parameter set with public key size about $57\,\mathrm{KB}$ and security level 2^{80}.

This paper is organized as follows. In Sect. 2, we review YTS' scheme, then construct an analogous scheme over even finite fields and propose an attack to it. In Sect. 3, we give our new variant of YTS' scheme mixed with HFEv. Security analysis of this new scheme is then presented in Sect. 4. Finally Sect. 5 concludes this paper.

2 Yasuda, Takagi and Sakurai's Signature Scheme and Its Analog over Finite Fields of Characteristic Two

In this section, we shall first briefly review Yasuda, Takagi and Sakurai's (YTS') signature scheme [YTS13], then present an analogous scheme over finite fields of characteristic two, and finally sketch an attack to this analog of YTS' scheme.

2.1 Yasuda, Takagi and Sakurai's Signature Scheme

YTS' scheme [YTS13] is constructed from the classification of quadratic forms over finite fields of odd characteristic. We give a brief review of it in the following.

Let q be a power of an *odd* prime p and δ a non-square element in $\mathbb{F}_q$. Moreover let $I_{n,\delta} = \begin{pmatrix} I_{n-1} & \\ & \delta \end{pmatrix}$, and I_{n-1} the $(n-1) \times (n-1)$ identity matrix. Then any $n \times n$ symmetric matrix A over $\mathbb{F}_q$ can be decomposed as either $A = X^T X$ or $A = X^T I_{n,\delta} X$ where X an $n \times n$ matrix.

Let $n = r^2$ and $m = r(r+1)/2$. Choose two one-one correspondences

$$\phi_1 : \text{the set of } r \times r \text{ matrices over } \mathbb{F}_q \overset{1-1}{\longleftrightarrow} \mathbb{F}_q^n,$$

$$\phi_2 : \text{the set of } r \times r \text{ symmetric matrices over } \mathbb{F}_q \overset{1-1}{\longleftrightarrow} \mathbb{F}_q^m.$$

Define two maps

$$F_1, F_2 : \mathbb{F}_q^n \to \mathbb{F}_q^m, \quad F_1(x) = \phi_2(X^T X), \quad F_2(x) = \phi_2(X^T I_{r,\delta} X).$$

Then the pair (F_1, F_2) can be used to construct a multivariate quadratic signature scheme as follows.

Let $R_1, R_2 : \mathbb{F}_q^n \to \mathbb{F}_q^n$ and $L : \mathbb{F}_q^m \to \mathbb{F}_q^m$ be three randomly chosen invertible affine transformations and

$$\bar{F}_1 = L \circ F_1 \circ R_1, \quad \bar{F}_2 = L \circ F_2 \circ R_2.$$

YTS' scheme can be described as follows.

Public Key. $\bar{F}_1, \bar{F}_2$.

Private Key. R_1, R_2, L.

Signature Generation. For a message $y \in \mathbb{F}_q^m$, first compute $y' = L^{-1}(y)$ and the corresponding symmetric matrix $Y = \phi_2^{-1}(y')$, then compute an $r \times r$ matrix X such that $Y = X^T X$ or $Y = X^T I_{r,\delta} X$, and the corresponding vector $x' = \phi_1(X)$, finally compute $x = R_1^{-1}(x')$ or $x = R_2^{-1}(x')$ correspondingly.

Verification. A signature x is accepted only if $\bar{F}_1(x) = y$ or $\bar{F}_2(x) = y$.

The public key size is $O(r^6)$, private key size is $O(r^4)$ and efficiency of signature generation is $O(r^4)$. The parameters $(q, r, n) = (6781, 11, 121)$ is proposed and claimed to have security of 140-bit in [YTS13].

After YTS' scheme was proposed in 2013, it was then quickly broken in 2014 by two different and independent attacks of Hashimoto [Has14], and of Zhang and Tan [ZT15a]. Hashimoto used an algebraic approach to recover the private key of YTS' scheme if R_1 has a special form, and then reduced the case of general R_1 to this special case. He implemented his attack and broke the parameters (6781, 11, 121) in hundreds of seconds. Zhang and Tan applied a geometric approach by first giving a simple matrix expression for the public map and discovering the underlying geometric structure of YTS' scheme in terms of invariant subspaces. They then converted the problem of recovering the private key into a geometric problem of decomposing the whole space into certain invariant subspaces and calculating their appropriate bases. Finally they applied the theory of invariant subspaces to develop an algorithm for recovering the private key.

Later on they extended their original work and implemented their attack successfully in [ZT15b] which totally and practically break YTS' scheme. In this extension paper, they tested various parameters and recovered all the private keys efficient. For example, the private key of (6781, 11, 121) was recovered in only about 14.77 s.

2.2 Analog of YTS' Scheme over Finite Fields of Characteristic Two

Let q be a positive power of 2 and δ an element in $\mathbb{F}_q$ such that $x^2 + x + \delta$ is irreducible over $\mathbb{F}_q$. There is also a classification for quadratic forms over finite fields of characteristic two, cf. pages 138–139 of [Tay92] or Theorem 6.30 of [LN97], which is more complicated than the case of odd characteristic. Due to limitation of space, we shall directly give a matrix form for this classification.

Recall that a representing matrix of a quadratic form $f \in \mathbb{F}_q[x_1, \ldots, x_r]$ is an $n \times n$ matrix A over $\mathbb{F}_q$ satisfying

$$f(\mathbf{x}) = \mathbf{x}^T A \mathbf{x}, \quad \mathbf{x} = (x_1, \ldots, x_n)^T.$$

There are many representing matrices for one f, but there is only one $n \times n$ upper triangular representing matrix $Q_f = (a_{ij})$ with $a_{ij} = 0$ for $i > j$ by rewriting $f = \sum_{1 \le i \le j \le n} a_{ij} x_i x_j$. Next we shall use upper triangular matrices to represent quadratic forms.

For even $r = 2t$ $(t \ge 1)$, define

$$Q_{2t} = \begin{pmatrix} 0 & 1 & & & \\ & 0 & & & \\ & & \ddots & & \\ & & & 0 & 1 \\ & & & & 0 \end{pmatrix}_{2t \times 2t}, \quad Q'_{2t} = \begin{pmatrix} Q_{2t-2} & & \\ & 1 & 1 \\ & & \delta \end{pmatrix}.$$

For odd $r = 2t + 1$ $(t \ge 1)$, define

$$Q_{2t+1} = \begin{pmatrix} Q_{2t} & \\ & 1 \end{pmatrix}, \quad Q'_{2t+1} = \begin{pmatrix} Q'_{2t} & \\ & 0 \end{pmatrix}.$$

We next define a helpful operation on square matrices, called folding operation. The folding matrix of an $r \times r$ matrix $A = (a_{ij})$ is $A^F = (a'_{ij})$ where $a'_{ii} = a_{ii}$ for all i, $a'_{ij} = a_{ij} + a_{ji}$ for (i, j) with $i < j$, and $a'_{ij} = 0$ for (i, j) with $i > j$.

Proposition 1. *For any $r \times r$ matrix C, $(C^T A^F C)^F = (C^T A C)^F$.*

We find that the folding operation disappears if A^F and $(A^F)^T$ are added together.

Proposition 2. $A^F + (A^F)^T = A + A^T$.

With the help of the above notations, we can now have the following matrix form for the classification.

Theorem 1. *For an $r \times r$ upper triangular matrix A over $\mathbb{F}_q$ (r can be even or odd), there is an $r \times r$ matrix X such that either $A = (X^T Q_r X)^F$ or $A = (X^T Q'_r X)^F$.* □

An algorithm for computing such a matrix X is sketched in [LN97], pages 286–287.

We next give our construction of a signature scheme analogous to YTS' signature scheme. Let $n = r^2$, $m = r(r+1)/2$, and another one-one correspondence

$$\phi_3 : \text{the set of } r \times r \text{ upper triangular matrices over } \mathbb{F}_q \xleftrightarrow{1-1} \mathbb{F}_q^m.$$

For $x \in \mathbb{F}_q^n$, $X = \phi_1^{-1}(x)$ is an $r \times r$ matrix. Define

$$F_1, F_2 : \mathbb{F}_q^n \to \mathbb{F}_q^m, \quad F_1(x) = \phi_3((X^T Q_r X)^F), \quad F_2(x) = \phi_3((X^T Q'_r X)^F).$$

Similar to YTS' scheme, the pair (F_1, F_2) is surjective and can be used as the central map of a multivariate signature scheme.

Let $R_1, R_2 : \mathbb{F}_q^n \to \mathbb{F}_q^n$ and $L : \mathbb{F}_q^m \to \mathbb{F}_q^m$ be three randomly chosen invertible affine transformations and

$$\bar{F}_1 = L \circ F_1 \circ R_1, \quad \bar{F}_2 = L \circ F_2 \circ R_2.$$

Our analog of YTS' scheme is described as follows.

Public Key. $\bar{F}_1, \bar{F}_2$.

Private Key. R_1, R_2, L.

Signature Generation. For a message $y \in \mathbb{F}_q^m$, first compute $y' = L^{-1}(y)$ and the corresponding upper triangular matrix $Y = \phi_3^{-1}(y')$, then compute an $r \times r$ matrix X such that $Y = (X^T Q_r X)^F$ or $Y = (X^T Q'_r X)^F$, and the corresponding vector $x' = \phi_1(X)$, finally compute $x = R_1^{-1}(x')$ or $x = R_2^{-1}(x')$ correspondingly.

Verification. A signature x is accepted if $\bar{F}_1(x) = y$ or $\bar{F}_2(x) = y$, otherwise rejected.

Some features of the scheme are given below.

Public Key Size. $\frac{1}{16} r(r+1)(r^2+1)(r^2+8)(\log_2 q)$ Bytes.

Private Key Size. $\frac{1}{4}(9r^4 + 2r^3 + 11r^2 + 2r)(\log_2 q)$ Bytes.

Efficiency. Algorithm of generating a signature is similar to YTS' scheme and has the same level of efficiency $O(r^4) = O(n^2)$.

Security. The security level against MinRank attack is at least $O(q^r) = O(q^{\sqrt{n}})$ for recovering L.

We remark that small q, such as $q = 2$, is impractical and we need only consider big q. This is because the public key size is very sensitive on r, i.e., $O(r^6)$. Thus r should be small and q should be big. For instance, if $q = 2$, to have security level even as low as 2^{20}, we should have $r \geq 20$, i.e., $n \geq 400$, then the public key size is already too huge, larger than 4 MB.

For $q > 2$, however, we find that the attacking method of [ZT15a, ZT15b] is still applicable but requires sophisticated modification. In the rest of this section, we briefly sketch such an attack to the above scheme with $q > 2$ which can totally break it.

2.3   Attack to the Analog of YTS' Scheme ($q > 2$)

It is sufficient to attack the first map of the public key. We shall omit the subscript and simply write $\bar{F}, F, R$ for $\bar{F}_1, F_1, R_1$ respectively. Since the affine parts of L, R can be recovered easily when $q > 2$, c.f. Appendix B of [ZT15b], we can consider only the case that both L, R are linear. Write $\bar{F} = (\bar{f}_1, \ldots, \bar{f}_m)$ with

$$\bar{f}_k(x) = x^T A_k x, \quad x \in \mathbb{F}_q^n$$

where A_k is an $n \times n$ upper triangular matrix publicly known.

We first give a simple matrix expression for the public map. Write the $m \times m$ matrix L in the following form

$$\begin{pmatrix} l_{1;11} & l_{1;12} & l_{1;22} & \cdots & l_{1;1r} & \cdots & l_{1;rr} \\ \vdots & \vdots & \vdots & & \vdots & & \vdots \\ l_{m;11} & l_{m;12} & l_{m;22} & \cdots & l_{m;1r} & \cdots & l_{m;rr} \end{pmatrix}.$$

In addition, let $l_{k;ji} = l_{k;ij}$ for $i < j$ and define the symmetric matrix

$$L_k = (l_{k;ij})_{r \times r}$$

corresponding to the kth row of L.

Then A_k has the following simple expression: if r is even, for $1 \leq k \leq m$,

$$A_k = \left(R^T \begin{pmatrix} \begin{matrix} 0 & L_k \\ & 0 \end{matrix} & & \\ & \ddots & \\ & & \begin{matrix} 0 & L_k \\ & 0 \end{matrix} \end{pmatrix} R \right)^F \tag{2.1}$$

with $r/2$ blocks $\begin{pmatrix} 0 & L_k \\ & 0 \end{pmatrix}$ on the diagonal, and if r is odd, for $1 \leq k \leq m$,

$$A_k = \left(R^T \begin{pmatrix} \begin{matrix} 0 & L_k \\ & 0 \end{matrix} & & & \\ & \ddots & & \\ & & \begin{matrix} 0 & L_k \\ & 0 \end{matrix} & \\ & & & D_k \end{pmatrix} R \right)^F, \quad \text{where } D_k = \begin{pmatrix} l_{k;11} & & \\ & \ddots & \\ & & l_{k;rr} \end{pmatrix} \tag{2.2}$$

with $(r-1)/2$ blocks $\begin{pmatrix} 0 & L_k \\ & 0 \end{pmatrix}$ on the diagonal.

Comparing with the case that q odd [ZT15a], there are two significant differences: (1) the middle matrix is singular here; (2) there is the folding operation on the right hand side. So the method for q odd cannot be directly applied to A_k here. Motivated by Kipnis and Shamir's attack to the oil-vinegar scheme in the case that the field is of characteristic two [KS98, DGS06], we try considering $B_k := A_k + A_k^T$ instead. For even r and odd r, B_k has the following expression respectively,

$$B_k = R^T \begin{pmatrix} \begin{smallmatrix} 0 & L_k \\ L_k & 0 \end{smallmatrix} & & & \\ & \ddots & & \\ & & \begin{smallmatrix} 0 & L_k \\ L_k & 0 \end{smallmatrix} \end{pmatrix} R, \quad B_k = R^T \begin{pmatrix} \begin{smallmatrix} 0 & L_k \\ L_k & 0 \end{smallmatrix} & & & \\ & \ddots & & \\ & & \begin{smallmatrix} 0 & L_k \\ L_k & 0 \end{smallmatrix} & \\ & & & 0 \end{pmatrix} R.$$

Compared to the left matrix, the right matrix has a zero block at the bottom diagonal. This difference makes the two situations very different and the latter situation is more complicated. Next we will discuss the two cases separately.

First Case: r Even. Since L is invertible, all L_k are linearly independent and thus form a basis for $r \times r$ symmetric matrices. Hence there is a linear combination of all L_k which is invertible and thus a linear combination of all B_k which is invertible. Pick such an invertible linear combination of all B_k and denote it B_0 and L_0 its corresponding linear combination of all L_k. Let $L_k' = L_0^{-1} L_k$ and $B_k' = B_0^{-1} B_k$. Then we have

$$B_k' = R^{-1} \begin{pmatrix} L_k' & & \\ & \ddots & \\ & & L_k' \end{pmatrix} R, \quad B_k' R^{-1} = R^{-1} \begin{pmatrix} L_k' & & \\ & \ddots & \\ & & L_k' \end{pmatrix}.$$

Based on the above identities for A_k, B_k, B_k', we can apply the method of [ZT15a, ZT15b] using invariant subspaces to recover equivalent private key R, L.

Second Case: r Odd. This case is troublesome. Since B_k has rank at most $n - r$, it can never be invertible. So the method for the case r even is not applicable directly here. Nevertheless we still can reduce it to the case r even as shown below.

Similar to the case r even, we pick a linear combination, B_0, of all B_k such that it is of rank $n - r$. Notice that the last r columns of $B_0 R^{-1}$ are zero which means that the last r columns of R^{-1} span the null space $N(B_0)$ of B_0. Since $RR^{-1} = I$, so $N(B_0)$, equivalently the subspace spanned by the last r columns of R^{-1}, is orthogonal to the subspace spanned by the first $n - r$ rows of R. Let R_1 be the submatrix consisting of the first $n - r$ rows of R, and R_2 the submatrix consisting of the last r rows of R. Then the row space $R(R_1)$ of R_1 is contained in $N(B_0)^{\perp}$. It should be noted that we may not have $R(R_1) = N(B_0)^{\perp}$ though $\dim R(R_1) + \dim N(B_0) = n$ due to the special theory of quadratic forms over finite fields of characteristic two. Practically we can pick a regular dim-$(n - r)$ subspace of $N(B_0)^{\perp}$ and treat it as $R(R_1)$ — try another B_0 if there is no regular dim-$(n - r)$ subspace. Since $R(R_1)$ is regular, there are two bases $\mathbf{a}_1, \ldots, \mathbf{a}_{n-r}$

and $\mathbf{b}_1, \ldots, \mathbf{b}_{n-r}$ such that the inner product $\mathbf{a}_i^T \cdot \mathbf{b}_j = 0$ if $i \neq j$ and 1 if $i = j$. Namely

$$(\mathbf{a}_1, \ldots, \mathbf{a}_{n-r})^T (\mathbf{b}_1, \ldots, \mathbf{b}_{n-r}) = I_{n-r}.$$

Next notice that the identity for B_k can be reduced to

$$B_k = R_1^T \begin{pmatrix} 0 & L_k & & & \\ L_k & 0 & & & \\ & & \ddots & & \\ & & & 0 & L_k \\ & & & L_k & 0 \end{pmatrix} R_1,$$

and R_1 can be written as

$$R_1 = R_1'(\mathbf{a}_1, \ldots, \mathbf{a}_{n-r})^T$$

where R_1' is an $(n-r) \times (n-r)$ matrix. Let

$$\bar{B}_k = (\mathbf{b}_1, \ldots, \mathbf{b}_{n-r})^T B_k (\mathbf{b}_1, \ldots, \mathbf{b}_{n-r}).$$

Then

$$\bar{B}_k = R_1'^T \begin{pmatrix} 0 & L_k & & & \\ L_k & 0 & & & \\ & & \ddots & & \\ & & & 0 & L_k \\ & & & L_k & 0 \end{pmatrix} R_1'.$$

Hence the problem of finding R_1 is then reduced to finding R_1', i.e., the case r even. After finding R_1, then R_2 can be found from the identity of A_k.

To summarize, the analog of YTS' scheme over finite field of characteristic two is somehow trickier than YTS' original scheme over finite field of odd characteristic. However, the method of the attack of [ZT15a, ZT15b] can still be applied to totally break this analog after a careful reduction.

3   A New Variant of YTS' Scheme Mixed with HFEv: YTS-HFEv

In this section, we shall construct a new variant of YTS' scheme, named YTS-HFEv, which is a mixture of YTS' scheme and the well known HFEv scheme. This construction is motivated by a most recent signature scheme called MI-T-HFE [ZT15c] in which a new idea is proposed to apply HFEv to enhance the security for signature schemes.

3.1   HFE and HFEv

Hidden field equations (HFE) was proposed by Patarin in 1996 [Pat96] as a candidate to repair the Matsumoto-Imai cryptosystem [MI88] after which was broken by his linearization equations in 1995 [Pat95].

Let q be a power of prime p which can be even or odd. Choose a degree s irreducible polynomial $g(x)$ over $\mathbb{F}_q$ and let $\mathbb{K} = \mathbb{F}_q[x]/(g(x))$ which is then a degree t extension of $\mathbb{F}_q$. Define the following isomorphism of vector spaces over $\mathbb{F}_q$,

$$\phi : \mathbb{K} \to \mathbb{F}_q^t, \quad \phi(a_0 + a_1 x + \cdots + a_{t-1}x^{t-1}) = (a_0, a_1, \ldots, a_{t-1})$$

An HFE polynomial with degree bound D is a polynomial over $\mathbb{K}$ of the following form

$$H(X) = \sum_{1 \le q^i + q^j \le D} a_{ij} X^{q^i + q^j} + \sum_{1 \le q^j \le D} b_j X^{q^j} + c.$$

The key here is that if D is relatively small, then $H(X) = Y$ can be solved efficiently using Berlekamp's algorithm with complexity $O(tD^2 \log_q D + D^3)$. Patarin's HFE encryption scheme has such an HFE polynomial as the core map and composes it with two invertible affine transformations.

Although HFE has been broken thoroughly [KS99, GJS06, BFP13], it has been developed into a most important family of multivariate public key schemes. An important variant of HFE is HFEv, an encryption scheme which was first presented in 1999 [KPG99] and remains secure until today. HFEv applies the idea of unbalanced oil and vinegar signature scheme [KPG99] and add a few new variables in $\mathbb{F}_q$, called vinegar variables, to HFE. These vinegar variables in $\mathbb{F}_q$ corresponds to a variable V in the extension field $\mathbb{K}$. The core map of HFEv is a polynomial of two variables X, V over $\mathbb{K}$ of the following form

$$H(X, V) = \sum_{q^i + q^j \le D} a_{ij} X^{q^i + q^j} + \sum_{q^i \le D} b_{ij} X^{q^i} V^{q^j} + \sum_{q^i \le D} d_i X^{q^i}$$
$$+ \sum c_{ij} V^{q^i + q^j} + \sum e_i V^{q^i} + f$$

where the degree of the vinegar variable V can be arbitrary high. To solve the equation $H(X, V) = Y$, one first assigns any value to V and then solve it as in HFE. A cryptanalysis of HFEv was given in [DS05] and showed that HFEv is secure if V is not very small. A famous example is QUARTZ [PCG01] which is a signature scheme constructed from HFEv simply with a few components deleted from the public map of HFEv.

3.2   The New Scheme YTS-HFEv Where YTS' Scheme Meets HFEv

In [ZT15c], the authors proposed an idea of using the HFEv encryption scheme to enhance the security of signature schemes. They applied the following special type of HFEv polynomials over $\mathbb{K}$:

$$H(X_1, X_2) = \sum_{0 \le i \le t} \sum_{1 \le q^j \le D} a_{ij} X_1^{q^i} X_2^{q^j} + \sum_{1 \le q^i + q^j \le D} b_{ij} X_2^{q^i} X_2^{q^j} + \sum_{1 \le q^j \le D} c_j X_2^{q^j},$$

where X_1 is the vinegar variable. Notice its difference with general HFEv polynomials: there are no terms $X_1^{q^i}$ and no constant term. The purpose of this specialty is to make equation $H(X_1, X_2) = 0$ always have a solution, at least the zero solution $X_2 = 0$. However, for a given X_1, a nonzero solution for X_2 is preferred among those solutions to $H(X_1, X_2) = 0$. We shall accept the zero solution $X_2 = 0$ if there is no nonzero solution. Using this type of HFEv polynomials to mix with signature schemes, we can not only enhance the security but also assure that every message has a valid signature, a feature somewhat desired by a signature scheme.

Following [ZT15c], define the following map for $\mathbf{x}_1, \mathbf{x}_2 \in \mathbb{F}_q^t$ to be used next

$$\bar{H} : \mathbb{F}_q^t \times \mathbb{F}_q^t \to \mathbb{F}_q^t, \quad \bar{H}(\mathbf{x}_1, \mathbf{x}_2) = \phi(H(\phi^{-1}(\mathbf{x}_1), \phi^{-1}(\mathbf{x}_2))).$$

Let $F_1, F_2 : \mathbb{F}_q^n \to \mathbb{F}_q^m$ be the pair of maps of YTS' scheme or its analog in the preceding section depending on whether q is odd or even. Then as it did in [ZT15c], define for $\mathbf{x}_1 \in \mathbb{F}_q^t$, $\mathbf{x}_2 \in \mathbb{F}_q^n$, $i = 1, 2$,

$$G_i(\mathbf{x}_1, \mathbf{x}_2) = F_i(\mathbf{x}_1) + S_i \cdot \bar{H}(T_i \cdot \mathbf{x}_1, \mathbf{x}_2) : \mathbb{F}_q^{t+n} \to \mathbb{F}_q^m$$

where S_1, S_2 are $m \times t$ matrices and T_1, T_2 are $t \times n$ matrices. Let

$$\bar{G}_1 = L \circ G_1 \circ R_1, \quad \bar{G}_2 = L \circ G_2 \circ R_2 : \mathbb{F}_q^{t+n} \to \mathbb{F}_q^m$$

where $R_1, R_2 : \mathbb{F}_q^{t+n} \to \mathbb{F}_q^{t+n}$ and $L : \mathbb{F}_q^m \to \mathbb{F}_q^m$ are three randomly chosen invertible linear maps. Then we have the following new signature scheme, called YTS-HFEv:

Public Key. $\bar{G}_1, \bar{G}_2$.
Private Key. $L, R_1, R_2, S_1, S_2, T_1, T_2$.
Signature Generation. A given message $\mathbf{y} \in \mathbb{F}_q^n$ is signed in the following way:
  1. Compute $\mathbf{y}' = L^{-1}(\mathbf{y})$.
  2. Solve $F_1(\mathbf{x}_1) = \mathbf{y}'$ or $F_2(\mathbf{x}_1) = \mathbf{y}'$ to get a solution $\mathbf{x}_2$.
  3. Substitute $\mathbf{x}_1$ into the corresponding $\bar{H}(T_i \cdot \mathbf{x}_1, \mathbf{x}_2) = 0$ and solve it by Berlekamp's algorithm. Among those solutions, pick a nonzero solution and assign it to $\mathbf{x}_2$. If there is only the zero solution, then let $\mathbf{x}_2 = 0$. Then $(\mathbf{x}_1, \mathbf{x}_2)$ is a solution to the corresponding $F_i'(\mathbf{x}) = \mathbf{y}'$.
  4. Compute the corresponding $\mathbf{x} = R_i^{-1}(\mathbf{x}_1, \mathbf{x}_2)$ which is then a signature.
Verification. A signature $\mathbf{x}$ is accepted if $\bar{F}_1'(\mathbf{x}) = \mathbf{y}$ or $\bar{F}_2'(\mathbf{x}) = \mathbf{y}$.

Key sizes and efficiency of the above scheme are given below.

Public Key Size. $\frac{1}{32} r(r+1)(r^2+1)(r^2+t+1)(\log_p q)\lceil \log_2 p \rceil$ Bytes. So r should be chosen small to have small public key size.
Private Key Size. $\frac{1}{32}(5r^4 + 2r^3 + 28r^2 t + r^2 + 4rt + 8t^2)(\log_p q)\lceil \log_2 p \rceil$ Bytes.
Efficiency of Signature Generation. $O((r^2 + t)^2 + tD^2 \log_q D + D^3)$.

Hence for practical reason, p should be close to $2^{\lceil \log_2 p \rceil}$ and r should be small to have small key size. In addition, D should be relatively small so that signature generation can be efficient. On the other hand, there should be as more terms of X_2 as possible so that the rank is not too small to maintain enough level of security, thus q should also be small since $q^i \leq D$.

Based on the above considerations and the security reason given in next section, we propose the following practical parameters

$$(q, r, t, D) = (4, 9, 40, 80).$$

Then the message length is $2m = r(r+1) = 90$ bits, signature length is $2(n+t) = 2(r^2 + t) = 242$ bits, public key size is $56.3\,\mathrm{KB}$, private key size is $8.7\,\mathrm{KB}$. We claim that the best attack to this new scheme is the High Rank Attack and the security level is 2^{80}. Detailed cryptanalysis is given in next section.

4 Security Analysis

In multivariate public key cryptography, it is generally difficult to prove the security of a scheme as there yet has been no provable security model in this subject. We will have to check the security of a scheme against all known attacks in this subject. In the future, if there were a provable security model, it would then be possible to formally prove the security of a scheme. In this section, we shall analyze the security of the new variant of YTS' scheme, YTS-HFEv, against attacks in multivariate public key cryptography. We shall omit those attacks obviously not applicable here, and take into account those attacks to YTS' scheme, High Rank Attack and attacks to HFEv.

Recall that the pair of the central map of YTS-HFEv are the following sums

$$G_i(\mathbf{x}_1, \mathbf{x}_2) = F_i(\mathbf{x}_1) + S_i \cdot \bar{H}(T_i \cdot \mathbf{x}_1, \mathbf{x}_2) : \mathbb{F}_q^{t+n} \to \mathbb{F}_q^m.$$

The purpose of adding $\bar{H}$ is to hide the structure of the F_i in YTS' scheme and it is unnecessary to have $t > m$. Notice that if $\mathbf{x}_2$ is always chosen to be 0 in the process of generating signatures, then a large collection of signatures would help identify the secret subspace of $(\mathbf{x}_1, 0)$. However, this would not happen in the design of signature generation, because after $\mathbf{x}_1$ is calculated, a nonzero solution to $\bar{H}(T_i \cdot \mathbf{x}_1, \mathbf{x}_2) = 0$ exists with high probability and is preferred. So leakage of the subspace of $(\mathbf{x}_1, 0)$ can be prevented.

4.1 Attacks to YTS' Scheme

First of all we consider those attacks threatening YTS' scheme and its analog over even finite fields, including MinRank attack and attacks in [Has14, ZT15a, ZT15b] and the one in Sect. 2 of this paper.

In YTS' scheme and its analog over even finite fields, the quadratic polynomial of each component of the central map corresponds to a matrix of rank r. This rank is preserved under linear combination. So MinRank attack is applicable to recover L based on this rank property, and the complexity is $O(q^r)$ [YTS13]. However, in YTS-HFEv, the adding of $S_i\bar{H}$ increases this rank to be no less than

$r + t$. So as long as t is big enough, the new scheme can then resist the MinRank attack. For example, to have security level of 2^{80}, we should have $q^{r+t} \geq 2^{80}$.

The attacks in [Has14, ZT15a, ZT15b] and the one in Sect. 2 of this paper all use the special structure of the central map and the public map. Namely the central map and the public map have too simple and too structured expressions. For example, as pointed out in [ZT15a], each component of YTS' public map can be expressed as

$$A_k = R^T \begin{pmatrix} L_k & & \\ & \ddots & \\ & & L_k \end{pmatrix} R,$$

and each component of the public map of our analog has an expression either (2.1) or (2.2). These expressions have very canonical structures making the private keys recoverable using specific methods. However, after being mixed with HFEv, this kind of special structure is totally destroyed so that these attacks are no longer applicable here.

4.2   High Rank Attack

High Rank Attack is to find linear combinations of the central map such that they have the most number of variables. This is equivalent to find the variables appearing the fewest times s in the central map. So high rank attack can recover L with complexity $O(q^s)$ and is powerful to break triangular schemes [CSV97, GC00, YC05]. For the new scheme here, this complexity is $O(q^t)$. If t is not big enough, $S_i \bar{H}$ may be removed and then attacks [Has14, ZT15a] can remain applicable. So it is necessary to protect $S_i \bar{H}$ from the High Rank Attack. To have security level 2^{80}, we should choose t such that $q^t \geq 2^{80}$.

4.3   Attacks to HFEv

This part is similar to the corresponding part of [ZT15c] as we apply the same idea of [ZT15c], i.e., using HFEv to mix with the central map. Notice that the new scheme YTS-HFEv is indeed of the type of HFEv, hence it is necessary to discuss those attacks to HFEv. This becomes clear if we lift the central map

$$G_i(\mathbf{x}_1, \mathbf{x}_2) = F_i(\mathbf{x}_1) + S_i \cdot \bar{H}(T_i \cdot \mathbf{x}_1, \mathbf{x}_2)$$

to the extension field $\mathbb{K}$,

$$G_i'(V, X) = \sum a_{ij}' V^{q^i + q^j} + \sum b_i' V^{q^i}$$
$$+ \sum_{0 \leq i \leq t} \sum_{1 \leq q^j \leq D} a_{ij} V^{q^i} X^{q^j} + \sum_{1 \leq q^i + q^j \leq D} b_{ij} V^{q^i} X^{q^j} + \sum_{1 \leq q^j \leq D} c_j X^{q^j}.$$

which obviously have the form of HFEv polynomials. The $\mathbf{x}_1$ of $G_i(\mathbf{x}_1, \mathbf{x}_2)$ corresponds to the vinegar variable V and $\mathbf{x}_2$ corresponds to variable X; F_i corresponds to the sum of the monomials $V^{q^i + q^j}, V^{q^i}$; and $T_i \bar{H}$ corresponds to the sum of the rest monomials.

The major attacks applicable to the HFE family are Kipnis-Shamir's attack [KS99] and direct algebraic attack [FJ03]. Kipnis-Shamir's attack relies on the MinRank problem and is improved by Ding and Schmidt [DS05] to attack HFEv as well. In their cryptanalysis, if v is very small, such as $v = 1$, then HFEv can be broken, but the complexity increases fast as v increases. Especially there would be no way to identify HFEv from a random system when the number of vinegar variables v and the extension degree of the field $\mathbb{K}$ over $\mathbb{F}_q$ are close.

For direct algebraic attack to HFEv and HFEv-, a solid theoretical estimation on the complexity is given by Ding and Yang in [DY13] by calculating the degree of regularity. They conclude that direct attack remains feasible for very small v but infeasible for big v. As an example, the famous QUARTZ signature scheme is an HFEv scheme with several components deleted. It has only 4 vinegar variables, its degree of regularity is bounded by 9 and its security level is estimated as 2^{92}.

Notice that YTS-HFEv has n vinegar variables which is bigger than the extension degree t. Since we have to choose t such as $q^t \geq 2^{80}$ due to the high rank attack, the public map would not be identifiable from a random system of quadratic polynomials against Kipnis-Shamir's MinRank attack. This choice of the parameter t also assures that its degree of regularity is very high according to [DY13], hence direct algebraic attack is also not applicable to this variant.

5   Conclusion

In this paper, we investigate the possibility of fixing the security problem of Yasuda, Takagi and Sakurai's interesting signature scheme. We first construct its analogous scheme over finite fields of characteristic two, but then we find that it is neither secure by developing an attack to it. To resolve this security issue, we apply the idea of HFEv and use a special type of HFEv polynomials to mix with Yasuda, Takagi and Sakurai's scheme and our analogous scheme. We then show that this new scheme, YTS-HFEv, can resist current attacks, and propose a parameter set with public key size 57 KB and security level 2^{80}. Future implementation is needed to verify the security claim made here.

Acknowledgments. The authors would like to thank the anonymous reviewers for their helpful comments on improving this paper. The first author would like to thank the financial support from the National Natural Science Foundation of China (Grant No. 61572189).

References

[BFP13] Bettale, L., Faugère, J.C., Perret, L.: Cryptanalysis of HFE, multi-HFE and variants for odd and even characteristic. Des. Codes Crypt. **69**(1), 1–52 (2013)

[Clo09] Clough, C.: Square: A New Family of Multivariate Encryption Schemes. Ph.D. thesis, University of Cincinnati (2009)

[CSV97] Coppersmith, D., Stern, J., Vaudenay, S.: The security of the birational permutation signature schemes. J. Crypt. **10**, 207–221 (1997)

[DGS06] Ding, J., Gower, J.E., Schmidt, D.S.: Multivariate Public Key Cryptosystems. Advances in Information Security, vol. 25. Springer, Heidelberg (2006)

[DS05] Ding, J., Schmidt, D.: Cryptanalysis of HFEv and internal perturbation of HFE. In: Vaudenay, S. (ed.) PKC 2005. LNCS, vol. 3386, pp. 288–301. Springer, Heidelberg (2005)

[DWY07] Ding, J., Wolf, C., Yang, B.-Y.: ℓ-invertible cycles for multivariate quadratic (MQ) public key cryptography. In: Okamoto, T., Wang, X. (eds.) PKC 2007. LNCS, vol. 4450, pp. 266–281. Springer, Heidelberg (2007)

[DY13] Ding, J., Yang, B.-Y.: Degree of regularity for HFEv and HFEv-. In: Gaborit, P. (ed.) PQCrypto 2013. LNCS, vol. 7932, pp. 52–66. Springer, Heidelberg (2013)

[FJ03] Faugère, J.-C., Joux, A.: Algebraic cryptanalysis of hidden field equation (HFE) cryptosystems using Gröbner bases. In: Boneh, D. (ed.) CRYPTO 2003. LNCS, vol. 2729, pp. 44–60. Springer, Heidelberg (2003)

[GC00] Goubin, L., Courtois, N.T.: Cryptanalysis of the TTM cryptosystem. In: Okamoto, T. (ed.) ASIACRYPT 2000. LNCS, vol. 1976, pp. 44–57. Springer, Heidelberg (2000)

[GJ79] Garey, M.R., Johnson, D.S.: Computers and intractability: A guide to the theory of NP-completeness. W.H. Freeman, New York (1979)

[GJS06] Granboulan, L., Joux, A., Stern, J.: Inverting HFE is quasipolynomial. In: Dwork, C. (ed.) CRYPTO 2006. LNCS, vol. 4117, pp. 345–356. Springer, Heidelberg (2006)

[Has14] Hashimoto, Y.: Cryptanalysis of the multivariate signature scheme proposed in PQCrypto 2013. In: Mosca, M. (ed.) PQCrypto 2014. LNCS, vol. 8772, pp. 108–125. Springer, Heidelberg (2014)

[KPG99] Kipnis, A., Patarin, J., Goubin, L.: Unbalanced oil and vinegar signature schemes. In: Stern, J. (ed.) EUROCRYPT 1999. LNCS, vol. 1592, pp. 206–222. Springer, Heidelberg (1999)

[KS98] Kipnis, A., Shamir, A.: Cryptanalysis of the oil and vinegar signature scheme. In: Krawczyk, H. (ed.) Advances in Cryptology – CRYPTO'98. LNCS, vol. 1462, pp. 257–266. Springer, Heidelberg (1998)

[KS99] Kipnis, A., Shamir, A.: Cryptanalysis of the HFE public key cryptosystem by relinearization. In: Wiener, M. (ed.) CRYPTO 1999. LNCS, vol. 1666, pp. 19–30. Springer, Heidelberg (1999)

[LN97] Lidl, R., Niederreiter, H.: Finite fields. Encyclopedia of Mathematics and Its Applications, vol. 20, 2nd edn. Cambridge University Press, Cambridge (1997)

[MI88] Matsumoto, T., Imai, H.: Public quadratic polynomial-tuples for efficient signature-verification and message-encryption. In: Günther, C.G. (ed.) EUROCRYPT 1988. LNCS, vol. 330, pp. 419–453. Springer, Heidelberg (1988)

[Pat95] Patarin, J.: Cryptanalysis of the Matsumoto and Imai public key scheme of Eurocrypt'88. In: Coppersmith, D. (ed.) CRYPTO 1995. LNCS, vol. 963, pp. 248–261. Springer, Heidelberg (1995)

[Pat96] Patarin, J.: Hidden fields equations (HFE) and isomorphisms of polynomials (IP): two new families of asymmetric algorithms. In: Maurer, U.M. (ed.) EUROCRYPT 1996. LNCS, vol. 1070, pp. 33–48. Springer, Heidelberg (1996)

[PBD14] Porras, J., Baena, J., Ding, J.: ZHFE, a new multivariate public key encryption scheme. In: Mosca, M. (ed.) PQCrypto 2014. LNCS, vol. 8772, pp. 229–245. Springer, Heidelberg (2014)

[PCG01] Patarin, J., Courtois, N.T., Goubin, L.: QUARTZ, 128-bit long digital signatures. In: Naccache, D. (ed.) CT-RSA 2001. LNCS, vol. 2020, pp. 282–288. Springer, Heidelberg (2001)

[Sho97] Shor, P.W.: Polynomial-time algorithms for prime factorization and discrete logarithms on a quantum computer. SIAM J. Comput. **26**(5), 1484–1509 (1997)

[Tay92] Taylor, D.E.: The Geometry of the Classical Groups. Sigma Series in Pure Mathematics. Heldermann Verlag, Berlin (1992)

[TDTD13] Tao, C., Diene, A., Tang, S., Ding, J.: Simple matrix scheme for encryption. In: Gaborit, P. (ed.) PQCrypto 2013. LNCS, vol. 7932, pp. 231–242. Springer, Heidelberg (2013)

[WBP05] Wolf, C., Braeken, A., Preneel, B.: Efficient cryptanalysis of RSE(2)PKC and RSSE(2)PKC. In: Blundo, C., Cimato, S. (eds.) SCN 2004. LNCS, vol. 3352, pp. 294–309. Springer, Heidelberg (2005)

[WBP06] Wolf, C., Braeken, A., Preneel, B.: On the security of stepwise triangular systems. Des. Codes Crypt. **40**, 285–302 (2006)

[WP05] Wolf, C, Preneel, B.: Taxonomy of public key schemes based on the problem of multivariatequadratic equations. Cryptology ePrint Archive, Report 2005/077 (2005). http://eprint.iacr.org/2005/077/

[WYHL06] Wang, L.-C., Yang, B.-Y., Hu, Y.-H., Lai, F.: A "medium-field" multivariate public-key encryption scheme. In: Pointcheval, D. (ed.) CT-RSA 2006. LNCS, vol. 3860, pp. 132–149. Springer, Heidelberg (2006)

[YC05] Yang, B.-Y., Chen, J.-M.: Building secure tame-like multivariate public-key cryptosystems: the new TTS. In: Boyd, C., González Nieto, J.M. (eds.) ACISP 2005. LNCS, vol. 3574, pp. 518–531. Springer, Heidelberg (2005)

[YTS13] Yasuda, T., Takagi, T., Sakurai, K.: Multivariate signature scheme using quadratic forms. In: Gaborit, P. (ed.) PQCrypto 2013. LNCS, vol. 7932, pp. 243–258. Springer, Heidelberg (2013)

[ZT15a] Zhang, W., Tan, C.H.: Algebraic cryptanalysis of Yasuda, Takagi and Sakurai's signature scheme. In: Lee, J., Kim, J. (eds.) Information Security and Cryptology - ICISC 2014. LNCS, vol. 8949, pp. 53–66. Springer, Switzerland (2014)

[ZT15b] Zhang, W., Tan, C.H.: Cryptanalysis of Yasuda, Takagi and Sakurai's Signature Scheme Using Invariant Subspaces. Cryptology ePrint Archive, Report 2015/1005 (2015). http://eprint.iacr.org/2015/1005

[ZT15c] Zhang, W., Tan, C.H.: MI-T-HFE, a new multivariate signature scheme. In: Groth, J. (ed.) IMACC 2015. LNCS, vol. 9496, pp. 43–56. Springer, Heidelberg (2015). doi:10.1007/978-3-319-27239-9_3

Symmetric Ciphers

Statistical and Algebraic Properties of DES

Stian Fauskanger[1(✉)] and Igor Semaev[2]

[1] Norwegian Defence Research Establishment (FFI), PB 25, 2027 Kjeller, Norway
stian.fauskanger@ffi.no
[2] Department of Informatics, University of Bergen, Bergen, Norway

Abstract. D. Davies and S. Murphy found that there are at most 660 different probability distributions on the output from any three adjacent S-boxes after 16 rounds of DES [5]. In this paper it is shown that there are only 72 different distributions for S-boxes 4, 5 and 6. The distributions from S-box triplets are linearly dependent and the dependencies are described. E.g. there are only 13 linearly independent distributions for S-boxes 4, 5 and 6. A coset representation of DES S-boxes which reveals their hidden linearity is studied. That may be used in algebraic attacks. S-box 4 can be represented by significantly fewer cosets than the other S-boxes and therefore has more linearity. Open cryptanalytic problems are stated.

Keywords: S-box · Output distributions · Linear dependencies · Coset representation

1 Introduction

The Data Encryption Standard (DES) is a symmetric block cipher from 1977. It has block size of 64 bits and a 56-bit key. DES in its original form is deprecated due to the short key. Triple DES [1] however, is still used in many applications (e.g. in chip-based payment cards). It is therefore still important to analyze its security. DES is probably the most analyzed cipher, and is broken by linear [8] and differential [3] cryptanalysis. Even so, the most effective method in practice is still exhaustive search for the key. There are also some algebraic attacks that can break 6-round DES [4].

Donald Davies and Sean Murphy described in [5] some statistical properties of the S-boxes in DES. They found that there are at most 660 different distributions on the output from any three adjacent S-boxes after 16 rounds. These distributions divide the key space into classes where equivalent keys make the output follow the same distributions. The correct class is found by identifying which distribution a set of plaintext/ciphertext pairs follow. They used this to give a known-plaintext attack. The time complexity of the attack is about the same as brute-force attack and requires approximately $2^{56.6}$ plaintext/ciphertext pairs. The attack was improved by Biham and Biryukov [2] where the key can be found with 2^{50} plaintext/ciphertext pairs with 2^{50} operations. Later, Kunz-Jacques and Muller [7] further improved the attack to a chosen-plaintext attack with time complexity 2^{45} using 2^{45} chosen plaintexts.

© Springer International Publishing Switzerland 2016
D. Lin et al. (Eds.): Inscrypt 2015, LNCS 9589, pp. 93–107, 2016.
DOI: 10.1007/978-3-319-38898-4_6

In this paper we study new statistical and algebraic properties of DES. In Sect. 2 we show Davies and Murphy's results, using different notations than theirs. We also show a new exceptional property of S_4, and use this to show that there are fewer different distributions on the output from $S_4 S_5 S_6$ compared to other triplets. The new properties are related to the forth S-box in DES, and is used to show that the number of different distributions on the output from S-box 4, 5 and 6 is at most 72 (after 16 rounds). This divides the key space into fewer, but larger, classes compared to Davies and Murphy's results.

The distributions from S-box triplets are linearly dependent. We give a description of the relations between the distributions, and upper bound the number of linearly independent distributions for each triplet. E.g. among the 72 different distributions for S-box 4, 5 and 6 there are only 13 linearly independent.

A coset representation of the DES S-boxes is suggested in Sect. 4. It is found that S-box 4 is abnormal again. It can be covered by 10 sub-cosets while the other S-boxes require at least 16. Also, the coset representation of S-box 4 contains 6 sub-cosets of size 8, while the other S-boxes contain at most one sub-coset of such size. The coset representation of S-boxes makes it possible to write the system of equations for DES in a more compact form than in [9,10].

Like the linear approximations discovered by Shamir [12] was later used by Matsui [8] to successfully break DES, these new properties might improve some attacks in the future. Two open problems are stated at the end of the paper. If solved that would improve statistical and algebraic attacks on DES.

1.1 Notations

Let X_{i-1}, X_i denote the input to the i-th round and X_i, X_{i+1} denote the i-th round output. So X_0, X_1 and X_{17}, X_{16} are plaintext and ciphertext blocks respectively, where the initial and final permutations are ignored. Let K_i be the 48-bit round key at round i. Then

$$X_{i-1} \oplus X_{i+1} = Y_i, \quad Y_i = P(S(\bar{X}_i \oplus K_i)), \tag{1}$$

where $\bar{X}_i$ is a 48-bit expansion of X_i, P denotes a permutation on 32 symbols, and S is a transform implemented by 8 S-boxes. Let S_j be a DES S-box, so

$$S_j(u_5, u_4, u_3, u_2, u_1, u_0) = (v_3, v_2, v_1, v_0), \tag{2}$$

where u_i and v_i are input and output bits respectively.

2 Results from Davies and Murphy

By (1), the XOR of the plaintext/ciphertext blocks are representable as follows

$$X_{17} \oplus X_1 = Y_2 \oplus Y_4 \oplus \ldots \oplus Y_{14} \oplus Y_{16}, \tag{3}$$

$$X_{16} \oplus X_0 = Y_1 \oplus Y_3 \oplus \ldots \oplus Y_{13} \oplus Y_{15}. \tag{4}$$

In this section we study the joint distribution of bits in $X_{17} \oplus X_1$ and in $X_{16} \oplus X_0$ which come from the output of 3 adjacent S-boxes in DES round function, and therefore in Y_i. These results are from [5], but presented using a different notation.

2.1 Definitions and a Basic Lemma

The output of 3 adjacent S-boxes is called (S_{i-1}, S_i, S_{i+1})-output when i is specified. When analysing (3) and (4) we assume the round function inputs $X_2, X_4, \ldots, X_{16}$ and $X_1, X_3, \ldots, X_{15}$ are uniformly random and independent respectively. Input to S_i is accordingly assumed to be uniformly random. These common assumptions were already in [5].

When we look at a reduced number of rounds in DES (k rounds), then $X_{k+1} \oplus X_1$ and $X_k \oplus X_0$ follows the distribution for the XOR of $k/2$ round-outputs (for even k). We will throughout this paper use $2n$ to denote the number of rounds. n is the number of outputs that are XORed, and full DES is represented by $n = 8$.

We define three distributions that are related to each S_i. We use notation (2).

1. The distribution of $(u_1, u_0, v_3, v_2, v_1, v_0)$ is called **right hand side distribution** and we denote $p_{y,r}^{(i)} = \mathbf{Pr}((u_1, u_0) = y$ and $(v_3, v_2, v_1, v_0) = r)$.
2. The distribution of $(u_5, u_4, v_3, v_2, v_1, v_0)$ is called **left hand side distribution** and we denote $q_{x,r}^{(i)} = \mathbf{Pr}((u_5, u_4) = x$ and $(v_3, v_2, v_1, v_0) = r)$.
3. The distribution of $(u_5, u_4, u_1, u_0, v_3, v_2, v_1, v_0)$ is called **LR distribution** and we denote

$$Q_{x,y,r}^{(i)} = \mathbf{Pr}((u_5, u_4) = x, \text{ and } (u_1, u_0) = y, \text{ and } (v_3, v_2, v_1, v_0) = r).$$

Obviously, $p_{y,r}^{(i)} = \sum_x Q_{x,y,r}^{(i)}$ and $q_{x,r}^{(i)} = \sum_y Q_{x,y,r}^{(i)}$, the sums are over 2-bit x, y respectively.

Lemma 1. *For any 2-bit x, y and any 4-bit r holds*

$$p_{y \oplus 2, r}^{(i)} + p_{y,r}^{(i)} = \frac{1}{32}, \tag{5}$$

$$q_{x \oplus 1, r}^{(i)} + q_{x,r}^{(i)} = \frac{1}{32}, \tag{6}$$

$$Q_{x,y,r}^{(i)} + Q_{x,y \oplus 2, r}^{(i)} + Q_{x \oplus 1, y, r}^{(i)} + Q_{x \oplus 1, y \oplus 2, r}^{(i)} = \frac{1}{64}. \tag{7}$$

Proof. The equalities (5) and (6) were found directly from the values of $p_{y,r}^{(i)}, q_{x,r}^{(i)}$, for instance, see those distributions listed for S_4 in Appendix A. Alternatively, by DES S-box definition, for any fixed (u_5, u_0) the distribution of (v_3, v_2, v_1, v_0) is uniform. So $(u_0, v_3, v_2, v_1, v_0)$ and $(u_5, v_3, v_2, v_1, v_0)$ are uniformly distributed and that implies (5) and (6) as Kholosha [6] later observed. The former implies (7) as well.

2.2 Output-Distributions on S-box Triplets

We study the distribution of the output from three adjacent S-boxes in DES round function. Let $(a_5, \ldots, a_0)$, $(b_5, \ldots, b_0)$ and $(c_5, \ldots, c_0)$ be the input to three adjacent S-boxes in one DES round. Then

$$(a_1, a_0) \oplus (b_5, b_4) = k \qquad \text{and} \qquad (b_1, b_0) \oplus (c_5, c_4) = k',$$

where k and k', the **common key bits**, are both 2-bit linear combinations of round-key-bits. By $k_j = (k_{j1}, k_{j0})$ and $k'_j = (k'_{j1}, k'_{j0})$ we denote the common key bits in round j.

Let (r, s, t) be a 12-bit output from S_{i-1}, S_i, S_{i+1} in one DES round. Then

$$\mathbf{Pr}(r, s, t \mid k, k') = 2^4 \times \sum_{x,y} p^{(i-1)}_{x \oplus k, r} \, Q^{(i)}_{x,y,s} \, q^{(i+1)}_{y \oplus k', t}. \tag{8}$$

The distribution of (r, s, t) after $2n$ rounds is the n-fold convolution of (8):

$$\mathbf{Pr}(r, s, t \mid k_1, k'_1, \ldots, k_n, k'_n) = \sum \prod_{i=1}^{n} \mathbf{Pr}(r_i, s_i, t_i \mid k_i, k'_i),$$

where the sum is over (r_i, s_i, t_i) such that $\bigoplus_i (r_i, s_i, t_i) = (r, s, t)$. By changing the order of summation and using (8) we get

$$\mathbf{Pr}(r, s, t \mid k_1, k'_1, \ldots, k_n, k'_n)$$
$$= 2^{4n} \times \sum p^{(i-1)}_{x_1 \oplus k_1, \ldots, x_n \oplus k_n, r} \times Q^{(i)}_{x_1, y_1, \ldots, x_n, y_n, s} \times q^{(i+1)}_{y_1 \oplus k'_1, \ldots, y_n \oplus k'_n, t}, \tag{9}$$

where the sum is over 2-bit $x_1, y_1, \ldots, x_n, y_n$, and

$$p^{(i)}_{x_1, \ldots, x_n, r} = \sum_{\oplus_j r_j = r} p^{(i)}_{x_1, r_1} \times \cdots \times p^{(i)}_{x_n, r_n},$$

$$q^{(i)}_{y_1, \ldots, y_n, t} = \sum_{\oplus_j t_j = t} q^{(i)}_{y_1, t_1} \times \cdots \times q^{(i)}_{y_n, t_n},$$

$$Q^{(i)}_{x_1, y_1, \ldots, x_n, y_n, s} = \sum_{\oplus_j s_j = s} Q^{(i)}_{x_1, y_1, s_1} \times \cdots \times Q^{(i)}_{x_n, y_n, s_n}.$$

Lemma 1 implies the following corollary.

Corollary 1. *For any 2-bit $x_1, y_1, \ldots, x_n, y_n$ and 4-bit r, t*

$$p^{(i)}_{x_1 \oplus k_1, \ldots, x_n \oplus k_n, r} = p^{(i)}_{x_1 \oplus k_{10}, \ldots, x_{n-1} \oplus k_{(n-1)0}, x_n \oplus 2\overline{k}, r},$$

$$q^{(i)}_{y_1 \oplus k'_1, \ldots, y_n \oplus k'_n, t} = q^{(i)}_{y_1 \oplus 2k'_{11}, \ldots, y_{n-1} \oplus 2k'_{(n-1)1}, y_n \oplus \overline{k'}, t},$$

where $\overline{k}$ and $\overline{k'}$ are the parity of $(k_{11}, \ldots, k_{n1})$ and $(k'_{10}, \ldots, k'_{n0})$.

Each value for the vector $(k_1, k'_1, \ldots, k_n, k'_n)$ can be mapped to a distribution on (r, s, t). Many of these distributions are equal to each other. Corollary 1 is now used to give an upper bound on the number of different distributions.

First, one can permute any (k_j, k'_j) and (k_i, k'_i) and get the same distribution. Also the distribution is defined by the parity of $(k_{11}, \ldots, k_{n1})$ and $(k'_{10}, \ldots, k'_{n0})$. There are 4 values for the two parity-bits, and there are $\binom{3+n}{n}$ combinations for the remaining $2n$ bits $(k_{10}, \ldots, k_{n0})$ and $(k'_{11}, \ldots, k'_{n1})$. Therefore there are at most $4 \times \binom{3+n}{n}$ different distributions on the output from three adjacent S-boxes. Table 1 lists the maximum number of different distributions after multiple rounds. Again, 16-round DES is specified by $n = 8$.

Table 1. Upper bound on number of different distributions for $2n$ rounds

n	1	2	3	4	5	6	7	8
Upper bound	16	40	80	140	224	336	480	660

3   New Statistical Property of S_4

In this section we find an exceptional property of S_4. In particular, we prove Lemma 2, and use it to show that there are fewer different output-distributions on $S_4 S_5 S_6$.

Lemma 2.   *For any 2-bit x, y, a and 4-bit r holds*

$$\sum_h p^{(4)}_{x \oplus a, h} \, p^{(4)}_{y \oplus a, h \oplus r} = \sum_h p^{(4)}_{x,h} \, p^{(4)}_{y, h \oplus r}.$$

Proof. By Lemma 1, $p^{(4)}_{x \oplus 2, h} + p^{(4)}_{x,h} = \frac{1}{32}$ for any 2-bit x and 4-bit h. It is easy to see the lemma is true for $a = 2$. All other cases are reduced to $a = 1$ and $x = y = 0$. Let

$$f(h) = \begin{cases} 0, & \text{if } h \notin \{0, 6, 9, 15\}; \\ 1, & \text{if } h \in \{0, 9\}; \\ -1, & \text{if } h \in \{6, 15\}. \end{cases}$$

From S_4 right hand side distribution values, see Table 5 in Appendix A, we find

$$p^{(4)}_{x \oplus 1, h} + p^{(4)}_{x, h} = \frac{1}{32} + \frac{(-1)^{x_1} f(h)}{64} \tag{10}$$

and then

$$\sum_h f(h) f(h \oplus r) = 4 f(r), \tag{11}$$

$$\sum_h p^{(4)}_{x,h} f(h \oplus r) = \frac{(-1)^{x_1} 2 f(r)}{64}, \tag{12}$$

for any 2-bit $x = (x_1, x_0)$ and any 4-bit r. Hence

$$\sum_h p^{(4)}_{1,h} \, p^{(4)}_{1, h \oplus r} = \sum_h \left(\frac{1}{32} + \frac{f(h)}{64} - p^{(4)}_{0,h} \right) \left(\frac{1}{32} + \frac{f(h \oplus r)}{64} - p^{(4)}_{0, h \oplus r} \right) =$$

$$\sum_h \frac{f(h) f(h \oplus r)}{64^2} - 2 \sum_h p^{(4)}_{0,h} \frac{f(h \oplus r)}{64} + \sum_h p^{(4)}_{0,h} \, p^{(4)}_{0, h \oplus r} = \sum_h p^{(4)}_{0,h} \, p^{(4)}_{0, h \oplus r}.$$

The lemma is proved.

This surprising property holds because (10), (11) and (12) are true simultaneously for the right hand side distribution $p^{(4)}_{x,h}$.

Corollary 2.   *For any 2-bit* $x_1, \ldots, x_n$ *and 4-bit* r *holds*

$$p^{(4)}_{x_1 \oplus k_1, \ldots, x_n \oplus k_n, r} = p^{(4)}_{x_1, \ldots, x_{n-1}, x_n \oplus \bar{k}, r},$$

where $\bar{k} = k_1 \oplus \cdots \oplus k_n$.

Proof.   By Lemma 2,

$$\sum_{h_1 \oplus h_2 = r} p^{(4)}_{x_1 \oplus k_1, h_1} p^{(4)}_{x_2 \oplus k_2, h_2} = \sum_{h_1 \oplus h_2 = r} p^{(4)}_{x_1, h_1} p^{(4)}_{x_2 \oplus (k_1 \oplus k_2), h_2}$$

for any x_1, x_2, k_1, k_2 and r. Therefore the corollary is true for $n = 2$. The general case follows recursively.

3.1   The Number of Different Output-Distributions

Davies and Murphy found that there are at most $4 \times \binom{3+n}{n}$ different distributions of the output from 3 adjacent S-boxes after $2n$ rounds. In this section we show (S_4, S_5, S_6)-output has at most $(8n + 8)$ different distributions.

Lemma 3. *Let* (r, s, t) *be* (S_4, S_5, S_6)-*output after* $2n$ *rounds. There are at most* $8n + 8$ *different distributions* (r, s, t) *can follow.*

Proof. By Corollaries 1 and 2 the distribution of (r, s, t) only depends on $\bigoplus_{j=1}^{n} k_j$, $\bigoplus_{j=1}^{n} k'_{j0}$ and common key bits $(k'_{11}, \ldots, k'_{n1})$, where the order of the last n bits is irrelevant. There are $n + 1$ combinations for $(k'_{11}, \ldots, k'_{n1})$ and 8 possible values for the three parity bits. The maximum number of different distributions is therefore at most $8n + 8$ as the lemma states.

We computed the actual number of different distributions for all 8 triplets. Table 2 lists the results for $n = 1, \ldots, 8$ together with the bound from Lemma 3 and Davies-Murphy's bound. Remark that 16-round DES is specified by $n = 8$.

It is not clear whether or not fewer different distribution can improve Davies-Murphy's attack. Intuitively, distinguishing between few distributions could be easier than distinguishing between many distributions (if the biases are approximately the same). At the same time, the number of keys in the class representing a given distribution is larger, so more work is required to identify the correct key in the class. Also, the triplet attack described by Davies and Murphy does not perform better than the attack based on the two S-box pairs in the triplet [5]. We do not know if it is possible to alter Davies-Murphy's attack so that fewer distribution would give an advantage.

Table 2. Number of different distributions for output of 3 adjacent S-boxes

n	1	2	3	4	5	6	7	8
D-M's bound for all triplets	16	40	80	140	224	336	480	660
New upper bound for (S_4, S_5, S_6)	16	24	32	40	48	56	64	72
Actual value for (S_4, S_5, S_6)	16	24	32	40	48	56	64	72
Actual value for other triplets	16	40	80	140	224	336	480	660

3.2   Linear Dependencies Between the Distributions

In this section we describe linear relations between distributions on the output from three adjacent S-boxes. We will see how (S_4, S_5, S_6) compares to the other triplets. A distribution can be represented by a row-vector $(v_0, \ldots, v_{2^{12}-1})$, where v_j is the probability of the output $j = (r, s, t)$.

Let M be a matrix whose rows are (S_{i-1}, S_i, S_{i+1})-output distributions. M is then called a **distribution matrix**. A non-zero vector r such that $rM = 0$ is called a linear relation for M. Let R be a matrix whose rows are linear relations for M, then R is called a **relation matrix** for M. Then

$$\mathrm{rank}(M) \leq k - \mathrm{rank}(R), \tag{13}$$

where k is the number of rows in M. There are five independent linear relations inside the right, LR and left distribution that can be used to find linear relation between the rows of M. By Lemma 1,

$$\sum_a C_a^1 \times p_{x \oplus a, r}^{(i)} = 0 \quad \text{and} \quad \sum_a C_a^2 \times q_{x \oplus a, r}^{(i)} = 0, \tag{14}$$

where $C^1 = (1, -1, 1, -1)$ and $C^2 = (1, 1, -1, -1)$. Also by Lemma 1, for any 2-bit x, y and 4-bit r

$$\sum_a Q_{x \oplus a, y, r}^{(i)} + Q_{x \oplus a, y \oplus 2, r}^{(i)} = \frac{1}{32}, \tag{15}$$

$$\sum_b Q_{x, y \oplus b, r}^{(i)} + Q_{x \oplus 1, y \oplus b, r}^{(i)} = \frac{1}{32}, \tag{16}$$

$$Q_{x, y, r}^{(i)} + Q_{x, y \oplus 2, r}^{(i)} + Q_{x \oplus 1, y, r}^{(i)} + Q_{x \oplus 1, y \oplus 2, r}^{(i)} = \frac{1}{64}. \tag{17}$$

One now subtracts (15) and (15) after changing $y \leftarrow y \oplus 1$, (16) and (16) after changing $x \leftarrow x \oplus 2$, then (17) and (17) after changing $y \leftarrow y \oplus 1$. So

$$\sum_{k, k'} C_{k, k'} \times Q_{x \oplus k, y \oplus k', r} = 0, \tag{18}$$

for any x, y and r, where C is any of

$$C^3 = (1, \ -1, \ 1, \ -1, \ 1, \ -1, \ 1, \ -1, \quad 1, \ -1, \quad 1, \ -1, \quad 1, \ -1, \quad 1, \ -1),$$
$$C^4 = (1, \quad 1, \ 1, \quad 1, \ 1, \quad 1, \ 1, \quad 1, \ -1, \ -1, \ -1, \ -1, \ -1, \ -1, \ -1, \ -1),$$
$$C^5 = (1, \ -1, \ 1, \ -1, \ 1, \ -1, \ 1, \ -1, \quad 0, \quad 0, \quad 0, \quad 0, \quad 0, \quad 0, \quad 0, \quad 0).$$

For instance, C^3 comes from

$$\sum_a Q_{x \oplus a, y, r}^{(i)} + Q_{x \oplus a, y \oplus 2, r}^{(i)} - \sum_a Q_{x \oplus a, y \oplus 1, r}^{(i)} + Q_{x \oplus a, y \oplus 3, r}^{(i)} = 0.$$

Both (14) and (18) are used to build linear relations between the distributions of (r, s, t), the output from three adjacent S-boxes after one round.

Lemma 4.

$$\textit{For any } k' \qquad\qquad \sum_k C_k^1 \times \boldsymbol{Pr}(r,s,t \mid k,k') = 0, \qquad (19)$$

$$\textit{for any } k \qquad\qquad \sum_{k'} C_{k'}^2 \times \boldsymbol{Pr}(r,s,t \mid k,k') = 0, \qquad (20)$$

$$\textit{for } C \in \{C^3, C^4, C^5\} \qquad \sum_{k,k'} C_{k,k'} \times \boldsymbol{Pr}(r,s,t \mid k,k') = 0. \qquad (21)$$

Proof. We will prove (19):

$$
\sum_k C_k^1 \times \boldsymbol{Pr}(r,s,t \mid k,k') = 2^4 \times \sum_k C_k^1 \times \left(\sum_{x,y} p_{x \oplus k,r}^{(i-1)} Q_{x,y,s}^{(i)} q_{y \oplus k',t}^{(i+1)} \right)
$$

$$
= 2^4 \times \sum_{x,y} \sum_k C_k^1 \times \left(p_{x \oplus k,r}^{(i-1)} Q_{x,y,s}^{(i)} q_{y \oplus k',t}^{(i+1)} \right)
$$

$$
= 2^4 \times \sum_{x,y} Q_{x,y,s}^{(i)} q_{y \oplus k',t}^{(i+1)} \times \left(\sum_k C_k^1 \times p_{x \oplus k,r}^{(i-1)} \right) = 0.
$$

Similarly (20) is proved. We will prove (21).

$$
\sum_{k,k'} C_{k,k'} \times \boldsymbol{Pr}(r,s,t \mid k,k') = 2^4 \times \sum_{k,k'} C_{k,k'} \times \left(\sum_{x,y} p_{x,r}^{(i-1)} Q_{x \oplus k, y \oplus k', s}^{(i)} q_{y,t}^{(i+1)} \right)
$$

$$
= 2^4 \times \sum_{x,y} \sum_{k,k'} C_{k,k'} \times \left(p_{x,r}^{(i-1)} Q_{x \oplus k, y \oplus k', s}^{(i)} q_{y,t}^{(i+1)} \right)
$$

$$
= 2^4 \times \sum_{x,y} p_{x,r}^{(i-1)} q_{y,t}^{(i+1)} \times \left(\sum_{k,k'} C_{k,k'} Q_{x \oplus k, y \oplus k', s}^{(i)} \right)
$$

$$
= 0.
$$

Lemma 4 implies there are 11 linear dependencies between rows of the distribution matrix after one round. The rank of the relation matrix is 10. We have also computed the rank of the distribution matrix which is 6. Since there are 16 distributions in total, we have found all 10 independent linear relations between the distributions. Lemma 4 is now used to build linear relations between the distributions after $2n$ rounds.

Lemma 5. *For any* $(k_1, \dots, k_n)$, $(k_1', \dots, k_n')$, *and* i

$$\sum_{k_i} C_{k_i}^1 \times \boldsymbol{Pr}(r,s,t \mid k_1, k_1', \dots, k_n, k_n') = 0, \qquad (22)$$

$$\sum_{k_i'} C_{k_i'}^2 \times \boldsymbol{Pr}(r,s,t \mid k_1, k_1', \dots, k_n, k_n') = 0, \qquad (23)$$

$$\sum_{k_i, k_i'} C_{k_i, k_i'} \times \boldsymbol{Pr}(r,s,t \mid k_1, k_1', \dots, k_n, k_n') = 0, \qquad (24)$$

where $C \in \{C^3, C^4, C^5\}$.

Proof. It is enough to prove (22) for $i = 1$.

$$\sum_{k_1} C_{k_1}^1 \times \mathbf{Pr}(r, s, t \mid k_1, k_1', \dots, k_n, k_n')$$

$$= \sum_{k_1} C_{k_1}^1 \times \sum{}' \prod_{j=1}^{n} \mathbf{Pr}(r_j, s_j, t_j \mid k_j, k_j')$$

$$= \sum{}' \prod_{j=2}^{n} \mathbf{Pr}(r_j, s_j, t_j \mid k_j, k_j') \sum_{k_1} C_{k_1}^1 \times \mathbf{Pr}(r_1, s_1, t_1 \mid k_1, k_1') = 0,$$

where $\sum{}'$ is over all (r_j, s_j, t_j) such that $\bigoplus_j (r_j, s_j, t_j) = (r, s, t)$. The proofs of (23) and (24) are similar.

Generating all relations from (22), (23) and (24) for all values of $(k_1, \dots, k_n)$, $(k_1', \dots, k_n')$, and i will make a relation matrix too large to calculate the rank when $n \geq 4$. We will instead consider a distribution matrix M, where each distribution occurs only once. We then generate a relation matrix for M. This way, by using (13), we find an upper bound on the rank of M for all triplets and $n \leq 8$, see row 2 and 3 in Table 3. Triplet $S_4 S_5 S_6$ have an upper bound on the rank which is lower than the other triplets. Full DES is specified by $n = 8$. We also computed the actual rank of M for each triplet, see row 4–11.

Each distribution is determined by a class of DES keys. Table 3 data suggests a strong statistical dependence between ciphertexts generated with representatives of such classes. An open problem is stated in the end of this paper, which if solved, could make use of these statistical dependencies to improve the probability of success on Davies-Murphy's attack.

Table 3. Rank of the distribution matrix for each triplet

n	1	2	3	4	5	6	7	8
Upper bound for $S_4 S_5 S_6$	6	7	8	9	10	11	12	13
Upper bound for other triplets	6	9	13	18	24	31	39	48
$S_1 S_2 S_3$	6	9	13	18	24	30	36	42
$S_2 S_3 S_4$	6	9	13	18	24	31	39	48
$S_3 S_4 S_5$	6	9	13	18	24	29	34	39
$S_4 S_5 S_6$	6	7	8	9	10	11	12	13
$S_5 S_6 S_7$	6	9	13	18	24	31	39	48
$S_6 S_7 S_8$	6	9	13	18	24	31	39	48
$S_7 S_8 S_1$	6	9	13	18	24	31	39	48
$S_8 S_1 S_2$	6	9	13	18	24	31	39	48

4   S-box Coset Representation and DES Equations

For each S_i by (2) a set T_i of 10-bit strings

$$(u_5, u_4, u_3, u_2, u_1, u_0, v_3, v_2, v_1, v_0) \tag{25}$$

is defined. They are vectors in a vector space of dimension 10 over field with two elements F_2 denoted F_2^{10}. Let V be any subspace of F_2^{10}. For any vector a the set $a \oplus V$ is called a coset in F_2^{10}. Let $\dim V = s$, then there are 2^{10-s} cosets associated with V. Also we say $a \oplus V$ has dimension s as well. Any coset of dimension s is a set of the solutions for a linear equation system

$$a \oplus V = \{x \mid xA = b\},$$

where A is a matrix of size $10 \times (10 - s)$, and $\operatorname{rank} A = 10 - s$, and b is a row vector of length $10 - s$.

Any set $T \subseteq F_2^{10}$ may be partitioned into a union of its sub-cosets. We try to partition into sub-cosets of largest possible dimension, in other words of largest size. Denote the set of such cosets by U, it is constructed by the following algorithm. One first constructs a list of all sub-cosets in T maximal by inclusion. Let C be a maximal in dimension coset from the list, then C is added to U and the Algorithm recursively applies to $T \backslash C$. Let

$$U = \{C_1, \ldots, C_r\}.$$

Therefore $x \in T$ if and only if x is a solution to the system $xA_k = b_k$ associated with $C_k \in U$.

The algorithm was applied to the vector sets T_i defined by DES S-boxes S_i. Let the sets of cosets U_i be produced. The results are summarised in Table 4, where $2^a \, 4^b \, 8^c$ means U_i contains a cosets of size 2, b cosets of size 4 and c cosets of size 8. The distribution is uneven. For instance, S_4 admits exceptionally many cosets of size 8. Disjoint sub-cosets which cover T_i for each $i = 1, \ldots, 8$ are listed in Appendix B, where strings (25) have integer number representation

$$u_5 2^9 + u_4 2^8 + u_3 2^7 + u_2 2^6 + u_1 2^5 + u_0 2^4 + v_3 2^3 + v_2 2^2 + v_1 2 + v_0.$$

4.1   More Compact DES Equations

Given one plaintext/ciphertext pair one constructs a system of equations in the key bits by introducing new variables after each S-box application, 128 equations for 16-round DES. By specifying S_i,

Table 4. Coset distribution for S-boxes

i	1	2	3	4	5	6	7	8
coset dist.	$2^6 \, 4^{13}$	$2^4 \, 4^{14}$	$2^6 \, 4^{11} \, 8$	$4^4 \, 8^6$	4^{16}	$2^6 \, 4^{13}$	$2^6 \, 4^{11} \, 8$	$2^4 \, 4^{12} \, 8$
# of cosets	19	18	18	10	16	19	18	17

$$\begin{array}{c} \bar{X}_{ji} \oplus K_{ji} \\ P^{-1}(X_{j-1\,i} \oplus X_{j-2\,i}) \end{array} = \begin{bmatrix} 0 & \dots & 63 \\ S_i(0) & \dots & S_i(63) \end{bmatrix}, \tag{26}$$

with 64 right hand sides, 10-bit vectors T_i written column-wise. Here $\bar{X}_{ji}$ and K_{ji} are 6-bit sub-blocks of $\bar{X}_j$ and K_j respectively. To find the key such equations are solved. That may be done with methods introduced in [10], see also [9]. The complexity heavily depends on the number of right hand sides.

We get a more compact representation, that is with lower number of sides. We use the previous section notation. Let U_i contain r cosets. So $x \in T_i$ if and only if x is a solution to exactly one of the linear equation systems

$$x A_k = b_k, \quad k = 1, \dots, r.$$

We cover the set of right hand side columns in (26) with sub-cosets from U_i and get (26) is equivalent to

$$\begin{bmatrix} \bar{X}_{ji} \oplus K_{ji} \\ P^{-1}(X_{j-1\,i} \oplus X_{j-2\,i}) \end{bmatrix} A_k = b_k, \quad k = 1, \dots, r \tag{27}$$

in sense that an assignment to the variables is a solution to (26) if and only if it is a solution to one of (27). The number of subsystems(also called sides) in (27), denoted by r, is between 10 and 19 depending on the S-box. For instance, in case of S_4 the Eq. (27) has only 10 subsystems, while (26) has 64. Such reduction generally allows a faster solution, see [11].

5   Conclusion and Open Problems

In the present paper new statistical and algebraic properties of the DES encryption were found. They may have cryptanalytic implications upon resolving the following theoretical questions.

The first problem is within the statistical cryptanalysis. Let the cipher key space be split into n classes $K_1, \dots, K_n$. Each class defines a multinomial distribution on some ≥ 2 outcomes, defined by plaintext and ciphertext bits. Let $P_1, \dots, P_n$ be all such distributions computed a priori. Let $\nu(k)$ denote a vector of observations on above outcomes for an unknown cipher key k. It is well known that the problem "decide $k \in K_i$" may be solved with maximum likelihood method as in [5]. For the classification of several observation vectors $\nu(k_1), \dots, \nu(k_s)$ the same method is applied.

Open problem is to improve the method (reduce error probabilities) given the vectors $P_1, \dots, P_n$ are linearly dependent. That would improve Davies-Murphy type attacks against 16-round DES as for 660 different distributions (72 for (S_4, S_5, S_6)) only ≤ 48 (13 for (S_4, S_5, S_6)) are linearly independent.

The second problem is related to algebraic attacks against ciphers. A new type time-memory trade-off for AES and DES was observed in [9,10]. Let m be the cipher key size. Let $\leq 2^l$ right hand sides be allowed in the combinations by Gluing of the MRHS equations [9,10] during solution. Gluing means writing several equations as one equation of the same type as (26). Then guessing

$\leq m - l$ key-bits is enough before the system of equations is solved by finding and removing contradictory right-hand sides in pairwise agreeing of the current equations. The overall time complexity is at least $2^{m-l} \times 2^l = 2^m$ operations as for each guess one needs to run over the right hand sides of at least one of the equations. However coset representation allows reducing the number of sides by writing them as (27). In case of DES the Eq. (26) for $i = 4$ is written with only 10 sides instead of 64. For AES instead of 256 right hand sides one can do 64 for each of the equations, see [11]. The combination of two Eq. (26) with Gluing has $\leq 2^{12}$ right hand sides. With coset representation the number of sides is at most 19^2 (at most 100 for the combination of two equations from S_4). Open problem is to reduce the time complexity of the above trade-off by using coset representation.

Acknowledgement. Stian Fauskanger is supported by the COINS Research School of Computer and Information Security.

A Appendix

A.1 S_4 Right, Left and LR Distribution

Section 2.1 define the right, left and LR distribution. Tables 5, 6 and 7 show the distributions for S-box 4.

Table 5. Right hand side distribution of S-box 4 (each entry $= 2^6 \times p_{x,r}^{(4)}$)

x\r	0	1	2	3	4	5	6	7	8	9	10	11	12	13	14	15
0	1	1	0	1	1	1	0	2	2	2	1	1	1	0	1	1
1	2	1	2	1	1	1	1	0	0	1	1	1	1	2	1	0
2	1	1	2	1	1	1	2	0	0	0	1	1	1	2	1	1
3	0	1	0	1	1	1	1	2	2	1	1	1	1	0	1	2

Table 6. Left hand side distribution of S-box 4 (each entry $= 2^6 \times q_{x,r}^{(4)}$)

x\r	0	1	2	3	4	5	6	7	8	9	10	11	12	13	14	15
0	2	0	0	2	0	1	2	1	1	1	1	1	0	2	1	1
1	0	2	2	0	2	1	0	1	1	1	1	1	2	0	1	1
2	2	1	0	1	0	0	2	1	1	1	2	1	1	2	0	1
3	0	1	2	1	2	2	0	1	1	1	0	1	1	0	2	1

Table 7. LR distribution of S-box 4 (each entry $= 2^6 \times Q^{(4)}_{x,y,r}$)

x y\r	0	1	2	3	4	5	6	7	8	9	10	11	12	13	14	15
0 0	1	0	0	0	0	0	0	1	0	1	0	0	0	0	1	0
0 1	1	0	0	0	0	0	1	0	0	0	0	1	0	1	0	0
0 2	0	0	0	1	0	0	1	0	0	0	1	0	0	1	0	0
0 3	0	0	0	1	0	1	0	0	1	0	0	0	0	0	0	1
1 0	0	1	0	0	1	0	0	0	1	0	0	1	0	0	0	0
1 1	0	1	1	0	1	0	0	0	0	0	0	0	0	0	1	0
1 2	0	0	1	0	0	1	0	0	0	0	0	0	1	0	0	1
1 3	0	0	0	0	0	0	0	1	0	1	1	0	1	0	0	0
2 0	0	0	0	0	0	0	0	1	0	1	1	0	1	0	0	0
2 1	1	0	0	1	0	0	0	0	0	0	1	0	0	1	0	0
2 2	1	0	0	0	0	0	1	0	0	0	0	1	0	1	0	0
2 3	0	1	0	0	0	0	1	0	1	0	0	0	0	0	0	1
3 0	0	0	0	1	0	1	0	0	1	0	0	0	0	0	0	1
3 1	0	0	1	0	0	1	0	0	0	1	0	0	1	0	0	0
3 2	0	1	1	0	1	0	0	0	0	0	0	0	0	0	1	0
3 3	0	0	0	0	1	0	0	1	0	0	0	1	0	0	1	0

B Appendix

B.1 Disjoint Sub-cosets for DES S-boxes

$U_1 = \{\{516, 626\}, \{678, 697\}, \{812, 827\}, \{841, 894\}, \{899, 922\}, \{944, 992\},$
$\quad \{14, 36, 326, 364\}, \{16, 87, 175, 232\}, \{63, 77, 572, 590\}, \{97, 130, 545, 706\},$
$\quad \{116, 158, 298, 448\}, \{178, 221, 938, 965\}, \{203, 241, 721, 747\},$
$\quad \{259, 282, 653, 660\}, \{310, 379, 437, 504\}, \{348, 389, 783, 982\},$
$\quad \{409, 425, 600, 616\}, \{467, 487, 851, 871\}, \{543, 759, 789, 1021\}\},$

$U_2 = \{\{365, 490\}, \{855, 870\}, \{892, 912\}, \{949, 1007\}, \{15, 19, 33, 61\},$
$\quad \{72, 84, 962, 990\}, \{110, 119, 134, 159\}, \{171, 178, 416, 441\},$
$\quad \{195, 216, 676, 703\}, \{228, 254, 396, 406\}, \{265, 295, 475, 501\},$
$\quad \{284, 304, 583, 619\}, \{322, 337, 737, 754\}, \{378, 453, 822, 905\},$
$\quad \{512, 602, 931, 1017\}, \{541, 558, 795, 808\}, \{568, 625, 773, 844\},$
$\quad \{650, 659, 717, 724\}\},$

$U_3 = \{\{341, 497\}, \{605, 624\}, \{648, 697\}, \{707, 759\}, \{876, 974\}, \{978, 1020\},$
$\{10, 29, 110, 121\}, \{32, 134, 301, 395\}, \{73, 80, 207, 214\},$
$\{163, 229, 312, 382\}, \{180, 250, 662, 728\}, \{257, 359, 420, 450\},$
$\{274, 412, 779, 901\}, \{443, 479, 525, 617\}, \{529, 687, 788, 938\},$
$\{550, 570, 834, 862\}, \{801, 883, 949, 999\},$
$\{55, 147, 332, 488, 580, 736, 831, 923\}\},$

$U_4 = \{\{45, 56, 290, 311\}, \{395, 401, 452, 478\}, \{711, 733, 968, 978\},$
$\{801, 820, 878, 891\}, \{7, 29, 328, 338, 683, 689, 996, 1022\},$
$\{78, 91, 257, 276, 749, 760, 930, 951\}, \{99, 117, 428, 442, 652, 666, 835, 853\},$
$\{128, 150, 495, 505, 608, 630, 783, 793\},$
$\{166, 191, 201, 208, 550, 575, 585, 592\},$
$\{234, 243, 357, 380, 522, 531, 901, 924\}\},$

$U_5 = \{\{2, 30, 323, 351\}, \{44, 59, 230, 241\}, \{68, 82, 203, 221\}, \{97, 124, 170, 183\},$
$\{135, 148, 685, 702\}, \{264, 277, 577, 604\}, \{293, 304, 367, 378\},$
$\{397, 462, 657, 722\}, \{403, 416, 960, 1011\}, \{441, 472, 948, 981\},$
$\{489, 502, 516, 539\}, \{546, 744, 844, 902\}, \{568, 711, 869, 922\},$
$\{619, 650, 783, 1006\}, \{631, 765, 809, 931\}, \{790, 831, 848, 889\}\},$

$U_6 = \{\{467, 504\}, \{591, 693\}, \{735, 762\}, \{795, 836\}, \{887, 897\}, \{918, 971\},$
$\{12, 26, 256, 278\}, \{33, 63, 74, 84\}, \{111, 114, 232, 245\},$
$\{137, 151, 162, 188\}, \{198, 301, 563, 984\}, \{217, 305, 642, 874\},$
$\{323, 349, 398, 400\}, \{356, 423, 830, 1021\}, \{382, 443, 521, 716\},$
$\{453, 491, 594, 636\}, \{532, 613, 800, 849\}, \{558, 665, 775, 944\},$
$\{680, 739, 941, 998\}\},$

$U_7 = \{\{402, 481\}, \{534, 587\}, \{621, 632\}, \{848, 872\}, \{926, 946\}, \{979, 1020\},$
$\{29, 43, 143, 185\}, \{48, 66, 426, 472\}, \{91, 110, 329, 380\},$
$\{148, 160, 730, 750\}, \{200, 237, 513, 548\}, \{209, 250, 969, 994\},$
$\{259, 286, 652, 657\}, \{300, 341, 447, 454\}, \{307, 359, 675, 759\},$
$\{571, 605, 793, 895\}, \{778, 815, 896, 933\},$
$\{4, 119, 389, 502, 692, 711, 821, 838\}\},$

$$U_8 = \{\{446, 498\}, \{519, 684\}, \{806, 911\}, \{949, 1019\}, \{13, 17, 100, 120\},$$
$$\{34, 63, 649, 660\}, \{72, 134, 297, 487\}, \{154, 179, 857, 880\},$$
$$\{175, 203, 266, 366\}, \{215, 244, 530, 561\}, \{309, 323, 828, 842\},$$
$$\{342, 379, 965, 1000\}, \{389, 400, 460, 473\}, \{555, 765, 768, 982\},$$
$$\{580, 698, 877, 915\}, \{609, 631, 718, 728\},$$
$$\{93, 225, 284, 416, 606, 738, 799, 931\}\}.$$

References

1. Barker, W.C., Barker, E.B.: SP 800–67 Rev. 1. Recommendation for the Triple Data Encryption Algorithm (TDEA) Block Cipher, January 2012
2. Biham, E., Biryukov, A.: An improvement of Davies' attack on DES. J. Cryptol. **10**(3), 195–205 (1997)
3. Biham, E., Shamir, A.: Differential cryptanalysis of the full 16-Round DES. In: Brickell, E.F. (ed.) CRYPTO 1992. LNCS, vol. 740, pp. 487–496. Springer, Heidelberg (1993)
4. Courtois, N.T., Bard, G.V.: Algebraic cryptanalysis of the data encryption standard. In: Galbraith, S.D. (ed.) Cryptography and Coding 2007. LNCS, vol. 4887, pp. 152–169. Springer, Heidelberg (2007)
5. Davies, D., Murphy, S.: Pairs and triplets of DES S-boxes. J. Cryptol. **8**(1), 1–25 (1995)
6. Kholosha, A.: Personal conversation with I. Semaev, September 2014
7. Kunz-Jacques, S., Muller, F.: New improvements of Davies-Murphy cryptanalysis. In: Roy, B. (ed.) ASIACRYPT 2005. LNCS, vol. 3788, pp. 425–442. Springer, Heidelberg (2005)
8. Matsui, M.: Linear cryptanalysis method for DES cipher. In: Helleseth, T. (ed.) EUROCRYPT 1993. LNCS, vol. 765, pp. 386–397. Springer, Heidelberg (1994)
9. Raddum, H.: MRHS equation systems. In: Adams, C., Miri, A., Wiener, M. (eds.) SAC 2007. LNCS, vol. 4876, pp. 232–245. Springer, Heidelberg (2007)
10. Raddum, H., Semaev, I.: Solving multiple right hand sides linear equations. Des. Codes Cryptogr. **49**(1), 147–160 (2008)
11. Semaev, I., Mikuš, M.: Methods to solve algebraic equations in cryptanalysis. Tatra Mt. Math. Publ. **45**(1), 107–136 (2010)
12. Shamir, A.: On the security of DES. In: Williams, H.C. (ed.) CRYPTO 1985. LNCS, vol. 218, pp. 280–281. Springer, Heidelberg (1986)

Accurate Estimation of the Full Differential Distribution for General Feistel Structures

Jiageng Chen[1]([⊠]), Atsuko Miyaji[2,3,4], Chunhua Su[2], and Je Sen Teh[5]

[1] Computer School, Central China Normal University, Wuhan 430079, China
chinkako@gmail.com
[2] School of Information Science,
Japan Advanced Institute of Science and Technology,
1-1 Asahidai, Nomi, Ishikawa 923-1292, Japan
{miyaji,chsu}@jaist.ac.jp
[3] Japan Science and Technology Agency (JST) CREST,
Kawaguchi Center Building 4-1-8, Honcho,
Kawaguchi-shi, Saitama 332-0012, Japan
[4] Graduate School of Engineering, Osaka University, Osaka, Japan
[5] School of Computer Sciences, Universiti Sains Malaysia, Gelugor, Malaysia
jesen_teh@hotmail.com

Abstract. Statistical cryptanalysis is one of the most powerful tools to analyze symmetric key cryptographic primitives such as block ciphers. One of these attacks, the differential attack has been demonstrated to break a wide range of block ciphers. Block cipher proposals previously obtain a rough estimate of their security margin against differential attacks by counting the number of active S-Box along a differential path. However this method does not take into account the complex clustering effect of multiple differential paths. Analysis under full differential distributions have been studied for some extremely lightweight block ciphers such as KATAN and SIMON, but is still unknown for ciphers with relatively large block sizes. In this paper, we provide a framework to accurately estimate the full differential distribution of General Feistel Structure (GFS) block ciphers with relatively large block sizes. This framework acts as a convenient tool for block cipher designers to determine the security margin of their ciphers against differential attacks. We describe our theoretical model and demonstrate its correctness by performing experimental verification on a toy GFS cipher. We then apply our framework to two concrete GFS ciphers, LBlock and TWINE to derive their full differential distribution by using super computer. Based on the results, we are able to attack 25 rounds of TWINE-128 using a distinguishing attack, which is comparable to the best attack to date. Besides that, we are able to depict a correlation between the hamming weight of an input differential characteristic and the complexity of the

J. Chen is partly supported by the National Natural Science Foundation of China under Grant 61302161.

A. Miyaji is partly supported by Grant-in-Aid for Scientific Research (C)(15K00183) and (15K00189).

C. Su is partly supported by JSPS KAKENHI 15K16005.

D. Lin et al. (Eds.): Inscrypt 2015, LNCS 9589, pp. 108–124, 2016.
DOI: 10.1007/978-3-319-38898-4_7

attack. Based on the proposed framework, LBlock and TWINE have shown to have 178 and 208-bit security respectively.

Keywords: Differential attack · GFS · Differential distribution · LBlock · TWINE

1 Introduction

Block ciphers have been playing an important role in information security to achieve confidentiality and integrity. Recently, block ciphers with lightweight designs start attracting research attention due to their wide range of potential applications such as RFID, wireless sensor networks and etcetera. These lightweight block ciphers usually have small block sizes which are less or equal to 64 bits and a smaller key size, filling in the gap where the traditional ciphers such as AES are not applicable anymore. The General Feistel Structure (GFS) is among one of the most popular designs that have received a lot of analysis. Recently proposed lightweight ciphers such as LBlock [22] and TWINE [21] belong to this design category.

Among all the methods to analyze block ciphers, differential attacks are one of the most powerful methods since its invention back in 1990 [5]. The attack is statistical in nature and its success relies on finding long differential paths with high probability. For a long time, one single ad hoc-found path is usually used in the differential cryptanalysis. Thus the study of the differential path has not received much attention until recently. First in papers [8,9], multiple differential cryptanalysis was theoretically analyzed to show that the attacker generally has more power in building the differential distinguisher if he or she has more knowledge in the differential distribution. Later in paper [1], the author analyzed an extremely lightweight block cipher, KATAN32 by computing the whole differential distribution, and indeed it further increased the number of rounds that can be attacked compared to the previous results. The downside of using the whole differential distribution is that the attacker is unable to filter subkey bits, which may cause the complexity to increase. Thus there exists another branch of research focusing more on the key recovery phase and key relation such as related key attacks. Representative results include [6,19] which will not be addressed further in this paper since our focus is only the single key model. The full differential distribution can be computed if the block size is less than 32 bits, as shown in [1]. However, for ciphers with large block sizes, it is currently computationally infeasible to construct the full distribution. Thus to a large extent, the method to derive an accurate full distribution remains unexploited.

From the provable security's point of view, it is desirable to derive a security bound on the number of rounds that is secure against differential attack. Currently for block ciphers with S-Box-based design, counting the number of active S-Box [18], which is the number of S-Box on the differential path, is the common way to evaluate the security. In the proposal of both LBlock and TWINE, the number of active S-Box multiplied by the largest differential probability of the S-Box is used to evaluate security margin. For more complicated designs which involves MixColumn operation as in AES, paper [17] provided a tight lower bound for the minimum number of active S-box for several GFS ciphers. Although counting the number of active S-Box may be a good approximation for one single path, the actual differential distribution involves complicated clustering effects which cannot be addressed by this model. Thus the security margin evaluated in this way may not be accurate, or in other words, the lower bound may be underestimated.

In this paper, we contribute mainly in two aspects. Firstly, we address the full differential distribution for GFS ciphers with relatively large block sizes by providing both theoretical and experimental frameworks. We partition the block according to the length of the S-Box input, which is the size of data blocks processed by these ciphers. Then we theoretically model the computation of the full differential distribution for any number of rounds and verify our evaluation by using a toy GFS cipher to show that the truncated differential distribution can be used to accurately evaluate the concrete differential distribution. Furthermore, due to the truncated differentials, the ability to store all the internal states allow us to perform quick computing of the distribution even for large rounds. By taking advantage of the supercomputer, we can perform the experiment to obtain full differential distributions for every input difference. As a result, our experiments have provided us with several new findings regarding the differential attack. Firstly, we discovered that input differences with relatively small hamming weights tend to lead to better distinguishers. Based on our framework, we evaluate two GFS ciphers LBlock and TWINE to derive the best differential attack so far. Especially for TWINE-128, we are able to obtain a comparable result by attacking 25 rounds. Also, we are able to provide the precise security margins against differential attacks for the full rounds of both LBlock and TWINE for the first time. This is by far the most accurate security proof for GFS designs to date.

Outline of the Paper. Section 2 provides the theoretical model to compute the complete differential distribution for truncated GFS with bijective S-box design. Experiments on the toy model are also provided in this Section to verify the correctness of the model. In Sect. 3, concrete evaluations on LBlock and TWINE are provided. Lastly, we conclude our paper with some final statements.

2   Differential Characteristic Revisited

Since the proposal of differential attack in [5], methods to find long differential paths with high probability becomes the key to the success of the attack. Matsui in [13] first proposed a branch and bound algorithm to efficiently search the high probability linear and differential path for cipher DES. The algorithm applies the greedy strategy to find the best single path with the highest probability. Since then, researchers began to follow this strategy when searching for good property paths. As an extension of the differential attack, the multi-differential attack tries to take advantage of multiple differential paths to further increase the attacker's advantage when distinguishing from random distribution. Works [8] and [9] are two of the representative ones. For block ciphers with S-Box based design, researchers count the number of active S-Box as a criteria to measure the security margin against differential attack. It is well known [11] that there usually exists more than one path that can lead from the same input α to the output β, so that the probability of the corresponding path is actually bigger. Unfortunately, researchers usually do not consider this differential cluster or linear hull effect when searching good paths. [7] recently took advantage of the differential cluster to further improve the rounds of the differential paths.

Let's assume a block cipher E is a markov cipher with n-bit block size and r_f rounds in total. Previously, researchers try to identify one single $r < r_f$ round path $\alpha_0 \to \beta_r$ with high probability $Prob(\alpha_0 \to \beta_r) > 2^{-n}$, so that the attacker does not use up the entire message space. Usually, r is far from the full rounds r_f if the cipher is well designed. If we continue the search for more rounds, we will end up with a single path with a tiny probability much smaller than 2^{-n}. On the other hand, if we assume all the differential paths are randomly distributed, for a full r_f-round cipher, the probability of any differential path $Prob(\alpha_0 \to \beta_{r_f})$ should be around 2^{-n}. Obviously, there is a gap between the two results. From the differential cluster or linear hull effect, we make the following assumption.

Lemma 1. *For an r-round ideal Markov block cipher E, a single r-round differential path is defined as $(\alpha_0 \to \beta_r)_{single} = (\alpha_0, \gamma_{1,i_1}, \gamma_{2,i_2}, ..., \gamma_{r-1,i_{r-1}}, \beta_r)$, where $I_t^{min} \le i_t \le I_t^{max}, 1 \le t \le r - 1$. Here I_t^{min} and I_t^{max} denote the smallest and largest differential values in round t respectively. Let's define its probability to be $Prob((\alpha_0 \to \beta_r)_{single}) = p_{i_1,i_2,...,i_{r-1}}$. Then the total probability of differential path $\alpha_0 \to \beta_r$ can be computed by*

$$Prob(\alpha_0 \to \beta_r) = \sum_{i_1=I_1^{min}}^{I_1^{max}} \cdots \sum_{i_{r-1}=I_{r-1}^{min}}^{I_{r-1}^{max}} p_{i_1,i_2,...,i_{r-1}} \approx 2^{-n}$$

which is approximately equal to 2^{-n}. And we call

$$CS_{(\alpha_0,\beta_r)} = \sum_{i_1=I_1^{min}}^{I_1^{max}} \cdots \sum_{i_{r-2}=I_{r-1}^{min}}^{I_{r-1}^{max}} 1$$

the corresponding cluster size $CS_{(\alpha_0,\beta_r)}$.

For large number of rounds r, we may assume $p_{i_1,i_2,...,i_{r-1}}$ to be tiny and have the relation $p_{i_1,i_2,...,i_{r-1}} \propto CS_{(\alpha_0,\beta_r)}^{-1}$. As a result, the complexity to find the real probability of some specific path is related to the corresponding cluster size $CS_{(\alpha_0,\beta_r)}$. As the number of rounds grow, cluster size becomes bigger which makes it more difficult to compute the real probability. Also notice that for real cipher, the probability varies for different paths and the cluster size is related to the input differential property. This relation will be discussed later in this paper. Next, we will discuss first how to theoretically evaluate the cluster size and the probability, and then efficiently compute the full clusters for GFS ciphers based on bijective S-Box design.

2.1   Theoretical Model to Evaluate the Cluster Size and Probability

General Feistel Structure (GFS) is one of the most popular and widely studied design strategies for constructing block ciphers. Recently in paper [20], the authors studied different permutations and derived the optimized ones for different parameter settings. Recently proposed lightweight block ciphers LBlock [22] and TWINE [21] belong to the GFS design.

In GFS, the plaintext is divided into d subblocks $P = (x_0^0, x_1^0, ..., x_{d-1}^0)$, where $|x_j^i| = 2^{n/d}$ bits in length. The output of the i-th round is derived as follows:

$$(x_0^i, x_1^i, ..., x_{d-1}^i) \leftarrow \pi(x_0^{i-1}, F^{i-1}(x_0^{i-1}) \oplus x_1^{i-1}, ..., F^{i-1}(x_{d-2}^{i-1}) \oplus x_{d-1}^{i-1})$$

where π is the permutation, and function $F : \{0,1\}^{n/d} \rightarrow \{0,1\}^{n/d}$ is the only non-linear function in GFS. For S-box based design with large subblock size n/d, usually MDS matrix is applied to provide further mixing within each subblock. However, in recent lightweight designs such as [21,22], n/d is small in size (usually 4 bits), and F is equivalent to a single S-Box. Figure 1 shows the GFS8 defined in [20] with two corresponding F functions. For the simplicity, in this paper we will stick to the lightweight version of GFS without the application of MDS.

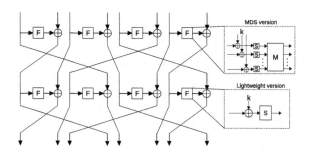

Fig. 1. GFS8 [20]

Below are some definitions that will be used for the theoretical evaluation. From now on, we use symbol α^C and α^T to denote a concrete differential and a truncated differential respectively.

Definition 1 (Structure, Branch Weight, Hamming Weight, Cancel Weight). *Let $\alpha^{C,i} = (\alpha_0^{C,i}, \alpha_1^{C,i}, ..., \alpha_{d-1}^{C,i})$ denote the concrete differential states for each of the rounds $0 \leq i \leq N-1$. Function Trunc maps the concrete differential state to the truncated differential state: $\alpha^{T,i} = (\alpha_0^{T,i}, \alpha_1^{T,i}, ..., \alpha_{d-1}^{T,i}) \leftarrow Trunc(\alpha_0^{C,i}, \alpha_1^{C,i}, ..., \alpha_{d-1}^{C,i})$, where $\alpha_j^{T,i} = 1$ if $\alpha_j^{C,i} \neq 0$, and $\alpha_j^{T,i} = 0$ if $\alpha_j^{C,i} = 0$. We call*

$$(\alpha^{T,0}, \alpha^{T,1}, ..., \alpha^{T,r})$$

a r-round truncated structure, or structure in short. We define the number of active S-Box of round i

$$B_i = B_i(\alpha^{T,i}) = \alpha_0^{T,i} + \alpha_2^{T,i} + \cdots + \alpha_{d-2}^{T,i}$$

*to be the **Branch Weight** of the corresponding round. We define the **Hamming Weight** of the i-th round differential state to be*

$$H_i = H_i(\alpha^{T,i}) = \sum_{j=0}^{d-1} \alpha_j^{T,i}$$

*Finally, we define the **Canceling Weight** G_i and **Non-Canceling Weight** W_i for round i to be*

$$G_i = \alpha_0^{T,i} \wedge \alpha_1^{T,i} \wedge \neg \alpha_1^{T-1,i+1} + \cdots + \alpha_{d-2}^{T,i} \wedge \alpha_{d-1}^{T,i} \wedge \neg \alpha_{d-1}^{T-1,i+1}$$

$$W_i = \alpha_0^{T,i} \wedge \alpha_1^{T,i} \wedge \alpha_1^{T-1,i+1} + \cdots + \alpha_{d-2}^{T,i} \wedge \alpha_{d-1}^{T,i} \wedge \alpha_{d-1}^{T-1,i+1}$$

where $(\alpha_0^{T-1,i+1}, \alpha_1^{T-1,i+1}, ..., \alpha_{d-1}^{T-1,i+1}) \leftarrow \pi^{-1}(\alpha_0^{T,i+1}, \alpha_1^{T,i+1}, ..., \alpha_{d-1}^{T,i+1})$

G_i counts the number of instances in round i where $\alpha_j^{T,i} = \alpha_{j+1}^{T,i} = 1$ while $\alpha_{j+1}^{T-1,i+1} = 0$, and W_i counts the number of instances in round i where $\alpha_j^{T,i} = \alpha_{j+1}^{T,i} = \alpha_{j+1}^{T-1,i+1} = 1$. Now we are ready to have the following theorem:

Lemma 2. *Let $\alpha_I^{C,0} \rightarrow \alpha_O^{C,r}$ be a r-round concrete differential path with $I \in \Omega_i$ and $O \in \Delta_o$. Ω_i and Δ_o denotes the concrete differential set following the i-th input and o-th output truncated difference. Assume we have in total m structures which have the same truncated input and output $\alpha_{\Omega_I}^{T,0}, \alpha_{\Delta_O}^{T,r}$ while differing in the middle, we call m the truncated cluster size of truncated path $(\alpha_{\Omega_I}^{T,0} \rightarrow \alpha_{\Delta_O}^{T,r})$. The jth structure can be presented as follows $(0 \leq j \leq m-1)$:*

$$(\alpha_{\Omega_I}^{T,0}, \alpha^{T,1,j}, ..., \alpha^{T,r-1,j}, \alpha_{\Delta_O}^{T,r})$$

Let's assume before proceeding round $0 \leq i \leq r-1$ in the jth structure, we have L_i^j concrete differential paths which are resulted from input differential $\alpha^{C,0}$. Then after i-th round, the number of total paths generated from $\alpha^{C,0}$ becomes

$$L_{i+1}^j = L_i^j \times R^{B_i^j} \times (2^{\frac{n}{d}} - 1)^{-G_i^j} \times \left(\frac{2^{\frac{n}{d}} - 1}{2^{\frac{n}{d}} - 2}\right)^{-W_i^j}$$

where R is the average branch number of the S-Box, and $L_0^j = 1$ (initially, there exists only one state). Then L_r^j can be denoted as

$$L_r^j = R^{\sum_{i=0}^{r-1} B_i^j} \cdot (2^{\frac{n}{d}} - 1)^{-\sum_{i=0}^{r-1} G_i^j} \cdot \left(\frac{2^{\frac{n}{d}} - 1}{2^{\frac{n}{d}} - 2}\right)^{-\sum_{i=0}^{r-1} W_i^j}.$$

Proof. For the jth structure $(\alpha_{\Omega_I}^{T,0}, \alpha^{T,1,j}, ..., \alpha^{T,r-1,j}, \alpha_{\Delta_O}^{T,r})$, we can easily compute parameters B_i^j, H_i^j, W_i^j and G_i^j for each round i. Assume before proceeding i-th round, we have L_i^j concrete differential paths which are derived from the input differential $\alpha^{C,0}$ which follows the truncated form $\alpha^{T,0}$. Since there are B_i^j active S-Box in this round, the increasing number of branches for each of the existed path can be computed as $R^{B_i^j}$. However, for each of the G_i^j XOR operation, we know from the next round truncated pattern, the two input differences will be canceled out. The probability for this event to happen is $(2^{\frac{n}{d}}-1)^{-G_i^j}$. Also for each of the W_i^j XOR operations, instead of probability 1, we need to exclude the cases where 0 may appear, thus the probability for this event to happen is $(\frac{2^{\frac{n}{d}}-1}{2^{\frac{n}{d}}-2})^{-W_i^j}$. Since we need the concrete paths to follow the truncated pattern, only the paths that follow the truncated pattern can survive. As a result, we have $L_{i+1}^j = L_i^j \times R^{B_i^j} \times (2^{\frac{n}{d}}-1)^{-G_i^j} \times (\frac{2^{\frac{n}{d}}-1}{2^{\frac{n}{d}}-2})^{-W_i^j}$ number of paths remaining. By computing this repeatedly, we can derive the total number of paths L_r^j after r-th round. $\square$

Theorem 1. *Assume we have 2^N concrete input differentials having the same truncated input difference, and the average single path probability for the truncated structure is $P_{ave}^{\sum_{i=0}^{r-1} B_i^j}$. Let the counter X^j denote the number of hits for any concrete output differences following the same output truncated difference $\alpha_{\Omega_I}^{T,r}$ in the j-th structure. Then*

$$X_{\alpha_{\Omega_I}^{C,0}, \alpha_{\Delta_O}^{C,r}}^j \sim \mathcal{B}(2^N \cdot L_r^j, \quad (2^{n/d}-1)^{-H_r} \cdot P_{ave}^{\sum_{i=0}^{r-1} B_i^j}) \approx$$

$$\mathcal{N}\left(2^N \cdot L_r^j \cdot (2^{\frac{n}{d}}-1)^{-H_r} \cdot P_{ave}^{\sum_{i=0}^{r-1} B_i^j}, \quad 2^N \cdot L_r^j \cdot \right.$$

$$\left. (2^{\frac{n}{d}}-1)^{-H_r} \cdot P_{ave}^{\sum_{i=0}^{r-1} B_i^j} \cdot (1-(2^{\frac{n}{d}}-1)^{-H_r} \cdot P_{ave}^{\sum_{i=0}^{r-1} B_i^j})\right)$$

Denote random variable $P^j = \frac{1}{2^N} \cdot X^j$ be the probability for the concrete path $\alpha_{\Omega_I}^{C,0} \rightarrow \alpha_{\Delta_O}^{C,r}$, and let $\Gamma_j^r = \frac{(2^{\frac{n}{d}}-1)^{-\sum_{i=0}^{r-1}(G_i^j+W_i^j)-H_r}}{(2^{\frac{n}{d}}-2)^{-\sum_{i=0}^{r-1} W_i^j}}$, then

$$P_{(\alpha_{\Omega_I}^{C,0} \rightarrow \alpha_{\Delta_O}^{C,r})}^j \sim \mathcal{N}\left(\Gamma_j^r, \quad (\Gamma_j^r \cdot (1-(2^{n/d}-1)^{-H_r} \cdot P_{ave}^{\sum_{i=0}^{r-1} B_i}))/2^N\right)$$

where P_{ave} is the average differential probability of the S-Box.

Proof. Since the truncated output difference has hamming weight H_r, the concrete differential space is $(2^{n/d}-1)^{H_r}$ (excluding the 0 case). For any $\alpha_{\Delta_{O,j}}^{C,r} \in \{0,1\}^{log(2^{n/d}-1)^{H_r}}$, the probability that it gets hit by the $2^N L_i^j$ paths x times follows the binomial distribution $\mathcal{B}(2^N \cdot L_r^j, \quad (2^{n/d}-1)^{-H_r} \cdot P_{ave}^{\sum_{i=0}^{r-1} B_i^j})$. Since $2^N L_i^j$ is large, we can approximate it by normal distribution as shown above.

To derive its probability distribution, we only need to divide by the number of total pairs 2^N. After extending L_r^j as above, branch number R is canceled by P_{ave} since for any S-Box, $R \cdot P_{ave} = 1$. Replace with Γ_j^r we derive the result. Notice that the mean of the distribution is not affected by the number of input pairs 2^N.

$$P^j_{(\alpha_{\Omega_I}^{C,0} \to \alpha_{\Delta_O}^{C,r})} \sim$$

$$\mathcal{N}\left(L_r^j \cdot (2^{\frac{n}{d}} - 1)^{-H_r^j} \cdot P_{ave}^{\sum_{i=0}^{r-1} B_i^j}, \quad (L_r^j \cdot (2^{\frac{n}{d}} - 1)^{-H_r^j} \cdot P_{ave}^{\sum_{i=0}^{r-1} B_i^j} \right.$$

$$\left. \cdot (1 - (2^{\frac{n}{d}} - 1)^{-H_r^j} \cdot P_{ave}^{\sum_{i=0}^{r-1} B_i^j}))/2^N \right)$$

$$= \mathcal{N}\left((R \cdot P_{ave})^{\sum_{i=0}^{r-1} B_i} \cdot \Gamma_j^r, \quad ((R \cdot P_{ave})^{\sum_{i=0}^{r-1} B_i} \cdot \Gamma_j^r \right.$$

$$\left. \cdot (1 - (2^{n/d} - 1)^{-H_r^j} \cdot P_{ave}^{\sum_{i=0}^{r-1} B_i}))/2^N \right)$$

$$= \mathcal{N}\left(\Gamma_j^r, \quad (\Gamma_j^r \cdot (1 - (2^{n/d} - 1)^{-H_r^j} \cdot P_{ave}^{\sum_{i=0}^{r-1} B_i}))/2^N \right) \qquad \square$$

Corollary 1. *The distribution of probability $(\alpha_{\Omega_I}^{C,0} \to \alpha_{\Delta_O}^{C,r})$ after considering the entire truncated cluster with size m has the following distribution.*

$$P_{(\alpha_{\Omega_I}^{C,0} \to \alpha_{\Delta_O}^{C,r})} \sim \mathcal{N}\left(\sum_{j=0}^{m-1} \Gamma_j^r, \quad \sum_{j=0}^{m-1} \Gamma_j^r / 2^N \right)$$

Corollary 1 is straightforward by taking the truncated cluster into consideration. Notice that for large number of rounds, $(1 - (2^{n/d} - 1)^{-H_r} \cdot P_{ave}^{\sum_{i=0}^{r-1} B_i})$ can be approximated to be one, and thus the distribution can be simplified as stated.

Since for any S-Box, we know that $R \cdot P_{ave} = 1$, thus the expect value will converge to some stable value $\sum \Gamma$ as the number of rounds become large. Actually, we can see that as the number of rounds becomes large, the probability of the paths tends to gather around the mean.

2.2   Experimental Verification

The evaluation of the probability for the concrete differential cluster is the key to the attack. Thus it is necessary to verify the correctness of the probability calculation, especially, the mean (Γ) of the probability distribution in Corollary 1. Our experiment has the following settings.

1. We design a toy version of GFS cipher. It has 32-bit block size with 8 4-bit subblocks. TWINE's S-Box is applied and we apply the optimal block shuffle No.2 for $k = 8$ from [20] as the permutation layer to guarantee good diffusion property. It can be seen as a smaller block size version of TWINE.

2. We target 7 rounds differential path and choose the truncated input difference $\alpha_{\Omega_I}^{T,0}$ and output difference $\alpha_{\Delta O}^{T,7}$, such that the concrete differential cluster size evaluated by the theoretical model is close to but less than 2^{30} so that we can practically collect enough sample data.

3. We compute 10^4 differential paths with randomly generated input and output concrete differences $\alpha_{\Omega_I}^{C,0}$ and $\alpha_{\Omega_O}^{C,7}$. The probability $Prob(\alpha_{\Omega_I}^{C,0} \to \alpha_{\Omega_O}^{C,7})$ is computed by considering every possible differential path from $\alpha_{\Omega_I}^{C,0}$ to $\alpha_{\Omega_O}^{C,7}$.

Even for 7 rounds, the computational cost is high when trying to find all the paths connecting some specific input and output difference $\alpha_{\Omega_I}^{C,0}$ and $\alpha_{\Omega_O}^{C,7}$. We apply the meet-in-the-middle approach when searching the path probability. First, we split the 7 rounds into two, 3 rounds + 4 rounds. Then starting from $\alpha_{\Omega_I}^{C,0}$, we compute every differential path till the middle point and save them in a hash table along with the corresponding probabilities. Then starting from $\alpha_{\Omega_O}^{C,7}$, we compute backwards for all the differential paths, and match the ones in the hash table. Once we find a match, update the total probability.

As a result, the computational cost is reduced from computing 7 rounds to computing the longer half, which is 4 rounds. The bottleneck is the memory storage, which is bounded by the hamming weight of the truncated difference in the matching round. The experimental results are summarized in Fig. 2. From the figure, it shows that the mean of the probability distribution is evaluated very accurately. The experimental mean is $2^{-31.9984}$ while the theoretical value is $2^{-31.9958}$. From the left figure, the histogram confirms the normal distribution of the probability. For this particular case, the normal approximation becomes rather accurate when the number of input pairs reaches around $2^N \approx 2^{37.4042}$. And this value also satisfies the condition in Theorem 1, which again confirms the accuracy of our model.

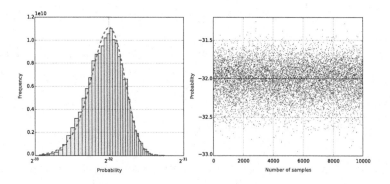

Fig. 2. Experimental result for toy cipher

3 Statistical Distinguisher and Some Observations for LBlock and TWINE

It it well known that when there are only two distributions to distinguish from, hypothesis testing based on Neyman-Pearson lemma [14] provides us with the most powerful test. [4] first provided a former analysis on how to build an optimal distinguisher between two sources, specifically one from random distribution and one from a real cipher distribution as in our context. They further derived the complexity to distinguish in the form of number of observable outputs or the input queries regarding the block cipher analysis based on the log-likelihood ratio statistics. Several following papers such as [1,9] take advantage of this distinguisher framework, and after combining with order statistics techniques addressed in [3], they were able to accurately evaluate the successful probability of a key recovery of the attack. Also, they were able to apply not only the traditional differential attack but also multiple, truncated and impossible differential attacks. The relation between a good statistical distinguisher and the number of rounds we can attack is pretty much straightforward. What may not seem to be trivial is the complexity of the key recovery, which will rely on the format of the output differential. However, it is known that if we use multiple differential outputs, the distinguisher behaves better and since we are especially interested in the extent to which we can distinguish theoretically for large rounds of GFS, we omit the key recovery discussion in this paper. We rearrange the core theorems from [4] that will be used in our evaluation as follows.

Theorem 2 ([4]). *Considering that $Z_1, Z_2, \ldots$ is a sequence of iid random variables of distribution D and that D_0 and D_1 share the same support, the log-likelihood ratio statistic follows normal distribution,*

$$Pr[\frac{LLR(Z^n) - n\mu}{\sigma\sqrt{n}} < t] \xrightarrow{n \to \infty} \Phi(t)$$

where $\mu = \mu_j$ with $\mu_0 = D(D_0\|D_1)$, $\mu_1 = -D(D_1\|D_0)$ and $\sigma_j^2 = \sum_{z \in \mathcal{Z}} Pr_{D_j}[z](\log\frac{Pr_{D_0}[z]}{Pr_{D_1}[z]}) - \mu_j^2$ for $j \in \{0,1\}$. And

$$LLR(Z^n) = \sum_{a \in \mathcal{Z}} N(a|Z^n) \log\frac{Pr_{D_0}[a]}{Pr_{D_1}[a]}$$

Denote v to be the number of samples need to distinguish between D_0 and D_1, then

$$v = \frac{4 \cdot \Phi^{-1}(P_e)^2}{\sum_{z \in \mathcal{Z}} \frac{(Pr_{D_0}[z] - Pr_{D_1}[z])^2}{Pr_{D_1}[z]}}$$

where P_e is the error probability, and D denotes the Kullback-Leibler distance

$$D(D_0\|D_1) = \sum_{z \in \mathcal{Z}} Pr_{D_0}[z] \log\frac{Pr_{D_0}[z]}{Pr_{D_1}[z]}$$

Here we assume D_1 has the uniform distribution, then $Pr_{D_1}[z] = 2^{-n}$ for $\forall z \in \{0,1\}^n$. From Corollary 1, we know that $Pr_{D_0}[z_i]$ follows different normal distributions. We know that the mean of the distribution is the unbiased point estimator for $Pr_{D_0}[z_i]$. Thus by replacing $Pr_{D_0}[z_i]$ with the corresponding mean derived by Corollary 1, we are able to compute the required number of samples v in order to distinguish.

3.1   Efficient Algorithm to Compute D_0

Deriving the full distribution D_0 is a practical issue. For GFS with 4-bit nibble and 64-bit block size, the truncated differential domain is shrunk down to 2^{16}. However, the computational cost will still grow exponentially as the number of rounds grows. Fortunately, we can store all the 2^{16} differential states for each of the rounds, which makes the computational cost grow linearly regarding the number of rounds. This will dramatically speed up the computing for D_0 regarding large number of rounds.

Algorithm 1. Searching D_0 for all input and output truncated differences

---

1: **Input:** Input truncated difference $\alpha^{T,0}$.
2: **Output:** Full distribution of D_0 given $\alpha^{T,0}$.
3: **procedure Dist_search**($r \leftarrow 0, \alpha^{T,0}$)
4:     $M = \{(s_i, p_i) | 0 \le i \le 2^{n/d}\} \leftarrow \emptyset$
5:     Append $(\alpha^{T,0}, 1.0)$ to M.
6:     **while** $r! = N - 1$ **do**
7:         $M_{out} \leftarrow M$
8:         **for** $\forall(s_i, p_i) \in M$ **do**
9:             // Given s_i, p_i, round function returns all the possible output diff and probabilities
10:             $\{(o_0, p_0'), ..., (o_{t-1}, p_{t-1}')\} \leftarrow round(s_i, p_i)$
11:             **for** $\forall(o_i, p_i')$ **do**
12:                 **if** $o_i \in M_{out}$ **then**
13:                     $p_i \leftarrow p_i + p_i'$
14:                 **else**
15:                     Append (o_i, p_i') to M_{out}
16:         $M \leftarrow M_{out}$
17:     Output $(s_i, p_i) \in M, 0 \le i \le 2^{n/d}$

For GFS with 4-bit sub-blocks and 64-bit block size, after around 7 rounds, M will include every truncated internal state. We apply the GMP library [10] when computing the probability so that we do not lose precision. However, as the number of rounds grow, the bias becomes miniscule, requiring large amounts of memory to store the precision. When we reach some large rounds, we cannot produce accurate result due to the memory limit. The algorithm is still very efficient considering that we need to perform the search for not only one but all the $2^{n/d} - 1$ input difference $\alpha^{T,0}$. The following experimental results show the number of rounds we have achieved with full precision as well as some rounds where precision was lost partially.

3.2 Observations on LBlock and TWINE

LBlock is a 32-round, 64-bit block cipher with Feistel structure proposed by Wenling Wu et al. in [22]. In each round after the left 32-bit side goes through a non-linear function F, it is XOR-ed with the right side that has performed an 8-bit left cyclic shift. TWINE is also a 64-bit block cipher with GFS structure proposed by Tomoyasu Suzaki, etc. in [21]. Different from LBlock, it supports 80 and 128 bits key length which both have the same 36 rounds. The F function of LBlock and round operation of TWINE are shown in Figs. 3 and 4.

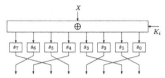

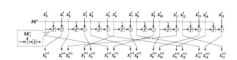

Fig. 3. F function for LBlock   **Fig. 4.** One round for TWINE

In [21], the authors already identified that both ciphers are very similar to each other regarding the Feistel structure and the permutation layer. This is also our motivation to study these two ciphers, first to compare the security margins and secondly, obtain the observataions for the behavior of GFS.

As we have pointed out, our framework can be used to exploit all the distributions under our theoretical model. In order to get a close look at the strength and weakness of the various differential paths given different input differences, we need to perform Algorithm 1 for all the $2^{16} - 1$ input differences for different number of rounds. Figures 5 and 6 show the experimental results of how many samples are required in order to distinguish the cipher from a uniformly distributed random source. Particularly, for each of the input differences (hamming weight), we consider all the possible output differences to derive the corresponding distinguisher. The experiment was performed on supercomputer Cray XC30 with 700 CPU cores (Intel Xeon E5-2690v3 2.6 GHz (Haswell)) running in parallel for around three days.

Both figures share some similarities which provide us with an insight into the properties of other GFS with bijective S-Box design. It also provides us with strategies on how to perform efficient cryptanalysis. Firstly, within the same number of rounds, we notice that the distinguisher will perform better as the hamming weight of the input differences decrements. Considering many previous researchers such as [16] favor the input difference with small hamming weight, this result seems to be straightforward. However, previous results did not consider the clustering effect where many small paths could eventually lead to

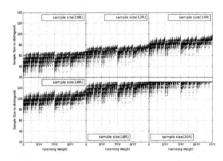 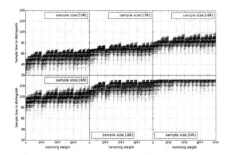

Fig. 5. Distinguisher for LBlock   **Fig. 6.** Distinguisher for TWINE

a better cluster. Here we clarify this situation by showing that input differences with large hamming weight tend to have better randomization property with respect to the differential distribution, thus an attacker should focus on searching the paths with small input hamming weight.

Secondly, this trend remains the same for different number of rounds, with the total number of pairs required to distinguish increasing as the number of rounds grows. This makes sense according to the Markov cipher model [15], which has been used to model modern block ciphers. Notice that for both LBlock and TWINE, starting from round 18, the number of pairs tends to converge to some threshold. This is due to the insufficient precision used in the GMP library. We expect that the original trend will persist no matter the number of rounds if we have enough memory space to store 2^{16} elements with large enough precision. In the current setting, we set the precision to be 10000 bits, which gives us a good balance between the precision of the results, and the experiment speed. Notice that even for 20 rounds, the results for the low hamming weight are still accurate and usable.

Distinguishing Attack. Now we give distinguishing attacks for LBlock and TWINE assuming the usage of the full code book. We have previously shown that input differences with small hamming weight tends to have better distinguishability. For any truncated input difference $\alpha^{T,0}$, the total number of differential pairs that conform to the input differential $\alpha^{T,0}$ is $2^{63+4 \times HW(\alpha^{T,0})}$, where $HW(\alpha^{T,0})$ denotes the hamming weight of $\alpha^{T,0}$. If the number of pairs v in order to distinguish derived from the statistical framework is smaller than $2^{63+4 \times HW(\alpha^{T,0})}$, then we are able to launch the distinguisher attack immediately. However, for larger rounds such as 18 rounds, the experimental result indicates that the input differential with the best distinguishing effect requires more pairs than the total amount that the cipher can provide. Therefore, instead of taking advantage of only one input difference, we can consider multiple input differences. One straightforward way is to store 2^{16} counters for each of the input difference, and we extend the distribution domain from 2^{16} to maximum 2^{32} counters. Let v_i denote the number of pairs required for input difference $\alpha_i^{T,0}$, then the number of pairs $v_{0...i}$ to distinguish can be computed as follows:

$$v_{0...i} = (\sum_{x=0}^{i} \frac{1}{v_x})^{-1}$$

This equation can be derived directly from Theorem 2. Notice that we will proceed with the input difference with small hamming weight first, thus v is sorted in ascending order based on hamming weight in order to provide which input difference to use first. In order to check the success of the attack, we need to be sure that

$$v_{0...i} < \sum_{x=0}^{i} 2^{63+4\times HW(\alpha_i^{T,0})}$$

For our distinguishing attack, the computational cost is the cost of the summing the counters, which requires $\sum_{x=0}^{i} 2^{63+4\times HW(\alpha_i^{T,0})}$ memory accesses. Under the conservative estimation that one memory access is equivalent to one round operation cost, which was also used in paper [12], the computational cost can be estimated as $\frac{1}{R} \times \sum_{x=0}^{i} 2^{63+4\times HW(\alpha_i^{T,0})}$ R-round computation, where R is the number of total rounds to attack.

Although for larger number rounds we currently do not have the accurate distribution for all the input differences due to the computational limitations, the input differences with small hamming weight are still accurate. Therefore, we can take advantage of this accurate region to launch the attack. For 21 rounds of LBlock, if we take the first 2^{11} input differences sorted according to v_i, then $v_{0...2^{11}} \approx 2^{97.69}$ which is less than the total available pairs $2^{100.67}$. This means we can actually perform the distinguishing attack as long as we have enough computing resources. The time complexity here is thus $2^{93.3}$ 21 rounds LBlock encryptions. TWINE behaves almost exactly the same as LBlock for the first 21 rounds. By applying our framework, we can provide an accurate security bound for different number of rounds. For example, a 21-round LBlock will theoretically fail to achieve the security level that we claim if we set the key size to be larger than 94 bits.

Next we summarize the security margin for both LBlock and TWINE regarding the distinguishing attack. Notice that we choose the distinguishing attack to bound the security since it is usually considered to be weaker than key recovery attack. So from a designer's point of view, we have to set the security parameter (key size) to be conservative in order to resist as many attacks as possible. Due to the limitation of computational resources, we can only derive the accurate values up to 21 rounds for both LBlock and TWINE accordingly. However, after observing the first 21 rounds for both LBlock and TWINE, the increase of the computational cost is log-linear with respect to the number of rounds. Thus the trend can be well extrapolated by using the least square methods. Figures 7 and 8 demonstrate the security level for full rounds of LBlock and TWINE, where the dotted line is the prediction while the solid line is the experimental results. Our analysis shows that if both ciphers use 80-bit key setting, then number of rounds considered to be secure is around 19. However, since TWINE also support 128-bit key, in order to satisfy the corresponding security, we will need at least 25 rounds. We notice that in [2], they can achieve 25-rounds key recovery attack for TWINE-128 by using MitM and impossible differential attack. By using truncated differential technique, however, they can only attack 23-rounds using dedicated techniques. Our result complements theirs by revealing a general pattern after an in-depth analysis of the differential distinguisher. From the differential characteristic's point of view, although Table 3 in [2] demonstrates several paths that are better than evaluated using active S-Box, they still cannot achieve more than 16 rounds for TWINE.

From the provable security's point of view, both full rounds LBlock and TWINE are secure, and our analysis can provide the accurate security margin which is around 178 bits and 208 bits for LBlock and TWINE respectively. The reason TWINE is more secure in this sense is that it has 4 more rounds than LBlock, and they are equivalently secure against differential attack if given the same number of rounds.

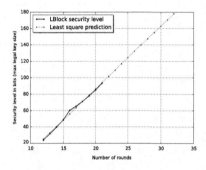
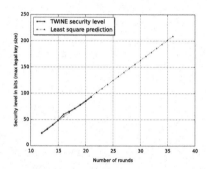

Fig. 7. Security level for LBlock   **Fig. 8.** Security level for TWINE

4   Conclusion

In this paper, we revisit the security of GFS with S-Box design regarding differential cryptanalysis. We evaluate the differential trails taking the full cluster into consideration by providing both theoretical and experimental results for the full distribution in truncated form. Our framework provides a solution for ciphers with relatively large block size to derive the full differential distribution. As a concrete application, we evaluate LBlock and TWINE to demonstrate the relationship between the hamming weight of the input difference and complexity of the attack. For TWINE-128, our attack can achieve 25 rounds, which is comparable to the best attacks up to date. More importantly, our framework enables us to compute the accurate security bound on full rounds LBlock and TWINE. As far as we know, this is the first achievement on security proof with exact security margin provided. This framework can be utilized by future cipher proposals to determine the minimum security margin of their designs.

References

1. Albrecht, M.R., Leander, G.: An all-in-one approach to differential cryptanalysis for small block ciphers. In: Knudsen, L.R., Wu, H. (eds.) SAC 2012. LNCS, vol. 7707, pp. 1–15. Springer, Heidelberg (2013)
2. Alex Biryukov, P.D., Perrin, L.: Differential analysis and meet-in-the-middle attack against round-reduced twine. Cryptology ePrint Archive, Report 2015/240 (2015)
3. Selçuk, A.A., Biçak, A.: On probability of success in linear and differential cryptanalysis. In: Cimato, S., Galdi, C., Persiano, G. (eds.) SCN 2002. LNCS, vol. 2576, pp. 174–185. Springer, Heidelberg (2003)
4. Baignères, T., Junod, P., Vaudenay, S.: How far can we go beyond linear cryptanalysis? In: Lee, P.J. (ed.) ASIACRYPT 2004. LNCS, vol. 3329, pp. 432–450. Springer, Heidelberg (2004)
5. Biham, E., Shamir, A.: Differential cryptanalysis of DES-like cryptosystems. In: Menezes, A., Vanstone, S.A. (eds.) CRYPTO 1990. LNCS, vol. 537, pp. 2–21. Springer, Heidelberg (1991)
6. Biryukov, A., Nikolić, I.: Automatic search for related-key differential characteristics in byte-oriented block ciphers: application to AES, Camellia, Khazad and others. In: Gilbert, H. (ed.) EUROCRYPT 2010. LNCS, vol. 6110, pp. 322–344. Springer, Heidelberg (2010)
7. Biryukov, A., Roy, A., Velichkov, V.: Differential analysis of block ciphers SIMON and SPECK. In: Cid, C., Rechberger, C. (eds.) FSE 2014. LNCS, vol. 8540, pp. 546–570. Springer, Heidelberg (2015)
8. Blondeau, C., Gérard, B.: Multiple differential cryptanalysis: theory and practice. In: Joux, A. (ed.) FSE 2011. LNCS, vol. 6733, pp. 35–54. Springer, Heidelberg (2011)
9. Blondeau, C., Gérard, B., Nyberg, K.: Multiple differential cryptanalysis using LLR and statistics. In: Visconti, I., De Prisco, R. (eds.) SCN 2012. LNCS, vol. 7485, pp. 343–360. Springer, Heidelberg (2012)
10. Granlund, T., et al.: The GNU Multiple Precision Arithmetic Library, 2.0.2 edn. TMG Datakonsult, Boston (1996)
11. Knudsen, L.R., Robshaw, M.: The Block Cipher Companion. Springer Science & Business Media, Heidelberg (2011)
12. Lu, J., Yap, W.-S., Wei, Y.: Weak keys of the full MISTY1 block cipher for related-key differential cryptanalysis. In: Dawson, E. (ed.) CT-RSA 2013. LNCS, vol. 7779, pp. 389–404. Springer, Heidelberg (2013)
13. Matsui, M.: On correlation between the order of S-Boxes and the strength of DES. In: De Santis, A. (ed.) EUROCRYPT 1994. LNCS, vol. 950, pp. 366–375. Springer, Heidelberg (1995)
14. Neyman, J., Pearson, E.S.: On the Problem of the Most Efficient Tests of Statistical Hypotheses. Springer, New York (1992)
15. O'Connor, L., Goli, J.: A unified Markov approach to differential and linear cryptanalysis. In: Safavi-Naini, R., Pieprzyk, J.P. (eds.) ASIACRYPT 1994. LNCS, vol. 917, pp. 385–397. Springer, Heidelberg (1995)
16. Özen, O., Varıcı, K., Tezcan, C., Kocair, Ç.: Lightweight block ciphers revisited: cryptanalysis of reduced round PRESENT and HIGHT. In: Boyd, C., González Nieto, J. (eds.) ACISP 2009. LNCS, vol. 5594, pp. 90–107. Springer, Heidelberg (2009)
17. Shibutani, K.: On the diffusion of generalized Feistel structures regarding differential and linear cryptanalysis. In: Biryukov, A., Gong, G., Stinson, D.R. (eds.) SAC 2010. LNCS, vol. 6544, pp. 211–228. Springer, Heidelberg (2011)

18. Shirai, T., Shibutani, K., Akishita, T., Moriai, S., Iwata, T.: The 128-bit blockcipher CLEFIA (Extended Abstract). In: Biryukov, A. (ed.) FSE 2007. LNCS, vol. 4593, pp. 181–195. Springer, Heidelberg (2007)
19. Sun, S., Hu, L., Wang, P., Qiao, K., Ma, X., Song, L.: Automatic security evaluation and (related-key) differential characteristic search: application to SIMON, PRESENT, LBlock, DES(L) and other bit-oriented block ciphers. In: Sarkar, P., Iwata, T. (eds.) ASIACRYPT 2014. LNCS, vol. 8873, pp. 158–178. Springer, Heidelberg (2014)
20. Suzaki, T., Minematsu, K.: Improving the generalized Feistel. In: Hong, S., Iwata, T. (eds.) FSE 2010. LNCS, vol. 6147, pp. 19–39. Springer, Heidelberg (2010)
21. Suzaki, T., Minematsu, K., Morioka, S., Kobayashi, E.: TWINE: a lightweight block cipher for multiple platforms. In: Knudsen, L.R., Wu, H. (eds.) SAC 2012. LNCS, vol. 7707, pp. 339–354. Springer, Heidelberg (2013)
22. Wu, W., Zhang, L.: LBlock: a lightweight block cipher. In: Lopez, J., Tsudik, G. (eds.) ACNS 2011. LNCS, vol. 6715, pp. 327–344. Springer, Heidelberg (2011)

Improved Zero-Correlation Cryptanalysis on SIMON

Ling Sun[1], Kai Fu[1], and Meiqin Wang[1,2(✉)]

[1] Key Laboratory of Cryptologic Technology and Information Security,
Ministry of Education, Shandong University, Jinan 250100, China
{lingsun,fukai6}@mail.sdu.edu.cn
[2] State Key Laboratory of Cryptology, P.O. Box 5159, Beijing 100878, China
mqwang@sdu.edu.cn

Abstract. SIMON is a family of lightweight block ciphers publicly released by the NSA. Up to now, there have been many cryptanalytic results on it by means of differential, linear, impossible differential, integral, zero-correlation linear cryptanalysis and so forth. At INDOCRYPT 2014, Wang *et al.* gave zero-correlation attacks for 20-round SIMON32, 20-round SIMON48/72 and 21-round SIMON48/96. We investigate the security of whole family of SIMON by using zero-correlation linear cryptanalysis in this paper. For SIMON32 and SIMON48, we can attack one more round than the previous zero-correlation attacks given by Wang *et al.* We are the first one to give zero-correlation linear approximations of SIMON64, SIMON96 and SIMON128. These approximations are also utilized to attack the corresponding ciphers.

Keywords: SIMON · Zero-correlation linear approximation · Cryptanalysis

1 Introduction

Lightweight primitives aim at finding an optimal compromise between efficiency, security and hardware performance. Lightweight ciphers have been used in many fields, such as RFID tags, smartcards, and FPGAs. The impact of lightweight cipher is likely to continue increasing in the future. In recent years, many lightweight ciphers have been developed, including KATAN [10], KLEIN [11], LED [12], Piccolo [15], PRESENT [8] and TWINE [17].

SIMON [6] is a family of lightweight block ciphers publicly released by the National Security Agency (NSA) in June 2013. NSA has developed three ciphers to date, including SIMON, SPECK and Skipjack. SIMON has been optimized for performance in hardware implementations, while its sister algorithm, SPECK [6], has been optimized for software implementations. SIMON and SPECK offer users a variety of block sizes and key sizes for different implementations.

Many cryptanalytic results have been published on SIMON. The first differential cryptanalysis on SIMON was presented by Abed *et al.* in [1].

© Springer International Publishing Switzerland 2016
D. Lin et al. (Eds.): Inscrypt 2015, LNCS 9589, pp. 125–143, 2016.
DOI: 10.1007/978-3-319-38898-4_8

Then, Biryukov *et al.* improved the differential cryptanalysis of SIMON32, SIMON48 and SIMON64 by searching better differential characteristics in [7]. Based on the differential distinguisher shown by Biryukov *et al.*, Wang *et al.* improved the key recovery attacks on SIMON32, SIMON48 and SIMON64 [18]. In [18], Wang *et al.* gave the attack on 21-round SIMON32, which is still the best attack up to now. In addition, Sun *et al.* identified better differential distinguisher for SIMON with MILP models in [16]. Impossible differential attack against SIMON was firstly presented in [2], then the improved impossible differential attacks on SIMON32 and SIMON48 were given in [19], which had been further improved by Boura *et al.* in [9].

For the integral attack, Wang *et al.* proposed the attack on 21-round SIMON32 in [19] based on a zero-sum integral distinguisher for 15-round SIMON32, which was obtained experimentally.

Zero-correlation linear attack is one of the recent cryptanalytic methods introduced by Bogdanov and Rijmen in [3]. This kind of attack is based on the linear approximation with correlation zero (*i.e.* the linear approximation with probability exactly $\frac{1}{2}$). The idea of multiple zero-correlation cryptanalysis was developed in recent years in [4] by Bogdanov and Wang. They proposed a new distinguisher by using the fact that there are numerous zero-correlation approximations in susceptible ciphers. In [5], a more powerful distinguisher called multidimensional zero-correlation distinguisher was introduced. Wang *et al.* also gave the zero correlation linear approximations for SIMON32 and SIMON48 in [19]. They employed these approximations to attack 20-round SIMON32, 20-round SIMON48/72 and 21-round SIMON48/96.

In this paper, we investigate the security of whole family of SIMON by using zero-correlation linear cryptanalysis. For SIMON32 and SIMON48, by using the technique of equivalent-key, our cryptanalysis can attack one more round than the previous zero-correlation attacks in [19]. We are the first ones to give zero-correlation linear approximations of SIMON64, SIMON96 and SIMON128. These approximations are also utilized to attack the corresponding ciphers.

Our Contributions. In this paper, we investigate the security of whole family of SIMON by using zero-correlation linear cryptanalysis. Our contributions can be summarized as follows:

– Based on the 11-round zero-correlation distinguisher for SIMON32 and 12-round zero-correlation distinguisher for SIMON48, we use the equivalent-key technique (*i.e.* by moving the subkey into the left-side of round function) to improve the key recovery attack on SIMON32 and SIMON48. Finally, we can attack 21-round SIMON32, 21-round SIMON48/72 and 22-round SIMON48/96. The equivalent-key technique has been widely used in various key-recovery attacks. This technique aims at reducing the number of guessed subkey by using equivalent subkeys to replace the original subkeys used in the cipher. This technique had been used in [13] by Isobe. But there exists a little difference. Because the subkey is XORed after non-linear function, the condition in [13] that some parts of plaintext should be fixed can be canceled.

- We provide 13-, 16- and 19- round zero-correlation linear approximations of SIMON64, SIMON96 and SIMON128, respectively. We also use them to analysis the security of the corresponding ciphers. We are the first one to give the zero-correlation linear cryptanalysis for SIMON64, SIMON96 and SIMON128.

Our results along with the previous zero-correlation attacks on SIMON32 and SIMON48 are listed in Table 1.

Table 1. Summary of zero-correlation attacks on SIMON

Cipher	Rounds	Time (ENs)	Data (KPs)	Memory (Bytes)	Ref.
SIMON32	20	$2^{59.9}$	2^{32}	$2^{41.4}$	[19]
SIMON32	21	$2^{59.4}$	2^{32}	$2^{31.0}$	**Sect. 4.1**
SIMON48/72	20	$2^{59.7}$	2^{48}	$2^{43.0}$	[19]
SIMON48/72	21	$2^{61.9}$	2^{48}	$2^{43.0}$	**Sect. 4.2**
SIMON48/96	21	$2^{72.6}$	2^{48}	$2^{46.7}$	[19]
SIMON48/96	22	$2^{80.5}$	2^{48}	$2^{43.0}$	**Sect. 4.2**
SIMON64/96	23	$2^{90.4}$	2^{64}	$2^{54.0}$	**Sect. 4.3**
SIMON64/128	24	$2^{116.8}$	2^{64}	$2^{54.0}$	**Sect. 4.3**
SIMON96/144	28	$2^{141.0}$	2^{96}	$2^{85.0}$	**Sect. 4.3**
SIMON128/192	32	$2^{156.8}$	2^{128}	$2^{117.0}$	**Sect. 4.3**
SIMON128/256	34	$2^{255.6}$	2^{128}	$2^{117.0}$	**Sect. 4.3**

KP: Known Plaintext; EN: Encryption.

Outline. The remainder of this paper is organized as follows. Section 2 gives a brief description of SIMON and a general introduction of zero-correlation linear cryptanalysis. Section 3 presents the zero-correlation linear distinguishers used in the following attacks. Section 4 covers the zero-correlation attacks on the whole family of SIMON. Finally, we conclude the paper in Sect. 5.

2 Preliminaries

2.1 Brief Description of SIMON

SIMON [6] is a family of lightweight block ciphers publicly released by the National Security Agency (NSA) in June 2013. SIMON offers users a variety of block sizes and key sizes for different implementations. Table 2 lists the different block and key sizes, in bits, for SIMON.

SIMON is a two-branch balanced Feistel network which consists of three operations: AND (&), XOR ($\oplus$) and rotation ($\lll$). We denote the input of the i-th round by $(L_i, R_i), i = 0, 1, \ldots, r - 1$. In round i, (L_i, R_i) is updated to (L_{i+1}, R_{i+1}) by using a function $F(x) = (x \lll 1) \& (x \lll 8) \oplus (x \lll 2)$ as follows:

Table 2. SIMON parameters

Block size	Key size
32	64
48	72, 96
64	96, 128
96	96, 144
128	128, 192, 256

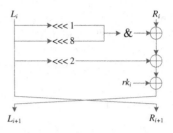

Fig. 1. Round function of SIMON

$$L_{i+1} = F(L_i) \oplus R_i \oplus rk_i,$$
$$R_{i+1} = L_i.$$

The output of the last round (L_r, R_r) is the ciphertext. An illustration of the round function is depicted in Fig. 1.

The key schedule of SIMON uses an LFSR-like procedure to generate r subkeys $rk_0, rk_1, \ldots, rk_{r-1}$. SIMON processes three slightly different key schedule procedures, depending on the number of word (ω) included in the master key. The first ω subkeys $rk_0, rk_1, \ldots, rk_{\omega-1}$ are initialized by the master key. The remaining subkeys are generated as follows:

$$rk_{i+m} = c \oplus (z_j)_i \oplus rk_i \oplus Y_m \oplus (Y_m \ggg 1),$$

$$Y_m = \begin{cases} rk_{i+1} \ggg 3 & \text{if } \omega = 2 \\ rk_{i+1} \oplus (rk_{i+2} \ggg 3) & \text{if } \omega = 3 \\ rk_{i+1} \oplus (rk_{i+3} \ggg 3) & \text{if } \omega = 4. \end{cases}$$

Here, the value c is constant `0xff...fc`, and $(z_j)_i$ denotes the i-th bit from one of the five constant sequences z_0, z_1, z_2, z_3 and z_4. The master key can be derived if any sequence of ω consecutive subkeys is known. For more information, please refer to [6].

2.2 Zero-Correlation Linear Cryptanalysis

Zero-correlation linear attack is one of the recent cryptanalytic methods introduced by Bogdanov and Rijmen in [3]. This kind of attack is based on the linear approximation with correlation zero (*i.e.* the linear approximation with probability exactly $\frac{1}{2}$). The idea of multiple zero-correlation cryptanalysis was developed in recent years in [4] by Bogdanov and Wang. They proposed a new distinguisher by using the fact that there are numerous zero-correlation approximations in susceptible ciphers. In [5], a more powerful distinguisher called multidimensional zero-correlation distinguisher was introduced.

Even though multiple zero-correlation cryptanalysis and multidimensional zero-correlation cryptanalysis perform better than zero-correlation linear cryptanalysis for various ciphers, we have to claim that they are not appropriate for SIMON. Multiple zero-correlation cryptanalysis and multidimensional zero-correlation cryptanalysis are more appropriate for word-level ciphers, such as AES, Skipjack and CAST-256.

The following Theorem is useful for computing the success probability of zero-correlation linear cryptanalysis.

Theorem 1 ([3, Proposition 3]). *The probability that the correlation value is 0 for a non-trivial linear approximation of a randomly drawn n-bit permutation can be approximated by $\frac{1}{\sqrt{2\pi}}2^{\frac{4-n}{2}}$ for $n \geq 5$.*

Based on the linear approximation of correlation zero, a technique similar to Matsui's Algorithm 2 [14] can be used for key recovery. Let the adversary have 2^n plaintext-ciphertext pairs and a zero-correlation linear approximation $\alpha \to \beta$ for a part of the cipher. The linear approximation is placed in the middle of the attacked cipher. Let E and D be the partial intermediate states of the data transform at the boundaries of the linear approximations (See Fig. 2). Then the key can be recovered using the following approach:

1. Guess the bits of the key needed to compute E and D. For each guess:
   (a) Partially encrypt the plaintexts and partially decrypt the ciphertexts up to the boundaries of the zero-correlation linear approximation $\alpha \to \beta$.
   (b) Estimate the correlation c of the linear approximation $\alpha \to \beta$ for the key guess using the partially encrypted and decrypted value E and D by counting how many times $\langle \alpha, E \rangle + \langle \beta, D \rangle$ is zero over 2^n plaintext-ciphertext pairs.
   (c) Perform a test on the estimated correlation c to tell of the estimated values of c is compatible with the hypothesis that the actual value of c is zero.

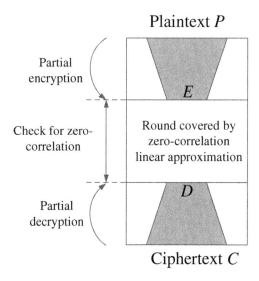

Fig. 2. Key recovery in zero-correlation linear cryptanalysis

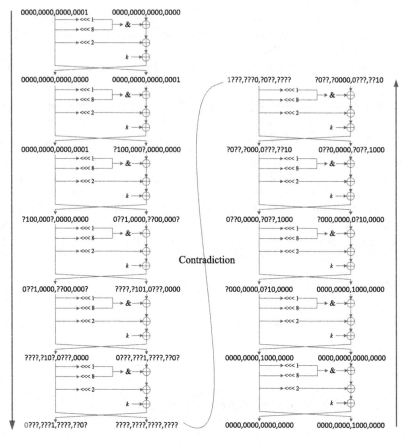

Fig. 3. Zero-correlation linear approximation of 11-round SIMON32. (Color figure online)

2. Test the surviving key candidates against a necessary number of plaintext-ciphertext pairs.

3   Zero-Correlation Linear Distinguishers of SIMON

3.1   Zero-Correlation Linear Distinguisher of SIMON32

For SIMON32, we use the 11-round zero-correlation linear distinguisher in [19], which is shown in Fig. 3. The input mask is (0x0001,0x0000) and the output mask is (0x0000,0x0080). The '0' at bottom left and the '1' at top right (in red) constitute the contradiction that ensures zero correlation.

3.2   Zero-Correlation Linear Distinguisher of SIMON48

Similarly, by using the 12-round zero-correlation linear distinguisher in [19], we can mount the key recovery attacks on 21-round SIMON48/72 and

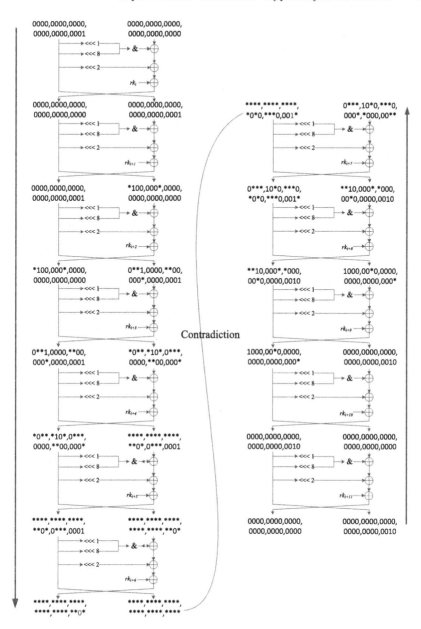

Fig. 4. Zero-correlation linear approximation of 12-round SIMON48. (Color figure online)

22-round SIMON48/96. The distinguisher used in the following attacks is shown in Fig. 4. The input mask is (0x000001,0x000000) and the output mask is (0x000000,0x000002). The '0' at bottom left and the '1' at top right (in red) constitute the contradiction that ensures zero correlation.

3.3 Zero-Correlation Linear Distinguishers of SIMON64, SIMON96 and SIMON128

In order to attack SIMON64/96/128, we first construct 13-, 16- and 19-round zero-correlation linear approximations for SIMON64, SIMON96 and SIMON128 by applying miss-in-the middle technique, which are shown in Figs. 5, 6 and 7, respectively.

	ROUND	LEFT	RIGHT
FORWARD	0	00000000000000000000000000000001	00000000000000000000000000000000
	1	00000000000000000000000000000000	00000000000000000000000000000001
	2	00000000000000000000000000000001	*100000*000000000000000000000000
	3	*100000*000000000000000000000000	0**10000**00000*0000000000000001
	4	0**10000**00000*0000000000000001	*0***10*0***0000**00000*00000000
	5	*0***10*0***0000**00000*00000000	0******1*******0*0***0000**00000*
	6	0******1*******0*0***0000**00000*	*****************************0*0***0000
	7	*****************************0*0***0000	0*******************************0*
	8	0*******************************0*	********************************
BACKWARD	8	1*******0*0***0000**00000*0*******	*0***0000**00000*00000000*0***10
	9	*0***0000**00000*00000000*0***10	0**00000*0000000000000010**1000
	10	0**00000*0000000000000010**1000	*0000000000000000000000*100000
	11	*0000000000000000000000*100000	00000000000000000000000010000000
	12	00000000000000000000000010000000	00000000000000000000000010000000
	13	00000000000000000000000000000000	00000000000000000000000010000000

Fig. 5. Zero-correlation linear approximation of 13-round SIMON64.

	ROUND	LEFT	RIGHT
FORWARD	0	000000000000000000000000000000000000000000000001	000000000000000000000000000000000000000000000000
	1	000000000000000000000000000000000000000000000000	000000000000000000000000000000000000000000000001
	2	000000000000000000000000000000000000000000000001	*100000*0000000000000000000000000000000000000000
	3	*100000*0000000000000000000000000000000000000000	0**10000**00000*000000000000000000000000000000001
	4	0**10000**00000*00000000000000000000000000000001	*0***10*0***0000**00000*000000000000000000000000
	5	*0***10*0***0000**00000*000000000000000000000000	0******1*******0*0***0000**00000*0000000000000001
	6	0******1*******0*0***0000**00000*0000000000000001	*1*****************0*0***0000**00000*00000000
	7	*1*****************0*0***0000**00000*00000000	0*****************************0*0***0000**00000*
	8	0*****************************0*0***0000**00000*	***0*0***0000
	9	***0*0***0000	0***0*
	10	0***0*	**
BACKWARD	6	*******************0*0***0000**00000*00000000*1*	****1******0*0***0000**00000*0000000000000010**
	5	****1******0*0***0000**00000*0000000000000010**	**10*0***0000**00000*00000000000000000000000*0*
	4	**10*0***0000**00000*000000000000000000000000*0*	10000*00000*000000000000000000000000000000010**
	3	10000*00000*000000000000000000000000000000010**	0000*0000000000000000000000000000000000000000*10
	2	0000*0000000000000000000000000000000000000000*10	000000000000000000000000000000000000000000001000
	1	000000000000000000000000000000000000000000001000	000000000000000000000000000000000000000000001000
	0	000000000000000000000000000000000000000000000000	000000000000000000000000000000000000000000001000

Fig. 6. Zero-correlation linear approximation of 16-round SIMON96.

ROUND	LEFT	RIGHT
FORWARD 0	00000000000000000000000000000000000000000000000000000000000000001	00000000000000000000000000000000000000000000000000000000000000000
1	00000000000000000000000000000000000000000000000000000000000000000	00000000000000000000000000000000000000000000000000000000000000001
2	00000000000000000000000000000000000000000000000000000000000000001	*100000*00000000000000000000000000000000000000000000000000000000
3	*100000*00000000000000000000000000000000000000000000000000000000	0**10000**00000000000000000000000000000000000000000000000000001
4	0**10000**00000*000000000000000000000000000000000000000000000001	*0***10*0**0000**00000*000000000000000000000000000000000000000
5	*0***10*0**0000**00000*000000000000000000000000000000000000001	0*****1*****0*0***0000**00000*00000000000000000000000000000000
6	0*****1*****0*0***0000**00000*000000000000000000000000000000001	*1****************0*0***0000**00000*0000000000000000000000001
7	*1****************0*0***0000**00000*00000000000000000000000000	0*****************************0*0***0000**00000*00000000000001
8	0*****************************0*0***0000**00000*000000000000001	*0***********************************0*0***0000**00000*00000000
9	*0***********************************0*0***0000**00000*00000000	0***0*0***0000**00000*
10	0***0*0***0000**00000*	***0*0***0000
11	***0*0***0000	0**
12	0**	**
BACKWARD 7	*************************0*0**0000*00000*000000000000000010	1******************0*0**0000*00000*00000000000000000000000*
6	1******************0*0**0000*00000*00000000000000000000010	******1*****0*0**0000*00000*000000000000000000000000000010
5	******1*****0*0**0000*00000*0000000000000000000000000010	0***10*0**0000*00000*0000000000000000000000000000000010
4	0***10*0**0000*00000*00000000000000000000000000000000010	**10000*00000*00000000000000000000000000000000000000010
3	**10000*00000*00000000000000000000000000000000000000010	100000*00000000000000000000000000000000000000000000010
2	100000*00000000000000000000000000000000000000000000010	00000000000000000000000000000000000000000000000000010
1	00000000000000000000000000000000000000000000000000010	00000000000000000000000000000000000000000000000000010
0	00000000000000000000000000000000000000000000000000000	00000000000000000000000000000000000000000000000000010

Fig. 7. Zero-correlation linear approximation of 19-round SIMON128.

4   Zero-Correlation Linear Cryptanalysis of SIMON

In this section, we investigate the security of whole family of SIMON by using zero-correlation linear cryptanalysis. We use 11- and 12-round zero-correlation linear approximations of SIMON32 and SIMON48 in [19] to present the key recovery attacks on 21-round SIMON32, 21-round SIMON48/72 and 22-round SIMON48/96. We also utilize the distinguishers presented in Sect. 3.3 to attack SIMON64, SIMON96 and SIMON128.

4.1   Zero-Correlation Linear Cryptanalysis of SIMON32

In this section, we use the 11-round zero-correlation linear distinguisher (See Fig. 3) in [19] to attack 21-round SIMON32. As shown in Fig. 8, we can add five rounds before the distinguisher and append five rounds after the distinguisher (*i.e.* the zero-correlation distinguisher starts from the 5-th round and ends at the 15-th round, with round number starting from 0). In this way, we can attack 21-round SIMON32.

Equivalent-Subkey Technique. The equivalent-subkey technique has been widely used in various key-recovery attacks. This technique aims at reducing the number of guessed subkey bits by replacing the equivalent subkeys with the original subkeys. This technique had been used in [13] by Isobe. But there exists a little difference. Because the subkey is XORed after non-linear function, the condition in [13] that some parts of plaintext should be fixed can be canceled.

In order to reduce the number of guessed subkey bits in the key recovery process, we move the subkey rk_i of the i-th round to the $(i + 1)$-th round, $(i = 0, 1, 2, 3, 4)$, to get the equivalent subkey K^i, see Fig. 8 (a). For example, K^0 in Fig. 8 (a) is equal to rk_0, and K^1 is equal to $(rk_0 \lll 2) \oplus rk_1$ and so forth. Note that K^4 is located in the distinguisher and doesn't need to be guessed. In Fig. 8 (a), we only list the guessed bits for K^i, $0 \leq i \leq 3$. Similarly, we can move the subkey rk_i of the i-th round to the $(i - 1)$-th round, $(i = 16, 17, 18, 19, 20)$, to get the equivalent subkey K^i, see Fig. 8 (b). Again, K^{16} is located in the distinguisher and doesn't need to be guessed. In Fig. 8 (b), we only list the guessed bits for K^i, $17 \leq i \leq 20$.

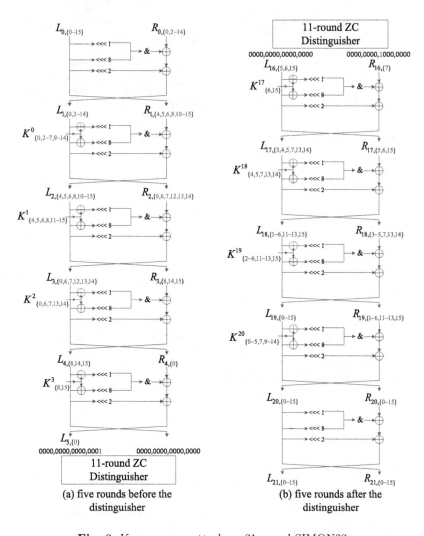

Fig. 8. Key recovery attack on 21-round SIMON32.

Key Recovery Process for SIMON32. In the following, R_i denotes the output of the i-th round. $R_{i,\{j\}}$ denotes the j-th bit of the R_i. $L_{i,\{j\}}$ is defined in a similar way. Note the bit position starts from '0'.

Firstly, we guess a part of the equivalent subkeys K^{17}, K^{18}, K^{19} and K^{20} (the concrete guessed key bits are shown in Fig. 8 (b)) and partially decrypt the ciphertext up to the state $R_{16,\{7\}}$. Next, we guess a part of the equivalent subkeys K^0, K^1, K^2, K^3 (the concrete guessed key bits are shown in Fig. 8 (a)) and partially encrypt the plaintext to the state $L_{5,\{0\}}$. We count the number of occurrences of the event that $L_{5,\{0\}} \| R_{16,\{7\}}$ is equal to "00" or "11". If the occurrence number is exactly equal to 2^{31}, we can keep the guessed 58-bit subkey as a possible subkey candidate, and discard it otherwise. To this end, 58-bit subkey is already guessed, which includes $K^0_{\{0,2-7,9-14\}}$, $K^2_{\{4-6,8,11-15\}}$, $K^3_{\{0,6,7,13,14\}}$, $K^4_{\{8,15\}}$, $K^{17}_{\{6,15\}}$, $K^{18}_{\{4,5,7,13,14\}}$, $K^{19}_{\{2-6,11-13,15\}}$ and $K^{20}_{\{0-5,7,9-14\}}$.

From Theorem 1, the probability that a wrong subkey guess is kept after the above procedure can be approximated by $\frac{1}{\sqrt{2\pi}} 2^{\frac{4-32}{2}} \approx 2^{-15.33}$. Thus, $2^{58} \times 2^{-15.33} = 2^{42.67}$ subkey candidates will be left. After that, we guess 6-bit subkey $K^0_{\{1,8,15\}} \| K^1_{\{0,1,2\}}$ and obtain 29 remaining bits of $K^1_{\{3,7,9,10\}} \| K^2_{\{1-5,8-12,15\}} \| K^3_{\{0-7,9-14\}}$ by solving the linear equations with Gaussian elimination. At last, we can compute all bits of the master key by inverting the key schedule, and check the correctness by using at most two plaintext-ciphertext pairs. We express this procedure in Algorithm 1.

Algorithm 1. Key Recovery Attack of SIMON32

---

1 Represent $K^{20}_{\{0-5,7,9-14\}} \| K^{19}_{\{2-6,11-13,15\}} \| K^{18}_{\{4,5,7,13,14\}} \| K^{17}_{\{6,15\}}$ by
$K^0 \| K^1 \| K^2 \| K^3$, and get 29 linear equations

2 **for** *all* $2^{42.67}$ *subkey candidates getting from the subkey recovery procedure (See Table 3)* **do**

3   **for** *all values of* $K^0_{\{1,8,15\}} \| K^1_{\{0,1,2\}}$ **do**

4   Get 29 linear equations with respect to
        $K^1_{\{3,7,9,10\}} \| K^2_{\{1-5,8-12,15\}} \| K^3_{\{0-7,9-14\}}$

5   Solve the linear equations by means of Gaussian elimination

6   **if** *solvable* **then**

7   Compute all bits of the master key according to the key schedule.

8   Verify the master key by using two plaintext-ciphertext pairs.

Complexity of Attack. The data complexity for the attack on SIMON32 is 2^{32} known plaintexts.

In this attack, the dominant term for the memory complexity is the term used to store 2^{31} 8-bit counters $T_0[\mathbf{X}_1^{32}]$, which makes the memory complexity be 2^{31} bytes.

The time complexity of each step in subkey recovery procedure is listed in Table 3. Overall, the time complexity in subkey recovery procedure is $2^{59.42}$ 21-round SIMON32 encryptions. In master key recovery phase, solving 29 linear equations with 29 variables by using Gaussian elimination needs about $\frac{1}{3} \cdot 29^3 \approx 8130$ bit-XOR operations, which can be measured by $\frac{8130}{16 \cdot 4 \cdot 21} \approx 2^{2.60}$ 21-round SIMON32 encryptions (Note that there are three XOR operations and

Table 3. Procedure of subkey recovery for SIMON32

Step	Input state	Guessed subkey (#Bits)	Computing (#Bits)	Counter (size)	Time complexity
0	X_0^{32}	$K_{\{0-5,7,9-14\}}^{20}$ $K_{\{2-6,11-13,15\}}^{19}$ $K_{\{4,5,7,13,14\}}^{18}$ $K_{\{6,15\}}^{17}$(29)	$R_{16,\{7\}}$ (36)*	$T_0[X_1^{32}]$(31)	$2^{32}\cdot2^{29}\cdot\frac{1+3+6+10+16}{16\times21}$ $\approx2^{55.78}$
1	X_1^{32}	None(0)	$L_{1,\{0,2-14\}}$ (14)	$T_1[X_2^{32}]$(25)	$2^{31}\cdot2^{29}\cdot\frac{14}{16\times21}\approx2^{55.41}$
2	X_2^{32}	$K_{\{0,3,5,7,10,12,14\}}^0$(7)	$L_{2,\{4,6,8,11,13,15\}}$ (6)	$T_2[X_3^{32}]$(24)	$2^{25}\cdot2^{36}\cdot\frac{6}{16\times21}\approx2^{55.19}$
3	X_3^{32}	$K_{\{4,6,11,13\}}^0$(4)	$L_{2,\{5,12,14\}}$ (3)	$T_3[X_4^{32}]$(20)	$2^{24}\cdot2^{40}\cdot\frac{3}{16\times21}\approx2^{57.19}$
4	X_4^{32}	$K_{\{2,9\}}^0$(2)	$L_{2,\{10\}}$ (1)	$T_4[X_5^{32}]$(17)	$2^{20}\cdot2^{42}\cdot\frac{1}{16\times21}\approx2^{53.61}$
5	X_5^{32}	$K_{\{6,8,13,15\}}^1$(4)	$L_{3,\{0,7,14\}}$ (3)	$T_5[X_6^{32}]$(15)	$2^{17}\cdot2^{46}\cdot\frac{3}{16\times21}\approx2^{56.19}$
6	X_6^{32}	$K_{\{5,12,14\}}^1$(3)	$L_{3,\{6,13\}}$ (2)	$T_6[X_7^{32}]$(13)	$2^{15}\cdot2^{49}\cdot\frac{2}{16\times21}\approx2^{56.61}$
7	X_7^{32}	$K_{\{4,11\}}^1$(2)	$L_{3,\{12\}}$ (1)	$T_7[X_8^{32}]$(10)	$2^{13}\cdot2^{51}\cdot\frac{1}{16\times21}\approx2^{55.61}$
8	X_8^{32}	$K_{\{0,7,14\}}^2$(3)	$L_{4,\{8,15\}}$ (2)	$T_8[X_9^{32}]$(8)	$2^{10}\cdot2^{54}\cdot\frac{2}{16\times21}\approx2^{56.61}$
9	X_9^{32}	$K_{\{6,13\}}^2$(2)	$L_{4,\{14\}}$ (1)	$T_9[X_{10}^{32}]$(5)	$2^8\cdot2^{56}\cdot\frac{1}{16\times21}\approx2^{55.61}$
10	X_{10}^{32}	$K_{\{8,15\}}^3$(2)	$L_{5,\{0\}}$ (1)	$T_{10}[X_{11}^{32}]$(2)	$2^5\cdot2^{58}\cdot\frac{1}{16\times21}\approx2^{54.61}$

Input State: input state of each step (See Table 4 for its concrete meaning);
Guessed Subkey: guessed subkey bits in each step;
Computing: state bits to be computed in each step;
Counter: counters to be constructed in each step;
Time Complexity: measured in 21-round SIMON32 encryption.
*: To compute $R_{16,\{7\}}$, we also need to compute $R_{17,\{5,6,15\}}$, $R_{18,\{3-5,7,13,14\}}$, $R_{19,\{1-6,11-13,15\}}$ and $R_{20,\{0-15\}}$, which are in total 36 bits.

Table 4. Explanation of symbols used in subkey recovery of SIMON32

Symbol	Meaning
X_0^{32}	$L_{0,\{0-15\}}\|R_{0,\{0,2-14\}}\|L_{21,\{0-15\}}\|R_{21,\{0-15\}}$
X_1^{32}	$L_{0,\{0-15\}}\ \|\ R_{0,\{0,2-14\}}\ \|\ R_{16,\{7\}}$
X_2^{32}	$L_{1,\{0,2-14\}}\|R_{1,\{4-6,8,10-15\}}\|R_{16,\{7\}}$
X_3^{32}	$L_{2,\{4,6,8,11,13,15\}}\|L_{1,\{0,2-4,6-14\}}\|R_{1,\{5,10,12,14\}}\|R_{16,\{7\}}$
X_4^{32}	$L_{2,\{4-6,8,11-15\}}\|L_{1,\{0,2,6-9,12-14\}}\|R_{1,\{10\}}\|R_{16,\{7\}}$
X_5^{32}	$L_{2,\{4-6,8,10-15\}}\|R_{2,\{0,6,7,12-14\}}\|R_{16,\{7\}}$
X_6^{32}	$L_{3,\{0,7,14\}}\|L_{2,\{4,5,8,10-12,14,15\}}\|R_{2,\{6,12,13\}}\|R_{16,\{7\}}$
X_7^{32}	$L_{3,\{0,6,7,13,14\}}\|L_{2,\{4,8,10,11,14,15\}}\|R_{2,\{12\}}\|R_{16,\{7\}}$
X_8^{32}	$L_{3,\{0,6,7,12-14\}}\|R_{3,\{8,14,15\}}\|R_{16,\{7\}}$
X_9^{32}	$L_{4,\{8,15\}}\|L_{3,\{0,6,12,13\}}\|R_{3,\{14\}}\|R_{16,\{7\}}$
X_{10}^{32}	$L_{4,\{8,14,15\}}\|R_{4,\{0\}}\|R_{16,\{7\}}$
X_{11}^{32}	$L_{5,\{0\}}\|R_{16,\{7\}}$

one AND operation in the round function of SIMON. For simplicity, we approximate them as four XOR operations in our analysis), thus the time complexity of master key recovery phase can be approximated as $2^{42.67}\times2^5\times2^{2.60}+2^{42.67}\times2^5\times(1+2^{-32})\approx2^{50.49}$ 21-round SIMON32 encryptions. Thus, the total time complexity of this attack is about $2^{59.42}$ 21-round SIMON32 encryptions.

4.2 Zero-Correlation Linear Cryptanalysis of SIMON48

Similarly, by using the 12-round zero-correlation linear distinguisher (See Fig. 4) in [19], we can mount key recovery attacks on 21-round SIMON48/72 and 22-round SIMON48/96.

Key Recovery Attack on 21-Round SIMON48/72. As shown in Fig. 9, we can add five rounds before the distinguisher and append four rounds after the distinguisher. In this way, we can attack 21-round SIMON48/72. We only list the guessed subkey bits in Fig. 9. The detailed attack procedure is proceeded in Algorithm 2.

The data complexity for the attack on SIMON48/72 is 2^{48} known plaintexts.

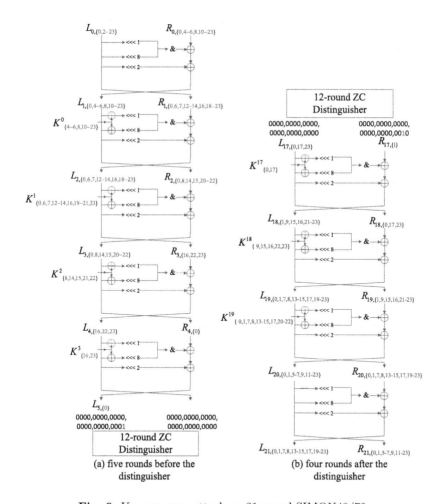

Fig. 9. Key recovery attack on 21-round SIMON48/72.

Algorithm 2. Key Recovery Attack of SIMON48/72

---

1 Represent $K^3_{\{16,23\}} \| K^{18}_{\{0,17\}} \| K^{19}_{\{9,15,16,22,23\}} \| K^{20}_{\{0,1,7,8,13-15,17,20-22\}}$ by $K^0 \| K^1 \| K^2$, and get 20 linear equations.

2 **for** all $2^{30.67}$ subkey candidates getting from the subkey recovery procedure (the concrete subkey recovery procedure is listed in Table 5) **do**

3   **for** all values of $K^0_{\{0-3,7,9\}} \| K^1_{\{1-5,8-11,15,17,18\}}$ **do**

4   Get 20 linear equations with respect to $K^1_{\{22\}} \| K^2_{\{0-7,9-13,16-20,23\}}$.

5   Solve the linear equations by means of Gaussian elimination

6   **if** solvable **then**

7   Compute all bits of the master key according to the key schedule.

8   Verify the master key by using two plaintext-ciphertext pairs.

Table 5. Procedure of subkey recovery for SIMON48/72[†]

Step	Input State	Guessed Subkey(♯Bits)	Computing(♯Bits)	Counter(Size)	Time Complexity
0	$X^{48,72}_0$	$K^{18}_{\{0,17\}} \| K^{19}_{\{9,15,16,22,23\}}$ $K^{20}_{\{0,1,7,8,13-15,17,20-22\}}$(18)	$R_{17,\{1\}}$ (24)*	$T_0[X^{48,72}_1]$(43)	$2^{48} \cdot 2^{18} \cdot \frac{1+3+7+13}{24\times21} \approx 2^{61.61}$
1	$X^{48,72}_1$	None(0)	$L_{1,\{0,4-6,8,10-23\}}$ (19)	$T_1[X^{48,72}_2]$(33)	$2^{43} \cdot 2^{18} \cdot \frac{19}{24\times21} \approx 2^{56.27}$
2	$X^{48,72}_2$	$K^0_{\{5,6,8,10,12,13,15-17,19,20,22,23\}}$(13)	$L_{2,\{0,6,7,13,14,16,18,20,21,23\}}$ (10)	$T_2[X^{48,72}_3]$(26)	$2^{33} \cdot 2^{31} \cdot \frac{10}{24\times21} \approx 2^{58.34}$
3	$X^{48,72}_3$	$K^0_{\{4,11,14,18,21\}}$(5)	$L_{2,\{12,19,22\}}$ (3)	$T_3[X^{48,72}_4]$(21)	$2^{26} \cdot 2^{36} \cdot \frac{3}{24\times21} \approx 2^{54.61}$
4	$X^{48,72}_4$	$K^1_{\{0,7,14,21\}}$(4)	$L_{3,\{8,15,22\}}$ (3)	$T_4[X^{48,72}_5]$(17)	$2^{21} \cdot 2^{40} \cdot \frac{3}{24\times21} \approx 2^{53.61}$
5	$X^{48,72}_5$	$K^1_{\{6,12,13,16,19,20,23\}}$(7)	$L_{3,\{0,20-22\}}$ (4)	$T_5[X^{48,72}_6]$(11)	$2^{17} \cdot 2^{47} \cdot \frac{4}{24\times21} \approx 2^{57.02}$
6	$X^{48,72}_6$	$K^2_{\{8,14,15,21,22\}}$(5)	$L_{4,\{16,22,23\}}$ (3)	$T_6[X^{48,72}_7]$(5)	$2^{11} \cdot 2^{52} \cdot \frac{3}{24\times21} \approx 2^{55.61}$
7	$X^{48,72}_7$	$K^3_{\{16,23\}}$(2)	$L_{5,\{0\}}$ (1)	$T_7[X^{48,72}_8]$(2)	$2^5 \cdot 2^{54} \cdot \frac{1}{24\times21} \approx 2^{50.02}$

Input State: input state of each step (See Table 6 for its concrete meaning);
Guessed Subkey: guessed subkey bits in each step;
Computing: state bits to be computed in each step;
Counter: counters to be constructed in each step;
Time Complexity: measured in 21-round SIMON48 encryption.

    * : To compute $R_{17,\{1\}}$, we also need to compute $R_{18,\{0,17,23\}}$, $R_{19,\{1,9,15,16,21-23\}}$ and $R_{20,\{0,1,7,8,13-15,17,19-23\}}$, which are in total 24 bits.

    † : The false positive probability of this attack is $\frac{1}{\sqrt{2\pi}} 2^{\frac{4-48}{2}} \approx 2^{-23.33}$ from Theorem 1.
The number of remaining subkey candidates is $2^{54} \cdot 2^{-23.33} \approx 2^{30.67}$ as we guess 54 subkey bits in total.

In this attack, the dominant term for the memory complexity is the term used to store 2^{43} 8-bit counters $T_0[X^{48,72}_1]$, which makes the memory complexity be 2^{43} bytes.

From Table 5, the time complexity for subkey recovery is about $2^{61.87}$ 21-round SIMON48/72 encryptions. In Algorithm 2, it will proceed Gaussian elimination process for $2^{30.67} \cdot 2^{18} = 2^{48.67}$ times, which can be ignored compared to $2^{61.87}$ 21-round encryptions. After that, the time complexity of checking the correctness of guess using two plaintext-ciphertext pairs also can be ignored compared to $2^{61.87}$ 21-round encryptions. Thus, the total time complexity is about $2^{61.87}$ 21-round SIMON48/72 encryptions.

Table 6. Explanation of symbols used in subkey recovery of SIMON48/72

Symbol	Meaning
$X_0^{48,72}$	$L_{0,\{0-23\}} \| R_{0,\{0-23\}} \| L_{21,\{0-23\}} \| R_{21,\{0-23\}}$
$X_1^{48,72}$	$L_{0,\{0,2-23\}} \| R_{0,\{4-6,8,10-23\}} \| R_{17,\{1\}}$
$X_2^{48,72}$	$L_{1,\{0,4-6,8,10-23\}} \| R_{1,\{0,6,7,12-14,16,18-23\}} \| R_{17,\{1\}}$
$X_3^{48,72}$	$L_{2,\{0,6,7,13,14,16,18,20,21,23\}} \| L_{1,\{0,4,8,10,11,14,15,17,18,20-22\}} \| R_{1,\{12,19,22\}} \| R_{17,\{1\}}$
$X_4^{48,72}$	$L_{2,\{0,6,7,12-14,16,18-23\}} \| R_{2,\{0,8,14,15,20-22\}} \| R_{17,\{1\}}$
$X_5^{48,72}$	$L_{3,\{8,15,22\}} \| L_{2,\{6,12,13,16,18-20,22,23\}} \| R_{2,\{0,14,20,21\}} \| R_{17,\{1\}}$
$X_6^{48,72}$	$L_{3,\{0,8,14,15,20-22\}} \| R_{3,\{16,22,23\}} \| R_{17,\{1\}}$
$X_7^{48,72}$	$L_{4,\{16,22,23\}} \| R_{4,\{0\}} \| R_{17,\{1\}}$
$X_8^{48,72}$	$L_{5,\{0\}} \| R_{17,\{1\}}$

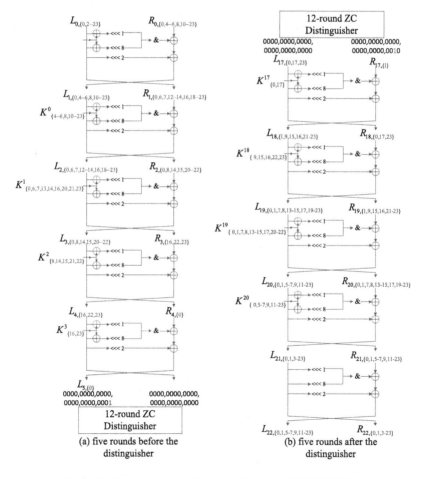

Fig. 10. Key recovery attack on 22-round SIMON48/96.

Algorithm 3. Key Recovery Attack of SIMON48/96

1 Represent $K^{17}_{\{0,17\}}\|K^{18}_{\{9,15,16,22,23\}}\|K^{19}_{\{0,1,7,8,13-15,17,20-22\}}\|K^{20}_{\{0,5-7,9,11-23\}}$ by $K^0\|K^1\|K^2\|K^3$, and get 36 linear equations.

2 **for** *all $2^{48.67}$ subkey candidates getting from the subkey recovery procedure (the concrete subkey recovery procedure is listed in Table 7)* **do**

3   **for** *all values of $K^0_{\{0-3,7,9\}}\|K^1_{\{1-5,8-11,15,17,18,22\}}\|K^2_{\{0-4\}}$* **do**

4   Get 36 linear equations with respect to $K^2_{\{5-7,9-13,16-20,23\}}\|K^3_{\{0-15,17-22\}}$.

5   Solve the linear equations by means of Gaussian elimination

6   **if** *solvable* **then**

7   Compute all bits of the master key according to the key schedule.

8   Verify the master key by using two plaintext-ciphertext pairs.

Table 7. Procedure of subkey recovery for SIMON48/96[†]

Step	Input State	Guessed Subkey($\sharp$Bits)	Computing($\sharp$Bits)	Counter(Size)	Time Complexity
0	$\boldsymbol{X}^{48,96}_0$	$K^{17}_{\{0,17\}}\|K^{18}_{\{9,15,16,22,23\}}$ $K^{19}_{\{0,1,7,8,13-15,17,20-22\}}$ $K^{20}_{\{0,5-7,9,11-23\}}$(36)	$R_{17,\{1\}}$ (43)*	$T_0[\boldsymbol{X}^{48,96}_1]$(43)	$2^{48}\cdot2^{36}\cdot\frac{43}{24\times22}\approx2^{80.38}$
1	$\boldsymbol{X}^{48,96}_1$	None(0)	$L_{1,\{0,4-6,8,10-23\}}$ (19)	$T_1[\boldsymbol{X}^{48,96}_2]$(33)	$2^{43}\cdot2^{36}\cdot\frac{19}{24\times22}\approx2^{74.20}$
2	$\boldsymbol{X}^{48,96}_2$	$K^0_{\{5,6,8,10,12,13,15-17,19,20,22,23\}}$(13)	$L_{2,\{0,6,7,13,14,16,18,20,21,23\}}$ (10)	$T_2[\boldsymbol{X}^{48,96}_3]$(26)	$2^{33}\cdot2^{49}\cdot\frac{10}{24\times22}\approx2^{76.28}$
3	$\boldsymbol{X}^{48,96}_3$	$K^0_{\{4,11,14,18,21\}}$(5)	$L_{2,\{12,19,22\}}$ (3)	$T_3[\boldsymbol{X}^{48,96}_4]$(21)	$2^{26}\cdot2^{54}\cdot\frac{3}{24\times22}\approx2^{72.54}$
4	$\boldsymbol{X}^{48,96}_4$	$K^1_{\{0,6,7,13,14,20,21\}}$(7)	$L_{3,\{8,14,15,21,22\}}$ (5)	$T_4[\boldsymbol{X}^{48,96}_5]$(14)	$2^{21}\cdot2^{61}\cdot\frac{5}{24\times22}\approx2^{75.28}$
5	$\boldsymbol{X}^{48,96}_5$	$K^1_{\{12,16,19,23\}}$(4)	$L_{3,\{0,20\}}$ (2)	$T_5[\boldsymbol{X}^{48,96}_6]$(11)	$2^{14}\cdot2^{65}\cdot\frac{2}{24\times22}\approx2^{70.96}$
6	$\boldsymbol{X}^{48,96}_6$	$K^2_{\{8,14,15,21,22\}}$(5)	$L_{4,\{16,22,23\}}$ (3)	$T_6[\boldsymbol{X}^{48,96}_7]$(5)	$2^{11}\cdot2^{70}\cdot\frac{3}{24\times22}\approx2^{73.54}$
7	$\boldsymbol{X}^{48,96}_7$	$K^3_{\{16,23\}}$(2)	$L_{5,\{0\}}$ (1)	$T_7[\boldsymbol{X}^{48,96}_8]$(2)	$2^5\cdot2^{72}\cdot\frac{1}{24\times22}\approx2^{67.96}$

Input State: input state of each step (See Table 8 for its concrete meaning);
Guessed Subkey: guessed subkey bits in each step;
Computing: state bits to be computed in each step;
Counter: counters to be constructed in each step;
Time Complexity: measured in 22-round SIMON48 encryption.

  * : To compute $R_{17,\{1\}}$, we also need to compute $R_{18,\{0,17,23\}}$, $R_{19,\{1,9,15,16,21-23\}}$, $R_{20,\{0,1,7,8,13-15,17,19-23\}}$ and $R_{21,\{0,1,5-7,9,11-23\}}$, which are in total 43 bits.

  † : The false positive probability of this attack is $\frac{1}{\sqrt{2\pi}}2^{\frac{4-48}{2}}\approx2^{-23.33}$ from Theorem 1. The number of remaining subkey candidates is $2^{72}\cdot2^{-23.33}\approx2^{48.67}$ for we guess 72 subkey bits in total.

Key Recovery Attack on 22-Round SIMON48/96. As shown in Fig. 10, we can add five rounds before the distinguisher and append five rounds after the distinguisher. In this way, we can attack 22-round SIMON48/96. We only list the guessed subkey bits in Fig. 10. The detailed attack procedure is proceeded in Algorithm 3.

The data complexity for the attack on SIMON48/96 is 2^{48} known plaintexts.

In this attack, the dominant term for the memory complexity is the term used to store 2^{43} 8-bit counters $T_0[\boldsymbol{X}^{48,96}_1]$, which makes the memory complexity to be 2^{43} bytes.

Table 8. Explanation of symbols used in subkey recovery of SIMON48/96

Symbol	Meaning
$X_0^{48,96}$	$L_{0,\{0-23\}} \| R_{0,\{0-23\}} \| L_{22,\{0-23\}} \| R_{22,\{0-23\}}$
$X_1^{48,96}$	$L_{0,\{2-23\}} \| R_{0,\{0,4-6,8,10-23\}} \| R_{17,\{1\}}$
$X_2^{48,96}$	$L_{1,\{0,4-6,8,10-23\}} \| R_{1,\{0,6,7,12-14,16,18-23\}} \| R_{17,\{1\}}$
$X_3^{48,96}$	$L_{2,\{0,6,7,13,14,16,18,20,21,23\}} \| L_{1,\{0,4,8,10,11,14,15,17,18,20-22\}} \| R_{1,\{12,19,22\}} \| R_{17,\{1\}}$
$X_4^{48,96}$	$L_{2,\{0,6,7,12-14,16,18-23\}} \| R_{2,\{0,8,14,15,20-22\}} \| R_{17,\{1\}}$
$X_5^{48,96}$	$L_{3,\{8,14,15,21,22\}} \| R_{2,\{0,20\}} \| L_{2,\{12,16,18,19,22,23\}} \| R_{17,\{1\}}$
$X_6^{48,96}$	$L_{3,\{0,8,14,15,20-22\}} \| R_{3,\{16,22,23\}} \| R_{17,\{1\}}$
$X_7^{48,96}$	$L_{4,\{16,22,23\}} \| R_{4,\{0\}} \| R_{17,\{1\}}$
$X_8^{48,96}$	$L_{5,\{0\}} \| R_{17,\{1\}}$

From Table 7, the time complexity for subkey recovery is about $2^{80.54}$ 22-round SIMON48/96 encryptions. In Algorithm 3, it will proceed Gaussian elimination process for $2^{48.67} \cdot 2^{24} = 2^{72.67}$ times, which can be ignored compared to $2^{80.54}$ 22-round encryptions. After that, the time complexity of checking the correctness of guess using two plaintext-ciphertext pairs also can be ignored compared to $2^{80.54}$ 22-round encryptions. Thus, the total time complexity is about $2^{80.54}$ 22-round SIMON48/96 encryptions.

4.3 Zero-Correlation Linear Cryptanalysis of SIMON64, SIMON96 and SIMON128

We can use the zero-correlation linear approximations showed in Figs. 5, 6 and 7 to attack SIMON64, SIMON96 and SIMON128, respectively. Since the attack procedures for them are similar, we only list the attack results in Table 9.

Table 9. Summary of ZC linear attack results on SIMON

Cipher	ZC linear distinguisher	Attacked rounds	Total rounds	Time (ENs)	Data (KPs)	Memory
SIMON64/96	13	23(5+13+5)*	42	$2^{90.4}$	2^{64}	2^{54} bytes
SIMON64/128	13	24(6+13+5)	44	$2^{116.8}$	2^{64}	2^{54} bytes
SIMON96/144	16	28(6+16+6)	54	$2^{141.0}$	2^{96}	2^{85} bytes
SIMON128/192	19	32(7+19+6)	69	$2^{156.8}$	2^{128}	2^{117} bytes
SIMON128/256	19	34(8+19+7)	72	$2^{255.6}$	2^{128}	2^{117} bytes

KP: Known Plaintext; EN: Encryption.

* : For $(a + b + c)$, a is the number of rounds before the distinguisher, b is the length of the distinguisher and c is the number of rounds after the distinguisher.

5 Conclusion

In this paper, we study the security of whole family of SIMON by using zero-correlation linear cryptanalysis. We improved the previous zero-correlation attacks for SIMON32 and SIMON48. Moreover, we present the 13-, 16- and 19-round zero correlation linear approximations of SIMON64, SIMON96 and SIMON128, respectively, and use them to attack the corresponding ciphers. We are the first one to give the zero-correlation linear cryptanalysis for SIMON 64, SIMON96 and SIMON128.

Acknowledgements. This work has been supported by 973 program (No. 2013CB834205), NSFC Projects (No. 61133013 and No. 61572293), Program for New Century Excellent Talents in University of China (No. NCET-13-0350), as well as Outstanding Young Scientists Foundation Grant of Shandong Province (No. BS2012DX018).

References

1. Abed, F., List, E., Lucks, S., Wenzel, J.: Differential cryptanalysis of round-reduced SIMON and SPECK. In: Cid, C., Rechberger, C. (eds.) FSE 2014. LNCS, vol. 8540, pp. 525–545. Springer, Heidelberg (2015)
2. Alkhzaimi, H., Lauridsen, M.: Cryptanalysis of the SIMON family of block ciphers. IACR Cryptology ePrint Archive, 2013/543 (2013)
3. Bogdanov, A., Rijmen, V.: Linear hulls with correlation zero and linear cryptanalysis of block ciphers. Designs, Codes and Cryptography **70**, 369–383 (2014). Springer, Heidelberg
4. Bogdanov, A., Wang, M.: Zero correlation linear cryptanalysis with reduced data complexity. In: Canteaut, A. (ed.) FSE 2012. LNCS, vol. 7549, pp. 29–48. Springer, Heidelberg (2012)
5. Bogdanov, A., Leander, G., Nyberg, K., Wang, M.: Integral and multidimensional linear distinguishers with correlation zero. In: Wang, X., Sako, K. (eds.) ASIACRYPT 2012. LNCS, vol. 7658, pp. 244–261. Springer, Heidelberg (2012)
6. Beaulieu, R., Shors, D., Smith, J., Treatman-Clark, S., Weeks, B., Wingers, L.: The SIMON and SPECK families of lightweight block ciphers. IACR Cryptology ePrint Archive, Report 2013/404 (2013)
7. Biryukov, A., Roy, A., Velichkov, V.: Differential analysis of block ciphers SIMON and SPECK. In: Cid, C., Rechberger, C. (eds.) FSE 2014. LNCS, vol. 8540, pp. 546–570. Springer, Heidelberg (2015)
8. Bogdanov, A.A., Knudsen, L.R., Leander, G., Paar, C., Poschmann, A., Robshaw, M., Seurin, Y., Vikkelsoe, C.: PRESENT: an ultra-lightweight block cipher. In: Paillier, P., Verbauwhede, I. (eds.) CHES 2007. LNCS, vol. 4727, pp. 450–466. Springer, Heidelberg (2007)
9. Boura, C., Naya-Plasencia, M., Suder, V.: Scrutinizing and improving impossible differential attacks: applications to CLEFIA, Camellia, LBlock and Simon. In: Sarkar, P., Iwata, T. (eds.) ASIACRYPT 2014. LNCS, vol. 8873, pp. 179–199. Springer, Heidelberg (2014)
10. Cannière, C., Dunkelman, O., Kneževiá, M.: KATAN and KTANTAN-a family of small and efficient hardware-oriented block ciphers. In: Clavier, C., Gaj, K. (eds.) CHES 2009. LNCS, vol. 5747, pp. 272–288. Springer, Heidelberg (2009)

11. Gong, Z., Nikova, S., Law, Y.W.: KLEIN: a new family of lightweight block ciphers. In: Juels, A., Paar, C. (eds.) RFIDSec 2011. LNCS, vol. 7055, pp. 1–18. Springer, Heidelberg (2012)

12. Guo, J., Peyrin, T., Poschmann, A., Robshaw, M.: The LED block cipher. In: Preneel, B., Takagi, T. (eds.) CHES 2011. LNCS, vol. 6917, pp. 326–341. Springer, Heidelberg (2011)

13. Isobe, T., Shibutani, K.: Generic key recovery attack on feistel scheme. In: Sako, K., Sarkar, P. (eds.) ASIACRYPT 2013, Part I. LNCS, vol. 8269, pp. 464–485. Springer, Heidelberg (2013)

14. Matsui, M.: Linear cryptanalysis method for DES cipher. In: Helleseth, T. (ed.) EUROCRYPT 1993. LNCS, vol. 765, pp. 386–397. Springer, Heidelberg (1994)

15. Shibutani, K., Isobe, T., Hiwatari, H., Mitsuda, A., Akishita, T., Shirai, T.: *Piccolo*: an ultra-lightweight blockcipher. In: Preneel, B., Takagi, T. (eds.) CHES 2011. LNCS, vol. 6917, pp. 342–357. Springer, Heidelberg (2011)

16. Sun, S., Hu, L., Wang, M., Wang, P., Qiao, K., Ma, X., et al.: Constructing mixed-integer programming models whose feasible region is exactly the set of all valid differential characteristics of SIMON. IACR Cryptology ePrint Archive, 2015/122 (2015)

17. Suzaki, T., Minematsu, K., Morioka, S., Kobayashi, E.: TWINE: a lightweight block cipher for multiple platforms. In: Knudsen, L.R., Wu, H. (eds.) SAC 2013. LNCS, vol. 7707, pp. 339–354. Springer, Heidelberg (2013)

18. Wang, N., Wang, X., Jia, K., Zhao, J.: Improved differential attacks on reduced SIMON versions. IACR Cryptology ePrint Archive, 2014/448 (2014)

19. Wang, Q., Liu, Z., Varici, K., Sasaki, Y., Rijmen, V., Todo, Y.: Cryptanalysis of reduced-round SIMON32 and SIMON48. In: Meier, W., Mukhopadhyay, D. (eds.) Progress in Cryptology – INDOCRYPT 2014. LNCS, vol. 8885, pp. 143–160. Springer, Heidelberg (2014)

A New Cryptographic Analysis of 4-bit S-Boxes

Ling Cheng$^{(\boxtimes)}$, Wentao Zhang, and Zejun Xiang

State Key Laboratory of Information Security, Institute of Information Engineering,
Chinese Academy of Sciences, Beijing, China
{chengling,zhangwentao,xiangzejun}@iie.ac.cn

Abstract. An exhaustive search of all 16! bijective 4-bit S-boxes has been conducted by Markku-Juhani et al. (SAC 2011). In this paper, we present an improved exhaustive search over all permutation-xor equivalence classes. We put forward some optimizing strategies and make some improvements on the basis of their work. For our program, it only takes about one-sixth of the time of the experiment by Markku-Juhani et al. to get the same results. Furthermore, we classify all those permutation-xor equivalence classes in terms of a new classification criterion, which has been come up with by Wentao Zhang et al. (FSE 2015). For some special cases, we calculate the distributions of permutation-xor equivalence classes with respect to their differential bound and linear bound. It turns out that only in three special cases, there exist S-boxes having a minimal differential bound $p = 1/4$ and a minimal linear bound $\epsilon = 1/4$, which imply the optimal S-boxes.

Keywords: 4-bit S-box · Classification · Exhaustive search · Differential cryptanalysis · Linear cryptanalysis · Time complexity

1 Introduction

S-boxes play an important role in block ciphers [1], which have been proposed for the first time in Lucifer cipher [2], whereafter are popularized by DES [3,4]. Since S-boxes act as the only non-linear part in many block ciphers, their cryptographic strength has a direct impact on the security of the whole block cipher. Two kinds of S-boxes with size of 4-bit and 8-bit are widely used. For example, AES [5] uses an 8-bit S-box, which has good performance against differential and linear cryptanalysis. However, AES is limited for some extremely resource-constrained environments [6,8]. The most compact hardware implementation of AES-128 still requires 2400GE [7]. In the past decade, with extensive deployment of tiny computing devices such as RFID and sensor network, many new lightweight block ciphers and hash functions have been proposed. The AES S-box needs more than 200GE, while a typical 4-bit S-box only needs about 20–30GE. Hence, to reduce the hardware area, 4-bit S-boxes are widely used in lightweight cryptographic primitives, such as LBlock [13], LED [12], PHOTON [14], PRESENT [9], PRIDE [16], PRINCE [15], RECTANGLE [11], SPONGENT [10], and so on. In such a situation, it is very important to have a better understanding of 4-bit S-boxes.

© Springer International Publishing Switzerland 2016
D. Lin et al. (Eds.): Inscrypt 2015, LNCS 9589, pp. 144–164, 2016.
DOI: 10.1007/978-3-319-38898-4_9

Even if the value of n is very small, the number of permutations over n-bit vectors is still too large. For example, the number of permutations over 4-bit vectors is 16! $\approx 2^{44.25}$, which is still huge. The number of permutations over 5-bit vectors is 32! $\approx 2^{117.66}$, which is too huge to be exhaustively searched. From this point of view, it is meaningful to study how to effectively reduce the time of the exhaustive search of 4-bit S-boxes. In [18], the author gave an exhaustive search over all bijective 4-bit S-boxes according to permutation-xor equivalence and gave an idea of *golden S-boxes* with ideal properties. The total time of creating a 1.4 GB file of the representatives of all permutation-xor equivalence classes is about half an hour with a 2011 consumer laptop. We will present some new results which can be used to improve the efficiency of the search algorithm greatly. Moreover, we will correct a few minor clerical errors appeared in the algorithm description in [18].

For 4-bit S-boxes, the optimal values are known with respect to differential and linear cryptanalysis (that is, differential uniformity and linearity). An S-box attaining these optimal values is called an optimal S-box. In [17], all optimal 4-bit S-boxes were classified according to affine equivalence, it is a surprising fact that there are only 16 different affine equivalence classes. In [19], a new classification of 4-bit optimal S-boxes have been conducted. Given an S-box, let $CarD1_S$ denote the number of times that 1-bit input difference causes a 1-bit output difference, and $CarL1_S$ denote the number of times that a 1-bit input mask causes a 1-bit output mask. The subset of 4-bit optimal S-boxes with the same values of $CarD1_S$ and $CarL1_S$ is called a category. All optimal 4-bit S-boxes were classified into 183 different categories. Among all the 183 categories, the authors specified 3 so-called platinum categories with the minimal value of $CarD1_S + CarL1_S$. In [19], the authors claimed that the category with $CarD1_S = 0$ and $CarL1_S = 0$ is the best case. However, they also proved that there is no such an optimal S-box. It is worth noting that only optimal S-boxes are considered in the work of [19]. A natural question is, whether there exists an S-box with $CarD1_S = 0$ and $CarL1_S = 0$ if we enlarge the scope from optimal 4-bit S-boxes to all bijective 4-bit S-boxes. This question is one motivation of our paper.

1.1   Contributions

In this paper, we give an improved exhaustive search over all permutation-xor equivalence classes of 16! bijective 4-bit boxes, which is one of our main contributions. Firstly we put forward three theorems, which are the core principles of our optimizing strategies. Next we give five lookup tables, which will be used to reduce some repeated calculations and speed up the search process. Based on the above description, we emphatically explain our optimizing strategies. Our experiments have been performed using one laptop with Intel Core i5 CPU. The total time of our program to create the same 1.4 GB file is about five minutes, which is one-sixth of the time spent in [18]. This improvement is potentially meaningful for future study of 5-bit S-boxes or 6-bit S-boxes, even for larger S-boxes.

Due to the fact that two permutation-xor equivalent S-boxes have the same values of CarD1$_S$ and CarL1$_S$ [19], we classify all permutation-xor equivalence classes of bijective 4-bit S-boxes in terms of the values of CarD1$_S$ and CarL1$_S$. For S-boxes with CarD1$_S$ + CarL1$_S$ ≤ 4, we calculate the distributions of S-boxes in relation to differential properties and linear properties. The results show that there exist 4-bit S-boxes satisfying CarD1$_S$ = 0 and CarL1$_S$ = 0. However, all the S-boxes with CarD1$_S$ = 0 and CarL1$_S$ = 0 are not good in relation to differential properties and linear properties, which are linear maps. In addition, all the S-boxes with (CarD1$_S$, CarL1$_S$) ∈ {(0, 1), (1, 0), (3, 0)} are linear maps too. Another fact which is of interest is that the minimal differential bound and linear bound of S-boxes with (CarD1$_S$, CarL1$_S$) ∈ {(1, 1), (1, 2), (2, 1)} both are 0.375. For PRESENT, RECTANGLE and SPONGENT, we expect the S-boxes with smaller values of CarD1$_S$ and CarL1$_S$. From this point, it is worth further studying S-boxes with (CarD1$_S$, CarL1$_S$) ∈ {(1, 1), (1, 2), (2, 1)}, which are potential classes for improving the security-performance tradeoff of PRESENT, RECTANGLE and SPONGENT.

1.2   Organization

This paper is organized as follows. Section 2 reviews some necessary definitions; Sect. 3 describes an improved exhaustive search over all permutation-xor equivalence classes; Sect. 4 revisits a new classification of 4-bit S-boxes according to CarD1$_S$ and CarL1$_S$; Sect. 5 concludes the paper; Appendix 5 presents some results of our experiments.

2   Preliminaries

2.1   Differential-Uniformity, Linearity, Optimal 4 bit S-Box

Definition 1. *Let S denote a* 4 × 4 *bijective S-box. Let* ΔX, ΔY *be two four-bit values, define* ND($\Delta X, \Delta Y$) *as:*

$$\text{ND}(\Delta X, \Delta Y) = \sharp\{x \in \mathbb{F}_2^4 | S(x) \oplus S(x \oplus \Delta X) = \Delta Y\}.$$

ND($\Delta X, \Delta Y$)/16 is the differential probability p of the characteristic ($\Delta X, \Delta Y$).

Definition 2. *Define* Diff(S) *as the **differential-uniformity** of S:*

$$\text{Diff}(S) = \max_{\Delta X \neq 0, \Delta Y} \text{ND}(\Delta X, \Delta Y).$$

Obviously, the differential-uniformity of S means the capacity for the resistance against differential cryptanalysis [4]. In general, the smaller the value of the differential-uniformity of an S-box, the more secure the S-box resists against differential cryptanalysis. It is known that for any 4 × 4 bijective S-box, Diff(S) ≥ 4 [17].

Definition 3. *Let S denote a 4×4 bijective S-box. Let ΓX, ΓY be two four-bit values, define* $\mathrm{Imb}(\Gamma X, \Gamma Y)$ *as the imbalance of the linear approximation:*

$$\mathrm{Imb}(\Gamma X, \Gamma Y) = |\sharp\{x \in \mathbb{F}_2^4 | \Gamma X \cdot x = \Gamma Y \cdot S(x)\} - 8|.$$

where $\cdot$ denotes the inner product on $\mathbb{F}_2^4$.

$\mathrm{Imb}(\Gamma X, \Gamma Y)/16$ is the bias ϵ of the linear approximation.

Definition 4. *Define* $\mathrm{Lin}(S)$ *as the **linearity** of S:*

$$\mathrm{Lin}(S) = \max_{\Gamma X, \Gamma Y \neq 0} \mathrm{Imb}(\Gamma X, \Gamma Y).$$

Obviously, The linearity of S means the capacity for the resistance against linear cryptanalysis [20]. Generally speaking, the smaller the value of the linearity of an S-box, the more secure the S-box resists against linear cryptanalysis. It is known that for any 4×4 bijective S-box, $\mathrm{Lin}(S) \geq 4$ [17].

Definition 5 ([17]). *Assume S is a 4×4 bijective S-box, which satisfies* $\mathrm{Diff}(S) = 4$ *and* $\mathrm{Lin}(S) = 4$, *then S is known as the **optimal S-box**.*

Let $\mathrm{wt}(x)$ denote the Hamming weight of bit vector x.

Definition 6 ([11]). *Define* $\mathrm{SetD1_S}$ *as follow:*

$$\mathrm{SetD1_S} = \{(\Delta X, \Delta Y) \in \mathbb{F}_2^4 \times \mathbb{F}_2^4 | \mathrm{wt}(\Delta X) = \mathrm{wt}(\Delta Y) = 1 \text{ and } \mathrm{ND}(\Delta X, \Delta Y) \neq 0\}.$$

Let $\mathrm{CarD1_S}$ *denote the cardinality of* $\mathrm{SetD1_S}$.

Definition 7 ([11]). *Define* $\mathrm{SetL1_S}$ *as follow:*

$$\mathrm{SetL1_S} = \{(\Gamma X, \Gamma Y) \in \mathbb{F}_2^4 \times \mathbb{F}_2^4 | \mathrm{wt}(\Gamma X) = \mathrm{wt}(\Gamma Y) = 1 \text{ and } \mathrm{Imb}(\Gamma X, \Gamma Y) \neq 0\}.$$

Let $\mathrm{CarL1_S}$ *denote the cardinality of* $\mathrm{SetL1_S}$.

2.2 Affine Equivalence and PE Equivalence

Definition 8 ([17]). *Let $A, B \in \mathrm{GL}(4, \mathbb{F}_2)$ be two invertible 4×4 matrices, and $a, b \in \mathbb{F}_2^4$ be two vectors. We call two S-boxes S and S' are **affine equivalent** if they satisfy:*

$$S'(x) = B(S(A(x) \oplus a)) \oplus b.$$

$A(x) \oplus a$ denote the inner affine transformation and $B(x) \oplus b$ denote the outer affine transformation.

It is well known that the values of $\mathrm{Diff}(S)$ and $\mathrm{Lin}(S)$ of an S-box both remain unchanged after applying an affine transformation [21,22]. In particular, when we apply an affine transformation to an optimal S-box, the new S-box we get is also an optimal S-box [17]. According to this property, all the optimal S-boxes can be classified into different affine equivalence classes. [17] gives all 16 affine equivalence classes of optimal S-boxes.

Definition 9 ([17]). *Let* $P_i, P_o \in \mathrm{GL}(4, \mathbb{F}_2)$ *be two* 4×4 *bit permutation matrices, and* $c_i, c_o \in \mathbb{F}_2^4$ *be two vectors. We call two S-boxes* S *and* S' *are* **permutation-xor equivalent(PE)** *if they satisfy:*

$$S'(x) = P_o(S(P_i(x) \oplus c_i)) \oplus c_o.$$

$P_i(x) \oplus c_i$ *denote the inner permutation-xor transformation and* $P_o(x) \oplus c_o$ *denote the outer permutation-xor transformation.*

In the following paper, we mainly concentrate on the PE classes. The inner permutation-xor transformation and outer permutation-xor transformation are called inner transformation and outer transformation for short.

3   An Improved Exhaustive Search over All PE Classes

In this section, we give an improved exhaustive search over all permutation-xor equivalence classes of all 16! bijective 4-bit boxes. Similar to the method in [18], our algorithm takes the least member of each PE class as the representative, then stores all the representatives on the disk together with the sizes of the PE classes.

A bijective 4-bit S-box can be expressed as a 4×16 bit matrix. Each column denotes a unique mapping of $0, 1, \cdots, 15$. Each row can be expressed as a 16-bit word.

Property 1 ([18]). Any 4×4 bijective S-box can be uniquely expressed as:

$$S(x) = \left(\sum_{i=0}^{3} 2^{P(i)} W_{i,(15-x)}\right) \oplus c.$$

P denotes the bit permutation of numbers $(0,1,2,3)$, $c \in \mathbb{F}_2^4$ denotes the xor constant. $W_i(i = 0, 1, 2, 3)$ is a 16-bit word which satisfies $0 < W_0 < W_1 < W_2 < W_3 < 2^{15}$. $W_i = \sum_{j=0}^{15} 2^j W_{i,j}$.

Due to the fact that S is bijective, it is required that $\mathrm{wt}(W_i) = 8$ for each W_i, and four W_i must be different from each other. There are $4! = 24$ values for P and $2^4 = 16$ values for the constant c. P and c make up $24 \times 16 = 384$ outer transformations. W_i defines the inner transformations. There are 384 inner transformations, which also consist of 24 bit permutations P_i and 16 constants c_i.

Given an S-box S, there are 384×384 S-boxes which are PE equivalent with S. We arrange them into a 384×384 matrix. Each S-box in the matrix is the result of applying an inner transformation and an outer transformation to S. Each row in the matrix corresponds to an inner transformation. Inner transformations are numbered from 0 to 383. Each column in the matrix corresponds to an outer transformation. Outer transformations are numbered from 0 to 383. The representative of the PE class is exactly the least member among all the 384×384 elements of the matrix. Now, we are ready to present three theorems.

Theorem 1. *Given an S-box S, and the corresponding matrix of PE equivalent S-boxes of S as described above, then for each row in the matrix, the 384 S-boxes within this row are distinct.*

Proof. Let $M_i(i = 0, 1, \cdots, 383)$ denote 384 outer transformations, $\varphi_i(i = 0, 1, \cdots, 383)$ denote inner transformation. $S_{i,0}, S_{i,1}, S_{i,2}, \cdots, S_{i,383}$ denote the S-boxes in the corresponding row in the matrix, i.e.

$$S_{i,0} = M_0(S(\varphi_i(x))), S_{i,1} = M_1(S(\varphi_i(x))), \cdots, S_{i,383} = M_{383}(S(\varphi_i(x))). \quad (1)$$

Assume that there exists $S_{i,j} = S_{i,k}$ with $S_{i,j} = M_j(S(\varphi_i(x))), S_{i,k} = M_k(S(\varphi_i(x)))$. Then, $M_j(S(\varphi_i(x))) = M_k(S(\varphi_i(x)))$. Since the outer transformations are invertible, then

$$S(\varphi_i(x)) = (M_j^{-1} \circ M_k)(S(\varphi_i(x))). \quad (2)$$

Thus $M_j = M_k$. Obviously this is contradict to the precondition $M_j \neq M_k$. So $S_{i,0}, S_{i,1}, S_{i,2}, \cdots, S_{i,383}$ are different from each other. □

Theorem 2. *Given an S-box S, and the corresponding matrix of PE equivalent S-boxes of S, for any two rows i, j in the matrix, let $\mathcal{A}_i(\mathcal{A}_j)$ denote the set of all 384 S-boxes in row $i(j)$, then either $\mathcal{A}_i = \mathcal{A}_j$, or $\mathcal{A}_i$ and $\mathcal{A}_j$ are disjoint.*

Proof. Let $M_i(i = 0, 1, \cdots, 383)$ denote 384 outer transformations, φ_i, φ_j denote any two inner transformations, $\mathcal{A}_i = (S_{i,0}, S_{i,1}, S_{i,2}, \cdots, S_{i,383})$ denote the set of 384 S-boxes in rows i of the matrix, and $\mathcal{A}_j = (S_{j,0}, S_{j,1}, S_{j,2}, \cdots, S_{j,383})$ denote the set of 384 S-boxes in row j of the matrix, i.e.

$$S_{i,0} = M_0(S(\varphi_i(x))), S_{i,1} = M_1(S(\varphi_i(x))), \cdots, S_{i,383} = M_{383}(S(\varphi_i(x))). \quad (3)$$

$$S_{j,0} = M_0(S(\varphi_j(x))), S_{j,1} = M_1(S(\varphi_j(x))), \cdots, S_{j,383} = M_{383}(S(\varphi_j(x))). \quad (4)$$

According to Theorem 1, $S_{i,0}, S_{i,1}, S_{i,2}, \cdots, S_{i,383}$ are different from each other, and $S_{j,0}, S_{j,1}, S_{j,2}, \cdots, S_{j,383}$ are different from each other.

Assume that there exists $S_{i,m} = S_{j,k}$ with $S_{i,m} = M_m(S(\varphi_i(x))), S_{j,k} = M_k(S(\varphi_j(x)))$. Then, $M_m(S(\varphi_i(x))) = M_k(S(\varphi_j(x)))$. Similarly, we get,

$$S(\varphi_i(x)) = (M_m^{-1} \circ M_k)(S(\varphi_j(x))). \quad (5)$$

From (3) and (5), then

$$S_{i,0} = (M_0 \circ M_m^{-1} \circ M_k)(S(\varphi_j(x))),$$
$$S_{i,1} = (M_1 \circ M_m^{-1} \circ M_k)(S(\varphi_j(x))),$$

$$\vdots \qquad\qquad (6)$$

$$S_{i,383} = (M_{383} \circ M_m^{-1} \circ M_k)(S(\varphi_j(x))).$$

From (4) and (6), it is deduced that $(S_{i,0}, S_{i,1}, \cdots, S_{i,383}) \subset (S_{j,0}, S_{j,1}, \cdots, S_{j,383})$. In turn, it can be proved that $(S_{j,0}, S_{j,1}, \cdots, S_{j,383}) \subset (S_{i,0}, S_{i,1}, \cdots, S_{i,383})$. Thus it means

$$(S_{i,0}, S_{i,1}, \cdots, S_{i,383}) = (S_{j,0}, S_{j,1}, \cdots, S_{j,383}). \tag{7}$$

That is to say, as long as there exists a same S-box in any two rows, the two corresponding sets of all 384 S-boxes in these two rows are exactly equal. Otherwise the two sets are disjoint. □

Theorem 3. *Given an S-box S, and the corresponding matrix of PE equivalent S-boxes of S, for any row i in the matrix, let $\mathcal{A}_i(i = 0, 1, \cdots, 383)$ denote the set of all 384 S-boxes in row i, and let $\mathcal{B} = \{\mathcal{A}_0, \mathcal{A}_1, \cdots, \mathcal{A}_{383}\}$ be a multiset. Then every element in $\mathcal{B}$ has a same cardinality. Furthermore, let n denote the number of distinct elements in $\mathcal{B}$, then any element $\mathcal{A}_i$ in $\mathcal{B}$ has a cardinality of $384/n$.*

Proof. According to Theorem 2, it is known that the two sets of all 384 S-boxes in any two rows in the matrix either are exactly equal, or are disjoint.

Firstly, we prove that every element in $\mathcal{B}$ has a same cardinality. Let $a_i(a_j)$ denote the cardinality of $\mathcal{A}_i(\mathcal{A}_j)$, $\mathcal{A}_i \neq \mathcal{A}_j$, then $\mathcal{A}_{i_0} = \mathcal{A}_{i_1} = \cdots, \mathcal{A}_{i_{a_i}-1}$, and $\mathcal{A}_{j_0} = \mathcal{A}_{j_1} = \cdots, \mathcal{A}_{j_{a_j}-1}$. We need to prove that $a_i = a_j$.

Let $\varphi_{i_k}(k = 0, 1, \cdots, a_i - 1)$ denote inner transformations that result in $\mathcal{A}_{i_k}(k = 0, 1, \cdots, a_i - 1)$. For each $\mathcal{A}_{i_k}$, let $S_{i,u}$ denote the least member among the 384 different S-boxes in $\mathcal{A}_{i_k}$, and $L_{i_k}(k = 0, 1, \cdots, a_i - 1)$ denote the outer transformation that lead to the least S-box $S_{i,u}$, i.e.

$$S_{i,u} = L_{i_0}(S(\varphi_{i_0}(x))), S_{i,u} = L_{i_1}(S(\varphi_{i_1}(x))), \cdots, S_{i,u} = L_{i_{a_i}-1}(S(\varphi_{i_{a_i}-1}(x))). \tag{8}$$

Similarly, let $\varphi_{j_k}(k = 0, 1, \cdots, a_j - 1)$ denote inner transformations that result in $\mathcal{A}_{j_k}(k = 0, 1, \cdots, a_j - 1)$. For each $\mathcal{A}_{j_k}$, let $S_{j,v}$ denote the least member among the 384 different S-boxes in $\mathcal{A}_{j_k}$, and $L_{j_k}(k = 0, 1, \cdots, a_j - 1)$ denote the outer transformation that lead to the least S-box $S_{j,v}$, i.e.

$$S_{j,v} = L_{j_0}(S(\varphi_{j_0}(x))), S_{j,v} = L_{j_1}(S(\varphi_{j_1}(x))), \cdots, S_{j,v} = L_{j_{a_j}-1}(S(\varphi_{j_{a_j}-1}(x))). \tag{9}$$

From the fact that $S_{i,u}$ and S are PE equivalent, $S_{j,v}$ and S are PE equivalent, it implies that $S_{i,u}$ and $S_{j,v}$ are PE equivalent as well. Let ϕ denote an inner transformation and M denote an outer transformation, then

$$S_{j,v} = M(S_{i,u}(\phi(x))). \tag{10}$$

From (8) and (10), then

$$S_{j,v} = (M \circ L_{i_0})(S((\varphi_{i_0} \circ \phi)(x))),$$
$$S_{j,v} = (M \circ L_{i_1})(S((\varphi_{i_1} \circ \phi)(x))),$$
$$\vdots \tag{11}$$
$$S_{j,v} = (M \circ L_{i_{a_i}-1})(S((\varphi_{i_{a_i}-1} \circ \phi)(x))).$$

From (9) and (11), it can be inferred,

$$(\varphi_{i_0} \circ \phi, \varphi_{i_1} \circ \phi, \cdots, \varphi_{i_{a_i-1}} \circ \phi) \subset (\varphi_{j_0}, \varphi_{j_1}, \cdots, \varphi_{j_{a_j-1}}). \tag{12}$$

Thus, it implies $a_i \leq a_j$. In turn, it can be proved that $a_j \leq a_i$. So $a_i = a_j$. If $\mathcal{A}_i = \mathcal{A}_j$, it is clear that $a_i = a_j$. Thus $a_i(i = 0, 1, \cdots, 383)$ are equal to each other.

Let n denote the number of distinct elements in $\mathcal{B}$. then

$$a_0 = a_1 = a_2 = \cdots = a_{383} = 384/n. \tag{13}$$

□

Theorems 1, 2, 3 have a great importance on counting the sizes of PE classes.

3.1   The Search Algorithm

The total number of bijective 4-bit S-boxes is $16! \approx 2^{44.25}$. Algorithm 1 describes the search of PE classes of all bijective 4-bit S-boxes, which is a modification of the algorithm in [18]. We correct a few minor clerical errors appeared in the algorithm description in [18]. Those highlighted in red in Algorithm 1 are the modified parts.

Now we will explain Algorithm 1 and the optimizing strategies in detail. Before performing the search, we establish five lookup tables to reduce some repeated calculations and speed up the search process. The five lookup tables are described as follow:

1. Table $wt8tab[6435]$: There are $\binom{16}{8} = 12,870$ 16-bit words with Hamming weight 8. Duo to the limit of $W_i < 2^{15}$, it needs to discard half of the words and remain 6435 candidates.
2. Table $word\_distribution[6435][384]$: For each word in table $wt8tab[6435]$, apply 384 inner transformations. To normalize the word, invert all bits of a word if the highest bit is set, then store the results in the corresponding position of table $word\_distribution[6435][384]$. 384 inner transformations consist of 24 bit permutations and 16 xor constants, where bit permutations are represented by P_i with index from 0 to 23, and xor constants are represented by c_i from 0 to 15. Each inner transformation is corresponding to a column in table $word\_distribution[6435][384]$ with index j, $j = P_i * 16 + c_i$. Thus when applying an inner transformation with index j (the index of bit permutation P_i is $j/16$, $c_i = j\%16$) to W_i, the result is just the value of $word\_distribution[i][j]$. This method avoids the repeated calculations when applying inner transformations to the same words.
3. Table $mw[6435]$: For each word W_i in table $wt8tab[6435]$, apply 384 inner transformations and normalize the results. The results are called the equivalents of W_i. Then select the minimal word among the 384 equivalents of W_i. The minimal equivalent of W_i is just the minimal member among the ith row in table $word\_distribution[6435][384]$. There are 58 different elements in table $mw[6435]$.

Algorithm 1. An improved search algorithm over PE classes.

1: **for** $i_0 = 0$ to 6434 **do**
2:     $W_0 = wt8tab[i_0]$
3:     **if** $mw(W_0) = W_0$ **then**
4:         **for** $i_1 = i_0 + 1$ to 6434 **do**
5:             $W_1 = wt8tab[i_1]$
6:             **if** $mw(W_1) \geq W_0$ and
                    $wt(t_2 = \neg W_0 \wedge W_1) = 4$ and $wt(t_3 = W_0 \wedge W_1) = 4$ and
                    $wt(t_1 = W_0 \wedge \neg W_1) = 4$ and $wt(t_0 = \neg W_0 \wedge \neg W_1) = 4$ **then**
7:                 **for** $i_2 = i_1 + 1$ to 6434 **do**
8:                     $W_2 = wt8tab[i_2]$
9:                     **if** $mw(W_2) \geq W_0$ and
                            $wt(u_0 = t_0 \wedge \neg W_2) = 2$ and $wt(u_4 = t_0 \wedge W_2) = 2$ and
                            $wt(u_1 = t_1 \wedge \neg W_2) = 2$ and $wt(u_5 = t_1 \wedge W_2) = 2$ and
                            $wt(u_2 = t_2 \wedge \neg W_2) = 2$ and $wt(u_6 = t_2 \wedge W_2) = 2$ and
                            $wt(u_3 = t_3 \wedge \neg W_2) = 2$ and $wt(u_7 = t_3 \wedge W_2) = 2$ **then**
10:                        **for** $j = 0$ to 7 **do**
11:                            $v_j = lsb(u_j)$
12:                        **end for**
13:                        **for** $b = 0$ to 255 **do**
14:                            $W_3 = \oplus_{j=0}^{7}(b_j u_j \oplus v_j)$
15:                            **if** $W_3 < 2^{15}$ **then**
16:                                **if** $W_3 > W_2$ and $mw(W_3) \geq W_0$ **then**
17:                                    **if** $(mw(W_1) > W_0$ and $mw(W_2) > W_0$ and $mw(W_3) > W_0)$
                                        then
18: $test1(W_0, W_1, W_2, W_3)$
19:                                    **else**
20: $test2(W_0, W_1, W_2, W_3)$
21:                                    **end if**
22:                                **end if**
23:                            **end if**
24:                        **end for**
25:                    **end if**
26:                **end for**
27:            **end if**
28:        **end for**
29:    **end if**
30: **end for**

---

4. Table $fix\_point[6435][384]$: If $word\_distribution[i][j] = mw[i]$, then fix_point $[i][j] = 1$, or else $fix\_point[i][j] = 0$. For each word in table $mw[6435]$, table $fix\_point[6435][384]$ records the inner transformations that make the word unchanged. Each minimal word maps to a set of inner transformations that make it unchanged, which are called $fix\_transformations$.

5. Table $position[32640]$: $position[word] = index$. It is an additional table to record the index of each word in the other four tables. Since the maximal value of a word is $32,640$, the size of table $position$ is defined as $32,640$.

The program mainly consists of four nested loops. We take many optimizing strategies to break the loop in advance. Hence, the total time complexity of Algorithm 1 is reduced greatly. For our program, we emphasize the following points.

1. The minimal words among the 384 equivalents of $W_i (i = 1, 2, 3)$ should be not less than W_0, i.e. $mw(W_1) \geq W_0$, $mw(W_2) \geq W_0$ and $mw(W_3) \geq W_0$. Let W_0^*, W_1^*, W_2^* and W_3^* denote the equivalents of W_0, W_1, W_2 and W_3, respectively. Due to the selection of W_0, the condition that $W_0^* \geq W_0$ is constant. Assume that there exists $W_1^* < W_0$, then sort W_0^*, W_1^*, W_2^* and W_3^* in ascending order and get the new S-box (W_0', W_1', W_2', W_3'). The new S-box (W_0', W_1', W_2', W_3') maybe equal to (W_0, W_1, W_2, W_3), which can result in duplicate S-boxes. Note that we add judgements for which the minimal words among the equivalents of $W_i (i = 1, 2, 3)$ may equal to W_0. Experiments show that there exactly exist S-boxes as the representatives of PE classes, satisfying that the minimal words among the equivalents of $W_i (i = 1, 2, 3)$ equal to W_0.
2. W_3 should be less than 2^{15} and we directly discard the candidates of W_3 which satisfy $W_3 \geq 2^{15}$. This solution is different from that in [18]. The solution in [18] is to invert all bits of W_3 if $W_3 \geq 2^{15}$. However, in our program, we find that the solution in [18] can lead to duplicate values for W_3. We don't know how the author of [18] solved this problem.
3. The method of checking that whether (W_0, W_1, W_2, W_3) is the least member of its PE class has a big influence on the time complexity of the algorithm. However, the author of [18] didn't explain the method of checking the least member. Now we will give our solution to the problem, which is to use two functions $test1$ and $test2$ in different cases, respectively.

Obviously, given an S-box S, and the corresponding matrix of PE equivalent S-boxes of S, the search of the least member of its PE class equals to the search of the least S-box in the matrix. However, it will take too much time if applying 384×384 transformations to any tested S-box and performing 384×384 comparisons. Based on the aforementioned three theorems, we find some optimization strategies for reducing the times of transformations and comparisons.

$test1(W_0, W_1, W_2, W_3)$: If the three minimal words among the equivalents of W_1, W_2, W_3 are bigger than W_0, i.e. $mw(W_1) > W_0$, $mw(W_2) > W_0$ and $mw(W_3) > W_0$, we use the function $test1$ to test whether (W_0, W_1, W_2, W_3) is the least member of its PE class or not. Since that W_0 is the minimal word among its equivalents $(mw(W_0) = W_0)$, the results after applying 384 transformations to W_0 must be no less than W_0, i.e. $W_0^* \geq W_0$. If $W_0^* > W_0$, the new S-box (W_0', W_1', W_2', W_3') must be larger than (W_0, W_1, W_2, W_3) no matter how $W_0^*, W_1^*, W_2^*, W_3^*$ are sorted, since there always exist $W_0^* > W_0$, $W_1^* > W_0$, $W_2^* > W_0$, and $W_3^* > W_0$. In this case, it only needs to traverse the transformations that make W_0 unchanged, instead of all 384 transformations. Details are as follows.

1. The $fix\_transformations$ of W_0 can be got from table fix_point. For each one of $fix\_transformations$, do

(a) Applying the transformation to W_1, W_2 and W_3, the corresponding new words W_1^*, W_2^* and W_3^* can be got by looking up table *word_distribution*.

(b) Sorting W_1^*, W_2^*, W_3^* and resulting in the final (W_0', W_1', W_2', W_3'), which satisfies $W_0 = W_0' < W_1' < W_2' < W_3'$. Note that each one of the *fix_transformations* to the four words corresponds to a row in the 384×384 matrix, sorting the new four words in ascending order and inverting all bits of a word if the highest bit is set correspond to the selection of the least S-boxes in the row.

(c) Comparing (W_1', W_2', W_3') with (W_1, W_2, W_3). If there exists $(W_1', W_2', W_3') < (W_1, W_2, W_3)$, (W_0, W_1, W_2, W_3) can't be the least member of its class, then the loop exits. Otherwise the loop continues.

2. For each one of the *fix_transformations* of W_0, we get a new S-box (W_0', W_1', W_2', W_3'). If (W_0, W_1, W_2, W_3) is no greater than any new S-box, we can say (W_0, W_1, W_2, W_3) is exactly the least member of its PE class. The pseudo code of *test1* refers to Algorithm 2 in Appendix 5.

Due to the fact that in most cases, it holds that $mw(W_1) > W_0$, $mw(W_2) > W_0$ and $mw(W_3) > W_0$. Moreover, there are only few transformations that make W_0 unchanged for most W_0, the method of using function *test1* greatly reduces the times of query and comparison for estimating the least member of PE classes.

Table 1 gives all minimal words with the numbers of transformations that make them unchanged. From Table 1, we can see that there is only one minimal word having *fix_transformations* with 384 values, while most of the rest have *fix_transformations* with a few values.

test2(W_0, W_1, W_2, W_3): As long as one of the three minimal words among the equivalents of W_1, W_2, W_3 equals to W_0, i.e. $mw(W_1) = W_0$ or $mw(W_2) = W_0$ or $mw(W_3) = W_0$, we use the function *test2* to test whether (W_0, W_1, W_2, W_3) is the least member of its PE class or not. Since that W_1^*, W_2^*, W_3^* may equal to W_0 and W_0^* also may be larger than W_1^*, W_2^*, W_3^*, the sizes and order of four words $W_0^*, W_1^*, W_2^*, W_3^*$ are uncertain. In this case, the function needs to traverse all the 384 transformations, and then perform as function *test1*. Only a few cases need to use function *test2*. It means that the numbers of cases that indeed need to traverse all the 384 transformations are few. The pseudo code of *test2* refers to Algorithm 3 in Appendix 5.

In function *test1* and *test2*, if (W_0, W_1, W_2, W_3) is the least member of the PE class, recording the numbers of inner transformations that make (W_0, W_1, W_2, W_3) stay unchanged, denoted as a_i, namely, the numbers of rows where the least member in the row is (W_0, W_1, W_2, W_3). According to Theorem 3, it is easy to calculate the expression $n = 384/a_i$, where n denote the number of distinct rows in the 384×384 matrix. According to Theorems 1, 2, each row in the 384×384 matrix have 384 different S-boxes, and any two rows either are exactly equal, or have no intersection. Thus the size of the PE class is $384/a_i \times 384$.

In conclusion, the method of using *test1* and *test2* in different cases can discard a part of unnecessary search branches, thereby greatly optimizes the

Table 1. Minimal words and the numbers of transformations that make them unchanged, denoted as T_n. The total number of minimal words is 58.

Min word	T_n	Min word	T_n	Min word	T_n	Min word	T_n
255	96	1020	8	2019	2	6030	8
383	12	1647	16	2022	4	6038	12
447	4	1659	2	2025	4	6042	2
495	4	1662	2	2033	4	6060	2
510	12	1695	16	2034	2	6120	24
831	24	1719	2	2040	4	6375	24
863	4	1725	2	4080	64	6627	4
879	2	1782	8	5739	12	6630	4
893	2	1785	8	5742	4	7128	32
894	2	1910	8	5783	12	7140	8
975	8	1913	4	5787	2	7905	16
983	2	1914	2	5790	4	15555	96
987	2	1973	2	5805	2	27030	384
989	2	1974	1	5820	2		
990	1	1980	2	5865	12		

time complexity of the search algorithm. The results of our search algorithm are in accordance with the results in [18].

4 Revisiting a New Classification of 4-Bit S-Boxes According to CarD1$_S$ and CarL1$_S$

Based on the results of Algorithm 1 and the fact that two PE equivalent S-boxes have the same values of CarD1$_S$ and CarL1$_S$ [19], all PE classes can be classified in terms of the values of (CarD1$_S$, CarL1$_S$). Table 2 gives the distribution of all PE classes in relation to the values of (CarD1$_S$, CarL1$_S$). It is clear that $0 \leq$ CarD1$_S \leq 16$ and $0 \leq$ CarL1$_S \leq 16$.

From [19], it is known that the larger the value of CarD1$_S$ or CarL1$_S$, the more likely there exists a weak differential or linear trail with only one active S-box in each round in PRESENT, RECTANGLE and SPONGENT. Thus we only consider the S-boxes satisfying CarD1$_S$ + CarL1$_S \leq 4$.

From Table 2, it shows that there exist S-boxes with (CarD1$_S$, CarL1$_S$) = $(0,0)$. At first sight, S-boxes with CarD1$_S = 0$ and CarL1$_S = 0$ should be the best cases. However, the experiments show that all of these S-boxes are linear, which have differential bound $p = 1$ and linear bound $\epsilon = 0.5$. Hence, the S-boxes with (CarD1$_S$, CarL1$_S$) = $(0,0)$ can not be used in a cryptographic primitive. Moreover, we find that all S-boxes with (CarD1$_S$, CarL1$_S$) $\in \{(0,1),(1,0),(3,0)\}$ are linear as well.

Table 2. Distribution of all PE classes in relation to CarD1s and CarL1s

CarD1s \ CarL1s	0	1	2	3	4	5	6	7	8	9	10	11	12	13	14	15	16
0	72	99	316	228	730	581	1063	394	408	42	52	2	11	0	0	0	1
1	86	315	948	2071	6114	9640	16859	15808	13915	6719	3961	946	447	31	12	0	3
2	63	460	1948	6827	22075	52044	101751	132926	138443	105953	65612	29192	11430	2670	721	123	45
3	10	280	2299	11578	47765	148909	341421	558452	692203	653972	476090	271967	121099	40048	11145	2073	421
4	11	141	1304	10713	62258	241428	673529	1345735	1997350	2217641	1898611	1258096	640436	245342	69734	13276	1753
5	0	50	562	6227	49741	252489	857875	2052257	3563727	4602531	4506477	3337822	1880941	777387	225303	43384	4407
6	0	14	275	3085	28073	175721	738143	2098840	4245378	6273427	6894880	5643446	3413680	1485437	447600	83252	7981
7	0	2	38	1169	13005	92958	454355	1515789	3528434	5901226	7191266	6412522	4135906	1884542	580195	109580	9576
8	0	0	22	416	4896	39500	214949	804756	2110860	3914983	5242315	5063217	3494708	1672586	530226	100526	8982
9	0	2	9	165	1749	14164	79351	323786	926955	1861446	2694300	2812294	2089841	1060151	350387	68372	6067
10	0	0	0	26	479	4287	24796	103979	302878	641126	976323	1087726	869398	483161	172739	35533	3187
11	0	0	0	5	124	1101	6153	25785	74957	158827	248427	289344	247300	151250	60161	13809	1282
12	0	0	0	8	45	326	1536	5279	14310	28329	43419	50159	46584	31344	14497	3729	455
13	0	0	0	0	18	83	378	966	2000	3377	4775	5711	5600	3995	2115	692	119
14	0	0	0	0	2	35	56	130	156	237	321	408	409	302	192	91	31
15	0	0	0	0	0	8	0	4	15	18	3	9	16	20	5	5	2
16	0	0	0	0	5	1	2	0	0	0	0	0	0	1	1	0	0

Note: The figure at position (i, j) denotes the number of PE classes satisfying CarD1s $= i$ and CarL1s $= j$. The sum of numbers in the table is 142,090,700, which is the total number of all PE classes. The product of the figure at position (i, j) and 384 means the number of S-boxes satisfying CarD1s $= i$ and CarL1s $= j$.

Table 3. The minimal values of differential bound and linear bound for each (CarD1$_S$, CarL1$_S$).

(CarD1$_S$, CarL1$_S$)	(p, ϵ)
(0,4), (1,3), (2,2)	(0.25, 0.25)
(0,2), (0,3)	(0.5, 0.375)
(1,1),(1,2),(2,1)	(0.375, 0.375)
(3,1)	(0.5, 0.25)
(2,0),(4,0)	(0.5, 0.5)

For each case with CarD1$_S$ + CarL1$_S$ $\leq$ 4 except $\{(0,0),\ (0,1),(1,0),(3,0)\}$, we calculate the distributions of S-boxes in relation to differential bound p and linear bound ϵ. Detailed distributions are presented in Appendix 5. From these results, we can get the minimal values of differential bound and linear bound for different cases, as Table 3 shows.

From Table 3, it can be seen that only if (CarD1$_S$, CarL1$_S$) $\in$ $\{(0,4),(1,3),(2,2)\}$, there exist S-boxes having a differential bound $p = 1/4$ and linear bound $\epsilon = 1/4$, which is in accordance with the results in [19]. Moreover, we can find that the S-boxes with (CarD1$_S$, CarL1$_S$) $\in$ $\{(1,1),(1,2),(2,1)\}$ have the minimal values of differential bound and linear bound both of 0.375. Although this kind of differential and linear bound is worse than the optimal S-boxes, the values of CarD1$_S$ + CarL1$_S$ are smaller than the optimal S-boxes. A natural question we pose here is: when considering the security of PRESENT, RECTANGLE and SPONGENT against differential and linear cryptanalysis, whether there exist better S-boxes among the S-boxes with (CarD1$_S$, CarL1$_S$) $\in$ $\{(1,1),(1,2),(2,1)\}$. We leave this question for future study.

5 Summary

We have presented some new results to greatly improve Saarinen's exhaustive search algorithm over all bijective 4-bit S-boxes. Our experimental results show that the efficiency of the search algorithm has been improved about 6 times. The exhaustive search of all bijective 4-bit S-boxes is relatively easy to achieve, but the exhaustive search of 5-bit S-boxes or beyond will be very difficult. Therefore optimization of the search algorithm is meaningful, which can be potentially used in the study of 5-bit, 6-bit or lager S-boxes.

Based on the results of the search over all PE classes of bijective 4-bit S-boxes, we classify all PE classes in terms of the values of CarD1$_S$ and CarL1$_S$. For S-boxes with CarD1$_S$ + CarL1$_S$ $\leq$ 4, we give the distributions in relation to differential and linear bounds. We verify the results in [19], and our results are in accordance with the results in [19]. In addition, we find that no good S-box satisfies that (CarD1$_S$, CarL1$_S$) $\in$ $\{(0,0),(0,1),(1,0),(3,0)\}$. Our results show that the S-boxes with (CarD1$_S$, CarL1$_S$) $\in$ $\{(0,0),(0,1),(1,0),(3,0)\}$ are all linear, which can not be used in a cryptographic primitive. Moreover, the S-boxes with (CarD1$_S$, CarL1$_S$) $\in$ $\{(1,1),(1,2),(2,1)\}$ have a minimal differential bound and linear bound both of 0.375, which can be further investigated to see if they can be used to improve the security-performance tradeoff of PRESENT, RECTANGLE and SPONGENT.

One of our future work is to study when $(CarD1_S, CarL1_S) \in \{(1,1),(1,2),(2,1)\}$, how the S-boxes influence the security of PRESENT, RECTANGLE and SPONGENT against differential and linear cryptanalysis.

Acknowledgements. The research presented in this paper is supported by the National Natural Science Foundation of China (No. 61379138), and the "Strategic Priority Research Program" of the Chinese Academy of Sciences (No. XDA06010701).

Appendix A

Algorithm 2. test1: estimate the least member of the PE class.

Input:

    The four words of an S-box: W_0, W_1, W_2, W_3

    The indexes of four words: id_0, id_1, id_2, id_3

Output:

    If (W_0, W_1, W_2, W_3) is exactly the least member of the PE class, output the numbers of transformations that make (W_0, W_1, W_2, W_3) unchanged. Otherwise output 0.

```
 1: result ← 0
 2: for i = 0 to 384 do
 3:    if fix_point[id₀][i] = 1 then
 4:       new_w[0] = word_distribution[id₁][i]
 5:       new_w[1] = word_distribution[id₂][i]
 6:       new_w[2] = word_distribution[id₃][i]
 7:       sort(new_w)
 8:       if new_w[0] < W₁ then
 9:          return 0
10:       else if new_w[0] = W₁ then
11:          if new_w[1] < W₂ then
12:             return 0
13:          else if new_w[1] = W₂ then
14:             if new_w[2] < W₃ then
15:                return 0
16:             else if new_w[2] = W₃ then
17:                result ++
18:             end if
19:          end if
20:       end if
21:    end if
22: end for
23: return result
```

See Tables 4, 5, 6, 7, 8, 9, 10, 11, 12, 13 and 14

Algorithm 3. test2: estimate the least member of the PE class.

Input:
>   The four words of an S-box: W_0, W_1, W_2, W_3
>   The indexes of four words: id_0, id_1, id_2, id_3

Output:
>   If (W_0, W_1, W_2, W_3) is exactly the least member of the PE class, output the numbers
>   of transformations that make (W_0, W_1, W_2, W_3) unchanged. Otherwise output 0.

```
 1: result ← 0
 2: for i = 0 to 384 do
 3:     new_w[0] = word_distribution[id₀][i]
 4:     new_w[1] = word_distribution[id₁][i]
 5:     new_w[2] = word_distribution[id₂][i]
 6:     new_w[3] = word_distribution[id₃][i]
 7:     sort(new_w)
 8:     if new_w[0] < W₀ then
 9:         return 0
10:     else if new_w[0] = W₀ then
11:         if new_w[1] < W₁ then
12:             return 0
13:         else if new_w[1] = W₁ then
14:             if new_w[2] < W₂ then
15:                 return 0
16:             else if new_w[2] = W₂ then
17:                 if new_w[3] < W₃ then
18:                     return 0
19:                 else if new_w[3] = W₃ then
20:                     result ++
21:                 end if
22:             end if
23:         end if
24:     end if
25: end for
26: return result
```

Table 4. Distribution of S-boxes with $CarD1_S = 0, CarL1_S = 2$ in relation to differential bound p and linear bound ϵ

LC →	$\epsilon \leq 1/4$		$1/4 < \epsilon \leq 3/8$		$3/8 < \epsilon \leq 1/2$	
DC ↓	n	%	n	%	n	%
$p \leq 1/4$	0	0.0000	0	0.0000	0	0.0000
$1/4 < p \leq 3/8$	0	0.0000	0	0.0000	0	0.0000
$3/8 < p \leq 1/2$	0	0.0000	1536	0.0127	21504	0.1772
$1/2 < p \leq 5/8$	0	0.0000	0	0.0000	0	0.0000
$5/8 < p \leq 3/4$	0	0.0000	1536	0.0127	36096	0.2975
$3/4 < p \leq 1$	0	0.0000	0	0.0000	60672	0.5000

Table 5. Distribution of S-boxes with $CarD1_S = 0, CarL1_S = 3$ in relation to differential bound p and linear bound ϵ

LC →	$\epsilon \leq 1/4$		$1/4 < \epsilon \leq 3/8$		$3/8 < \epsilon \leq 1/2$	
DC ↓	n	%	n	%	n	%
$p \leq 1/4$	0	0.0000	0	0.0000	0	0.0000
$1/4 < p \leq 3/8$	0	0.0000	0	0.0000	0	0.0000
$3/8 < p \leq 1/2$	0	0.0000	6144	0.0702	33792	0.3860
$1/2 < p \leq 5/8$	0	0.0000	0	0.0000	0	0.0000
$5/8 < p \leq 3/4$	0	0.0000	0	0.0000	23040	0.2632
$3/4 < p \leq 1$	0	0.0000	0	0.0000	24576	0.2807

Table 6. Distribution of S-boxes with $CarD1_S = 0, CarL1_S = 4$ in relation to differential bound p and linear bound ϵ

LC →	$\epsilon \leq 1/4$		$1/4 < \epsilon \leq 3/8$		$3/8 < \epsilon \leq 1/2$	
DC ↓	n	%	n	%	n	%
$p \leq 1/4$	768	0.0027	0	0.0000	0	0.0000
$1/4 < p \leq 3/8$	0	0.0000	18048	0.0644	0	0.0000
$3/8 < p \leq 1/2$	0	0.0000	40704	0.1452	98688	0.3521
$1/2 < p \leq 5/8$	0	0.0000	11520	0.0411	20736	0.0740
$5/8 < p \leq 3/4$	0	0.0000	1536	0.0055	58368	0.2082
$3/4 < p \leq 1$	0	0.0000	2304	0.0082	27648	0.0986

Table 7. Distribution of S-boxes with $CarD1_S = 1, CarL1_S = 1$ in relation to differential bound p and linear bound ϵ

LC →	$\epsilon \leq 1/4$		$1/4 < \epsilon \leq 3/8$		$3/8 < \epsilon \leq 1/2$	
DC ↓	n	%	n	%	n	%
$p \leq 1/4$	0	0.0000	0	0.0000	0	0.0000
$1/4 < p \leq 3/8$	0	0.0000	384	0.0032	0	0.0000
$3/8 < p \leq 1/2$	0	0.0000	0	0.0000	5760	0.0476
$1/2 < p \leq 5/8$	0	0.0000	0	0.0000	0	0.0000
$5/8 < p \leq 3/4$	0	0.0000	0	0.0000	17280	0.1429
$3/4 < p \leq 1$	0	0.0000	0	0.0000	97536	0.8063

Table 8. Distribution of S-boxes with $CarD1_S = 1, CarL1_S = 2$ in relation to differential bound p and linear bound ϵ

LC →	$\epsilon \leq 1/4$		$1/4 < \epsilon \leq 3/8$		$3/8 < \epsilon \leq 1/2$	
DC ↓	n	%	n	%	n	%
$p \leq 1/4$	0	0.0000	0	0.0000	0	0.0000
$1/4 < p \leq 3/8$	0	0.0000	2304	0.0063	0	0.0000
$3/8 < p \leq 1/2$	2304	0.0063	16896	0.0464	72960	0.2004
$1/2 < p \leq 5/8$	0	0.0000	0	0.0000	0	0.0000
$5/8 < p \leq 3/4$	0	0.0000	5376	0.0148	77568	0.2131
$3/4 < p \leq 1$	0	0.0000	0	0.0000	186624	0.5127

Table 9. Distribution of S-boxes with $CarD1_S = 1, CarL1_S = 3$ in relation to differential bound p and linear bound ϵ

LC →	$\epsilon \leq 1/4$		$1/4 < \epsilon \leq 3/8$		$3/8 < \epsilon \leq 1/2$	
DC ↓	n	%	n	%	n	%
$p \leq 1/4$	1536	0.0019	0	0.0000	0	0.0000
$1/4 < p \leq 3/8$	3027	0.0039	16896	0.0212	0	0.0000
$3/8 < p \leq 1/2$	6144	0.0077	82944	0.1043	165120	0.2076
$1/2 < p \leq 5/8$	0	0.0000	5376	0.0068	13824	0.0174
$5/8 < p \leq 3/4$	0	0.0000	10752	0.0135	262272	0.3298
$3/4 < p \leq 1$	0	0.0000	0	0.0000	227328	0.2859

Table 10. Distribution of S-boxes with $CarD1_S = 2, CarL1_S = 0$ in relation to differential bound p and linear bound ϵ

LC →	$\epsilon \leq 1/4$		$1/4 < \epsilon \leq 3/8$		$3/8 < \epsilon \leq 1/2$	
DC ↓	n	%	n	%	n	%
$p \leq 1/4$	0	0.0000	0	0.0000	0	0.0000
$1/4 < p \leq 3/8$	0	0.0000	0	0.0000	0	0.0000
$3/8 < p \leq 1/2$	0	0.0000	0	0.0000	2304	0.0952
$1/2 < p \leq 5/8$	0	0.0000	0	0.0000	0	0.0000
$5/8 < p \leq 3/4$	0	0.0000	0	0.0000	0	0.0000
$3/4 < p \leq 1$	0	0.0000	0	0.0000	21888	0.9048

Table 11. Distribution of S-boxes with $CarD1_S = 2, CarL1_S = 1$ in relation to differential bound p and linear bound ϵ

LC →	$\epsilon \leq 1/4$		$1/4 < \epsilon \leq 3/8$		$3/8 < \epsilon \leq 1/2$	
DC ↓	n	%	n	%	n	%
$p \leq 1/4$	0	0.0000	0	0.0000	0	0.0000
$1/4 < p \leq 3/8$	0	0.0000	2304	0.0130	0	0.0000
$3/8 < p \leq 1/2$	3072	0.0174	0	0.0000	16512	0.0935
$1/2 < p \leq 5/8$	0	0.0000	0	0.0000	0	0.0000
$5/8 < pp \leq 3/4$	0	0.0000	0	0.0000	16896	0.0957
$3/4 < pp \leq 1$	0	0.0000	0	0.0000	137856	0.7804

Table 12. Distribution of S-boxes with $CarD1_S = 2, CarL1_S = 2$ in relation to differential bound p and linear bound ϵ

LC →	$\epsilon \leq 1/4$		$1/4 < \epsilon \leq 3/8$		$3/8 < \epsilon \leq 1/2$	
DC ↓	n	%	n	%	n	%
$p \leq 1/4$	1536	0.0021	0	0.0000	0	0.0000
$1/4 < p \leq 3/8$	2304	0.0031	14592	0.0195	0	0.0000
$3/8 < p \leq 1/2$	21504	0.0287	29184	0.0390	147456	0.1971
$1/2 < p \leq 5/8$	0	0.0000	0	0.0000	0	0.0000
$5/8 < p \leq 3/4$	0	0.0000	18432	0.0246	198144	0.2649
$3/4 < p \leq 1$	0	0.0000	0	0.0000	314880	0.4209

Table 13. Distribution of S-boxes with $CarD1_S = 3, CarL1_S = 1$ in relation to differential bound p and linear bound ϵ

LC →	$\epsilon \leq 1/4$		$1/4 < \epsilon \leq 3/8$		$3/8 < \epsilon \leq 1/2$	
DC ↓	n	%	n	%	n	%
$p \leq 1/4$	0	0.0000	0	0.0000	0	0.0000
$1/4 < p \leq 3/8$	0	0.0000	0	0.0000	0	0.0000
$3/8 < p \leq 1/2$	3072	0.0286	0	0.0000	16896	0.1571
$1/2 < p \leq 5/8$	0	0.0000	0	0.0000	0	0.0000
$5/8 < p \leq 3/4$	0	0.0000	0	0.0000	3840	0.0357
$3/4 < p \leq 1$	0	0.0000	0	0.0000	83712	0.7786

Table 14. Distribution of S-boxes with $CarD1_S = 4, CarL1_S = 0$ in relation to differential bound p and linear bound ϵ

LC →	$\epsilon \leq 1/4$		$1/4 < \epsilon \leq 3/8$		$3/8 < \epsilon \leq 1/2$	
DC ↓	n	%	n	%	n	%
$p \leq 1/4$	0	0.0000	0	0.0000	0	0.0000
$1/4 < p \leq 3/8$	0	0.0000	0	0.0000	0	0.0000
$3/8 < p \leq 1/2$	0	0.0000	0	0.0000	3456	0.8182
$1/2 < p \leq 5/8$	0	0.0000	0	0.0000	0	0.0000
$5/8 < p \leq 3/4$	0	0.0000	0	0.0000	0	0.0000
$3/4 < p \leq 1$	0	0.0000	0	0.0000	768	0.1818

References

1. Feistel, H.: Block Cipher Cryptographic System. U.S. Patent 3,798,359, Filed 30 June 1971
2. Ben-Aroya, I., Biham, E.: Differential cryptanalysis of lucifer. In: Stinson, D.R. (ed.) CRYPTO 1993. LNCS, vol. 773, pp. 187–199. Springer, Heidelberg (1994)
3. National Bureau of Standards: Data Encryption Standard. FIPS PUB 46. National Bureau of Standards, U.S. Department of Commerce, Washington D.C., 15 January 1977
4. Biham, E., Shamir, A.: Differential cryptanalysis of DES-like cryptosystems. In: Menezes, A., Vanstone, S.A. (eds.) CRYPTO 1990. LNCS, vol. 537, pp. 2–21. Springer, Heidelberg (1991)
5. Daemen, J., Rijmen, V.: The Design of Rijndael: AES - The Advanced Encryption Standard. Springer, Heidelberg (2002)
6. Yamamoto, D., Hospodar, G., Maes, R., Verbauwhede, I.: Performance and security evaluation of AES S-box-based glitch PUFs on FPGAs. In: Bogdanov, A.; Sanadhya, S. (eds.) SPACE 2012. LNCS, vol. 7644, pp. 45–62. Springer, Heidelberg (2012)
7. Moradi, A., Poschmann, A., Ling, S., Paar, C., Wang, H.: Pushing the limits: a very compact and a threshold implementation of AES. In: Paterson, K.G. (ed.) EUROCRYPT 2011. LNCS, vol. 6632, pp. 69–88. Springer, Heidelberg (2011)
8. Engels, D., Fan, X., Gong, G., Hu, H., Smith, E.M.: Hummingbird: ultra-lightweight cryptography for resource-constrained devices. In: Sion, R., Curtmola, R., Dietrich, S., Kiayias, A., Miret, J.M., Sako, K., Sebé, F. (eds.) FC 2010 Workshops. LNCS, vol. 6054, pp. 3–18. Springer, Heidelberg (2010)
9. Bogdanov, A.A., Knudsen, L.R., Leander, G., Paar, C., Poschmann, A., Robshaw, M., Seurin, Y., Vikkelsoe, C.: PRESENT: an ultra-lightweight block cipher. In: Paillier, P., Verbauwhede, I. (eds.) CHES 2007. LNCS, vol. 4727, pp. 450–466. Springer, Heidelberg (2007)
10. Bogdanov, A., Knežević, M., Leander, G., Toz, D., Varıcı, K., Verbauwhede, I.: SPONGENT: a lightweight hash function. In: Preneel, B., Takagi, T. (eds.) CHES 2011. LNCS, vol. 6917, pp. 312–325. Springer, Heidelberg (2011)
11. Zhang, W., Bao, Z., Lin, D., Rijmen, V., Yang, B., Verbauwhede, I.: RECTANGLE: A Bit-slice Ultra-Lightweight Block Cipher Suitable for Multiple Platforms. Cryptology ePrint Archive: Report 2014/084. http://eprint.iacr.org/2014/084

12. Guo, J., Peyrin, T., Poschmann, A., Robshaw, M.: The LED block cipher. In: Preneel, B., Takagi, T. (eds.) CHES 2011. LNCS, vol. 6917, pp. 326–341. Springer, Heidelberg (2011)
13. Wu, W., Zhang, L.: LBlock: a lightweight block cipher. In: Lopez, J., Tsudik, G. (eds.) ACNS 2011. LNCS, vol. 6715, pp. 327–344. Springer, Heidelberg (2011)
14. Guo, J., Peyrin, T., Poschmann, A.: The PHOTON family of lightweight hash functions. In: Rogaway, P. (ed.) CRYPTO 2011. LNCS, vol. 6841, pp. 222–239. Springer, Heidelberg (2011)
15. Borghoff, J., et al.: PRINCE – a low-latency block cipher for pervasive computing applications. In: Wang, X., Sako, K. (eds.) ASIACRYPT 2012. LNCS, vol. 7658, pp. 208–225. Springer, Heidelberg (2012)
16. Albrecht, M.R., Driessen, B., Kavun, E.B., Leander, G., Paar, C., Yalçın, T.: Block ciphers – focus on the linear layer (feat. PRIDE). In: Garay, J.A., Gennaro, R. (eds.) CRYPTO 2014, Part I. LNCS, vol. 8616, pp. 57–76. Springer, Heidelberg (2014)
17. Leander, G., Poschmann, A.: On the classification of 4 bit S-boxes. In: Carlet, C., Sunar, B. (eds.) WAIFI 2007. LNCS, vol. 4547, pp. 159–176. Springer, Heidelberg (2007)
18. Saarinen, M.-J.O.: Cryptographic analysis of all 4 × 4-bit S-boxes. In: Miri, A., Vaudenay, S. (eds.) SAC 2011. LNCS, vol. 7118, pp. 118–133. Springer, Heidelberg (2012)
19. Zhang, W., Bao, Z., Rijmen, V., Liu, M.: A new classification of 4-bit optimal S-boxes and its application to PRESENT, RECTANGLE and SPONGENT. In: FSE 2015. Cryptology ePrint Archive: Report 2015/433 (2015)
20. Matsui, M.: Linear cryptanalysis method for DES cipher. In: Helleseth, T. (ed.) EUROCRYPT 1993. LNCS, vol. 765, pp. 386–397. Springer, Heidelberg (1994)
21. Carlet, C., Charpin, P., Zinoviev, V.: Codes, bent functions and permutations suitable for DES-like cryptosystems. In: Carlet, C., Charpin, P., Zinoviev, V. (eds.) Designs, Codes and Cryptography, vol. 15, pp. 125–156. Springer, Heidelberg (1998)
22. Nyberg, K.: Differentially uniform mappings for cryptography. In: Helleseth, T. (ed.) EUROCRYPT 1993. LNCS, vol. 765, pp. 55–64. Springer, Heidelberg (1994)

Elipptic Curve and Cryptographic Fundamentals

On Generating Coset Representatives of $PGL_2(\mathbb{F}_q)$ in $PGL_2(\mathbb{F}_{q^2})$

Jincheng Zhuang[1,2(✉)] and Qi Cheng[3]

[1] State Key Laboratory of Information Security, Institute of Information Engineering
Chinese Academy of Sciences, Beijing 100093, China
zhuangjincheng@iie.ac.cn
[2] State Key Laboratory of Mathematical Engineering and Advanced Computing,
Wuxi 214125, China
[3] School of Computer Science, The University of Oklahoma,
Norman, OK 73019, USA
qcheng@ou.edu

Abstract. There are $q^3 + q$ right $PGL_2(\mathbb{F}_q)-$cosets in the group $PGL_2(\mathbb{F}_{q^2})$. In this paper, we present a method of generating all the coset representatives, which runs in time $\tilde{O}(q^3)$, thus achieves the optimal time complexity up to a constant factor. Our algorithm has applications in solving discrete logarithms and finding primitive elements in finite fields of small characteristic.

Keywords: Projective linear group · Cosets · Discrete logarithm · Primitive elements

1 Introduction

The discrete logarithm problem (DLP) over finite fields underpins the security of many cryptographic systems. Since 2013, dramatic progresses have been made to solve the DLP when the characteristic is small [1–13,15–21]. Particularly, for a finite field $\mathbb{F}_{q^n}$, Joux [19] proposed the first algorithm with heuristic running time at most $q^{n^{1/4+o(1)}}$. Subsequently, Barbulescu et al. [3] proposed the first algorithm with heuristic quasi-polynomial running time $q^{(\log n)^{O(1)}}$. In [20], these algorithms are coined as Frobenius representation algorithms. One key component of algorithms in [3,19] is the relation generation, which requires enumerating the cosets of $PGL_2(\mathbb{F}_q)$ in $PGL_2(\mathbb{F}_{q^d})$, where d is a small integer, e.g. $d = 2$ [19].

J. Zhuang—This work was partially supported by the National Natural Science Foundation of China under Grant 61502481, the Strategic Priority Research Program of the Chinese Academy of Sciences under Grant XDA06010701, and the Open Project Program of the State Key Laboratory of Mathematical Engineering and Advanced Computing for Jincheng Zhuang.
Q. Cheng—This work was partially supported by China 973 Program under Grant 2013CB834201 and by US NSF under Grant CCF-1409294 for Qi Cheng.

© Springer International Publishing Switzerland 2016
D. Lin et al. (Eds.): Inscrypt 2015, LNCS 9589, pp. 167–177, 2016.
DOI: 10.1007/978-3-319-38898-4_10

Huang and Narayanan [14] have applied Joux's relation generation method for finding primitive elements of finite fields of small characteristic. There is another method of generating relations, see [7].

To illustrate the application of enumerating cosets of $PGL_2(\mathbb{F}_q)$ in $PGL_2(\mathbb{F}_{q^2})$, we briefly recall Joux's method [19] of generating relations among linear polynomials of a small characteristic finite field $\mathbb{F}_{q^{2k}} = \mathbb{F}_{q^2}[X]/(I(X))$, where $I(X) \in \mathbb{F}_{q^2}[X]$ is an irreducible factor of $h_1(X)X^q - h_0(X)$ with the requirement that the degrees of $h_0(X), h_1(X)$ are small. Let x be the image of $X \mod (I(X))$. Such Frobenius representation has the crucial property that $x^q = \frac{h_0(x)}{h_1(x)}$. It is well known that:

$$\prod_{\alpha \in \mathbb{F}_q} (y - \alpha) = y^q - y.$$

Applying the Mobius transformation

$$y \mapsto \frac{ax + b}{cx + d}$$

where the matrix $m = \begin{pmatrix} a & b \\ c & d \end{pmatrix} \in \mathbb{F}_{q^2}^{2 \times 2}$ is nonsingular, we get

$$\prod_{\alpha \in \mathbb{F}_q} (\frac{ax + b}{cx + d} - \alpha) = (\frac{ax + b}{cx + d})^q - \frac{ax + b}{cx + d}.$$

We deduce [4]:

$$h_1(x)(cx + d) \prod_{\alpha \in \mathbb{F}_q} ((ax + b) - \alpha(cx + d))$$
$$= (a^q h_0(x) + b^q h_1(x))(cx + d) - (ax + b)(c^q h_0(x) + d^q h_1(x))$$
$$(\mod x^q h_1(x) - h_0(x)).$$

If the right-hand side can be factored into a product of linear factors over $\mathbb{F}_{q^2}$, we obtain a relation of the form

$$\lambda^{e_0} \prod_{i=1}^{q^2} (x + \alpha_i)^{e_i} = \prod_{i=1}^{q^2} (x + \alpha_i)^{e'_i} \quad (\mod x^q h_1(x) - h_0(x)), \tag{1}$$

where λ is a multiplicative generator of $\mathbb{F}_{q^2}$, $\alpha_1 = 0, \alpha_2, \alpha_3, \ldots, \alpha_{q^2}$ is a natural ordering of elements in $\mathbb{F}_{q^2}$, and e_i's and e'_i's are non-negative integers.

Recall that for a given finite field $\mathbb{F}_q$, the projective general linear group $PGL_2(\mathbb{F}_q) = GL_2(\mathbb{F}_q)/E$, where E is the subgroup of $GL_2(\mathbb{F}_q)$ consisting of non-zero scalar matrices. Following the notion in [3], we denote $\mathcal{P}_q$ as a set of the right cosets of $PGL_2(\mathbb{F}_q)$ in $PGL_2(\mathbb{F}_{q^2})$, namely,

$$\mathcal{P}_q = \{PGL_2(\mathbb{F}_q)t | t \in PGL_2(\mathbb{F}_{q^2})\}.$$

Note that the cardinality of $\mathcal{P}_q$ is $q^3 + q$. It was shown in [3,19] that the matrices in the same right coset produce the same relation. In [19], Joux suggested two ways to generate relations: the first is to investigate the structure of cosets of $PGL_2(\mathbb{F}_q)$ in $PGL_2(\mathbb{F}_{q^2})$, and the second is to use hash values to remove duplicate relations. The second approach needs to enumerate the elements in $PGL_2(\mathbb{F}_{q^2})$ that has cardinality about q^6, hence has time complexity at least q^6. It may not be the most time-consuming part inside a subexponential algorithm. However, if we want a more efficient algorithm to compute the discrete logarithms of elements, or to construct a primitive element, this complexity can be a bottleneck. In this paper, we develop the first approach to generate cosets representatives efficiently.

1.1   Our Result

In this work, we give an almost complete characterization of $\mathcal{P}_q$. The case of determining left cosets is similar. Our main result is the following:

Theorem 1. *There exists a deterministic algorithm that runs in time $\tilde{O}(q^3)$ and computes a set $S \subseteq PGL_2(\mathbb{F}_{q^2})$ such that*

1. $|S| \leq q^3 + 2q^2 - q + 2$;
2. $\mathcal{P}_q = \{PGL_2(\mathbb{F}_q)t | t \in S\}$.

Here we follow the convention that uses the notation $\tilde{O}(f(q))$ to stand for $O(f(q) \log^{O(1)} f(q))$. Note that the time complexity of our algorithm is optimal up to a constant factor, since the $\mathcal{P}_q$ has size $q^3 + q$.

2   A Preliminary Classification

We deduce our main result by two steps. Firstly, we describe a preliminary classification. Then, we deal with the dominating case. In this section, the main technical tool we use is the fact that the following operations on a matrix over $\mathbb{F}_{q^2}$ will not change the membership in a right coset of $PGL_2(\mathbb{F}_q)$ in $PGL_2(\mathbb{F}_{q^2})$:

- Multiply the matrix by an element in $\mathbb{F}_{q^2}^*$;
- Multiply a row by an element in $\mathbb{F}_q^*$;
- Add a multiple of one row with an element in $\mathbb{F}_q$ into another row;
- Swap two rows.

Proposition 1. *Let g be an element in $\mathbb{F}_{q^2} \setminus \mathbb{F}_q$. Each right coset of $PGL_2(\mathbb{F}_q)$ in $PGL_2(\mathbb{F}_{q^2})$ is equal to $PGL_2(\mathbb{F}_q)t$, where t is one of the following four types:*

(I) $\begin{pmatrix} 1 & b \\ c & 1 \end{pmatrix}$, *where $b, c \in \mathbb{F}_{q^2} \setminus \mathbb{F}_q, bc \neq 1$.*

(II) $\begin{pmatrix} 1 & b_1 \\ g & d_2 g \end{pmatrix}$, *where $b_1, d_2 \in \mathbb{F}_q^*, b_1 \neq d_2$.*

(III) $\begin{pmatrix} 0 & 1 \\ c & d \end{pmatrix}$, where $c \in \mathbb{F}_{q^2}^*, d \in \mathbb{F}_{q^2}$.

(IV) $\begin{pmatrix} 1 & 0 \\ c & d \end{pmatrix}$, where $c \in \mathbb{F}_{q^2}, d \in \mathbb{F}_{q^2}^*$.

Proof. Let $\begin{pmatrix} a & b \\ c & d \end{pmatrix}$ be a representative of a right coset of $PGL_2(\mathbb{F}_q)$ in $PGL_2(\mathbb{F}_{q^2})$. If any of a, b, c, d is zero, then we divide them by the other non-zero element in the same row, and swap rows if necessary, we will find a representative of type (III) or (IV). So we may assume that none of the entries are zero. Dividing the whole matrix by a, we can assume $a = 1$. Consider the nonsingular matrix

$$\begin{pmatrix} 1 & b_1 + b_2 g \\ c_1 + c_2 g & d_1 + d_2 g \end{pmatrix},$$

where $b_i, c_i, d_i \in \mathbb{F}_q$ for $1 \le i \le 2$. We distinguish the following cases. Note that we may also assume $c_1 = 0$, since we can add the multiple of the first row with $-c_1$ into the second. We start with the matrix

$$\begin{pmatrix} 1 & b_1 + b_2 g \\ c_2 g & d_1 + d_2 g \end{pmatrix}$$

where $c_2 \ne 0$.

Case 1. $b_2 \ne 0$

Subtracting $\frac{d_2}{b_2}$ times the first row from the second row, the matrix becomes

$$\begin{pmatrix} 1 & b_1 + b_2 g \\ -\frac{d_2}{b_2} + c_2 g & d_1 - \frac{b_1 d_2}{b_2} \end{pmatrix}.$$

We can assume that $d_1 - \frac{b_1 d_2}{b_2} \ne 0$. The matrix is in the same coset with a matrix of type (I) since we can divide the second row by $d_1 - \frac{b_1 d_2}{b_2}$, and b_2 and c_2 are not zero.

Case 2. $b_2 = 0$

We will assume $b_1 \ne 0$. After subtracting $\frac{d_1}{b_1}$ times the first row from the second row, the matrix becomes

$$\begin{pmatrix} 1 & b_1 \\ -\frac{d_1}{b_1} + c_2 g & d_2 g \end{pmatrix}$$

Assume $d_2 \ne 0$.

1. If $d_1 = 0$, then the matrix can be reduced to type (II) by dividing the second row by c_2.
2. If $d_1 \ne 0$, adding the product of the second row with $\frac{b_1}{d_1}$ into the first row, we get

$$\begin{pmatrix} \frac{b_1 c_2}{d_1} g & b_1 + \frac{b_1 d_2}{d_1} g \\ -\frac{d_1}{b_1} + c_2 g & d_2 g \end{pmatrix}.$$

Dividing all the entries in the matrix by g, we get

$$\begin{pmatrix} \frac{b_1 c_2}{d_1} & \frac{b_1 d_2}{d_1} + b_1 g^{-1} \\ c_2 - \frac{d_1}{b_1} g^{-1} & d_2 \end{pmatrix}.$$

Dividing the first row by $\frac{b_1 c_2}{d_1}$ and the second row by d_2, the matrix is reduced to type (I), since $\frac{b_1 d_2}{d_1} + b_1 g^{-1}$ and $c_2 - \frac{d_1}{b_1} g^{-1}$ are in $\mathbb{F}_{q^2} \setminus \mathbb{F}_q$. □

There are only $O(q^2)$ many possibilities for Case (II). Next, we simplify Cases (III) and (IV) further. As a conclusion, we can see that there are only $O(q^2)$ many possibilities in Case (III) and (IV) as well.

Proposition 2. *Let*

$$\begin{pmatrix} 0 & 1 \\ c & d \end{pmatrix} = \begin{pmatrix} 0 & 1 \\ c_1 + c_2 g & d_1 + d_2 g \end{pmatrix}$$

be one representative of a right coset of $PGL_2(\mathbb{F}_q)$ in $PGL_2(\mathbb{F}_{q^2})$, where $c_1, c_2, d_1, d_2 \in \mathbb{F}_q$. Then it belongs to $PGL_2(\mathbb{F}_q)t$, where t is of the following two types:

(III-a): $\begin{pmatrix} 0 & 1 \\ g & d_2 g \end{pmatrix}$, *where $d_2 \in \mathbb{F}_q$.*

(III-b): $\begin{pmatrix} 0 & 1 \\ 1 + c_2 g & d_2 g \end{pmatrix}$, *where $c_2 \in \mathbb{F}_q, d_2 \in \mathbb{F}_q$.*

Proof. There are two cases to consider.

1. Assume $c_1 = 0$. Subtracting the second row by the first row times d_1, we get

$$\begin{pmatrix} 0 & 1 \\ c_2 g & d_2 g \end{pmatrix}.$$

Since $c_2 \neq 0$, after dividing the second row by c_2, the matrix is reduced to type (III-a).

2. Assume $c_1 \neq 0$. Subtracting the second row by the first row times d_1, we get

$$\begin{pmatrix} 0 & 1 \\ c_1 + c_2 g & d_2 g \end{pmatrix}.$$

Dividing the second row by c_1, we get

$$\begin{pmatrix} 0 & 1 \\ 1 + c_2 g & \frac{d_2}{c_1} g \end{pmatrix}.$$

Thus the matrix is reduced to type (III-b), which completes the proof.   □

Similarly, we have the following proposition.

Proposition 3. *Let*

$$\begin{pmatrix} 1 & 0 \\ c & d \end{pmatrix} = \begin{pmatrix} 1 & 0 \\ c_1 + c_2 g & d_1 + d_2 g \end{pmatrix}$$

be one representative of a right coset of $PGL_2(\mathbb{F}_q)$ in $PGL_2(\mathbb{F}_{q^2})$. Then it belongs to $PGL_2(\mathbb{F}_q)t$, where t is of the following two types:

(IV-a): $\begin{pmatrix} 1 & 0 \\ c_2 g & g \end{pmatrix}$, *where $c_2 \in \mathbb{F}_q$.*

(IV-b): $\begin{pmatrix} 1 & 0 \\ c_2 g & 1 + d_2 g \end{pmatrix}$, *where $c_2 \in \mathbb{F}_q, d_2 \in \mathbb{F}_q$.*

3   The Dominating Case

In this section, we show how to reduce the cardinality of type (I) in Proposition 1 from $O(q^4)$ to $O(q^3)$, which is the main case of representative of cosets. The following proposition shows that if

$$A_1 = \begin{pmatrix} 1 & b \\ c & 1 \end{pmatrix}, A_2 = \begin{pmatrix} 1 & b' \\ c' & 1 \end{pmatrix}$$

are of type (I) and

$$\frac{b^q - b}{c - c^q} = \frac{b'^q - b'}{c' - c'^q}, \frac{1 - bc^q}{b - c^q} = \frac{1 - b'c'^q}{b' - c'^q},$$

then A_1 and A_2 are in the same coset. Note that the first value is in $\mathbb{F}_q$. Considering parameters of the above special format is inspired by the equations appeared in [19].

Proposition 4. *Fix $v \in \mathbb{F}_q^*$ and $w \in \mathbb{F}_{q^2}$. Suppose that we solve the equations*

$$\begin{cases} \frac{x^q - x}{y - y^q} = v, \\ \frac{1 - xy^q}{y - y^q} = w, \end{cases} \tag{2}$$

under conditions $x, y \in \mathbb{F}_{q^2} \setminus \mathbb{F}_q$ and $xy \neq 1$, and find two pairs of solutions $(b, c), (b', c')$, then A_1 and A_2 are in the same right coset of $PGL_2(\mathbb{F}_q)$ in $PGL_2(\mathbb{F}_{q^2})$, where

$$A_1 = \begin{pmatrix} 1 & b \\ c & 1 \end{pmatrix}, A_2 = \begin{pmatrix} 1 & b' \\ c' & 1 \end{pmatrix}.$$

Proof. The proof consists of two steps. Firstly, we will parametrize the variety corresponding to solutions of $(x, y)'s$ to Eq. (2). Then we will deduce the desired result.

Note that x, y are in $\mathbb{F}_{q^2}$, we have $x^{q^2} = x$ and $y^{q^2} = y$. From Eq. (2), it follows that

$$w^q = \left(\frac{1 - xy^q}{y - y^q}\right)^q = \frac{1 - x^q y}{y^q - y}.$$

So $\frac{w^q}{v} = \frac{1-x^q y}{x-x^q}$ and $y - \frac{w^q}{v} = \frac{xy-1}{x-x^q}$. Thus

$$(y - \frac{w^q}{v})^{q+1} = \frac{(xy-1)(x^q y^q - 1)}{(x^q - x)(x - x^q)} = \frac{(1-xy^q)(1-x^q y)}{(x^q - x)(x - x^q)} - \frac{y-y^q}{x^q - x},$$

which equals $(\frac{w^q}{v})^{q+1} - \frac{1}{v}$. Besides, we have

$$-vy + w + w^q = \frac{y(x-x^q)}{y-y^q} + \frac{1-xy^q}{y-y^q} + \frac{x^q y - 1}{y-y^q} = x.$$

Hence Eq. (2) imply the following

$$\begin{cases} (y - \frac{w^q}{v})^{q+1} = (\frac{w^q}{v})^{q+1} - \frac{1}{v} \in \mathbb{F}_q, \\ x = -vy + w + w^q. \end{cases} \tag{3}$$

Let γ be one of the $(q+1)$-th roots of $(\frac{w^q}{v})^{q+1} - \frac{1}{v}$. Suppose that

$$c = \frac{w^q}{v} + \zeta_1 \gamma, c' = \frac{w^q}{v} + \zeta_2 \gamma,$$

where ζ_1, ζ_2 are two distinct $(q+1)$-th roots of unity, and

$$b = -vc + w + w^q = w - v\zeta_1\gamma,$$

$$b' = -vc' + w + w^q = w - v\zeta_2\gamma.$$

It follows that

$$A_1 = \begin{pmatrix} 1 & w - v\zeta_1\gamma \\ \frac{w^q}{v} + \zeta_1\gamma & 1 \end{pmatrix}, A_2 = \begin{pmatrix} 1 & w - v\zeta_2\gamma \\ \frac{w^q}{v} + \zeta_2\gamma & 1 \end{pmatrix}.$$

Since A_2 is not singular, we deduce

$$A_2^{-1} = \frac{1}{\det(A_2)} \begin{pmatrix} 1 & -w + v\zeta_2\gamma \\ -\frac{w^q}{v} - \zeta_2\gamma & 1 \end{pmatrix}.$$

Thus,

$$A_1 A_2^{-1} = \frac{1}{\det(A_2)} \begin{pmatrix} (v\zeta_1\gamma - w)(\frac{w^q}{v} + \zeta_2\gamma) + 1 & -v(\zeta_1\gamma - \zeta_2\gamma) \\ \zeta_1\gamma - \zeta_2\gamma & (v\zeta_2\gamma - w)(\frac{w^q}{v} + \zeta_1\gamma) + 1 \end{pmatrix}$$

$$= \frac{1}{\det(A_2)} \begin{pmatrix} m_{11} & m_{12} \\ m_{21} & m_{22} \end{pmatrix}.$$

Note that $m_{12} = -vm_{21}, m_{11} - m_{22} = (w^q + w)m_{21}$. They imply that $\frac{m_{12}}{m_{21}} \in \mathbb{F}_q$ and $\frac{m_{11}-m_{22}}{m_{21}} \in \mathbb{F}_q$. It remains to prove $\frac{m_{11}}{m_{21}} \in \mathbb{F}_q$. Let $\delta = \frac{m_{11}}{m_{21}}$. Note that

$$\delta \in \mathbb{F}_q \iff \delta = \delta^q$$
$$\iff m_{11}m_{21}^q = m_{11}^q m_{21}$$
$$\iff m_{11}m_{21}^q \in \mathbb{F}_q.$$

Since $\gamma^{q+1} = (\frac{w^q}{v})^{q+1} - \frac{1}{v} = \frac{w^{q+1}-v}{v^2}$, we have $\frac{w^{q+1}}{v} = v\gamma^{q+1} + 1$. Hence

$$m_{11} = w^q\zeta_1\gamma + v\zeta_1\gamma\zeta_2\gamma - w\zeta_2\gamma - v\gamma^{q+1}.$$

Thus

$$m_{11}m_{21}^q = \gamma^{q+1}\{(w^q + w) - (w\zeta_1^q\zeta_2 + w^q\zeta_1\zeta_2^q) + v(\zeta_2\gamma + \zeta_2^q\gamma^q) - v(\zeta_1\gamma + \zeta_1^q\gamma^q)\}.$$

Since

$$\gamma^{q+1} \in \mathbb{F}_q,$$

$$w^q + w \in \mathbb{F}_q, w\zeta_1^q\zeta_2 + w^q\zeta_1\zeta_2^q \in \mathbb{F}_q,$$

$$\zeta_2\gamma + \zeta_2^q\gamma^q \in \mathbb{F}_q, \zeta_1\gamma + \zeta_1^q\gamma^q \in \mathbb{F}_q,$$

we deduce $m_{11}m_{21}^q \in \mathbb{F}_q$, which implies $\frac{m_{11}}{m_{21}} \in \mathbb{F}_q$ and $\frac{m_{22}}{m_{21}} \in \mathbb{F}_q$. Thus

$$A_1A_2^{-1} = \frac{\zeta_1\gamma - \zeta_2\gamma}{\det(A_2)}\begin{pmatrix} \frac{m_{11}}{m_{21}} & -v \\ 1 & \frac{m_{22}}{m_{21}} \end{pmatrix}$$

$$\in PGL_2(q),$$

which implies that A_1 and A_2 are in the same right coset of $PGL_2(\mathbb{F}_q)$ in $PGL_2(\mathbb{F}_{q^2})$. This completes the proof. $\qquad\square$

Remark 1. Following a similar approach, it can be shown that A_1 and A_2 are also in the same left coset of $PGL_2(\mathbb{F}_q)$ in $PGL_2(\mathbb{F}_{q^2})$.

The map sending x to x^{q+1} is a group endomorphism from $\mathbb{F}_{q^2}^*$ to $\mathbb{F}_q^*$. Observe that $(\frac{w^q}{v})^{q+1} - \frac{1}{v}$ is in $\mathbb{F}_q$. If it is not zero, then

$$(y - \frac{w^q}{v})^{q+1} = (\frac{w^q}{v})^{q+1} - \frac{1}{v} \tag{4}$$

has $q + 1$ distinct solutions in $\mathbb{F}_{q^2}$. Out of these solutions, at most two of them satisfy $(-vy + w + w^q)y = 1$ because the degree on y is two. All the other solutions satisfy $xy \neq 1$.

Lemma 2. *Of all the solutions of Eq. (4), at most two of them are in $\mathbb{F}_q$.*

Proof. The number of solution in $\mathbb{F}_q$ is equal to the degree of $gcd(y^q - y, (y - \frac{w^q}{v})^{q+1} - (\frac{w^q}{v})^{q+1} + \frac{1}{v})$. And

$$(y - \frac{w^q}{v})^{q+1} - (\frac{w^q}{v})^{q+1} + \frac{1}{v}$$

$$= (y^q - \frac{w}{v^q})(y - \frac{w^q}{v}) - (\frac{w^q}{v})^{q+1} + \frac{1}{v}$$

$$\equiv (y - \frac{w}{v^q})(y - \frac{w^q}{v}) - (\frac{w^q}{v})^{q+1} + \frac{1}{v} \pmod{y^q - y}.$$

The last polynomial has degree 2. $\qquad\square$

Algorithm 1. Algorithm of generating right coset representatives of $PGL_2(\mathbb{F}_q)$ in $PGL_2(\mathbb{F}_{q^2})$

Input: A prime power $q \geq 4$ and an element $g \in \mathbb{F}_{q^2} - \mathbb{F}_q$
Output: A set S including all right coset representatives of $PGL_2(\mathbb{F}_q)$ in $PGL_2(\mathbb{F}_{q^2})$.

1: **for** $\alpha \in \mathbb{F}_q$ **do**
2:     $R[\alpha] \leftarrow \emptyset$
3: **end for**
4: **for** $\beta \in \mathbb{F}_{q^2}$ **do**
5:     $\alpha \leftarrow \beta^{q+1}$
6:     **if** the cardinality of $R[\alpha]$ is < 5 **then**
7:         $R[\alpha] \leftarrow R[\alpha] \cup \{\beta\}$
8:     **end if**
9: **end for** $\triangleright$ Now $R[\alpha]$ is a set consisting of at most 5 $(q+1)$-th root of α.
10: $S \leftarrow \emptyset$ $\triangleright$ Initialize S
11: **for** $(v, w) \in \mathbb{F}_q^* \times \mathbb{F}_{q^2}$ **do** $\triangleright$ Adding elements of type (I) in Proposition 1
12:     $\alpha \leftarrow (\frac{w^q}{v})^{q+1} - \frac{1}{v}$
13:     **for** $r \in R[\alpha]$ **do**
14:         $y \leftarrow \frac{w^q}{v} + r$
15:         $x \leftarrow -vy + w + w^q$
16:         **if** $xy \neq 1$ and $x \notin \mathbb{F}_q$ and $y \notin \mathbb{F}_q$ **then**
17:             $S \leftarrow S \cup \{\begin{pmatrix} 1 & x \\ y & 1 \end{pmatrix}\}$
18:             **break**
19:         **end if**
20:     **end for**
21: **end for**
22: **for** $(b_1, d_2) \in \mathbb{F}_q^* \times \mathbb{F}_q^*$ **do** $\triangleright$ Adding elements of type (II) in Proposition 1
23:     **if** $b_1 \neq d_2$ **then**
24:         $S \leftarrow S \cup \{\begin{pmatrix} 1 & b_1 \\ g & d_2 g \end{pmatrix}\}$
25:     **end if**
26: **end for**
27: **for** $d_2 \in \mathbb{F}_q$ **do** $\triangleright$ Adding elements of type (III) in Proposition 1
28:     $S \leftarrow S \cup \{\begin{pmatrix} 0 & 1 \\ g & d_2 g \end{pmatrix}\}$
29: **end for**
30: **for** $(c_2, d_2) \in \mathbb{F}_q \times \mathbb{F}_q$ **do**
31:     $S \leftarrow S \cup \{\begin{pmatrix} 0 & 1 \\ 1 + c_2 g & d_2 g \end{pmatrix}\}$
32: **end for**
33: **for** $c_2 \in \mathbb{F}_q$ **do** $\triangleright$ Adding elements of type (IV) in Proposition 1
34:     $S \leftarrow S \cup \{\begin{pmatrix} 1 & 0 \\ c_2 g & g \end{pmatrix}\}$
35: **end for**
36: **for** $(c_2, d_2) \in \mathbb{F}_q \times \mathbb{F}_q$ **do**
37:     $S \leftarrow S \cup \{\begin{pmatrix} 1 & 0 \\ c_2 g & 1 + d_2 g \end{pmatrix}\}$
38: **end for**
39: **return** S;

We observe that $-vy + w + w^q$ is in $\mathbb{F}_q$ if and only if y is in $\mathbb{F}_q$. Thus we have

Corollary 3. *Suppose that $q \geq 4$, and $(\frac{w^q}{v})^{q+1} - \frac{1}{v} \neq 0$. There must exist one solution of Eq. (3) that satisfy $x, y \in \mathbb{F}_{q^2} \setminus \mathbb{F}_q$ and $xy \neq 1$.*

Remark 2. To list all coset representatives of type (I) in Proposition 1, one can find one pair of $(b, c) \in (\mathbb{F}_{q^2} \setminus \mathbb{F}_q) \times (\mathbb{F}_{q^2} \setminus \mathbb{F}_q)$ for every $(v, w) \in \mathbb{F}_q^* \times \mathbb{F}_{q^2}$ by solving Eq. (3). Assume that $q \geq 4$. In order to solve Eq. (3), one can build a table indexed by elements in $\mathbb{F}_q^*$. In the entry of index $\alpha \in \mathbb{F}_q^*$, we store 5 distinct $(q+1)$-th roots of α in $\mathbb{F}_{q^2}$. The table will be built in advance, in time at most $\tilde{O}(q^2)$. For given $v \in \mathbb{F}_q^*$ and $w \in \mathbb{F}_{q^2}$, one can find $y \in \mathbb{F}_{q^2}$ satisfying Eq. (4) and x as $-vy + w + w^q$ in time $\log^{O(1)} q$ such that $xy \neq 1$ and $x, y \in \mathbb{F}_{q^2} \setminus \mathbb{F}_q$ since there are at most 4 such pairs from the discussion above. Thus, determining the dominating case can be done in time $\tilde{O}(q^3)$.

4   Concluding Remarks

We summarise our algorithm in Algorithm 1. Based on the discussions above, the number of representatives of types (I), (II), (III) and (IV) is no more than $q^3 - q^2, q^2 - 3q + 2, q^2 + q$ and $q^2 + q$ respectively, thus the total number of representatives of all four types (counting repetitions) is no more than $q^3 + 2q^2 - q + 2$. From Remark 2, we can see that the time complexity is $\tilde{O}(q^3)$. Hence Theorem 1 follows.

Acknowledgements. The authors would like to thank anonymous reviewers, Eleazar Leal, Robert Granger and Frederik Vercauteren for helpful comments and discussions.

References

1. Adj, G., Menezes, A., Oliveira, T., Rodríguez-Henríquez, F.: Computing discrete logarithms in $\mathbb{F}_{3^{6*137}}$ and $\mathbb{F}_{3^{6*163}}$ using Magma. In: Arithmetic of Finite Fields - 5th International Workshop, WAIFI 2014, pp. 3–22 (2014)
2. Adj, G., Menezes, A., Oliveira, T., Rodríguez-Henríquez, F.: Weakness of $\mathbb{F}_{3^{6*1429}}$ and $\mathbb{F}_{2^{4*3041}}$ for discrete logarithm cryptography. Finite Fields Appl. **32**, 148–170 (2015)
3. Barbulescu, R., Gaudry, P., Joux, A., Thomé, E.: A heuristic quasi-polynomial algorithm for discrete logarithm in finite fields of small characteristic. In: Nguyen, P.Q., Oswald, E. (eds.) EUROCRYPT 2014. LNCS, vol. 8441, pp. 1–16. Springer, Heidelberg (2014)
4. Cheng, Q., Wan, D., Zhuang, J.: Traps to the BGJT-algorithm for discrete logarithms. LMS J. Comput. Math. **17**, 218–229 (2014). (Special issue for ANTS 2014)
5. Göloglu, F., Granger, R., McGuire, G., Zumbrägel, J.: Discrete logarithms in GF(2^{1971}). NMBRTHRY list, 19 February 2013
6. Göloglu, F., Granger, R., McGuire, G., Zumbrägel, J.: Discrete logarithms in GF(2^{6120}). NMBRTHRY list, 11 April 2013

7. Göloğlu, F., Granger, R., McGuire, G., Zumbrägel, J.: On the function field sieve and the impact of higher splitting probabilities. In: Canetti, R., Garay, J.A. (eds.) CRYPTO 2013, Part II. LNCS, vol. 8043, pp. 109–128. Springer, Heidelberg (2013)
8. Göloğlu, F., Granger, R., McGuire, G., Zumbrägel, J.: Solving a 6120-bit DLP on a desktop computer. In: Lange, T., Lauter, K., Lisoněk, P. (eds.) SAC 2013. LNCS, vol. 8282, pp. 136–152. Springer, Heidelberg (2014)
9. Granger, R., Kleinjung, T., Zumbrägel, J.: Discrete logarithms in $GF(2^{9234})$. NMBRTHRY list, 31 January 2014
10. Granger, R., Kleinjung, T., Zumbrägel, J.: Discrete logarithms in the Jacobian of a genus 2 supersingular curve over $GF(2^{367})$. NMBRTHRY list, 30 January 2014
11. Granger, R., Kleinjung, T., Zumbrägel, J.: Breaking '128-bit secure' supersingular binary curves-(or how to solve discrete logarithms in $\mathbb{F}_{2^{4*1223}}$ and $\mathbb{F}_{2^{12*367}}$). In: Garay, J.A., Gennaro, R. (eds.) CRYPTO 2014, Part II. LNCS, vol. 8617, pp. 126–145. Springer, Heidelberg (2014)
12. Granger, R., Kleinjung, T., Zumbrägel, J.: On the powers of 2. Cryptology ePrint Archive, Report 2014/300 (2014). http://eprint.iacr.org/
13. Granger, R., Kleinjung, T., Zumbrägel, J.: On the discrete logarithm problem in finite fields of fixed characteristic (2015). arXiv preprint arXiv:1507.01495v1
14. Huang, M.-D., Narayanan, A.K.: Finding primitive elements in finite fields of small characteristic. In: Proceedings of the 11th International Conference on Finite Fields and Their Applications, Topics in Finite Fields, AMS Contemporary Mathematics Series (2013)
15. Joux, A.: Discrete logarithms in $GF(2^{1778})$. NMBRTHRY list, 11 February 2013
16. Joux, A.: Discrete logarithms in $GF(2^{4080})$. NMBRTHRY list, 22 March 2013
17. Joux, A.: Discrete logarithms in $GF(2^{6168})$. NMBRTHRY list, 21 May 2013
18. Joux, A.: Faster index calculus for the medium prime case application to 1175-bit and 1425-bit finite fields. In: Johansson, T., Nguyen, P.Q. (eds.) EUROCRYPT 2013. LNCS, vol. 7881, pp. 177–193. Springer, Heidelberg (2013)
19. Joux, A.: A new index calculus algorithm with complexity $L(1/4 + o(1))$ in small characteristic. In: Lange, T., Lauter, K., Lisoněk, P. (eds.) SAC 2013. LNCS, vol. 8282, pp. 355–379. Springer, Heidelberg (2014)
20. Joux, A., Pierrot, C.: Improving the polynomial time precomputation of Frobenius representation discrete logarithm algorithms - simplified setting for small characteristic finite fields. In: Sarkar, P., Iwata, T. (eds.) ASIACRYPT 2014. LNCS, vol. 8873, pp. 378–397. Springer, Heidelberg (2014)
21. Kleinjung, T.: Discrete logarithms in $GF(2^{1279})$. NMBRTHRY list, 17 October 2014

Recovering a Sum of Two Squares Decomposition Revisited

Xiaona Zhang[1,2,3], Li-Ping Wang[1,2(✉)], Jun Xu[1,2], Lei Hu[1,2],
Liqiang Peng[1,2,3], Zhangjie Huang[1,2], and Zeyi Liu[1,2,3]

[1] State Key Laboratory of Information Security,
Institute of Information Engineering, Chinese Academy of Sciences,
Beijing 100093, China
{zhangxiaona,wangliping,xujun,hulei,pengliqiang,
huangzhangjie,liuzeyi}@iie.ac.cn
[2] Data Assurance and Communications Security Research Center,
Chinese Academy of Sciences, Beijing 100093, China
[3] University of Chinese Academy of Sciences, Beijing, China

Abstract. Recently, in [6] Gomez et al. presented algorithms to recover a decomposition of an integer $N = rA^2 + sB^2$, where N, r, s are positive integers, and A, B are the wanted unknowns. Their first algorithm recovers two addends by directly using rigorous Coppersmith's bivariate integer method when the error bounds of given approximations to A and B are less than $N^{\frac{1}{6}}$. Then by combining with the linearization technique, they improved this theoretical bound to $N^{\frac{1}{4}}$. In this paper, we heuristically reach the bound $N^{\frac{1}{4}}$ with experimental supports by transforming the integer polynomial concerned in their first algorithm into a modular one. Then we describe a better heuristic algorithm, the dimension of the lattice involved in this improved method is much smaller under the same error bounds.

Keywords: Sum of squares · Lattice · LLL algorithm · Coppersmith's method

1 Introduction

Coppersmith's method to solve univariate modular polynomial [5] and bivariate integer polynomial [4] enjoys prevalent cryptographic applications, such as breaking the RSA crypto system as well as many of its variant schemes [1,12,14,16,18–20], cracking the validity of the multi-prime Φ-hiding assumptions [9,21], revealing the secret information of kinds of pseudorandom generators [2,6,10], and analyzing the security of some homomorphic encryption schemes [22]. The essence of this famed algorithm is to find integer linear combinations of polynomials which share a common root modulo a certain integer. These derived polynomials possess small coefficients and can be transformed into ones holding true over integers. Thus one can extract the desired roots using standard root-finding algorithms.

© Springer International Publishing Switzerland 2016
D. Lin et al. (Eds.): Inscrypt 2015, LNCS 9589, pp. 178–192, 2016.
DOI: 10.1007/978-3-319-38898-4_11

A noted theorem of Fermat addresses those integers which can be expressed as the sum of two squares. This property relies on the factorization of the integer, from which a sum of two squares decomposition (if exists) can be efficiently computed [8]. Recently, Gutierrez et al. [7] gave an algorithm to recover a decomposition of an integer $N = rA^2 + sB^2$, where r, s are known integers, and A, B are the wanted unknowns. When approximations A_0, B_0 to A, B are given, their first algorithm can recover the two addends under the condition that the approximation errors $|A - A_0|, |B - B_0|$ are no bigger than $N^{\frac{1}{6}}$.

In this paper, we first illustrate a method to solve a certain bivariate modular polynomial $f_N(x, y) = a_1 x^2 + a_2 x + a_3 y^2 + a_4 y + a_0$ based on Coppersmith's method. The trick to solve this kind of polynomial can be directly used to recover the two addends A, B of $N = rA^2 + sB^2$ from their approximations with an error tolerance $N^{\frac{1}{4}}$. The least significant bits exposure attacks on A and B can also be quickly executed by applying the method to solve this certain type polynomial. Next, we present a better method for recovering A, B from its approximations A_0, B_0. This improved approach transforms the problem into seeking the coordinates of a certain vector in our built lattice. The problem of finding these coordinates can be reduced to extracting the small roots of a different bivariate modular polynomial $f'_N(x, y) = b_1 x^2 + b_2 x + b_3 y^2 + b_4 y + b_5 xy + b_0$. The derived error bound is $N^{\frac{1}{3}}$ in this way.

The rest of this paper is organized as follows. In Sect. 2, we recall some preliminaries. In Sect. 3, we first describe the method to solve $f_N(x, y) = a_1 x^2 + a_2 x + a_3 y^2 + a_4 y + a_0$ and then give our deduction on error bound $N^{\frac{1}{4}}$ as well as the least significant bits exposure attacks on A, B, both of which are based on finding the small roots of $f_N(x, y)$. In Sect. 4, we elaborate a better method for recovering the addends of a sum of two squares. The theoretical error bound derived by this approach is $N^{\frac{1}{3}}$. Finally, we give some conclusions in Sect. 5.

2 Preliminaries

2.1 Lattices

Let $\mathbf{b_1}, \ldots, \mathbf{b_\omega}$ be linear independent row vectors in $\mathbb{R}^n$, and a lattice $\mathcal{L}$ spanned by them is

$$\mathcal{L} = \{\sum_{i=1}^{\omega} k_i \mathbf{b_i} \mid k_i \in \mathbb{Z}\},$$

where $\{\mathbf{b_1}, \ldots, \mathbf{b_\omega}\}$ is a basis of $\mathcal{L}$ and $B = [\mathbf{b_1}^T, \ldots, \mathbf{b_\omega}^T]^T$ is the corresponding basis matrix. The dimension and determinant of $\mathcal{L}$ are respectively

$$\dim(\mathcal{L}) = \omega, \det(\mathcal{L}) = \sqrt{\det(BB^T)}.$$

For any two-dimensional lattice $\mathcal{L}$, the Gauss algorithm can find out the reduced basis vectors $\mathbf{v_1}$ and $\mathbf{v_2}$ satisfying

$$\|\mathbf{v_1}\| \leq \|\mathbf{v_2}\| \leq \|\mathbf{v_1} \pm \mathbf{v_2}\|$$

in polynomial time. One can deduce that $\mathbf{v_1}$ is the shortest nonzero vector in $\mathcal{L}$ and $\mathbf{v_2}$ is the shortest vector in $\mathcal{L} \setminus \{k\mathbf{v_1} \mid k \in \mathbb{Z}\}$. Moreover, there are following results, which will be used in Sect. 4.

Lemma 1 (See Gómez et al., 2006 [6], Lemma 3). *Let $\mathbf{v_1}$ and $\mathbf{v_2}$ be the reduced basis vectors of $\mathcal{L}$ by the Gauss algorithm and $\mathbf{x} \in \mathcal{L}$. For the unique pair of integers (α, β) that satisfies $\mathbf{x} = \alpha\mathbf{v_1} + \beta\mathbf{v_2}$, we have*

$$\|\alpha\mathbf{v_1}\| \leq \frac{2}{\sqrt{3}}\|\mathbf{x}\|, \ \|\beta\mathbf{v_2}\| \leq \frac{2}{\sqrt{3}}\|\mathbf{x}\|.$$

Lemma 2 (See Gómez et al., 2006 [6], Lemma 5). *Let $\{\mathbf{u}, \mathbf{v}\}$ be a reduced basis of a 2-rank lattice $\mathcal{L}$ in $\mathbb{R}^r$. Then we have*

$$det(\mathcal{L}) \leq \| \mathbf{u} \|\| \mathbf{v} \| \leq \frac{2}{\sqrt{3}}det(\mathcal{L}).$$

The reduced basis calculation in two-rank lattices is far from being obtained for general lattices. The subsequently proposed reduction definitions all have to make a choice between computational efficiency and good reduction performances. The distinguished LLL algorithm takes a good balance, outputting a basis reduced enough for many applications in polynomial time.

Lemma 3 [17]. *Let $\mathcal{L}$ be a lattice. In polynomial time, the LLL algorithm outputs reduced basis vectors $\mathbf{v_1}, \ldots, \mathbf{v_\omega}$ that satisfy*

$$\|\mathbf{v_1}\| \leq \|\mathbf{v_2}\| \leq \cdots \leq \|\mathbf{v_i}\| \leq 2^{\frac{\omega(\omega-1)}{4(\omega+1-i)}} det(\mathcal{L})^{\frac{1}{\omega+1-i}}, 1 \leq i \leq \omega.$$

2.2 Finding Small Roots

Coppersmith gave rigorous methods for extracting small roots of modular univariate polynomials and bivariate integer polynomials. These methods can be heuristically extended to multivariate cases. Howgrave-Graham's [11] reformulation to Coppersmith's method is widely adopted by researchers for cryptanalysis.

Lemma 4 [11]. *Let $g(x_1, x_2) \in \mathbb{Z}[x_1, x_2]$ be an integer polynomial that consists of at most ω nonzero monomials. Define the norm of $g(x_1, x_2) =: \sum b_{i_1, i_2} x_1^{i_1} x_2^{i_2}$ as the Euclidean norm of its coefficient vector, namely,*

$$\|g(x_1, x_2)\| = \sqrt{\sum b_{i_1, i_2}^2}.$$

Suppose that

1. $g(x_1^{(0)}, x_2^{(0)}) = 0 \pmod{N}$, for $|x_1^{(0)}| < X_1$, $|x_2^{(0)}| < X_2$;
2. $\|g(X_1 x_1, X_2 x_2)\| < \frac{N}{\sqrt{\omega}}$.

Then $g(x_1^{(0)}, x_2^{(0)}) = 0$ holds over integers.

Combining Howgrave-Graham's lemma with the LLL algorithm, one can deduce that if

$$2^{\frac{\omega(\omega-1)}{4(\omega+1-i)}} \det(\mathcal{L})^{\frac{1}{\omega+1-i}} < \frac{N}{\sqrt{\omega}},$$

the polynomials corresponding to the shortest i reduced basis vectors hold over integers. Neglecting the low order terms which are independent on N, the above condition can be simplified as

$$\det(\mathcal{L}) < N^{\omega+1-i}. \tag{1}$$

After obtaining enough equations over integers, one can extract the shared roots by either resultant computation or Gröbner basis technique.

We need the following assumption through our analyses, which is widely adopted in previous works.

Assumption 1. *The Gröbner basis computations for the polynomials corresponding to the first few LLL-reduced basis vectors produce non-zero polynomials.*

3   Recovering the Addends from $N = rA^2 + sB^2$

In this section, we first describe the trick for finding the small roots of polynomial $f_N(x, y) = a_1 x^2 + a_2 y^2 + a_3 x + a_4 y + a_0$. Next, we address the problem of recovering the decomposition of a given number $N = rA^2 + sB^2$ only from its approximations to its addends A, B, where N, r, s are public positive integers. Then, we discuss how to achieve A and B when the least significant bits of them are revealed. Both of these two attacks can be transformed into solving the studied polynomial $f_N(x, y)$.

3.1   Solving Polynomial $f_N(x, y)$

Without loss of generosity, we assume $a_1 = 1$ since we can make it by multiplying f_N with $a_1^{-1} \bmod N$. If this inverse does not exist, one can factorize N. Set

$$f(x, y) = a_1^{-1} f_N(x, y) \bmod N.$$

Next, we find the small roots of $f(x, y)$ by Coppersmith's method. Build shifting polynomials

$$g_{k,i,j}(x, y) = x^i y^j f^k(x, y) N^{m-k},$$

where $i = 0, 1; k = 0, ..., m - i; j = 0, ..., 2(m - k - i)$. Obviously,

$$g_{k,i,j}(x, y) \equiv 0 \bmod N^m.$$

Construct a lattice $\mathcal{L}$ using the coefficient vectors of $g_{k,i,j}(xX, yY)$ as basis vectors. We sort the polynomials $g_{k,i,j}(xX, yY)$ and $g_{k',i',j'}(xX, yY)$ according to the lexicographical order of vectors (k, i, j) and (k', i', j'). In this way, we can

Table 1. Example of the lattice formed by vectors $g_{k,i,j}(xX, yY)$ when $m = 2$. The upper triangular part of this matrix is all zero, so omitted here, and the non-zero items below the diagonal are marked by $*$.

	1	y	y^2	y^3	y^4	x	xy	xy^2	x^2	x^2y	x^2y^2	x^3	x^4
$g_{0,0,0}$	N^2												
$g_{0,0,1}$		YN^2											
$g_{0,0,2}$			Y^2N^2										
$g_{0,0,3}$				Y^3N^2									
$g_{0,0,4}$					Y^4N^2								
$g_{0,1,0}$						XN^2							
$g_{0,1,1}$							XYN^2						
$g_{0,1,2}$								XY^2N^2					
$g_{1,0,0}$	$*$	$*$	$*$			$*$			X^2N				
$g_{1,0,1}$		$*$	$*$	$*$			$*$			X^2YN			
$g_{1,0,2}$			$*$	$*$	$*$			$*$	$*$		X^2Y^2N		
$g_{1,1,0}$						$*$	$*$	$*$	$*$			X^3N	
$g_{2,0,0}$	$*$	$*$	$*$	$*$	$*$	$*$	$*$	$*$	$*$	$*$	$*$	$*$	X^4

ensure that each of our shifting polynomials introduces one and only one new monomial, which gives a lower triangular structure for $\mathcal{L}$. We give an example for $m = 2$ in the following Table 1.

Then its determinant can be easily calculated as products of the entries on the diagonal as $det(\mathcal{L}) = X^{S_X} Y^{S_Y} N^{S_N}$ as well as its dimension ω where

$$\omega = \sum_{i=0}^{1} \sum_{k=0}^{m-i} \sum_{j=0}^{2(m-k-i)} 1 = 2m^2 + 2m + 1 = 2m^2 + o(m^2).$$

$$S_x = \sum_{i=0}^{1} \sum_{k=0}^{m-i} \sum_{j=0}^{2(m-k-i)} (2k + i) = \frac{1}{3}m(4m^2 + 3m + 2) = \frac{4}{3}m^3 + o(m^3).$$

$$S_y = \sum_{i=0}^{1} \sum_{k=0}^{m-i} \sum_{j=0}^{2(m-k-i)} j = \frac{1}{3}m(4m^2 + 3m + 2) = \frac{4}{3}m^3 + o(m^3).$$

$$S_N = \sum_{i=0}^{1} \sum_{k=0}^{m-i} \sum_{j=0}^{2(m-k-i)} (m - k) = \frac{2}{3}m(2m^2 + 3m + 1) = \frac{4}{3}m^3 + o(m^3).$$

Put these relevant values into inequality $det(\mathcal{L}) < N^{m\omega}$. After some basic calculations, we gain the bound

$$XY < N^{\frac{1}{2}}.$$

When $X = Y$, which means the two unknowns are balanced, the above result is

$$X = Y < N^{\frac{1}{4}}.$$

We summarize our result in the following theorem.

Theorem 1. *Let N be a sufficiently large composite integer of unknown factorization. Given a bivariate polynomial $f_N(x, y) = a_1x^2 + a_2x + a_3y^2 + a_4y + a_0 \mod N$, where $|x| \leq X$, $|y| \leq Y$. Under Assumption 1, if*

$$XY < N^{\frac{1}{2}},$$

one can extract all the solutions (x, y) of equation $f_N(x, y) \equiv 0 \pmod{N}$ in polynomial time.

3.2   Recovering a Decomposition from Approximations

In this subsection, we describe the method to recover A, B of $N = rA^2 + sB^2$ from their approximations.

Supposing that positive integers r and s are given. Set $N = rA^2 + sB^2$, where A, B are balanced addends, and A_0, B_0 are the approximations to A, B, that is $A = A_0 + x$ and $B = B_0 + y$, where x, y are bounded by Δ. Then, one can recover A and B according to Theorem 1 when

$$\Delta < N^{\frac{1}{4}}.$$

The concrete analysis is as follows. Note that

$$N = r(A_0 + x)^2 + s(B_0 + y)^2, \tag{2}$$

which gives rise to a bivariate modular polynomial

$$f_1(x, y) = rx^2 + sy^2 + 2A_0rx + 2B_0sy + rA_0^2 + sB_0^2 \equiv 0 \mod N,$$

this is exactly the same type of the polynomial we discussed in Sect. 3.1. So we gain the result $\Delta < N^{\frac{1}{4}}$ simply by substituting both X and Y appeared in Theorem 1 to Δ.

The experimental results to support the above analysis is displayed in Table 2, which matches well with the derived theoretical bound.

Table 2. Experimental results for error bound $\Delta = \frac{1}{4}$ with 512 bit N

N (bits)	m	dim	$log_N \Delta$	LLL (seconds)	Gröbner (seconds)
512	5	61	0.227	12.901	15.631
	6	85	0.230	49.172	606.360
	7	113	0.233	187.076	517.549
	8	145	0.235	566.471	3204.339
	9	181	0.236	1512.586	5538.002
	10	221	0.237	3430.463	out of memory

Table 3. Experimental results for Remark 1 with 512 bit N

N (bits)	m	dim	$log_N \Delta$	LLL (seconds)	Gröbner (seconds)
512	4	28	0.130	0.842	0.265
	5	36	0.132	3.806	0.842
	6	45	0.133	14.914	1.420
	8	66	0.135	143.349	11.532

Table 4. Experimental results for different modulus with 1024 bit N

N (bits)	M	m	dim	$log_N \Delta$	LLL (seconds)	Gröbner (seconds)
1024	$N-1$	6	85	0.23	582.258	144.005
	$2N-1$	6	85	0.23	587.046	145.440
	N^2-1	6	85	0.23	5917.165	1159.431

Remark 1. Gutierrez et al. discussed the same problem in [7]. They arranged Eq. (2) to a bivariate integer polynomial as follows,

$$f_1'(x,y) = rx^2 + sy^2 + 2A_0 rx + 2B_0 sy + rA_0^2 + sB_0^2 - N. \tag{3}$$

By directly applying Coppersmith's theorem [3], their derived error bound is only $N^{1/6}$. We do experiments for their method, part of the results are displayed in Table 3. The experimental results show that our method works much better.

Coppersmith's original method [3] for solving bivariate integer polynomial is difficult to understand. Coron [13] first reformulated Coppersmith's work and the key idea of which can be described as follows, choosing a proper integer R, and transforming the situation into finding a small root modulo R. Then, by applying LLL algorithm, a polynomial with small coefficients can be found out, which is proved to be algebraically independent with the original equation.

Our approach described above also transforms the integer equation into a modular polynomial. The difference between our method and Coppersmith's theorem [3] lies in the construction of shifting polynomials. We take use of the information of the power of the original polynomial. Although we didn't prove that the obtained polynomial with small coefficients is algebraically independent with the original polynomial, which is true in most cases during the experiments.

Remark 2. We studied different situations to transform Eq. (3) into modular ones as the modulus varies. For instance $q(x,y) = f_1(x,y) + M \equiv 0 \ mod \ (N+M)$. The experimental results for different M are shown in Table 4.

Specifically, we also consider non-constant modular polynomial

$$f_2(x,y) = rx^2 + sy^2 + 2A_0 rx + 2B_0 sy \equiv 0 \ mod \ (N - rA_0^2 - sB_0^2). \tag{4}$$

In this way, the corresponding theoretical error bound for recovering the addends from their approximations is $N^{1/6}$(please refer to Appendix A for the detailed analyses). However, the experimental results show a much better performance, which is displayed in Table 5.

3.3   Recovering a Decomposition from Non-approximations

Actually, the most significant bits exposure attack of A and B can be viewed as a special case of the above problem (recovering a a sum of two squares from its approximations). In this subsection, we consider the case when the least significant bits of A, B are leaked.

Given r, s are positive integers, set $N = rA^2 + sB^2$, where A, B are balanced addends. When half bits of A and B in the LSBs are intercepted, one can recover A, B according to Theorem 1.

Suppose $A = xM + A_0$, $B = yM + B_0$, where M, A_0 and B_0 are the gained integers, and x, y refers to the unknown parts. Then we have the following relation

$$N = r(xM + A_0)^2 + s(yM + B_0)^2,$$

which can be expanded to a bivariate modular polynomial

$$f_3(x, y) = rM^2x^2 + sM^2y^2 + 2rA_0Mx + 2sB_0My + rA_0^2 + sB_0^2 \equiv 0 \ mod \ N.$$

Set the upper bound for x and y as Δ_1 and put it into Theorem 1, we get $\Delta_1 < N^{\frac{1}{4}}$. Since

$$M = \frac{A - A_0}{x} > \frac{A - A_0}{N^{\frac{1}{4}}} \approx \frac{A}{N^{\frac{1}{4}}} \approx \frac{N^{\frac{1}{2}}}{N^{\frac{1}{4}}} = N^{\frac{1}{4}},$$

From these analyses, we get that half information from A and B can reveal the whole knowledge of both addends, no matter the leaked bits are LSBs or MSBs.

Table 5. Experimental results for Remark 2 with 512 bit N

N (bits)	m	dim	$log_N\Delta$	LLL (seconds)	Gröbner (seconds)
512	2	12	0.16	0.001	0.001
	3	24	0.19	0.016	0.14
	4	40	0.20	0.406	1.888
	5	60	0.21	2.558	45.490
	7	112	0.22	57.954	2028.294

4   A Better Method for Recovering the Addends

In this section, we reduce the problem of recovering a sum of two squares decomposition to seeking the coordinates of a desired vector in a certain lattice. Then we can find these coordinates by applying Coppersmith's method to solve a type of modular polynomials where the concerned monomials are x^2, y^2, xy, x, y and 1. Dealt this way, the theoretical error tolerance can be improved to $N^{1/3}$, and the involved lattices in this approach possess much smaller dimensions compared to the ones in Sect. 3.

4.1   The Reduction of Recovering the Addends

From the initial key relation $N = r(A_0 + x)^2 + s(B_0 + y)^2$ we have

$$2rA_0x + 2sB_0y + rx^2 + sy^2 = N - rA_0^2 - sB_0^2. \tag{5}$$

Hence, the recovery of vector

$$\mathbf{e} := (X_1, X_2, X_3) = ((r + s)\Delta x, (r + s)\Delta y, rx^2 + sy^2)$$

solves the problem. Here Δ represents the upper bound for x and y. It is not hard to see that vector $\mathbf{e}$ is in a shifted lattice $\mathbf{c} + S$, $\mathbf{c} = (c_1, c_2, c_3) \in \mathbb{Z}^3$, where $(\frac{c_1}{(r+s)\Delta}, \frac{c_2}{(r+s)\Delta}, c_3)$ is a particular solution of (5) and S is a two-dimensional lattice

$$\begin{pmatrix} (r + s)\Delta & 0 & -2A_0r \\ 0 & (r + s)\Delta & -2B_0s \end{pmatrix}.$$

According to Minkowski's theorem [15], when $||\mathbf{e}|| < \sqrt{2}\sqrt{det(S)}$, one can recover $\mathbf{e}$ by solving the closet vector problem. Further, the norm of $\mathbf{e}$ satisfies $||\mathbf{e}|| \leq \sqrt{3}(r + s)\Delta^2$, and $det(S) \geq 2(r + s)\Delta\sqrt{\frac{min(r,s)*N}{2}}$ with condition $min(r, s) * N \geq 4\sqrt{N}\Delta(r^{3/2} + s^{3/2})$. These constraints give rise to the error bound $\Delta < N^{1/6}$, as discussed in [7].

Next, we present our analysis for the case when $\Delta > N^{1/6}$. Here, we tag $\mathbf{f} = ((r + s)\Delta f_1, (r + s)\Delta f_2, f_3)$ as the output of the CVP algorithm on S, and use $\{\mathbf{u} = ((r+s)\Delta u_1, (r+s)\Delta u_2, u_3), \mathbf{v} = ((r+s)\Delta v_1, (r+s)\Delta v_2, v_3)\}$ to denote the Gauss reduced basis for S. Then $\mathbf{e} = \mathbf{f} + \alpha\mathbf{u} + \beta\mathbf{v}$, where α, β represent the corresponding coordinates of vector $\mathbf{e} - \mathbf{f}$ in lattice S. Thus, the problem is converted to finding the parameters α and β, which satisfy equation

$$\begin{aligned} 2A_0r(f_1 + \alpha u_1 + \beta v_1) + 2B_0s(f_2 + \alpha u_2 + \beta v_2) \\ + r(f_1 + \alpha u_1 + \beta v_1)^2 + s(f_2 + \alpha u_2 + \beta v_2)^2 + rA_0^2 + sB_0^2 - N = 0. \end{aligned} \tag{6}$$

We first derive the upper bounds for the unknowns α, β. Since $\mathbf{e} - \mathbf{f} = \alpha\mathbf{u} + \beta\mathbf{v}$, from Lemma 1, we get

$$||\alpha\mathbf{u}||||\beta\mathbf{v}|| \leq \frac{2}{\sqrt{3}}||\mathbf{e} - \mathbf{f}|| \leq 4(r + s)\Delta^2.$$

Thus, $|\alpha| \leq \frac{4(r+s)\Delta^2}{||\mathbf{u}||}$, $|\beta| \leq \frac{4(r+s)\Delta^2}{||\mathbf{v}||}$. Further, according to Lemma 2, there is $det(\mathcal{S}) \leq ||\mathbf{u}||\,||\mathbf{v}|| \leq \frac{2}{\sqrt{3}}det(\mathcal{S})$. Then we have

$$|\alpha||\beta| \leq \frac{4(r+s)\Delta^2}{det(\mathcal{S})} \leq c_1\Delta^{3/2}N^{-1/4},$$

where $c_1 = 2^{7/4}(r+s)^{1/2}min(r,s)^{-1/4}$ is a constant.

Notice that Eq. (6) can be arranged to

$$
\begin{aligned}
(ru_1^2 + su_2^2)\alpha^2 &+ (rv_1^2 + sv_2^2)\beta^2 + 2(ru_1v_1 + su_2v_2)\alpha\beta + 2(A_0ru_1 \\
&+ B_0su_2 + rf_1u_1 + sf_2u_2)\alpha + 2(A_0rv_1 + B_0sv_2 + rf_1v_1 + sf_2v_2)\beta \quad (7) \\
&+ 2A_0rf_1 + 2B_0sf_2 + rf_1^2 + sf_2^2 + rA_0^2 + sB_0^2 \equiv 0 \bmod N,
\end{aligned}
$$

which represents a certain type of modular polynomials consisting of monomials x^2, y^2, xy, x, y and 1. Next, we describe our analysis for solving such polynomials.

4.2 Solving a Certain Type of Modular Polynomials

Let $f'_N(x,y) = b_1x^2 + b_2y^2 + b_3xy + b_4x + b_5y + b_0 \bmod N$. Assume $b_1 = 1$, otherwise, set

$$f'(x,y) = b_1^{-1}f'_N(x,y) \bmod N.$$

If the inverse $b_1^{-1} \bmod N$ does not exist, one can factorize N. Next, we use Coppersmith's method to find the small roots of this polynomial. Build shifting polynomials $h_{k,i,j}(x,y)$ which possess the same roots modular N^m with $f'(x,y) \equiv 0 \bmod N$ as follows:

$$h_{k,i,j}(x,y) = x^iy^j f'^k(x,y)N^{m-k},$$

where $i = 0,1; k = 0,...,m-i; j = 0,...,2(m-k)-i$.

Construct a lattice $\mathcal{L}'$ using the coefficient vectors of $h_{k,i,j}(xX, yY)$ as basis vectors. We sort the polynomials $h_{k,i,j}(xX, yY)$ and $h_{k',i',j'}(xX, yY)$ according to lexicographical order of vectors (k, i, j) and (k', i', j'). Therefore, we can ensure that each of our shifting polynomials introduces one and only one new monomial, which gives a triangular structure for $\mathcal{L}'$.

Then the determinant of $\mathcal{L}'$ can be easily calculated as products of the entries on the diagonal as $det(\mathcal{L}') = X^{S_X}Y^{S_Y}N^{S_N}$ as well as its dimension ω where

$$\omega = \sum_{k=0}^{m-i}\sum_{i=0}^{1}\sum_{j=0}^{2(m-k)-i} 1 = 2m^2 + o(m^2),$$

$$S_X = \sum_{k=0}^{m-i}\sum_{i=0}^{1}\sum_{j=0}^{2(m-k)-i} (2k+i) = \frac{4}{3}m^3 + o(m^3),$$

$$S_Y = \sum_{k=0}^{m-i}\sum_{i=0}^{1}\sum_{j=0}^{2(m-k)-i} j = \frac{4}{3}m^3 + o(m^3),$$

$$S_N = \sum_{k=0}^{m-i}\sum_{i=0}^{1}\sum_{j=0}^{2(m-k)-i} (m-k) = \frac{4}{3}m^3 + o(m^3).$$

Put these relevant values into inequality $\det(\mathcal{L}') < N^{m\omega}$. After some basic calculations, we gain the bound

$$XY < N^{\frac{1}{2}}.$$

We summarize our result in the following theorem.

Theorem 2. *Let N be a sufficiently large composite integer of unknown factorization and $f'_N(x,y) = b_1 x^2 + b_2 x + b_3 y^2 + b_4 y + b_5 xy + b_0 \mod N$ be a bivariate modular polynomial, where $|x| \leq X$, $|y| \leq Y$. Under Assumption 1, if*

$$XY < N^{\frac{1}{2}},$$

one can extract all the solutions (x,y) of equation $f'_N(x,y) \equiv 0 \pmod{N}$ in polynomial time.

Next, we use the above method to solve Eq. (7), and then recover the unknown addends.

4.3   Recover the Addends

Notice that Eq. (7) is exactly the same type of polynomial discussed in Sect. 4.2. Put the derived upper bounds for $|\alpha||\beta|$ in Sect. 4.1 into Theorem 2,

$$|\alpha||\beta| \leq c_1 \Delta^{3/2} N^{-1/4} \leq N^{1/2}.$$

Solve this inequality, omit the constant terms, and we obtain the optimized bound for the approximation error terms

$$\Delta < N^{\frac{1}{3}}. \tag{8}$$

Compared to Sect. 3, this method performs much better in practice since the dimensions of the involved lattices are much smaller when the error bounds are the same. We present the comparison results in Table 6, where one can see a remarkable improvement in the performing efficiency.

Remark 3. As in Sect. 3, we also analyzed the case when transforming Eq. (6) into a non-constant modular polynomial. The corresponding error bound is then $N^{1/4}$. Table 7 is the experimental results for this situation. Please refer to Appendix B for the detailed analysis.

5   Conclusions and Discussions

We revisit the problem of recovering the two addends in this paper. Our first algorithm improves Gutierrez et al.'s first result $N^{1/6}$ to $N^{1/4}$ by transforming the derived polynomial into a modular one. Then we improve this bound to $N^{1/3}$ in theory by reducing the problem of recovering a sum of two squares decomposition to seeking the coordinates of a desired vector in a certain lattice.

Table 6. A comparison between Sect. 4 (the left part datas) and Sect. 3 (the right part datas)

N (bits)	$log_N \Delta$	m	dim	LLL (seconds)	Gröbner (seconds)	m'	dim'	LLL' (seconds)	Gröbner' (seconds)
1024	0.19	1	6	0.016	0.001	2	13	0.047	0.031
	0.20	2	15	0.187	0.109	3	25	1.248	0.406
	0.21	2	15	0.172	0.109	3	25	1.030	0.967
	0.22	2	15	0.187	0.140	4	41	14.383	3.416
512	0.23	4	45	6.334	11.591	6	85	49.172	606.360
	0.235	5	66	47.612	68.391	8	145	566.471	3204.339
	0.236	6	91	229.789	579.091	9	181	1512.586	5538.002
	0.237	7	120	949.094	3410.151	10	221	3430.463	out of memory
	0.238	7	120	855.868	1696.823	–	–	–	-
	0.239	8	153	2852.619	out of memory	–	–	–	-

Table 7. Experimental results for Remark 3 with 512 bit N

N (bits)	m	dim	$log_N \Delta$	LLL (seconds)	Gröbner (seconds)
512	2	14	0.21	0.031	0.016
	3	27	0.22	0.328	0.187
	6	90	0.23	180.930	188.434

J.Gutierrez et al. did similarly in [7], and their optimized bound is $N^{1/4}$. Our second approach performs much better than the first one since the dimension of the required lattice is much smaller when the same error bounds are considered. The tricks to solve the derived polynomials in Sects. 3 and 4 are similar, both of which transform integer relations to modular polynomials. We study four kinds of modular polynomials in our work (two types are discussed in Remarks 2 and 3). The tricks for solving these polynomials may find other applications in cryptanalysis.

We do experiments to testify the deduced results. The tests are done in Magma on a PC with Intel(R) Core(TM) Quad CPU (3.20 GHz, 4.00 GB RAM, Windows 7). These datas well support our analyses, however, as the error terms go larger, the dimensions of the required lattices are huger. The time, memory costs also increase greatly, which stops our experiment at a not good enough point. Hope people who are interested in this problem can bring us further supports for the experiments.

Acknowledgements. The authors would like to thank anonymous reviewers for their helpful comments and suggestions. The work of this paper was partially supported by National Natural Science Foundation of China (No. 61170289) and the National Key Basic Research Program of China (2013CB834203).

A   Analysis for Remark 2

In this part, we give the details to show that when dealing with Eq. (3) as a non-constant modular polynomial (4), the corresponding error bound is $N^{1/6}$.

First, we display the trick for finding the small roots of $f_2(x,y) = rx^2 + sy^2 + 2A_0rx + 2B_0sy \equiv 0 \bmod (N - rA_0^2 - sB_0^2)$. Set $M = N - rA_0^2 - sB_0^2$ as the modulus. The shifting polynomials for this equation can be constructed as

$$
\begin{cases}
g_{k,i}^1(x,y) = y^i M^m, \\
i = 1, ..., 2m; \\
g_{k,i}^2(x,y) = x^j y^i f_3^k(x,y) M^{m-k}, \\
k = 0, ..., m-1; j = 1, 2; i = 0, ..., 2(m-k-1);
\end{cases}
$$

Suppose $|x| \leq X = N^\delta, |y| \leq Y = N^\delta$, then $M \approx N^{\frac{1}{2}+\delta}$. Similarly, the coefficients of $g^1(xX, yY), g^2(xX, yY)$ can be arranged as a lower triangular lattice $\mathcal{L}_1$, whose determinant can be easily calculated as $det(\mathcal{L}_1) = X^{S_X} Y^{S_Y} M^{S_M}$, where

$$\omega = 2m^2 + 2m = 2m^2 + o(m^2).$$

$$S_X = \frac{1}{3}m(4m^2 + 3m + 2) = \frac{4}{3}m^3 + o(m^3).$$

$$S_Y = \frac{1}{3}m(4m^2 + 3m + 2) = \frac{4}{3}m^3 + o(m^3).$$

$$S_M = \frac{1}{3}m(4m^2 + 9m - 1) = \frac{4}{3}m^3 + o(m^3).$$

Put these values into inequality $det(\mathcal{L}_1) \leq M^{m\omega}$, we obtain $\delta \leq \frac{1}{6}$, which means that the error bound derived by this method is

$$\Delta \leq N^{\frac{1}{6}},$$

a poorer bound compared to $N^{\frac{1}{4}}$. The experimental results in Table 5 show that this method works much better in practice than in theoretic analysis, although still weaker than the result in Sect. 3.2.

B   Analysis for Remark 3

Notice that the problem of finding coordinates for vector $\mathbf{e} - \mathbf{f}$ can also be transformed into solving a non-constant modular equation

$$
\begin{aligned}
q(\alpha, \beta) = {} & (ru_1^2 + su_2^2)\alpha^2 + (rv_1^2 + sv_2^2)\beta^2 + 2(ru_1v_1 + su_2v_2)\alpha\beta \\
& + (2rf_1u_1 + 2sf_2u_2 - u_3)\alpha + (2rf_1v_1 + 2rf_2v_2 - v_3)\beta \\
& \equiv 0 \bmod (N - 2rA_0f_1 - 2sB_0f_2 - rf_1^2 - sf_2^2 - rA_0^2 - sB_0^2)
\end{aligned}
$$

Set $M = |N - 2rA_0f_1 - 2sB_0f_2 - rf_1^2 - sf_2^2 - rA_0^2 - sB_0^2|$ as the modulus. Then the problem reduced to solving

$$q'(x,y) = x^2 + b_2y^2 + b_3xy + b_4x + b_5y \equiv 0 \bmod M.$$

Here we assume that $q'(x, y)$ is a monic irreducible polynomial, since we can make it satisfied by multiplying the modular inverse term. We apply Coppersmith's method to solve this polynomial. The shifting polynomials can be constructed as

$$\begin{cases} g_{k,i}^1(x,y) = y^i M^m, \\ i = 1, ..., 2m; \\ g_{k,i}^2(x,y) = y^i q'^k(x,y) M^{m-k}, \\ k = 1, ..., m, i = 0, ..., 2(m-k); \\ g_{k,i}^3(x,y) = xy^i q'^k(x,y) M^{m-k}, \\ k = 0, ..., m-1, i = 0, ..., 2(m-k) - 1; \end{cases}$$

From the former analysis, we know that $|x|, |y| \le \Delta^{3/2}N^{-1/4} = X = Y$, and $M \approx \Delta^2$. Similarly, the coefficients of $g^1(xX, yY), g^2(xX, yY)$ and $g^3(xX, yY)$ can be arranged as a lower triangular lattice $\mathcal{L}_2$, whose determinant can be easily calculated as $det(\mathcal{L}_2) = X^{S_X} Y^{S_Y} M^{S_M}$, where

$$\omega = 2m^2 + 3m = 2m^2 + o(m^2).$$
$$S_X = \frac{2}{3}m(2m^2 + 3m + 1) = \frac{4}{3}m^3 + o(m^3).$$
$$S_Y = \frac{2}{3}m(2m^2 + 3m + 1) = \frac{4}{3}m^3 + o(m^3).$$
$$S_M = \frac{1}{6}m(8m^2 + 15m + 1) = \frac{4}{3}m^3 + o(m^3).$$

Put these values into inequality $det(\mathcal{L}_2) \le M^{m\omega}$, we gain the corresponding error bound

$$\Delta \le N^{\frac{1}{4}}.$$

References

1. Aono, Y.: A new lattice construction for partial key exposure attack for RSA. In: Jarecki, S., Tsudik, G. (eds.) PKC 2009. LNCS, vol. 5443, pp. 34–53. Springer, Heidelberg (2009)
2. Bauer, A., Vergnaud, D., Zapalowicz, J.-C.: Inferring sequences produced by nonlinear pseudorandom number generators using Coppersmith's methods. In: Fischlin, M., Buchmann, J., Manulis, M. (eds.) PKC 2012. LNCS, vol. 7293, pp. 609–626. Springer, Heidelberg (2012)
3. Coppersmith, D.: Small solutions to polynomial equations, and low exponent RSA vulnerabilities. J. Cryptol. **10**(4), 233–260 (1997)

4. Coppersmith, D.: Finding a small root of a bivariate integer equation; factoring with high bits known. In: Maurer, U.M. (ed.) EUROCRYPT 1996. LNCS, vol. 1070, pp. 178–189. Springer, Heidelberg (1996)
5. Coppersmith, D.: Finding a small root of a univariate modular equation. In: Maurer, U.M. (ed.) EUROCRYPT 1996. LNCS, vol. 1070, pp. 155–165. Springer, Heidelberg (1996)
6. Gomez, D., Gutierrez, J., Ibeas, A.: Attacking the pollard generator. IEEE Trans. Inf. Theor. $52(12)$, 5518–5523 (2006)
7. Gutierrez, J., Ibeas, Á., Joux, A.: Recovering a sum of two squares decomposition. J. Symb. Comput. 64, 16–21 (2014)
8. Hardy, K., Muskat, J.B., Williams, K.S.: A deterministic algorithm for solving $n = fu^2 + gv^2$ in coprime integers u and v. J. Math. Comput. 55, 327–343 (1990)
9. Herrmann, M.: Improved cryptanalysis of the multi-prime ϕ - hiding assumption. In: Nitaj, A., Pointcheval, D. (eds.) AFRICACRYPT 2011. LNCS, vol. 6737, pp. 92–99. Springer, Heidelberg (2011)
10. Herrmann, M., May, A.: Attacking power generators using unravelled linearization: when do we output too much? In: Matsui, M. (ed.) ASIACRYPT 2009. LNCS, vol. 5912, pp. 487–504. Springer, Heidelberg (2009)
11. Howgrave-Graham, N.: Finding small roots of univariate modular equations revisited. In: Darnell, M. (ed.) Crytography and Coding. LNCS, vol. 1355, pp. 131–142. Springer, Heidelberg (1997)
12. Jochemsz, E., May, A.: A strategy for finding roots of multivariate polynomials with new applications in attacking RSA variants. In: Lai, X., Chen, K. (eds.) ASIACRYPT 2006. LNCS, vol. 4284, pp. 267–282. Springer, Heidelberg (2006)
13. Coron, J.-S.: Finding small roots of bivariate integer polynomial equations revisited. In: Cachin, C., Camenisch, J.L. (eds.) EUROCRYPT 2004. LNCS, vol. 3027, pp. 492–505. Springer, Heidelberg (2004)
14. Kakvi, S.A., Kiltz, E., May, A.: Certifying RSA. In: Wang, X., Sako, K. (eds.) ASIACRYPT 2012. LNCS, vol. 7658, pp. 404–414. Springer, Heidelberg (2012)
15. Kannan, R.: Minkowski's convex body theorem and integer programming. Math. Oper. Res. $12(3)$, 415–440 (1987)
16. Kiltz, E., O'Neill, A., Smith, A.: Instantiability of RSA-OAEP under chosen-plaintext attack. In: Rabin, T. (ed.) CRYPTO 2010. LNCS, vol. 6223, pp. 295–313. Springer, Heidelberg (2010)
17. Lenstra, A.K., Lenstra, H.W., Lovász, L.: Factoring polynomials with rational coefficients. Math. Ann. $261(4)$, 515–534 (1982)
18. May, A.: Using LLL-reduction for solving RSA and factorization problems. In: Nguyen, P.Q., Vallée, B. (eds.) The LLL Algorithm: Survey and Applications. ISC, pp. 315–348. Springer, Heidelberg (2010)
19. Sarkar, S.: Reduction in lossiness of RSA trapdoor permutation. In: Bogdanov, A., Sanadhya, S. (eds.) SPACE 2012. LNCS, vol. 7644, pp. 144–152. Springer, Heidelberg (2012)
20. Sarkar, S., Maitra, S.: Cryptanalysis of RSA with two decryption exponents. Inf. Process. Lett. 110, 178–181 (2010)
21. Tosu, K., Kunihiro, N.: Optimal bounds for multi-prime ϕ-hiding assumption. In: Mu, Y., Seberry, J., Susilo, W. (eds.) ACISP 2012. LNCS, vol. 7372, pp. 1–14. Springer, Heidelberg (2012)
22. van Dijk, M., Gentry, C., Halevi, S., Vaikuntanathan, V.: Fully homomorphic encryption over the integers. In: Gilbert, H. (ed.) EUROCRYPT 2010. LNCS, vol. 6110, pp. 24–43. Springer, Heidelberg (2010)

Improved Tripling on Elliptic Curves

Weixuan Li[1,2,3], Wei Yu[1,2(✉)], and Kunpeng Wang[1,2]

[1] State Key Laboratory of Information Security,
Institute of Information Engineering, Chinese Academy of Sciences, Beijing, China
yuwei_1_yw@163.com
[2] Data Assurance and Communication Security Research Center,
Chinese Academy of Sciences, Beijing, China
[3] University of Chinese Academy of Sciences, Beijing, China

Abstract. We propose efficient strategies for calculating point tripling on Hessian ($8M+5S$), Jacobi-intersection ($7M+5S$), Edwards ($8M+5S$) and Huff ($10M+5S$) curves, together with a fast quintupling formula on Edwards curves. M is the cost of a field multiplication and S is the cost of a field squaring. To get the best speeds for single-scalar multiplication without regarding perstored points, computational cost between different double-base representation algorithms with various forms of curves is analyzed. Generally speaking, tree-based approach achieves best timings on inverted Edwards curves; yet under exceptional environment, near optimal controlled approach also worths being considered.

Keywords: Elliptic curves · Scalar multiplication · Point arithmetic · Double-base number system

1 Introduction

Compared with finite fields $\mathbb{F}_q$, solving the elliptic curve discrete logarithm problem (ECDLP) in $E(\mathbb{F}_q)$ is much harder. For example, index calculus is a subexponential algorithm that solves DLP for the multiplicative group of a finite field $\mathbb{F}_q^*$, yet the best known countermeasures against ECDLP take exponential time. It means that when security level is equivalent, elliptic curve cryptosystem (ECC) has key and message sizes that are at least $5 - 10$ times smaller than those for other public-key cryptosystems, including RSA and $\mathbb{F}_q$-based DLP systems. This superiority promotes the implementation of ECC in resource limited equipments, like smart cards and cellular phones.

Practical efficiency of curve-based cryptographic system is significantly influenced by the speed of fundamental operation: scalar multiplication $[k]P$ of an integer $k \in \mathbb{Z}$ by a generic elliptic point P. A wide range of advances has been established to improve the efficiency of scalar multiplication.

This work is supported in part by National Research Foundation of China under Grant No. 61502487, 61272040, and in part by National Basic Research Program of China (973) under Grant No. 2013CB338001.

D. Lin et al. (Eds.): Inscrypt 2015, LNCS 9589, pp. 193–205, 2016.
DOI: 10.1007/978-3-319-38898-4_12

On the one hand, since the introduction of double-base chain (DBC) by Dimitrov, Imbert and Mishra [1], new and optimized scalar-recoding algorithms have attracted considerable attention to speed up single-scalar multiplication. See [2–6] for extensive progress on the subject. In spite of algorithmic properties, implementation complexity of double-base ($\{2,3\}$ or $\{2,5\}$ as usually used) number system relies on the cost of basic operations on elliptic groups—addition or mixed-addition, doubling, tripling, quintupling—at the same time. For instance, by signed DBC with $\{2,3\}$-base, any integer can be rewritten as a sequence of bits in $\{0^{(2)}, 0^{(3)}, \pm1\}$ in scalar evaluation phase, meanwhile the number of $0^{(2)}$, $0^{(3)}$ and ±1 is exactly the amount of doublings, triplings, additions in point multiplication phase respectively.

On the other hand, new elliptic curve forms with *unified* addition[1] formula were successively investigated in literature, aiming to resist side-channel attacks as well. From a security standpoint, unified addition formulas of Jacobi intersection [7], Hessian [8,9], Jacobi quartic [10], Edwards [11] and Huff [12] can execute doubling operations the same way as additions in insecure environments. It leaks no side channel information on scalars, and provides a simplified protection against *simple power analysis* (SPA). From an efficiency standpoint, arithmetic on various curves establish new speed records for single-scalar multiplication. In particular, twisted curves with different coordinate systems also draw some interest, *cf.* [8,13–15].

This paper is on optimizing point operations of several previously mentioned elliptic curves, on which we study the performance of different DBC algorithms. We introduce background knowledge of DBC in Sect. 2. Then we show faster tripling formulas for Hessian, Jacobi-intersection, Edwards and Huff curves in Sect. 3. Remarkably, our new Jacobi-intersection tripling formula is competitive with that of tripling-oriented Doche-Icart-Kohel curves [16], which is the fastest one at present. Cost of point tripling on various elliptic curves will be shown in Table 1. Section 4 contains a comprehensive comparison of total complexities between three efficient scalar multiplication methods, including recently proposed Near Optimal Controlled (NOC) [4], greedy algorithm [1] and tree [19] approaches. We conclude this paper in Sect. 5.

We emphasize that with several choices of coordinate systems, the usefulness of Edwards curves in establishing speed records for single-scalar multiplication make it valuable to develope further improvements. Nevertheless our new tripling formula doesn't gain enough improvement. In compensation, we give a new strategy for computing point quintupling on Edwards curve in appendix.

2 Preliminary

2.1 Double-Base Number System

The use of *double-base number system* (DBNS) in cryptographic systems is proposed by Dimitrov, Jullien and Miller [17]. By DBNS, any positive integer n

---

[1] An addition formula is advertised as *unified* if it can handle generic doubling, that is, the two addends are identical.

is represented as a sum or difference of $\{2,3\}$-integer (number of the form $2^b 3^t$), i.e.,

$$n = \sum_{i=1}^{l} c_i 2^{a_i} 3^{b_i},$$

where $c_i \in \{1, -1\}$ and $c_1 = 1$. l is called the length of expansion.

DBNS can largely reduce the total complexities of scalar multiplication than 2-radix based representation systems, e.g. *non-adjacent-form* (NAF) family [18, Chap. 9]. However, for DBNS there are at least two salient weaknesses. One ingredient is on search problem: although this system is highly redundant, how to find the shortest representation is still an open problem now. The other is on trade-offs between storage and efficiency: to achieve best timings of DBNS, extra storage space is needed, see [4, Example 1].

Of these two problems, the former one is quite tough—it is conjectured to be NP complete. Yet the latter is relatively easy to solve by making a compromise between storage and efficiency, known as *double-base chain*.

2.2 Double-Base Chain

Introduced as a special form of DBNS, DBC [1] translates any integer n into a DBNS representation with restricted exponents, satisfying:

$$n = \sum_{i=1}^{l} c_i 2^{a_i} 3^{b_i},$$

where $c_i \in \{1, -1\}$, and $a_1 \geq a_2 \geq \cdots \geq a_l$, $b_1 \geq b_2 \geq \cdots \geq b_l$.

It's feasible to apply Horner-like fashion in point multiplication phase due to decreasing characteristic of $\{a_i\}_{1 \leq i \leq l}$, $\{b_i\}_{1 \leq i \leq l}$. $2^{a_1} 3^{b_1}$ is called the leading term of expansion, and it's easy to see that we need no less than a_1 doublings, b_1 triplings and $l - 1$ additions to perform such scalar multiplication $[n]P$.

Although how to find the shortest double-base chain (a.k.a canonic DBC) for random integers remains unsolved, there are many efficiently computable algorithms that can compute DBCs with low Hamming weight, for example, binary/ternary [2], modified greedy algorithm, tree approach, multi-base NAF [20], Near Optimal Controlled DBC.

3   Improved Tripling Formulas

In this section, we introduce improved step-by-step computation of tripling formulas ($P_3 = [3]P_1$) for Hessian, Jacobi-intersection, Edwards and Huff curves over $\mathbb{F}_q$, with $\mathrm{Char}(\mathbb{F}_q) \neq 2, 3$. We omit affine coordinates, because basic arithmetic (e.g. point addition, doubling) in them inevitably involve expensive field inversions. When field multiplication doesn't gain many time penalties, projective coordinates are frequently used instead, usually trading an inversion to several multiplications and reducing the total cost as a consequence. We drop the cost of additions, subtractions, and multiplications by small constants in underlying fields as well. Group operations are expressed in terms of multiplication(M), squaring(S) and multiplying by constant(D) in the sequel.

3.1   Tripling Formula on Hessian Curves

The use of Hessian curves in scalar multiplication was introduced by Smart [21] and Joye [9]. A homogeneous projective Hessian curve over $\mathbb{F}_q$ is defined by

$$X^3 + Y^3 + Z^3 = dXYZ,$$

where $d \in \mathbb{F}_q$ and $d^3 \neq 27$. The neutral element is $(1, -1, 0)$.

A family of *generalized* Hessian curves was investigated in [8]. Efficient *unified* addition formulas for it were presented, which are *complete*[2] too.

Introduced by Hisil, Carter and Dawson [22], the original formula of inversion-free tripling on Hessian curves is shown as follows. Notice that choosing another curve parameter doesn't influence the computation of doubling and addition. Reset curve parameter $k = d^{-1}$. Inverting a constant on $\mathbb{F}_q$ can be computed in advance before scalar multiplicaion, without affecting the cost of the tripling formula.

$$X_3 = X_1^3(Y_1^3 - Z_1^3)(Y_1^3 - Z_1^3) + Y_1^3(X_1^3 - Y_1^3)(X_1^3 - Z_1^3)$$
$$Y_3 = Y_1^3(X_1^3 - Z_1^3)(X_1^3 - Z_1^3) - X_1^3(X_1^3 - Y_1^3)(Y_1^3 - Z_1^3)$$
$$Z_3 = k(X_1^3 + Y_1^3 + Z_1^3)((X_1^3 - Y_1^3)^2 + (X_1^3 - Z_1^3)(Y_1^3 - Z_1^3))$$

The tripling operation can be computed by:

$$A \leftarrow X_1^3, B \leftarrow Y_1^3, C \leftarrow Z_1^3, D \leftarrow (A - B)(C - A),$$

$$E \leftarrow (B - C)^2, F \leftarrow D(B - C), G \leftarrow (2A - B - C)^2 + 2D + E$$

$$X_3 \leftarrow 2(A \cdot E - B \cdot D), Y_3 \leftarrow X_3 + 2F, Z_3 \leftarrow k(A + B + C)G.$$

The best known explicit algorithm for Hessian tripling costs $8M + 6S + 1D$ [22,23]. Our new tripling formula is valid by performing $8M + 5S + 1D$ and $1S$ is saved. We point out that an extended projective coordinate system $(X, Y, Z, X^2, Y^2, Z^2, 2XY, 2XZ, 2YZ)$ for Hessian introduced in [24] reduces the total cost of addition formula and is beneficial for side-channel attack resistance. However this system is not suitable for tripling operations, so we ignore it in efficiency-oriented comparison in Sect. 4.

---

[2] As defined in [11] an addition formula is complete if it works for all pairs of inputs without exceptional cases.

3.2 Tripling Formula on Jacobi-Intersection Curves

Any elliptic curve over $\mathbb{F}_q$ is birationally equivalent to an intersection of two quadric surfaces in $\mathbb{P}^3(\mathbb{F}_q)$. Recall from [7,10], a projective point (S, C, D, T) in the Jacobi-intersection form satisfies

$$\begin{cases} S^2 + C^2 = T^2 \\ aS^2 + D^2 = T^2, \end{cases}$$

where $a \in \mathbb{F}_q$, $a(1 - a) \neq 0$. The identity point is $(0, 1, 1, 1)$.

Explicit inversion-free tripling formula [22] is as follows:

$$S_3 = S_1(k(kS_1^8 + 6S_1^4C_1^4 + 4S_1^2C_1^6) - 4S_1^2C_1^6 - 3C_1^8)$$
$$C_3 = C_1(k(3kS_1^8 + 4S_1^6C_1^2 - 4S_1^6C_1^2 - 6S_1^4C_1^4) - C_1^8)$$
$$D_3 = D_1(k(-kS_1^8 + 4S_1^6C_1^2 + 6S_1^4C_1^4 + 4S_1^2C_1^6) - C_1^8)$$
$$T_3 = T_1(k(-kS_1^8 - 4S_1^6C_1^2 + 6S_1^4C_1^4) - 4S_1^2C_1^6 - C_1^8),$$

where curve parameter $k = a - 1$. The terms can be organized as:

$$A \leftarrow S_1^2, B \leftarrow C_1^2, C \leftarrow A^2, D \leftarrow B^2, E \leftarrow (A + B)^2 - C - D,$$

$$F \leftarrow kC - D, G \leftarrow kC + D, H \leftarrow ((k - 1)E + 2F)G,$$

$$I \leftarrow (E - F)(F + kE), J \leftarrow (k + 1)EG,$$

$$S_3 \leftarrow S_1(H + I), C_3 \leftarrow C_1(H - I), D_3 \leftarrow D_1(I + J), T_3 \leftarrow T_1(I - J).$$

The above formula costs $7M + 5S + 2D$. Hisil et al. [22] proposed two versions of tripling formula on Jacobi-intersection curves, which are known best, one costs $4M + 10S + 5D$, and the other costs $7M + 7S + 3D$.

Moreover in [24], they showed how a redundant extended coordinate system can remarkably reduce the cost of addition on Jacobi-intersection curves by at least $2M$. The homogeneous projective coordinate system is named as "modified Jacobi-intersection", by which a point is represented as the sextuplet (S, C, D, T, U, V) with $U = SC$, $V = DT$. If we use it to perform tripling operation, we get $2M$ punishment and the total cost becomes $9M + 5S + 2D^3$. Yet in this case, the new tripling formula of modified Jacobi-intersection coordinate system is still faster than that of [23, Jacobi intersections].

3.3 Tripling Formula on Edwards Curves

In [25], Harold Edwards proposed a new form of elliptic curves and thoroughly investigated its mathematical aspects. Later Bernstein and Lange [11] established fast explicit formulas for elliptic group on Edwards curves. They also proposed inverted coordinate system that allows reduced additions in [26].

---

[3] The computation of E in the first line can be done as $E \leftarrow 2U^2$ alternatively. It saves 2 field additions.

The reader is referred to [13,15] for arithmetic on twisted Edwards curves and further improvements with different coordinate systems.

A Edwards curve over $\mathbb{F}_q$ defined by homogeneous projective coordinate (X, Y, Z) is

$$E : (X^2 + Y^2)Z^2 = c^2(Z^4 + dX^2Y^2),$$

where $c, d \in \mathbb{F}_q$ such that $dc^4 \neq 0, 1$. The identity for elliptic group is $(0, c, 1)$.

The tripling formula on Edwards curves [27] is shown as follows.

$$X_3 = X_1(X_1^4 + 2X_1^2Y_1^2 - 4c^2Y_1^2Z_1^2 + Y_1^4)(X_1^4 - 2X_1^2Y_1^2 + 4c^2Y_1^2Z_1^2 - 3Y_1^4)$$
$$Y_3 = Y_1(X_1^4 + 2X_1^2Y_1^2 - 4c^2X_1^2Z_1^2 + Y_1^4)(3X_1^4 + 2X_1^2Y_1^2 - 4c^2X_1^2Z_1^2 - Y_1^4)$$
$$X_3 = Z_1(X_1^4 - 2X_1^2Y_1^2 + 4c^2Y_1^2Z_1^2 - 3Y_1^4)(3X_1^4 + 2X_1^2Y_1^2 - 4c^2X_1^2Z_1^2 - Y_1^4)$$

The formula can be organized as:

$$A \leftarrow X_1^2, B \leftarrow Y_1^2, C \leftarrow (2c \cdot Z_1)^2, D \leftarrow A^2, E \leftarrow B^2,$$

$$F \leftarrow B(2A - C), G \leftarrow A(2B - C), H \leftarrow D + E + F,$$

$$I \leftarrow D - 3E - F, J \leftarrow D + E + G, K \leftarrow H + I + J,$$

$$X_3 \leftarrow X_1 \cdot HI, Y_3 \leftarrow Y_1 \cdot JK, Z_3 \leftarrow Z_1 \cdot IK.$$

This operation costs $8M + 5S + 1D$. To our knowledge, the previously best known tripling formulas cost $9M + 4S + 1D$ or $7M + 7S + 1D$ [22,27]. New formula given above trades $1M$ to $1S$, and gets several advantages because squaring costs less than multiplication in most cases. Similar routine can be applied to inverted Edwards coordinate.

Among Edwards curves family, the fastest addition is derived from its twisted form $-x^2 + y^2 = 1 + dx^2y^2$ with $(X, Y, \frac{XY}{Z}, Z)$ coordinate system. This redundant representation system save $1M$ for addition compared with inverted Edwards, leading to extra $1M$ for doubling though. So, it isn't suitable for DBC and we don't discuss further application of this coordinate system, because the amount of required doubling is usually more than that of addition.

3.4   Tripling Formula on Huff Curves

Joye, Tibouchi and Vergnaud presented unified and parameter-independent addition formulas for Huff's form elliptic curves [12], and studied its cryptographic application especially for pairing computations. The set of points on Huff satisfy

$$aX(Y^2 - Z^2) = bY(X^2 - Z^2),$$

where $a, b \in \mathbb{F}_q$ and $a^2 \neq b^2$. The identity element for the additive group on Huff's is $(0, 0, 1)$.

The tripling formula is shown as:

$$X_3 = X_1(X_1^2Y_1^2 - X_1^2Z_1^2 - Y_1^2Z_1^2 - 3Z_1^4)(X_1^2Y_1^2 - X_1^2Z_1^2 + 3Y_1^2Z_1^2 + Z_1^4)^2$$
$$Y_3 = Y_1(X_1^2Y_1^2 - X_1^2Z_1^2 - Y_1^2Z_1^2 - 3Z_1^4)(X_1^2Y_1^2 + 3X_1^2Z_1^2 - Y_1^2Z_1^2 + Z_1^4)^2$$
$$Z_3 = Z_1(X_1^2Y_1^2 - X_1^2Z_1^2 + 3Y_1^2Z_1^2 + Z_1^4)(X_1^2Y_1^2 + 3X_1^2Z_1^2 - Y_1^2Z_1^2 + Z_1^4)$$
$$\cdot (3X_1^2Y_1^2 + X_1^2Z_1^2 + Y_1^2Z_1^2 - Z_1^4)$$

The terms can be organized as follows:

$$A \leftarrow X_1^2, B \leftarrow Y_1^2, C \leftarrow Z_1^2, D \leftarrow A(B+C), E \leftarrow C(B+C),$$

$$F \leftarrow C(A+C), G \leftarrow F-E, H \leftarrow D-E-2F, I \leftarrow 2F+D-E,$$

$$J \leftarrow D+E-2G, K \leftarrow H+I+J, X_3 \leftarrow X_1 \cdot HJ^2, Y_3 \leftarrow Y_1 \cdot HI^2, Z_3 \leftarrow Z_1 \cdot IJK.$$

Above formula costs $10M + 5S$, and is independent of curve parameters. We don't give detailed improvement of tripling formulas on its generalized forms, due to limited benefits from the arithmetic on Huff's model, even comparing with Weierstrass curves in Jacobian coordinates. Interested readers are referred to [28] for discussion.

3.5 Cost Comparison Between Tripling Operations

The rest of this section includes a cost comparison between tripling formulas of various elliptic curves, see Table 1. Total complexities are counted for both

Table 1. Cost comparison between tripling formulas of different coordinate systems.

Systems	Tripling cost	Total cost	
		$1S = 0.8M$	$1S = 0.75M$
Huff (OLD)[28]	$10M + 6S$	$14.8M$	$14.5M$
Huff	**10M + 5S**	**14M**	**13.75M**
Jacobian [23]	$5M + 10S + 1D$	$13M$	$12.5M$
Hessian (OLD)	$8M + 6S + 1D$	$12.8M$	$12.5M$
Jacobi-quartic	$4M + 11S + 2D$	$12.8M$	$12.25M$
Jacobi-intersection-2 (OLD)	$7M + 7S + 3D$	$12.6M$	$12.25M$
Jacobian, $a = -3$ [23]	$7M + 7S$	$12.6M$	$12.25M$
Edwards (OLD)	$9M + 4S + 1D$	$12.2M$	$12M$
Edwards	**8M + 5S + 1D**	**12M**	**11.75M**
Hessian	**8M + 5S + 1D**	**12M**	**11.75M**
Jacobi-intersection-1 (OLD)	$4M + 10S + 5D$	$12M$	$11.5M$
Jacobi-intersection	**7M + 5S + 2D**	**11M**	**10.75M**
3DIK [16]	$6M + 6S + 2D$	$10.8M$	$10.5M$

$1S = 0.8M$ and $1S = 0.75M$ cases. We assume $1D = 0M$. It makes sense if chosen curve constants are of small values, or with extremely low (or high) hamming weight, so that the cost of D is equal that of several negligible additions on underlying field. Contributions introduced in this section are highlighted in **bold**.

As shown in Table 1, tripling formula on previously mentioned curves gains further improvement compared with the original ones (labelled with OLD in bracket). In particular, our new Jacobi-intersection tripling formula is competitive with that of tripling-oriented Doche/Icart/Kohel curves (denoted as 3DIK), by a difference of $0.2M$ in "$1S = 0.8M$" case.

4   Experiments

In this section, we are interested in how different options of curve shapes and scalar-recoding algorithms influence the speeds of scalar multiplication for generic elliptic point P. Before starting speed records for implementing DBC on various curves, basic arithmetic of involved curves is listed in Table 2, including optimizations given in this paper and latest results in literature. Our analysis is efficiency-oriented rather than simple power attack resistance, so several coordinate systems with reduced addition formula but expensive doubling, tripling operations are excluded in our consideration, as has been demonstrated in Sect. 3.

Table 2. Basic operations on various curves.

Curve shapes	mADD	DBL	TRL
3DIK	$7M + 4S + 1D$	$2M + 7S + 2D$	$6M + 6S + 2D$
Jacob	$7M + 4S$	$1M + 8S + 1D$	$5M + 10S + 1D$
Jacob-3	$7M + 4S$	$3M + 5S$	$7M + 7S$
ExtJacQuartic	$6M + 3S + 1D$	$2M + 5S$	$4M + 11S + 2D$
JacoIntersection	$10M + 1S + 2D$	$2M + 5S + 1D$	$7M + 5S + 2D$
ExtJacIntersection	$10M + 1S + 2D$	$2M + 5S + 1D$	$9M + 5S + 2D$
Hessian	$10M$	$7M + 1S$	$8M + 5S + 1D$
Huff	$10M$	$6M + 5S$	$10M + 5S$
InvEdw	$8M + 1S + 1D$	$3M + 4S + 1D$	$8M + 5S + 1D$

In Table 2, "Jacob" is referred to short Weierstrass curves $y^2 = x^3 + ax + b$ with projective Jacobian coordinate $(x, y) = (\frac{X}{Z^3}, \frac{Y}{Z^2})$, and "Jacob-3" is referred to the special case when $a = -3$. Moreover we consider a faster representation system (X, Y, Z, X^2, Z^2) of Jacobi-quartic form $y^2 = x^4 + 2ax^2 + 1$, whose detailed description and explicit formulas can be seen in [22–24]. "ExtJacIntersection" is referred to modified coordinate system for Jacobi-intersection as has been discussed in Sect. 3.

As for algorithmic aspect, we select Near Optimal Controlled DBC, greedy and tree approaches to generate DBCs for integers of 256, 320 and 512 bits. What should be pointed out is that as analyzed in [19], the average length of DBCs returned by tree approach tends to decrease when the size of coefficient set grows. For fair comparison and erasing precomputation, the coefficient set of tree approach is restricted to $\{1, -1\}$.

Total cost are counted disregarding necessary time to find optimal DBC, merely includes: 1. the amount of mixed addition (mADD), corresponding to l—the length of DBC; 2. the amount of doubling (DBL), corresponding to a_1—the power of 2 in the leading term of DBC; 3. the amount of tripling (TPL), corresponding to b_1—the power of 3 in the leading term of DBC. Average numbers of required mADD, DBL, TPL of these algorithms are shown in Table 3.

Table 3. Theoretical operations consumption of different algorithms.

Bits	NOC			Greedy			Tree		
	mADD	DBL	TPL	mADD	DBL	TPL	mADD	DBL	TPL
256	48	198	37	58.73	153	65	55.15	142.57	71.55
320	62	260	38	70.80	180	89	68.94	178.21	89.44
512	95	406	67	112.07	286	143	110.30	285.13	143.10

To allow easy comparison, we assume $1S = 0.8M$ as customary. Precomputation that derived from transforming affine points into extended projective coordinate system is also disregarded, due to limited influence on total complexities. For example, additional cost of transforming affine point (x, y) to ExtJacQuartic, JacIntersection and ExtJacIntersection are both $1M$.

Table 4. Total cost of NOC, greedy, tree approaches.

Curve shapes	NOC			Greedy			Tree		
	256	320	512	256	320	512	256	320	512
3DIK	2394	3018.8	4778.2	2463.846	3051.36	4861.114	2418.802	3023.536	4837.53
Jacob	2435.8	3050.4	4844.4	2576.246	3211.16	5118.514	2547.698	3184.662	5095.32
Jacob-3	2341.8	2931.2	4655.2	2489.046	3103.56	4946.914	2462.05	3077.602	4924.03
ExtJacQuartic	2064.8	2567.2	4091.6	2243.332	2813.92	4487.788	2234.52	2793.188	4468.98
JacIntersection	2113.4	2647.6	4199	2267.284	2823.64	4499.356	2238.09	2797.652	4476.12
ExtJacIntersection	2187.4	2723.6	4333	2397.284	3001.64	4785.356	2381.19	2976.532	4762.32
Hessian	2468.4	3104	4920.8	2560.7	3180	5067.5	2522.146	3152.718	5044.21
Huff	2978	3752	5948	3027.3	3754	5982.7	2978.9	3723.66	5957.7
InvEdw	2094	2613.6	4157.2	2245.424	2807.04	4475.416	2227.854	2784.854	4455.646

As can be seen from Table 4, NOC is the fastest one among these three algorithms, and it provides speed-ups for greedy and tree approaches by a factor of

5.4 % and 4.62 % approximately. Theoretically, superiority of NOC is yielded as a result of two advantages. First, smaller Hamming weight of returned DBCs by NOC leads to less addition operations during performing scalar multiplication, making it particularly beneficial for reducing algorithmic complexity. Second, for scalars of size t, it's easy verifiable that leading terms in Table 4 all satisfy $2^{a_1}3^{b_1} \approx 2^t$. Owing to high a_1/b_1 ratio, NOC algorithm is extremely suitable on elliptic curves with lower cost ratio of doubling over tripling, like inverted Edwards, extended Jacobi quartic.

But how to find its optimal expansion by NOC algorithm is disappointedly troublesome. Results in [4] reveal that this approach is practical to handle integers of size around 60 to 70 bits only. An alternatively applicative condition for this approach is in cryptographic protocols with fixed-scalar multiplication, like key-agreement. When handling scalar multiplication with generic scalars and elliptic points, tree-based search is optimal.

We now turn to curve selections. For NOC, extended Jacobi-quartic is the speed leader of both 256, 320 and 512 bits integers with necessary 5 registers to represent a projective point. When using tree approach, inverted Edwards coordinate system provides best performance with 3 registers to represent a projective point during scalar multiplication. Besides, Jacobi intersection form also behaves well, slightly slower than Inverted Edwards.

5   Conclusion

We have shown several optimizations for point tripling formulas on different elliptic curves, largely improving the efficiency of double-base chains for scalar multiplication. Moreover we provide an alternative efficient formula to calculate point quintupling on Edwards curves in appendix, a potential usefulness of which exists in establishing new speed records of quintupling-involved double-base number system.

We point out that what we did on elliptic curves mainly focuses on the arithmetic of their standard projective coordinates. We don't give detailed optimizations for all known extended coordinate systems. Some of them have been discussed in this work; the others are quite redundant and required to be further optimized in future work.

Taking everything into account, among discussed DBC algorithms, tree approach is the optimal one for practical implementation on inverted Edwards coordinate system for both 256, 320 and 512 bits integers. In some limited conditions like small or fixed-scalar multiplication, NOC can be used as an alternative.

A   Quintupling Formula on Edwards

We show a new formula to calculate the 5-fold of a point P on Edwards in this section. Let $(X_5, Y_5, Z_5) = 5(X_1, Y_1, Z_1)$. Explicit expression of (X_5, Y_5, Z_5) is quite involved so we exclude it in this context. Yet it's straightforward computable using curve equation and addition formula, one can accomplish it with the help of Magma or SageMath. An alternative algorithm for computing (X_5, Y_5, Z_5) is as follows:

$$A \leftarrow X_1^2, B \leftarrow Y_1^2, C \leftarrow Z_1^2, D \leftarrow A^2, E \leftarrow B^2, F \leftarrow C^2,$$

$$G \leftarrow (A+C)^2 - D - F, H \leftarrow (B+C)^2 - E - F, I \leftarrow (A+B)^2, J \leftarrow I - D - E,$$

$$K \leftarrow I^2, L \leftarrow I - G - H, M \leftarrow (D-E)^2, N \leftarrow J^2,$$

$$O \leftarrow (D-E)(K - 2d(K - M - 2N)), P \leftarrow 2M(I + 4F - G - H),$$

$$Q \leftarrow K - 4d \cdot N, R \leftarrow (D - E - G + H)Q, S \leftarrow L(2M - Q),$$

$$T \leftarrow O + P, U \leftarrow P - Q, V \leftarrow R + S, W \leftarrow R - S,$$

$$X_5 \leftarrow X_1(U+W)(U-W), Y_5 \leftarrow Y_1(T+V)(T-V), Z_5 \leftarrow Z_1(T+V)(U-W).$$

The above algorithm derives an efficient quintupling formula that costs $10M + 12S + 2D$. Including previous work reported in [27], cost of different strategies for computing projective quintupling formula on Edwards curves is listed as Table 5. It turns out that the new formula is preferred in most practical environments when $D\backslash M$, $S\backslash M$-ratio are less than 1.

Table 5. Different quintupling formulas on Edwards curves.

	Cost analysis
Bernstein et al.	$17M + 7S$
Bernstein et al.	$14M + 11S$
This work	$10M + 12S + 2D$

References

1. Dimitrov, V., Imbert, L., Mishra, P.K.: Efficient and secure elliptic curve point multiplication using double-base chains. In: Roy, B. (ed.) ASIACRYPT 2005. LNCS, vol. 3788, pp. 59–78. Springer, Heidelberg (2005)
2. Adikari, J., Dimitrov, V.S., Imbert, L.: Hybrid binary-ternary number system for elliptic curve cryptosystems. IEEE Trans. Comput. **60**(2), 254–265 (2011)
3. Dimitrov, V., Howe, E.: Lower bounds on the lengths of double-base representations. Proc. Am. Math. Soc. **139**(10), 3423–3430 (2011)
4. Doche, C.: On the enumeration of double-base chains with applications to elliptic curve cryptography. In: Sarkar, P., Iwata, T. (eds.) ASIACRYPT 2014. LNCS, vol. 8873, pp. 297–316. Springer, Heidelberg (2014)
5. Doche, C., Kohel, D.R., Sica, F.: Double-base number system for multi-scalar multiplications. In: Joux, A. (ed.) EUROCRYPT 2009. LNCS, vol. 5479, pp. 502–517. Springer, Heidelberg (2009)
6. Doche, C., Sutantyo, D.: New and improved methods to analyze and compute double-scalar multiplications. IEEE Trans. Comput. **63**(1), 230–242 (2014)

7. Liardet, P.-Y., Smart, N.P.: Preventing SPA/DPA in ECC systems using the jacobi form. In: Koç, Ç.K., Naccache, D., Paar, C. (eds.) CHES 2001. LNCS, vol. 2162, pp. 391–401. Springer, Heidelberg (2001)

8. Farashahi, R.R., Joye, M.: Efficient arithmetic on hessian curves. In: Nguyen, P.Q., Pointcheval, D. (eds.) PKC 2010. LNCS, vol. 6056, pp. 243–260. Springer, Heidelberg (2010)

9. Joye, M., Quisquater, J.-J.: Hessian elliptic curves and side-channel attacks. In: Koç, Ç.K., Naccache, D., Paar, C. (eds.) CHES 2001. LNCS, vol. 2162, pp. 402–410. Springer, Heidelberg (2001)

10. Billet, O., Joye, M.: The Jacobi model of an elliptic curve and side-channel analysis. In: Fossorier, M.P.C., Høholdt, T., Poli, A. (eds.) AAECC 2003. LNCS, vol. 2643, pp. 34–42. Springer, Heidelberg (2003)

11. Bernstein, D.J., Lange, T.: Faster addition and doubling on elliptic curves. In: Kurosawa, K. (ed.) ASIACRYPT 2007. LNCS, vol. 4833, pp. 29–50. Springer, Heidelberg (2007)

12. Joye, M., Tibouchi, M., Vergnaud, D.: Huffs model for elliptic curves. In: Algorithmic Number Theory, pp. 234–250. Springer, Heidelberg (2010)

13. Bernstein, D.J., Birkner, P., Joye, M., Lange, T., Peters, C.: Twisted edwards curves. In: Vaudenay, S. (ed.) AFRICACRYPT 2008. LNCS, vol. 5023, pp. 389–405. Springer, Heidelberg (2008)

14. Feng, R., Nie, M., Wu, H.: Twisted Jacobi intersections curves. In: Kratochvíl, J., Li, A., Fiala, J., Kolman, P. (eds.) TAMC 2010. LNCS, vol. 6108, pp. 199–210. Springer, Heidelberg (2010)

15. Hisil, H., Wong, K.K.-H., Carter, G., Dawson, E.: Twisted edwards curves revisited. In: Pieprzyk, J. (ed.) ASIACRYPT 2008. LNCS, vol. 5350, pp. 326–343. Springer, Heidelberg (2008)

16. Doche, C., Icart, T., Kohel, D.R.: Efficient scalar multiplication by isogeny decompositions. In: Yung, M., Dodis, Y., Kiayias, A., Malkin, T. (eds.) PKC 2006. LNCS, vol. 3958, pp. 191–206. Springer, Heidelberg (2006)

17. Dimitrov, V.S., Jullien, G.A., Miller, W.C.: Theory and applications of the double-base number system. IEEE Trans. Comput. **48**(10), 1098–1106 (1999)

18. Handbook of elliptic and hyperelliptic curve cryptography. CRC Press (2005)

19. Doche, C., Habsieger, L.: A tree-based approach for computing double-base chains. In: Mu, Y., Susilo, W., Seberry, J. (eds.) ACISP 2008. LNCS, vol. 5107, pp. 433–446. Springer, Heidelberg (2008)

20. Longa, P., Gebotys, C.: Fast multibase methods and other several optimizations for elliptic curve scalar multiplication. In: Jarecki, S., Tsudik, G. (eds.) PKC 2009. LNCS, vol. 5443, pp. 443–462. Springer, Heidelberg (2009)

21. Smart, N.P.: The Hessian form of an elliptic curve. In: Koç, Ç.K., Naccache, D., Paar, C. (eds.) CHES 2001. LNCS, vol. 2162, pp. 118–125. Springer, Heidelberg (2001)

22. Hisil, H., Carter, G., Dawson, E.: New formulae for efficient elliptic curve arithmetic. In: Srinathan, K., Rangan, C.P., Yung, M. (eds.) INDOCRYPT 2007. LNCS, vol. 4859, pp. 138–151. Springer, Heidelberg (2007)

23. Bernstein, D.J., Lange, T.: Explicit-formulas database (2007)

24. Hisil, H., Wong, K.K.H., Carter, G., et al.: Faster group operations on elliptic curves. In: Proceedings of the Seventh Australasian Conference on Information Security, vol. 98, pp. 7–20. Australian Computer Society Inc. (2009)

25. Edwards, H.: A normal form for elliptic curves. Bull. Am. Math. Soc. **44**(3), 393–422 (2007)

26. Bernstein, D.J., Lange, T.: Inverted edwards coordinates. In: Boztaş, S., Lu, H.-F.F. (eds.) AAECC 2007. LNCS, vol. 4851, pp. 20–27. Springer, Heidelberg (2007)
27. Bernstein, D.J., Birkner, P., Lange, T., Peters, C.: Optimizing double-base elliptic-curve single-scalar multiplication. In: Srinathan, K., Rangan, C.P., Yung, M. (eds.) INDOCRYPT 2007. LNCS, vol. 4859, pp. 167–182. Springer, Heidelberg (2007)
28. Wu, H., Feng, R.: Elliptic curves in Huffs model. Wuhan Univ. J. Nat. Sci. **17**(6), 473–480 (2012)

Web and Application Security

An Approach for Mitigating Potential Threats in Practical SSO Systems

Menghao Li[1,2], Liang Yang[1,3], Zimu Yuan[1(✉)], Rui Zhang[1], and Rui Xue[1]

[1] Institute of Information Engineering, Chinese Academy of Sciences, Beijing, China
{limenghao,yangliang,yuanzimu,zhangrui,xuerui}@iie.ac.cn
[2] University of Chinese Academy of Sciences, Beijing, China
[3] School of Information Engineering, Tianjin University of Commerce, Tianjin, China

Abstract. With the prosperity of social networking, it becomes much more convenient for a user to sign onto multiple websites with a web-based single sign-on (SSO) account of an identity provider website. According to the implementation of these SSO system, we classify their patterns into two general abstract models: independent SSO model and standard SSO model. In our research, we find both models contain serious vulnerabilities in their credential exchange protocols. By examining five most famous identity provider websites (e.g. Google.com and Weibo.com) and 17 famous practical service provider websites, we confirm that these potential vulnerabilities of the abstract models can be exploited in the practical SSO systems. With testing on about 1,000 websites in the wild, we are sure that the problem that we find is widely existing in the real world. These vulnerabilities can be attributed to the lack of integrity protection of login credentials. In order to mitigate these threats, we provide an integral protection prototype which help keeping the credential in a secure environment. After finishing the designation, we implement this prototype in our laboratory environment. Furthermore, we deploy extensive experiments for illustrating the protection prototype is effective and efficient.

Keywords: Single Sign-on · Web security · Integrity

1 Introduction

As a convenient and popular authorization method, single sign-on (SSO) is widely deployed by multiple websites as a way for logging in with a third-party account. For example, you can easily log into Smartsheet.com and Rememberthemilk.com using your Google account instead of individual accounts from each of them. It means that your Google account is authorized to access their resources by both websites. SSO reduces password fatigue from different username and password combinations and time spent on re-entering passwords for the same identity.

Thanks to the prosperity of social networking, multiple SSO systems, such as OpenID [4], Google AuthSub [20], SAML [7], and OAuth [5, 13], have been widely deployed on commercial websites. The SSO system works through the interactions among three parties: a client browser (the user), the identity provider (IDP, e.g.

© Springer International Publishing Switzerland 2016
D. Lin et al. (Eds.): Inscrypt 2015, LNCS 9589, pp. 209–226, 2016.
DOI: 10.1007/978-3-319-38898-4_13

Google.com), and service provider (SP, e.g. Smartsheet.com). The security of an SSO system is expected to prevent an unauthorized client from accessing to a legitimate user's account on the SP side. Given the fact that more and more high-value personal data are stored on the Internet, such as cloud websites, the flaws in SSO systems can completely expose the private information assets to the hackers. It forces SSO system developers to try their best to patch the flaws or build up a safer SSO system. However, in recent years, more and more logic flaws and vulnerabilities have been discovered.

By analyzing many popular commercial websites, we abstract the practical SSO systems into two categories. The first category of SSO systems is deployed with OAuth2.0 protocol, which is standardized by RFC 6749 [11] and is used to replace the previous SSO systems such as OpenID and AuthSub. The previous work on OAuth2.0 mostly focuses on the formal analysis [2, 15, 29] and auto detection of the vulnerabilities [2, 39]. But they do not come up with practical solutions. We focuses on the practical OAuth2.0 SSO systems deployed on the commercial websites, such as Google and Weibo, then extracts the workflows of the practical SSO OAuth2.0 systems. Besides, we also analyze the independent developed SSO systems. We find that those independent developed SSO systems follow a simple communication model which has only three steps. Without doubt, we find that both of these categories of SSO models have vulnerabilities.

By rechecking the commercial websites under our built general SSO models, we find that almost all of them obey the models and the vulnerabilities are similar on each website. Moreover we also find that some websites deploy SSO systems that mix the two general model together. This mixed model makes the analysis a bit complex. But we still find the integrity problems in the mixed model. We give a real world example of the mixed model SSO system in Sect. 4.

As the vulnerabilities can all be attributed to the lack of integrity protection on the login credential, we attempt to protect the credential's integrity with cryptographic method and try to not affect the original performance of the SSO system. In this paper, we propose protection prototype in Sect. 5. Our prototype can prevent the attackers from stealing the victim's credential and logging into victim's account with the entire access rights as the original victim.

Contributions. We first classify current popular SSO systems into two categories and build two abstract SSO models for analyzing the security of practical SSO systems. Then we parse the workflow of two kinds of SSO models in depth and find the vulnerabilities in those models.

Second, we verify that the vulnerabilities which pervasively existing in practical SSO websites obey the logic vulnerabilities we discovered in the abstract models.

Our third contribution is attempting to design a protection prototype. For mitigating the vulnerabilities, we focus on the integrity protection of the credentials by binding them with a protected parameter. As the channel that has the user browser's participation is not secure enough, our protection prototype exploit a direct channel (or private channel) between IDP and SP to deliver the binding parameter. The prototype can guarantee the integrity of the credentials and mitigate the threats from the network attacker and web attacker. The evaluation also shows that the overhead of prototype's performance is low comparing with the original SSO model.

2   Abstract Models of SSO Protocols

In this section, we discuss about our abstract models which are extracted from the practical SSO systems. We parse these practical systems in our research and focus on the information and data exchange workflows in them. In order to construct the models, we first investigate those websites that provide SSO login method and parse the login APIs of these websites with practical login actions. We manually analyze the massive SSO login documentations and extract the key parameters that should be pay much more attention during the parse of practical SSO login actions. As a result, we classify our models into two categories, which are named independent SSO model and standard SSO model. The independent model reflects the SSO models which the websites developed independently. The standard model represent those websites who follow the standard SSO information exchange protocols such as [11].

In our analysis, we summarize that a basic SSO system contains three entities, which are named IDP (Identity Provider), SP (Service Provider) and Client (Users), and the communication channels that connect each of the three entities together. The IDP is a server or a service cloud that stores user's account and password. It provides authentication of the identity of an individual user and authorizes the SP to access user's account on the IDP side. The SP, which is also called RP(resource provider) in some previous researches, is also a server or cloud that provides application services, such as a forum website, a cloud storage or a news subscription website. The client, in our research, represents a web browser that is connected to the internet which plays both as a redirection device and a resource visitor.

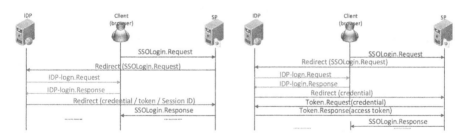

Fig. 1.   Independent SSO model   **Fig. 2.**   Standard SSO Model

2.1   Independent SSO Model

In the independent SSO model, we find that the IDP and SP only exchange data or messages through the Client (which is specifically a web browser). The Client acts as redirect party who can get all the messages and data between the IDP and SP. In Fig. 1, we show the detail workflow of the independent SSO model and the key parameters delivered in the communication channels. In the model, we mark out three channels in

3 different colors. We call the 3 channels as SSO-login channel, redirect channel and IDP-side verification channel. The SSO-login channel is only between the Client and the SP(the purple part of Fig. 1). It represents the SSO login request and response round trip in the model, and it stands at the first and last steps in the workflow. The redirect channel exploits the redirect functionality of the Client's browser (the green part of Fig. 1). In this part, the Client works as the redirect device who has the ability to receive and forward the messages between IDP and SP. The verification channel is used to deliver the messages between IDP and Client for verifying the user's identity who is on the Client-side (the orange part).

Now, we depict the workflow of SSO login and authentication in this model step by step.

- **Step 1**: When the Client want to log in the SP using the SSO method, it generates an *SSOlogin.Request* and delivers the login request to the SP server through the SSO-login channel.
- **Step 2**: When the SP receives this SSO login request, a redirect channel is generated among IDP, SP and Client. Then the SP redirects Client's SSO login request to the IDP through the Client's browser which acts as a relayed device.
- **Step 3**: After the IDP gets the redirected SSO login request, The IDP firstly need to sponsor a verification channel with the Client directly. Then the IDP verifies the identity of the user by checking the user's username and password which is supplied from the Client.
- **Step 4**: Once the verification is successfully accomplished, the IDP responses a credential (it could also be a token or a session ID) to the SP using the redirect channel.
- **Step 5**: After the SP gets the redirected credential, it responses the Client with an *SSOlogin.Response* under the *SSOlogin* channel.

When the user on the Client side receives this *SSOlogin.Response*, the user is capable to browse the custom content on the SP server, such as the news subscription.

Security Analysis. First of all, we review the model from the communication entities' perspective. There are three entities on the inter-connected channels (IDP, SP and Client), we discuss the security capability of them respectively. As the IDP and SP are represented as the servers in the model, they could be mass-flowed Internet websites in the real world, such as Google and NetEase. These websites have large quantity of sensitive data, which need to be protected, and enough financial investment on the security part. So the IDP and SP have much stronger security capability than just a personal PC or laptop. However, on the opponent side, the Client could just be a computer or smart mobile device. The investment on these personal devices security is limited, many malwares and Trojans focus on exploiting the personal devices other than a website.

Next, we review the model from the communication channels' perspective. With the TLS/SSL encryption technics used in the Internet communication, it shows that an encrypted channel are safer than an unencrypted channel. However, our research shows that only a few practical SSO systems in this model used HTTPS (which supports TLS/SSL) as one of their communication channels.

From the security analysis on the two aspects, we can conclude that the messages which are redirected by the Client on the redirect channels could expose the content into insecure environment. The key point of the independent SSO model's security should be focus on the step 4 of the model's workflow. In other words, this model's security depends on the confidentiality and integrity of the significant parameters, such as credentials, tokens or sessionIDs in the redirect channel through in step 4.

2.2 Standard SSO Model

The IDP and SP exchange messages not only through the Client as the redirect party, but also through a direct connection between them. In Fig. 2, we show the detail of this model's workflow. Comparing with Fig. 1, it has 4 channels: *SSOlogin* channel, the redirect channel, the verification channel and the direct channel. As the first three channels have been described in Sect. 2.1, we skip the discussion on them. Here we focus on the fourth channel – the direct channel (the red part). This channel is built between the IDP and SP directly without the participation of the Client. The functionality of this channel is to check whether the credential is generated by the same IDP and exchange for the second credential– access token.

Now we depict the details of the login workflows in the standard OAuth2.0 SSO model. The first 4 steps are similar with the independent model, and the step 5 and step 6 shows the additional token exchange in this SSO model.

- **Step 1**: When the Client starts a login request to the SP using the SSO method, it generates an *SSOlogin.Request* and send it to the SP through the *SSOlogin* channel.
- **Step 2**: Then the SP redirects Client's SSO login request to the IDP through the Client's browser which acts as a relayed device.
- **Step 3**: After the IDP gets the redirected SSO login request in step 2, the IDP sponsors a verification channel with the Client directly. Then the IDP verifies the identity of the user by checking the user's username and password which is supplied from the Client. The step is shown as IDP-login.Request and IDP-login.Response in the orange part.
- **Step 4**: Once the verification is successfully accomplished, the IDP responses a primary credential to the SP using the redirect channel as the response to Redirect(SSOlogin.request).
- **Step 5**: When the SP gets the redirected credential, it does not directly response the Client on the SSOlogin channel. What the SP has to do is to resend the credential back to the IDP to get the access token on the direct channel, which is used to allow the user on the Client to access the resources on the SP. This step is shown as the Token.Request(*credential*) and Token.Response(access token) in Fig. 2.
- **Step 6**: After the SP gets the access token, it response the Client with an SSOlogin response through the firstly established channel.

Now if the user successfully passed all the 6 steps, he should be able to visit the special subscription recourses on the SP.

Security Analysis. We still analyze the standard model from two perspectives. From the perspective of communication entities, the vulnerability in the three entities lies on the Client side which has the weakest protection technic. From the perspective of communication channels, the vulnerability exists in the insecure channel. Here it refers to the redirected channel where the Client takes part in.

Combining these two aspects, our analysis focuses on the Client side and the communication channels nearby it. It means that the redirect channel is still significant in our security analysis.

As is shown in Fig. 2, the standard SSO model extends the independent model with extra credential exchange steps. These steps are used for checking the correctness and availability of the credential and exchange for the real token. In order to keep these steps secure, this model uses the private direct connection between the IDP and SP without the participation of the Client and the redirect channels. It makes the attackers on the redirect channel environment have no chance to get the access token for login. From this point, this model is much safer than the independent model.

But when we go further, we find that the standard model still has its vulnerability which is analogous to the independent model. The integrity of the credential in step 4 is still not well-protected. Even though the following steps provide the direct channel for the security, the attacker can still stealthily get the content that contains the victim's credential on the redirect channel. Neither the SP nor the IDP checks whether the credential matches the Client's identity.

3   Adversary Models

We consider two different adversary models called network attacker [2, 29] and web attacker [21] which have the potential capability to exploit the vulnerabilities of practical SSO systems.

3.1   Network Attacker

Network attacker can be separated into two categories: active attacker and passive attacker. The active attacker is capable to intercept and modify the packages in the channel where it lies. The passive attacker is only capable to eavesdrop the packages on the channel, but cannot intercept or modify them. We consider man-in-the-middle attacker as our network attacker model, which belongs to one of the active attacker patterns. The man-in-the-middle attacker can intercept the messages on the channel between Client and the IDP or on the channel between Client and the SP. The credentials redirected by the Client could be intercepted and modified by this attacker.

In practice, for mitigating the threats from the man-in-the-middle attack, many web-based data transfers are available only under secured channels (for example, HTTPS). The encrypted channel makes the man-in-the-middle attack becomes unavailable because the attacker cannot tell which parameter is the correct credential from the cipher text. However, recent researches have indicated that the encrypted channel cannot completely stop the man-in-the-middle attack on the Internet. The attacker is able to deploy some HTTPS proxies [33–37] on the channel between the Client and Server to intercept the encrypted data stream and modify them on the proxy. On those proxies, the messages are decrypted, the attackers can understand the messages and pick out the credentials in the data stream. The trick of these HTTPS proxies is to pretend to be the forged server to the real client or forged client to the real server. These proxies just sit in the middle, decrypting traffic from both sides. Here how to trick the victim to install these HTTPS proxies is a kind of social engineering attack projects, and it is out of the scope of our paper.

Figure 3 shows the two roles the attacker is able to play in the communication between client and server.

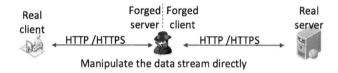

Fig. 3. Network Attacker

3.2 Web Attacker

Web attacker refers to those who control a malicious website on the Internet. The web attacker first lures the victim to visit this malicious website by following a malicious URI in a hyper-linked image or a malicious link address, such as a misleading link or image. When victim visits the malicious website, the attacker injects malicious code into victim's browser (e.g. XSS attack [30]) or replace victim's credential with attacker's (e.g. CSRF attack [28]). In the SSO login situation, the web attacker can require the victim delivering the credential to the malicious website under his control (XSS attack) or pushing the attacker's credential on the victim's browser for cheating the victim to login the SP as the attacker (CSRF attack).

Figure 4 shows the capability of the web attacker.

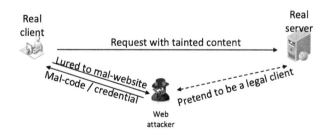

Fig. 4. Web Attacker

Our practical attack experiments (Sect. 4) and our protection prototype (Sect. 5) consider the threats under these two adversary models.

4   Case Study of Practical SSO Websites

In this section, we discuss our practical attack experiments on some of those famous websites in China, including Google, Weibo [22, 24], Tencent QQ [14], Alipay [17, 27], Taobao [26]. These five websites that we picked out all play the role of the IDP. Besides the Alipay websites deploys as our independent SSO model, the rest implement the standard OAuth2.0 SSO model we summarized in Sect. 2. For each IDP, we register two test account, namely Alice and Bob, and test whether the vulnerabilities work when logging into a practical SP. In our experiments, we login Bob's account with Alice's username and password by stealthily getting Bob's credential when Bob starts his login workflow.

Our experiment environment is as follow. First of all, we build up a local area network (LAN) to impersonate our test environment and connect two computers to the LAN. Then we deploy windows 7 as the operating system and play the role of victim (which means to be Alice) on one of the computers. We deploy Ubuntu14.10 as the attacker (which means to be Bob). On the Alice's computer, we install a web debugger tool – fiddler [9] for analyzing the web packages the victim gets and sends. On the Bob's computer, we install mitm-proxy [33], which is able to intercept the HTTPS data stream traffic on it, to filter the victim's SSO login messages for intercepting the Alice's login credentials.

4.1   Google Account

There are many service provider websites deploy Google account as one of their login method. In this part, we choose an online project management software – *smartsheet.com* [23] as our test SP. Although there are some SSO flaws have been reported in the previous research [3], their research focuses on the logic flaws on the smartsheet.com that the developers do not consider carefully and talks little about the vulnerabilities in the SSO protocol which is implemented between Google and Smartsheet. Besides, when we begin our study, Google has changed its SSO protocol from OpenID to OAuth2.0. So we cannot directly get experience from the previous research.

Fortunately, our study shows that the Google SSO login model follows our standard SSO model in Sect. 2.2. In our experiments, we register two new Google accounts, for example, *Alice@gmail.com* and *Bob@gmail.com*, and login smartsheet.com.

We search Alice's decrypted messages on the proxy and find the credential is named as *code*. Then we let Bob intercept Alice's following data traffic and stealthily keep Alice's code value in Bob's proxy. Now we start Bob's login workflow and also block the data stream when Bob gets his own *code*. Then Bob replaces his own *code* with Alice's, which is cut from her login workflow, and releases the modified redirect data stream to *smartsheet.com*. Without doubt, Bob successfully logs into Alice's account and controls the whole content of Alice's. Now Bob can do whatever he want to on the Alice's account.

During our impersonated attack, the only protection on this redirect message depends on the HTTPS protocol. But the integrity of this *code* is not protected. That is why Bob can exploit Alice's account without being detected by either *Google* or *smatsheet.com*.

4.2   Weibo.Com

Weibo.com also depends on standard OAuth2.0 SSO framework. It redirects the login credential through user's browser to the SP and it also calls this credential as *code*. However, different from the Google SSO login method, *Weibo* does not implement encrypted channels among the three abstract entities. Both network attacker and web attacker can be able to easily steal the victim's login credential.

In our experiment, we choose *Baidu* [38], a famous search engine service and cloud storage service provider in China, as the instance of the SP server. Like what we do in the Google case, we also register two *Weibo* accounts, which we still call them Alice and Bob, and confirm the availability of each account. Then we start our vulnerability exploit test. We put Bob on the proxy which Alice's login messages have to go through. On the proxy, we filter Alice's traffic data stream and search for the login credential which *Weibo* redirects to *Baidu*. As the channels are not encrypted every network package on the internet is displayed in plaintext. Bob is able to read Alice's packages directly and gets the login *code* of Alice's *Weibo* account.

Weibo redirects the *code* through a piece of JavaScript code in the response to the Alice's browser. The JavaScript code of Alice and Bob are shown as below:

On Alice's side, the code is as follows:

```
<script language=`javascript'>
callbackfunc({
http://baidu.com/.../afterauth?mkey=xxx
&code=code-of-alice});
</script>
```

On Bob's side, the code is as follows:

```
<script language=`javascript'>
callbackfunc({
http://baidu.com/.../afterauth?mkey=yyy
&code=code-of-bob});
</script>
```

Comparing the JavaScript code of two accounts, we find that the only difference of the redirect URI is the parameters: *code* and *mkey*, where the code is the login credential and the mkey is a ticket for preventing the CSRF attack. On the browsers, we intercept the redirection of the credentials of both Alice and Bob and replace Bob's *code* with Alice's. Then we redirect the modified Bob's URI back to *Baidu*. As a consequence, *Baidu* accepts the modified URI and regards Bob as Alice because Bob gives *Baidu* Alice's credential.

4.3   Alipay.Com

Alipay.com is an online payment and e-commerce management website (like PayPal) hosted by the Alibaba Group, a very famous Chinese online trade company. In practice, Alipay accounts can be used to login some other popular websites in China, such as

Xunlei and *Youku*. In our test, we choose *Xunlei* as the test SP and login it with Alipay accounts. Alice still plays the role of victim and Bob is the attacker.

In our test, we find that the Alipay is not following our standard SSO model, it is constructed under the independent SSO model which is discussed in Sect. 2.1. The SP does not resend the credential back to IDP for checking the validity. So we focus on the credential, which has been redirected through the user's browser, and detect whether it could be modified without being known by the SP.

Unfortunately, our test shows that the credential is composed with three parameters which is very different from the only one parameter in the standard OAuth2.0 model. These three parameters are *User_ID*, *token* and *sign*.

Although there exist a signature to protect the credential, we still find a way to let Bob hack into Alice's Alipay account. We test the Alipay SSO login method a lot of times, and find that the signature *sign* only protect the parameter of *token*.

It means that we can modify the *User_ID* to any value we want without being detected by *Xunlei.com*. Furthermore, we discover that the *User_ID* is a constant and plaintext. Each time we login no matter Alice's account or Bob account, the User_ID is an invariant. It means that the User_ID is guessable which is similar to the vulnerabilities in [2, 3, 15]. What the attacker need to do is to follow some rules to guess a legal User_ID. With this guessed User_ID attacker can log into any legitimate user's *Xunlei* account and get their sensitive data.

The Alipay SSO system also deploy a piece of javascript code as the redirect method. At the same time, its redirect messages only depend on HTTP which is insecure for delivering URL and significant parameters. The redirection URI is like: `http://xunlei.com/…/entrance.php?…token=xxx&user_id=USERID&sign=xxx&…`

Unlike the vulnerability in the standard OAuth2.0 SSO model, this vulnerability can be attributed to the logic flaws when the developers design the entire system. So it only suit for the Alipay SSO system and is not universal.

4.4 Taobao.Com

Taobao.com [26] is the most famous online shopping website in China. It also provides SSO login method, which is called AliSSO system. AliSSO system mixes the features of both independent SSO model and standard model together. From the perspective of the three entities of IDP, SP and Client, AliSSO follows the independent SSO model. When the credential is got by the SP, it does not need to send it back to IDP for checking the validity.

However, the SP does not directly accepts this credential. AliSSO separates the SP into two parts, in which one is a resource server and the other is an authentication server. The resource server stores the user's data and information and provides services to the user. The authentication server is in charge of certificating the identity of the legitimate user. When the SP gets the credential, it firstly generates another access token and redirects the token to the authentication server through user's browser after the authentication server gets the second access token, it generates a ticket and directly send to the resource server without the participation of user's browser. These steps are much more like the standard OAuth2.0 SSO model.

In our experiment, we choose *weibo* as an instance of our SP websites. Then we register two *taobao* accounts, namely Alice and Bob, and confirm the availability of each account. After that we begin our vulnerability exploit test. We suppose Bob as the attacker and put it on a proxy which Alice has to go through.

When we catch the data stream of Alice between taobao and weibo, we find that it is hard to modify the credential, which is named as *tbp*. As this parameter is protected by a signature, any change of the *tbp* will not be accepted by weibo. Then we let Alice's login workflow continues. After weibo gets the credential tbp and check the signature, it generates a second credential and redirects it to the authentication sub-server, login.weibo.com. This redirection also goes through Alice's browser, we can catch it on the proxy. When the sub-server gets the second credential, *alt*, it directly send *alt* to resource.weibo.com following the standard OAuth2.0 SSO model. After resource.weibo.com gets the *alt*, it responses Alice with her personal content.

In this login workflow, we find the second credential, alt, is not well protected. As Bob is on the proxy that Alice has to go through, he can replace his alt with Alice's and login Alice's account on weibo.com without any prevention from either weibo.com or taobao.com.

We have reported this vulnerability to the technic support group of *Weibo*, and got their thanks email in two days. Before we write our paper, this vulnerability has been patched.

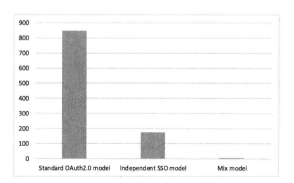

Fig. 5. Classified SSO Models

In practice, we have tested 1,037 websites manually. Most websites, except Google, in our experiment are located in China because some most famous websites, such as Facebook and Twitter, cannot visit in China mainland. But this problem does not affect our research. The conclusion of our tests is that most websites deploy the standard OAuth2.0 SSO model. The rest are independent SSO model and mixed SSO model (such as the taobao.com). The mixed model is not a new model, it is just combined from the two abstract SSO models together. The classified model graph is shown in Fig. 5. Then we pick up 9 typical SP websites and 5 IDP websites from our tested SSO websites. And we list the vulnerabilities and flaws of them in Table 1.

Table 1. SSO threats in real-world websites

SP / IDP	Smart-sheet	Remember-themilk	Weibo	Baidu	Youku	Sohu	Xunlei	Iqiyi	JD
Google	△	△							
Weibo				△	△	△	△	△	
QQ			△	△	△	△		△	△
Alipay					□	□	□	□	
Taobao			○				○		

Note: △ – Standard OAuth2.0 SSO model; □ – Independent SSO model; ○ – MixedSSO model;

5   Integrity Protection and Threat Mitigation

We can attribute the vulnerabilities we discuss in previous sections to the lack of the login credentials' integrity protection. In this section, we give out our prototype scheme for protecting the login credentials integrity. Our prototype can mitigate the threats from the network attack and web attack which are under the adversary models in Sect. 3. We build up our test environment in our lab with a LAN and two servers which play the roles of IDP and SP. Then we implement our prototype on those two servers and test it through another computer which acts as the Client. Finally, we compare the performance of our prototype and the original SSO system. The consequence shows that the performance of our prototype is acceptable.

5.1   Prototype Design

Our basic purpose is to avoid web attackers or network attackers stealing the legitimate user's login credentials and protect the credentials integrity. In this part, we first describe how our prototype prevents the web attackers and then we talk about how it prevents the network attackers. The workflow of our prototype is shown in Fig. 6.

Protection from Web Attackers. We use Same Origin Policy (SOP) [32] and HTTPOnly Policy [31] on the SP side to perform the protection. This protection can avoid attacker luring victims to login attacker's account unconsciously.

On the SP side, we add a parameter, *stat*, in the SSO redirect URL and set the browser's *cookie* with a parameter, *signstat*, which is a signature of stat and label this cookie as HttpOnly. When the IDP gets the redirect URL, it regards the parameter of *stat* as a component of the URL and append the credential after it. Then the IDP delivers it to the Client's browser. When the redirection URL that contains the credential and *stat* comes into the Client's browser, the browser redirects the credential to the SP with *cookie* back. When the SP gets the credential, *stat* and *cookie* back, it first computes whether the signature of *stat* in the URI matches the signature value in the *cookie*. If the signature of *stat* matches the value in the *cookie*, it means that this URL is not from the web attacker. The SP believes the user on the Client is a legitimate user.

The security of this design of *stat* depends on SOP and HTTPOnly which need the participation of the cookie. As the web attacker lures the victim to visit a malicious website under his control, the attacker prefers to put his own credential as a redirect URL in the response and send back to victim browser. When the victim gets the redirect URL that contain attacker credential, the browser wants to send the URL to the SP. If there is not protection, the attackers credentials would be send to SP and the SP would regard the victim as the attacker. In case the victim does not notice that he has logged into a wrong account and upload some significant files in this account, attacker can get those files a few minutes later just by legally login his account. However, with the help of SOP and HTTPOnly, this threat is blocked.

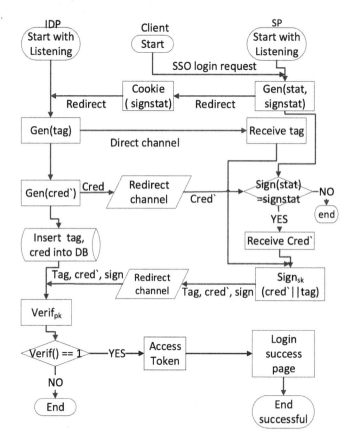

Fig. 6. WorkFlow of the Protection Prototype

Protection from Network Attackers. In order to mitigate the threats from network attackers, we need the participation of both IDP and SP. Besides, we also need two different channels: one is the redirect channel through the Client, the other is the direct or private channel between the IDP and SP.

In our adversary models, the network attacker can hack into an encrypted channel with the help of the SSL-proxy tools (such as mitmproxy). What the attacker need to do is to stealthily install a HTTPS proxy certificate on the victim's computer. But this work is out of our scope, we do not discuss it in our paper. This strong capability makes the confidentiality invalid on the HTTPS channels. In this situation, the integrity of the credential becomes a very significant point in the SSO system. But neither standard OAuth2.0 framework nor independent developed SSO system protect the integrity very well. We have easily logged into another user's account without knowing his or her username and password (Sect. 4). For mitigating the threats from the network attackers, we use the direct channel between IDP and SP to deliver a binding parameter, which we call it *tag*, for verifying the credential's integrity. Supposed that this direct channel is invisible in the attacker's view. So the *tag* is delivered securely between IDP and SP. After IDP delivers the *tag* directly to SP, it generates a corresponding credential which is bonded to the tag. And we let the IDP keep the pair of the original (*tag, credential*) in its database for checking the integrity of credential that delivered back from the SP. Then the IDP redirect the credential to Client's browser. On the SP side, it gets the tag from the direct channel and gets the credential from the redirect channel. Once the SP gets the login credential, we call credential' from the redirect channel, it binds the credential and the tag with a signature function $sign_{sk}(credential'||tag)$. The *sk* is the secrete key which is negotiated between IDP and SP. It is used for signing the value of $credential'||tag$. Then SP delivers the signature back to IDP through the direct channel with the (*tag, credential'*) pair. Correspondingly, the IDP has a public key *pk* for verifying the signature. After the IDP gets the signature and (*tag, credential'*) pair, it first searches the database with the value of tag. Then IDP verifies the signature of $sign_{sk}(credential'||tag)$ with the verify function $verif_{pk}(tag, credential, sign_{ak})$. If the verification successes ($verif_{pk} = 1$), it means that the attacker does not modify the credential when redirecting it. At this time, the IDP sends the access token directly to the SP, then SP notices the Client it has logged in SP successfully. If the verification fails, IDP reports an error and drop the (*tag, credential*) pair in the database.

5.2 Implementation

We deploy two desktop computers to impersonate the real SP and IDP called s-SP and s-IDP. Both of the computers have an Intel Core i7-3770 3.4 GHz CPU and 4 GB memory. The operation system is Ubuntu 14.10 LTS. We install the service software, including PHP 5.5.11, Apache 2.4.9 and MySQL server 5.6, and configure the web environment on both computers.

In our implementation, we deploy our prototype on the standard OAuth2.0 SSO framework and we call the login credential as *code*. In order to simplify the workflow of the impersonated SSO system, we omit the user's IDP-login steps. When an SSO login request comes from s-SP, s-IDP circumvents the verification steps and directly begins the authorization and login operations. During the authentication and authorization steps, we give s-SP a secrete key, *sk,* for signing the *code* with a binding parameter, *tag,* which is got through the direct channel from s-SP, and we give s-IDP a public key *pk* for verifying the signature of *code* that is given by the s-SP.

On the s-SP side, we add a parameter, stat, for preventing the attack from a malicious website. This parameter not only exists in the redirect URL but also has a signature in the user browser cookie. With the help of the SOP and HTTPOnly policies, the web attackers cannot get the signature of *stat* in the cookie between browser and the real SP. Once the forged *stat* is delivered back to s-SP, the server finds that the *stat* does not match the signature in the cookie and it will stop the following login workflow. This parameter can perfectly prevent the CSRF and XSS attacks that are sponsored by the web attackers.

Another thing need to pay attention on the s-SP is the synchronization of the parameters for generating the signature. Here they refer to tag and code specifically. It should be careful to handle this problem, because tag and code come from different channels. The tag comes from the direct channel between the s-IDP and s-SP and it is delivered to s-SP before the code. But the code comes through the redirect channel which is relayed from the user's browser. These two parameters cannot arrive at s-SP at the same time. If we do not consider the synchronization of these two parameters, s-SP may put Alice's code and Bob's tag together and compute a signature of the mixed-user parameters which is not correct for the s-IDP for verification. This problem might cause Bob logs into Alice's account. Our solution on this problem is simple. We build a concurrence lock on the s-SP side, which makes the s-SP can only deal with one user's login request.

5.3 Evaluation

Our implementation is about 100 lines of PHP and JavaScript code. Our evaluation depends on the execution time of the code. We set two timestamps in the entire login workflow. The first one is set at the SSO login page, when the user clicks the SSO login button, we get a timestamp. The second one is set on the login success page, if the user login successful, we record the second timestamp. The execution time is the difference of the two timestamps. Then we execute 400 times, and get the average time as the general execution time. The comparison between the original SSO model and our protection prototype is shown in Fig. 7.

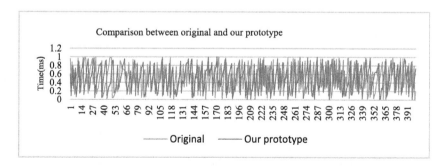

Fig. 7. Time spending comparison between original SSO model and our protection prototype

For the performance, we compare our prototype with the original SSO model which do not show any protections on the integrity of the credentials. Averaged 400 independent executions of each model, the overhead of the protection prototype is only increased by 0.418 % compared with the original SSO model. It means that the performance of our prototype is acceptable.

6 Related Work

Many previous works have been done to study the security of SSO systems. Wang et al. [3] discovered the SSO flaws in OpenID [4] and Flash. The flaws of OpenID cause the IDP to exclude the email element from the list of element it signs, which is sent back to the SP through a BRM. When the flaws of OpenID are reported to Google by the authors, Google replaces OpenID with OAuth2.0 as the SSO system [18, 19]. Armando et al. [10] studied on SAML-based SSO for Google Apps and gave the formal analysis of SAML 2.0 [6, 7] web browser SSO system. They used formal method to extract the abstract protocol in SAML 2.0 and built up the formal model of SAML. Somorovsky et al. [1] did a lot of researches in revealing vulnerabilities in formal SAML SSO systems. They revealed the threat from XML signature wrapping attacks is a big problem in the systems.

Bansal et al. [15] and Sun et al. [29] discovered the attacks on OAuth2.0 by formal analysis of the basic document of RFC 6749 [11]. They analyzed the formalized OAuth2.0 protocol and revealed that the potential threats coming from CSRF attack or token stolen during the redirection.

Before we finish our work, a vulnerability named *Covert Redirect* [16, 25] was reported about the OAuth2.0 on the Internet. It describes a process where a malicious attacker intercepts a request from an SP to an IDP and changes the parameter called "redirect_uri" with the intention of causing the IDP to direct the authorization credentials to a malicious location rather than to the original SP, thus exposing any returned secrets (e.g. credentials) to the attacker.

Zhou et al. [39] have built an automated SSO vulnerabilities test tool. This tool can detect whether a commercial website exists popular vulnerabilities, such as access_token misuse or OAuth credentials leak. But they only deploy the Facebook as the IDP site.

7 Conclusion

In this paper, we disclose the reason of the vulnerabilities that exist in commercial web SSO systems. We studied the SSO systems on 17 popular websites and classified them into two abstract models. Then we verify our models on about 1,000 SSO supported websites in the wild. Most websites follow the standard OAuth2.0 SSO model but there still some other websites prefer developing their own SSO system that depends on the independent model. We also elaborate our security analysis on these practical commercial websites that deploy different SSO models. That is the credentials could be intercepted by the attackers to log into the SP as the victim. For mitigating the threats focus on the credential's integrity, we give our protection prototype on guaranteeing the integrity of the credentials which is simple and efficient to deploy in practice. It not only fixes the vulnerabilities of the two abstract SSO models and the mixed model, but also mitigates the threats from the two adversary models mentioned in Sect. 3. However, our prototype also has its limitation. For example, on the SP side, it does not support concurrent SSO requests so far. Our prototype has to deploy on both IDP and SP server-sides. That is a trivial and cumbersome work. In the future work, we want to improve our prototype on these two problems and try our best to make our protection prototype to be a convenient independent third party middle-ware which can be deployed on any IDP or SP websites.

Acknowledgement. This work is supported by the "Strategic Priority Research Program" of the Chinese Academy of Sciences, Grants No. XDA06010701, National Natural Science Foundation of China (No.61402471, 61472414, 61170280), and IIE's Cryptography Research Project. Thanks to Wei Yang for helping recording the experiments. Thanks to a number of anonymous reviewers and Prof. Jian Liu who gave us very useful feedback on a previous version of this paper.

References

1. Juraj, S., Andreas, M., Jörg, S., Marco, K., Meiko, J.: On breaking SAML: be whoever you want to be. In: USENIX Security (2012)
2. Bai, G., Lei, J., Meng, G., Venkatraman, S.S., Saxena, P., Sun, J., Liu, Y., Dong, J.S.: AUTHSCAN: automatic extraction of web authentication protocols from implementations. In: NDSS (2013)
3. Wang, R., Chen, S., Wang, X.: Signing me onto your accounts through facebook and google: a traffic-guided security study of commercially deployed. In: IEEE S&P (2012)
4. OpenID. http://openid.net/
5. OAuth Protocols. http://oauth.net/
6. Technology report SAML protocol. http://xml.coverpages.org/saml.html
7. SAML2.0 Wikipedia. http://en.wikipedia.org/wiki/SAML 2.0
8. Wang, R., Chen, S., Wang, X., Qadeer, S.: How to shop for free online security analysis of cashier-as-a-service based web stores. In: IEEE S&P (2011)
9. Fiddler–The free web debugging proxy. http://www.telerik.com/fiddler
10. Armando, A., Carbone, R., Compagna, L., Cuellar, J., Abad, L.: Formal analysis of SAML 2.0 web browser single sign-on: breaking the SAML-based single sign-on for google apps. In: ACM FMSE (2008)
11. OAuth2.0 Authorization Framework. http://tools.ietf.org/html/rfc6749
12. Google Accounts Authentication and Authorization. https://developers.google.com/accounts/docs/OAuth2
13. OAuth2.0 documentation. http://oauth.net/documentation/
14. Wikipedia Tencent. http://en.wikipedia.org/wiki/Tencent
15. Bansal, C., Bhargavan, K., Maffeis, S.: Discovering concrete attacks on website authorization by formal analysis. In: IEEE CSF (2012)
16. Covert Redirect. http://tetraph.com/covert_redirect/
17. AlipayOpenAPI. https://openhome.alipay.com/doc/docIndex.htm
18. Google Accounts authorization and authentication Open ID 2.0 migration. https://developers.google.com/accounts/docs/OpenID?hl=en-US
19. Google Accounts authorization and authentication Using OAuth2.0 for login (OpenID Connect). https://developers.google.com/accounts/docs/OAuth2Login?hl=en-US
20. Google AuthSub. https://developers.google.com/accounts/docs/AuthSub
21. Akhawe, D., Barth, A., Lam, P.E., Mitchell, J., Song, D.: Towards a formal foundation of web security. In: CSF (2010)
22. Sinaweibo, Wikipedia. http://en.wikipedia.org/wiki/SinaWeibo

23. Smartsheet.com, one online project management software. https://www.smartsheet.com/
24. Weibo openAPI. http://open.weibo.com/wiki/
25. Covert Redirect Vulnerability Related to OAuth 2.0 and OpenID. http://tetraph.com/covert_redirect/oauth2_openid_covert_redirect.html
26. Taobao, Wikipedia. http://en.wikipedia.org/wiki/Taobao
27. AlipayWikipedia. http://en.wikipedia.org/wiki/Alibaba_Groupn#Alipay
28. Cross-Site Request Forgery (CSRF), The Open Web Application Security Project (OWASP). https://www.owasp.org/index.php/Cross-Site_Request_Forgery_(CSRF)
29. Sun, S.T., Beznosov. K.: The devil is in the (implementation) details: an empirical analysis of OAuth SSO systems. In: ACM CCS (2012)
30. Cross-Site Scripting (XSS), The Open Web Application Security Project (OWASP). https://www.owasp.org/index.php/XSS
31. HttpOnly, The Open Web Application Security Project (OWASP). https://www.owasp.org/index.php/HttpOnly
32. Same Origin Policy, W3C Web Security. https://www.w3.org/Security/wiki/Same_Origin_Policy
33. MitmProxy, An interactive console program that allows traffic flows to be intercepted, inspected, modified and replayed. https://mitmproxy.org/
34. SSL Man in the Middle Proxy. http://crypto.stanford.edu/ssl-mitm/
35. Cloudshark Appliance. https://appliance.cloudshark.org/
36. SSLsplit - transparent and scalable SSL/TLS interception. https://www.roe.ch/SSLsplit
37. Sslsniff, A tool for automated MITM attacks on SSL connections. http://www.thoughtcrime.org/software/sslsniff/
38. Baidu, Wikipedia. http://en.wikipedia.org/wiki/Baidu
39. Zhou, Y., Evans, D.: SSOScan: automates testing of web applications for single sign on vulnerabilities. In: 23rd USENIX Security Symposium (2014)

EQPO: Obscuring Encrypted Web Traffic with Equal-Sized Pseudo-Objects

Yi Tang[1,2(✉)] and Manjia Lin[1]

[1] School of Mathematics and Information Science, Guangzhou University,
Guangzhou 510006, China
ytang@gzhu.edu.cn
[2] Key Laboratory of Mathematics and Interdisciplinary Sciences of Guangdong,
Higher Education Institutes, Guangzhou University, Guangzhou 510006, China

Abstract. Internet users are concerned with their private web browsing behaviors. Browsing a webpage introduces a typical request-response-based network traffic which is associated with the structure of corresponding HTML document. This may make the traffic of a specified webpage demonstrate different features from others even when the traffic is encrypted. Traffic analysis techniques can be used to extract those features to identify that webpage, and hence the webpages the user visited could be disclosed though they might be encrypted. In this paper, we propose EQPO, a method to defend against traffic analysis by obscuring web traffic with EQual-sized Pseudo-Objects. A pseudo-object is composed by some original objects, object fragments, or padding octets. We define a structure of EQPO-enabled HTML document to force object requests and responses be on pseudo-objects. For a webpage set, by equalizing the sizes of pseudo-objects and the numbers of pseudo-objects requests in each webpage, we can make the traffic for those webpages with no identifiable features. We have implemented a proof of concept prototype and validate the proposed countermeasure with some state of the art traffic analysis techniques.

Keywords: Encrypted web traffic · Webpage identification · Traffic analysis · Equal-sized pseudo-object

1 Introduction

Browsing webpages privately has attracted much attention in recent years due to the increasing awareness of privacy protection. Internet users want to preserve the privacy of not only what content they have browsed but also which specified webpage they have visited. Encryption is effective in protecting the privacy of data contents transferred in networks, but it is not a winner-take-all method in distinguishing different webpages. Traffic features demonstrated by different webpages could be used to identify them page by page even if they are transmitted in encrypted form.

© Springer International Publishing Switzerland 2016
D. Lin et al. (Eds.): Inscrypt 2015, LNCS 9589, pp. 227–245, 2016.
DOI: 10.1007/978-3-319-38898-4_14

The network traffic introduced by browsing webpages is on request-response transactions. The request profile, such as the number and the sequence, is on the webpage structure, and the response amount sizes are on the sizes of corresponding resources (objects) embedded in basic HTML document. Current popular secure suites, such as SSL, SSH, IPSec, and Tor, etc., are focused on encrypting data contents and do not alter the number of object requests. Even for the encrypted contents, their sizes are not changed significantly comparing to the related original plain ones. This makes the traffic of different webpages demonstrate distinguishable features even if the traffic is encrypted. The typical traffic features, such as the order, number, length, or timing of packets, etc., can be extracted by traffic analysis (TA) and may lead to identifying the webpages precisely the user visited, or even inferring the data the user privately input [1–3,5,9,10].

Proposals against TA analysis are on changing traffic features. They can be operated at server side, client side, or on client-server cooperation, and worked on network level, transport level, and application level [3]. Padding extra bytes into transmitting data is the most general method. The padding procedure can be executed at server side before or after encryption [2,5]. An improved strategy on padding is traffic morphing, which makes a specified webpage traffic similar to another predefined traffic distribution [15]. These efforts are on fine-grained single object analysis and they are not efficient against the coarse-grained aggregated statistics [3]. The BuFLO method intends to cut off the aggregated associations among packet sizes, packet directions, and time costs [3] by sending specified packets in a given rate during a given time period. Some other techniques on higher level, such as HTTPOS [7], try to influence the packet generation at server side by customizing specified HTTP requests or TCP headers at client side.

Most of the TA analysis target on the identifying webpages in different websites. It seems that webpages in the same website may challenge the effectiveness of TA analysis because of their structure and resource similarities. However, recent researches also show that those webpages cannot escape from TA based identification [8]. For example, the technique discussed in [8] can identify specified webpages in the same website with up to 90 % accuracy. Partitularly, if it is used in the website related to healthcare successfully, the subsequent inference could be launched to reveal the reason why someone went to consult a doctor.

In this paper, we propose EQPO, a TA defence method on web application level, which focuses on preventing webpage identification in a same website. Motivated by the k-anonymity technique in database community [11], we intend to make the traffic of any page in a website similar with each other by introducing same numbers of equal-sized pseudo-object in pages. We define a pseudo-object as an object fragment combination which is composed by a set of object fragments. To make each pseudo-object with the same size, some pseudo-objects may be appended with padding octets. To make each webpage with the same number of pseudo-objects, some extra pseudo-objects may be filled up with padding octets. Our proposed method is on client-server cooperation. We translate a common traditional webpage into equal-sized-pseudo-object-enabled (EQPO-enabled) webpage. When the browser renders the EQPO-enabled webpage, the embedded script is invoked and initiates requests for those pseudo-objects. Cooperatively, a script running on server will produce pseudo-objects with given size and given number. This kind of object generating and fetching procedure generalizes the traffic features of the webpages in that webpage set and hence may be used to defend against the traffic analysis.

The contribution of this paper can be enumerated as follows.

1. We introduce the notion of equal-sized pseudo-object to design a new defence method against traffic analysis. The key idea is to generate as same as possible network traffic for webpages in a page set.
2. We develop the EQPO-enabled webpage structure to support the requests and responses for equal-sized pseudo-objects. Given a page set with size k, by composing the pseudo-objects with the same size and the same number, the traffic feature of any EQPO-enabled webpage in that set is similar with other $k-1$ webpages. And hence it is hard to identify a specified webpage in that set.
3. We have implemented a proof of concept prototype with data URI scheme and the AJAX technique, and we demonstrate the effectiveness of EQPO on defending against some typical TA attacks.

The rest of this paper is structured as follows. In Sect. 2, we overview some works on traffic analysis. In Sect. 3, we introduce the notion of equal-sized pseudo-objects. In Sect. 4, we discuss the method to construct the pseudo-objects. In Sect. 5, we conduct some experiments to validate our proposed method. And finally, the conclusion is drawn in Sect. 6.

2   Traffic Analysis in Encrypted Web Flows

2.1   Web Traffic

HTTP protocol is a typical request-response based protocol. To retrieve a document resource (object) from a web server, a browser first initiates a request for that object according to the corresponding URI (Uniform Resource Identifier), and then the server responses the request with required object contents. A webpage can be viewed as a set of objects that can be visited in a sequence. When visiting a webpage, the browser first fetches the basic HTML document from the web server who hosts that document, and then, issues HTTP requests to fetch other objects in sequence. Although the object requests could be on different connections, the requests order is logically depended on the structure of retrieved HTML document.

It is generally well known that webpages from different organizations have distinctly different structures, and hence could introduce distinguishable traffic features. However, webpages in a same website could also introduce distinguishable web traffic [8]. As an instance, Table 1 demonstrates the numbers and sizes of objects related to a small website, *maths.gzhu.edu.cn*.

Table 1. Features of some webpages in *maths.gzhu.edu.cn* (Retrieved July 3, 2015)

Object	default.asp		about.asp		news.asp	
	Number	Size (kB)	Number	Size (kB)	Number	Size (kB)
HTML	1	25.7	1	16.5	1	22.7
CSS	1	2.6	1	2.6	1	2.6
Image	67	940.0	15	379.7	38	557.3
Others	3	38.7	1	8.7	2	11.3
Total	72	1,007.0	18	407.5	42	593.9

According to this table, all of the three webpages have distinguishable object numbers and object sizes although they have the same CSS object. Counting object requests and aggregating traffic amounts can fingerprint these webpages easily.

2.2 Encrypted Web Traffic

It is well known that the HTTP protocol is not secure because of the data transmission in plain. A simple man-in-the-middle (MITM) attack could easily eavesdrop and intercept the HTTP conversations. HTTPS is designed to mitigate such MITM attacks by providing bidirectional encrypted transmissions. According to HTTPS protocol, the HTTPS payloads are encrypted but the TCP and IP headers are preserved. Noticed that the payloads are encrypted by a block cipher, such as the AES algorithm, and the lengths of encrypted payloads are almost the same as the plain ones except some octets are padded into a single block. This means that the secrets of payload contents are protected, but the real communicating address pairs and connections are easy to identify in HTTPS traffic.

The tunnel-based transmissions is used to hide real communicating IP address. It encapsulates entire original IP packet into a new IP packet. If the tunnel is encrypted, the encapsulated packet is also encrypted. A typical encrypted tunnel is the secure shell (SSH) tunnel, which means that a user may visit an external web server in private if he can connect to an external SSH server to create an SSH tunnel. However, the payloads are also encrypted by block ciphers and the encrypted payloads demonstrate almost same sizes as the corresponding plain versions. In general, a client is only communicating with an SSH proxy and the server behind the proxy is protected.

Tor is a special system for anonymization communication. Not only the communicating payloads but also the communicating pairs are protected. Different from the padding strategy in HTTPS and SSH, each Tor packet is padded to the size of MTU (Maximum Transmission Unit). It implies that all the packet cells in Tor have the same size.

As discussed above, current secure suites for Web browsing are focused on protecting the communicating contents, and protecting communication pairs in different level according to the security requirements. However, the traffic amounts, direction, and intervals can be sniffed by an adversary in the middle. The webpages could be identified even if they are transmitted in encrypted.

2.3 Traffic Analysis

When the traffic is encrypted and the encryption is perfect, analyzing the packet payload is meaningless. However, the encrypted payload size and the packet direction can be recognized clearly from the encrypted traffic. Consider that the current cipher suites cannot significantly enlarge the difference between the size of encrypted payload and its corresponding plain version, we can assume that the encryption is approximatively size-preserved. This implies that the size of an encryption object is similar to the size of object in plain form. Combining with the order of objects in transmitting, the structure of a specified webpage could be identified even if it is transmitted in encrypted form.

Table 2. Traffic analysis attack instances

Method	Classifier	Features considered
LL [5]	naïve Bayes	packet lengths
HWF [4]	multinomial naïve Bayes	packet lengths
LCC [6]	edit distance	packet lengths, order
DCRS [3]	naïve Bayes	total trace time,
		bidirectional total bytes,
		bytes in traffic bursts

Traffic analysis plays a key role in identifying a webpage in a webpage set. The core technique for traffic analysis is machine learning. There are two operation steps included in learning procedure, one for model training and the other for data classifying. A model is first trained by sampling data to extract generalized data features, and then it is incorporated into a classifier to distinguish new coming data. In particular, TA classifiers are constructed with supervised machine learning algorithms. It means that a classifier is trained on sets of traces that are labeled with k different webpages, and then it is used to determine whether or not a new set of traces is from a given webpage. Formally, the TA classifier is trained to a given labeled feature set $\{(\boldsymbol{F}_1, page_1), (\boldsymbol{F}_2, page_2), ..., (\boldsymbol{F}_k, page_k)\}$, where each $\boldsymbol{F}_i$ is a feature vector and $page_i$ is a webpage label. And then, a new set of traces with feature $\boldsymbol{F}'$ is input and the classifier will decide which label $page_i$ that the $\boldsymbol{F}'$ is attached. Some typical traffic analysis methods are enumerated in Table 2.

Liberatore and Levine [5] developed a webpage identification method (LL) by using naïve Bayes (NB) classifier. The LL method uses the packet direction and the packet length as feature vector. According to this method, NB is used to predict a label *page*: $page = \arg\max_i P(page_i|\boldsymbol{F}')$ for a given feature vector $\boldsymbol{F}'$ using Bayes rule $P(page_i|\boldsymbol{F}') = \frac{P(\boldsymbol{F}'|page_i)P(page_i)}{P(\boldsymbol{F}')}$, where $i \in \{1, 2, ..., k\}$. The LL method adopts the kernel density estimation to estimate the probability $P(\boldsymbol{F}'|page_i)$ over the example vector during the training phase, and the $P(page_i)$ is set to k^{-1}. The normalization constant $P(\boldsymbol{F}')$ is computed as $\sum_{i=1}^{k} P(\boldsymbol{F}'|page_i) \cdot P(page_i)$.

Herrmann, Wendolsky, and Federrath [4] proposed a method (HWF) by using a multinomial naïve Bayes (MNB) classifier. Both LL and HWF methods use the same basic learning method with the same traffic features. The difference is in the computation of $P(\boldsymbol{F}'|page_i)$. The HWF method determines the $P(\boldsymbol{F}'|page_i)$ with normalized numbers of occurrences of features while the LL method determines with corresponding raw numbers.

Observing that the order of non-MTU packets is almost invariable between packet sequences from the same webpage, Lu, Chang, and Chan [6] proposed a method (LCC) on the Levenshtein distance. The outgoing and incoming non-MTU packet length of $page_i$, $L_{out,i}$ and $L_{in,i}$, are obtained through learning. For the new traces t, the corresponding length pair, $L_{out,t}$ and $L_{in,t}$, are computed. The formula, $1 - \alpha \cdot D(L_{out,i}, L_{out,t}) - (1-\alpha) \cdot D(L_{in,i}, L_{in,t})$, is used to evaluate the difference, where α is the bias factor, which is set to 0.6 in their experiments, and D is the Levenshtein distance, which is equal to the number of insertions, deletions and substitutions of packet lengths to transform one packet sequence into another.

Most of the works are on single fine-grained packet analysis. In [3], Dyer, Coull, Ristenpart, and Shrimpton proposed an identification method (DCRS) based on coarse trace attributes, including total transmission time, total per-direction bandwidth, and traffic burstiness (total length of non ack packets sent in a direction between two packets sent in another direction). They used NB as the underlying machine learning algorithm and build the VNG++ classifier. Their results show that TA methods can reach a high identification accuracy against existed countermeasures without using individual packet lengths. It implies that the chosen feature attributes may be the most important factor in identifying webpages.

2.4   The Assumption

We follow the general scenario assumed in webpage identifying methods [4]: a user, say *Alice*, wants to protect which webpages she browsed from a web server against third parties. She can use popular secure suites, such as SSL, SSH, and IPSec, etc., to make the webpage contents transmit in encrypted form. The attacker, *Mallory*, is located between *Alice* and the web server and he can record the traffic between the two entities. Although the traffic may be encrypted, *Mallory* can retrieve the source and destination of the traffic, and also the sizes of packet payloads. He can identify *Alice* and the website based on the retrieved IP address. He intends to use traffic analysis techniques to identify which webpage in that website that *Alice* just visited.

2.5 Motivation of the EQPO Method

We first overview some methods against traffic analysis. The basic idea against TA analysis is to change traffic features. A typical method is using padding octets in packets to change the distributions of traffic packets. We consider the following defence countermeasures.

1. **PadMTU.** All packet sizes are increased to MTU. This method makes the length of each packet in session reach to the length of maximum transmission unit (MTU).
2. **PadRand.** For each packet in session, its size is increased to $len + r$, where $r, r \in \{0, 8, ..., \mathsf{MTU} - len\}$, is a random chosen number and len is the original packet length.
3. **Morphing.** Change packet length distribution in a webpage and make it look like another webpage.
4. **BuFLO.** The abbreviation of Buffered Fixed-Length Obfuscator, denoted by (d, ρ, t), where d is the size of fixed-length packets, ρ denotes the rate or frequency (in milliseconds) of packets sending, and t denotes the minimum amount of time (in milliseconds) for sending packets. The BuFLO counter-measure is different from the former 3 padding based methods for changing packet lengths, it tries to mitigate the effectiveness of coarse trace attributes by adjusting the traffic burstiness. A BuFLO implementation will send a packet of length d every ρ milliseconds until communications are stopped and at least t milliseconds of time have elapsed.

The features of web traffic are depended on the request-response traffic introduced by object retrieving transactions when webpages are rendering. Current techniques against traffic analysis are intended to make traffic features change dynamically. Observing that the same structure of webpages will generate indistinguishable traffic, if we can browse different webpages with a same number of equal-sized objects, we could make the traffic indistinguishable.

In the rest of this paper, we will propose EQPO, a method against traffic analysis. When visiting webpages in a page set, we intend to generate same number of requests for objects at client side, where each of those objects has the same size. Thus the traffic for object request-response transactions can be makd indistinguishable.

3   The Equal-Sized Pseudo-Object

We now turn our attention to how to construct equal-sized objects in a set of objects. We will introduce the notion of pseudo-object. We first review the scheme of data URI which provides a method to translate binary objects into text-based objects. And then we will propose the text-based pseudo-object.

3.1   Data URI

The dynamic programming language, JavaScript, is commonly viewed as part of web browsers. It has many routines to support the text string operations. The HTTP is an application protocol to exchange or transfer hypertext. When rendering a webpage, many types of objects are needed to retrieve from servers. Not all objects are text-based. For the non-text-based objects, such as the image files, splitting them directly with JavaScript language into fragments is not easy. We consider adopting the data URI scheme to handle the non-text-based objects [16].

The data URI scheme allows inclusion of small media type data as immediate data inline. It has the form of data:[<mediatype>][;base64],<encoded-data> where the mediatype part specifies the Internet media type and the ;base64 indicates that the data is encoded as base64. If both options are omitted, default to text/plain;charset=US-ASCII.

For example, the segment <img src="data:image/png;base64,iVBORw0..."> could be used to define an inline image embedded in HTML document. Considering that the base64-code is text-based, the base64-encoded objects can be easily cut into fragments and translated into pseudo-objects. It is trivial to compose any sizes of pseudo-objects at client side.

3.2   The Pseudo-Object

Let obj be an object in a webpage and $\|$ be the concatenation operator.

Definition 1. *A fragmentation of obj with length $m(m \geq 1)$ is a piece set, $F(obj) = \{f_1, f_2, ..., f_m\}$, such that $obj = \|_{i=1}^{m} f_i$ where $\forall f_i : f_i \neq \phi$.*

For example, let jso be a script object in webpage *page* with content is <script>alert("Hello World!");</script>. A fragmentation of object jso with length 2 is the piece set $\{jso\_f_1, jso\_f_2\}$, where the pieces jso_f_1 and jso_f_2 is <script>alert("Hello World"); and </script>, respectively. It is obviously that we can reassemble the object jso by simply concatenating the fragments in sequence, i.e., $jso = jso\_f_1 \| jso\_f_2$. It is noted that an object itself is also a fragmentation of this object with length 1. For example, if $csso$ is the css object in *page* whose content is hr {color:sienna;}, $csso$ can also be viewed as a fragmentation.

Suppose there is an object set S, $S = \{obj_1, obj_2, ..., obj_n\}$, where each obj_i is in a webpage *page* and $1 \leq i \leq n$. Let $F(obj_i) = \{obj\_f_{i,1}, obj\_f_{i,2}, ..., obj\_f_{i,m_i}\}$ be a fragmentation of obj_i with length m_i.

Definition 2. *A pseudo-object po with length l is $\|_{j=1}^{l} obj\_f_{i_j, n_i}$, where each component object $obj_{i_j} \in S$, $obj\_f_{i_j, n_i} \in F(obj_i)$, and for any two component object obj_{i_j} and $obj_{i_{j'}}$, $obj_{i_j} \neq obj_{i_{j'}}$ if $i_j \neq i_{j'}$.*

For the two objects, jso and $csso$, we discussed before, we can construct some pseudo-objects. For example, the pseudo-object, $po_1 = jso\_f_1$, is only constructed by the first fragment of jso, while the object, $po_2 = jso\_f_2||csso$, is concatenated by the second fragment of jso and $csso$.

Follow the Definition 2, we call the objects associated with a pseudo-object as the component objects, abbreviated as *components*. We also call the fragments associated with a pseudo-object as the component pseudo-fragment. When without causing confusion, we abbreviate the pseudo-fragment as *fragment*. As an example, the components of po_2 are jso and $csso$, and the corresponding fragments are jso_f_2 and $csso$.

We call the number of bytes contained in a fragment as the size of the fragment. Correspondingly, the sum of the sizes of all fragments in a pseudo-object is called the size of this pseudo-object. For example, the size of the fragment jso_f_1 and jso_f_2 is 30 and 9, respectively, and the size of the pseudo-object po_1 and po_2 is 30 and 27, respectively.

We say that two pseudo-objects are equal-sized if they have the same size. In general, given any two pseudo-objects, they are generally not equal-sized. However, we can append some padding octets to equalize them. We use the notation $Padding(n)$ to denote a string with n padding octets. As an example, if $po_2' = po_2||Padding(3)$, po_1 and po_2' are equal-sized.

Suppose we have another webpage *page**. Besides the object jso and $csso$ in *page*, it has a third object jso*, `<script>alert("Hi!");</script>`. When the browser needs to render *page* and *page**, besides the request for the basic HTML file, it will issue other 2 and 3 requests for objects, respectively. It is easy to distinguish *page* and *page** by using traffic analysis because of the different object request numbers and different object response amounts.

We consider the defence method with the same traffic features, i.e., we intend to browse the two pages with the same request numbers and the same object response amounts. As an example, we can require the browser to issue 3 object requests, with 30 bytes of response data each, to download all the objects in these two pages, respectively. To do so, we only need the requests and responses to be on equal-sized pseudo-objects. For the webpage *page*, we can define the 3 pseudo-objects as:

(a) $po_1 = jso\_f_1$;
(b) $po_2 = jso\_f_2||css||Padding(3)$;
(c) $po_3 = Padding(30)$,

while for *page*$'$, we define

(a) $po_1^* = jso\_f_1$;
(b) $po_2^* = jso\_f_2||csso||jso^*\_f_1$;
(c) $po_3^* = jso^*\_f_2||Padding(3)$,

where $jso^*\_f_1$ is `<sc` and $jso^*\_f_2$ is `ript>alert("Hi!");</script>`. Since these six pseudo-objects are equal-sized with each other, the traffic features for pseudo-objects in these two pages are the same if the requests and responses are on pseudo-objects. It implies that it is difficult to distinguish the two pages according to the encrypted web traffic.

4   Proof of Concept Implementation

In this section, we will provide our proof of concept EQPO implementation. We will discuss how to represent pseudo-objects in an HTML document, and how to retrieve those objects from web servers. We will present the structure of equal-sized pseudo-object-enabled (EQPO-enabled) HTML document to support retrieving the pseudo-object and assembling the original objects.

4.1   The EQPO-enabled HTML Document

In order to support retrieving pseudo-objects, it needs to redefine the structure of traditional HTML document. We call the HTML document that can support accessing equal-sized pseudo-objects as the equal-sized-pseudo-object-enabled (EQPO-enabled) HTML document. The following demonstrates an instance structure for EQPO-enabled HTML document with two img tags.

```
<html>
    <head> ...... </head>
    <script> ......
      function EQPOObject()
      ......
    </script>
    <body onload="EQPOObject()">
        ......
        <img id = objID1>
        ......
        <img id = objID2>
        ......
    </body>
</html>
```

To retrieve the pseudo-objects, the scripts for pseudo-objects must be included in HTML document and the URIs for extern objects are also needed to change. The document fragment <body onload="EQPOObject()"> implies that when the basic HTML document has been loaded, the onload event triggers the embedded script for EQPO objects. Note that the contents within img tags are referred to object identifiers (objIDs), it makes browser do not issue request for individual image file. When the script EQPOObject() is initiated, an XML-HttpRequest (XHR) object is created. The XHR object provides an easy way to retrieve data from a URI without having to refresh a full webpage. This means that some parts of the webpage could be updated while not downloading the whole. We use this XHR object to download EQPO objects.

The open() and send() methods in XHR object are used to require EQPO objects. The server generates EQPO objects and returns them to client. The construction of EQPO objects is on the order of the original objects in HTML document. The property of responseText in the XHR object is used to read the response content from server. When the browser handles the response, the onreadystatechange event listener is invoked and the property of readyState the current state of the request for the XHR objects. Particularly, when readyState is 4, which means done, and the status property is 200, which means ok, the response content has been retrieved successfully. A predefined procedure is invoked to call the success method in callback object to deal the server response. The response string responseText contains the EQPO objects. We then can decompose the string text according to the predefined syntax and obtain renderable object contents.

The parameter document includes the pseudo-object request number, the pseudo-object size, the original object number, and original object header sequences. This sequence can be expressed as the regular expression

`<objID@objlen@<mediatype>(;base64)?,|>{n},`

where n is the number of objects in this webpage, objlen is the size of the original object, which is denoted by objID, encoded in text or base64, and each object item is separated by |.

4.2 The Communications for Equal-Sized Pseudo-Objects

Figure 1 demonstrates the communications between browser and web server for pseudo-objects. When the browser initiates the request for basic HTML document, the server returns the EQPO-enabled HTML document. The browser renders it, and then requires parameters for equal-sized pseudo-objects.

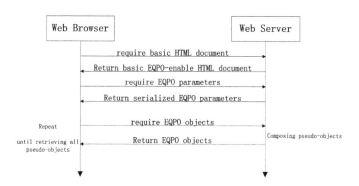

Fig. 1. Communications for pseudo-Objects

The pseudo-objects are fetched according to a predefined order and composed at server side on the structure of original webpage. The composed pseudo-object is in the form of `<objID@encoded-data>(|<objID@encoded-data>)*` with given size. With the received encoded object sizes, the browser maintains a buffer to store the downloaded fragment contents for each object. A simple comparing operation could be used to decide whether or not a given object has been downloaded. The returned encoded pseudo-objects are then decomposed and dispatched to the browser for rendering. The require-compose-decompose-dispatch procedure will be continue until all of equal-sized pseudo-objects has been downloaded.

It is noted that the notion of pseudo-object does not address the EQPO-enabled document and the EQPO parameter document. In fact, both of the two documents are text-based, it is easy to change their sizes for each webpage by appending HTML comments or padding octets. With the defined EQPO-enabled HTML document and the AJAX based communications, each translated webpage will demonstrate very similar traffic, especially when the traffic is encrypted.

5   Experiments and Discussions

5.1   The Experiment Setup

Our experiments are on artificial webpages with only image objects. We create an image library by picking some image files whose sizes are ranged from $5k$ to $25k$ from Internet. We then randomly select n ($n \in [n_L, n_H]$) image files to construct 200 traditional webpages and corresponding EQPO-enabled webpages, respectively. For each webpage, we visit 100 times via HTTPS and SSH tunnel, respectively. We record the traces, strip packet payloads with the TCPurify tool [18], and construct four types of trace set, EQPOHTTPS, EQPOSSH, TrHTTPS, and TrSSH, as listed in Table 3.

Table 3. Types of trace set

Type	Description
EQPOHTTPS	traces for visiting EQPO-enabled webpages via HTTPS
EQPOSSH	traces for visiting EQPO-enabled webpages via SSH
TrHTTPS	traces for visiting original webpages via HTTPS
TrSSH	traces for visiting original webpages via SSH

Note that the EQPOHTTPS dataset are on traces for visiting EQPO-enabled webpages via HTTPS. Analyzing those traces based on a certain traffic analysis technique can validate the capabilities of our proposed EQPO method against that specified traffic classifier. The comparison tests will be on TrHTTPS, the traces for visiting original webpages via HTTPS. With the tool from [17], we can test some popular defence countermeasures such as those described in Sect. 2, i.e., the PadMTU, PadRand, Morphing, and the BuFLO countermeasures. The usages of dataset EQPOSSH and TrSSH are analogous, the differences are that these two datasets are recorded from SSH tunnel.

In our conducted experiments, we set $[n_L, n_H]$ in 3 cases, i.e., [20, 40], [40, 60], and [60, 80]. These cases are denoted by $case_{[20,40]}$, $case_{[40,60]}$, and $case_{[60,80]}$, respectively. For each case, we set the pseudo-object request number as 35, 45, and 50, respectively. The sizes of each pseudo-object are equalized to 30,000 bytes.

To test the performances against traffic analysis, we run the code from [17] with classifiers and countermeasures we discussed in Sect. 2. The number of web-pages, k, is set to 2^i, where $1 \leq i \leq 7$, and 200, respectively. We use the default parameters in original code configuration. For each k with different classifiers and countermeasures, we run the test 10 times and average the accuracy as the ratio of successful identification.

5.2   Visiting Webpages with EQPO Method via HTTPS

HTTPS is a common secure protocol to resist MITM attacks in HTTP communication. It provides bidirectional encrypted communications between clients and servers. The HTTP payloads are encrypted but the TCP and IP headers are preserved.

We conduct a set of experiments to test our proposed method against the 4 discussed classifiers on EQPOHTTPS and compare with the results with other 4 countermeasures on TrHTTPS dataset. We figure the comparison results in following figures. It is noted that as the evaluation metric for the countermeasure against traffic analysis, the lower the identification accuarcy, the higher capabilities the countermeasure against analysis.

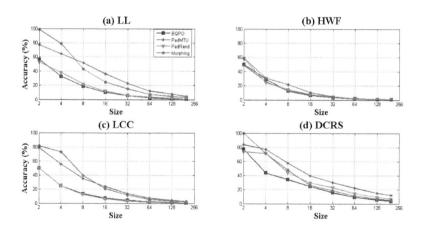

Fig. 2. Comparisons for $case_{[40,60]}$ with HTTPS traffic, EQPO vs PadMTU, PadRand, and Morphing.

Figure 2 demonstrates the comparison results with PadMTU, PadRand, and Morphing countermeasures for transmissions via HTTPS. As demonstrated in Fig. 2, the performances of the EQPO are different in the 4 addressed classifiers. It is the most effective countermeasure to defend against the LL, LCC, and

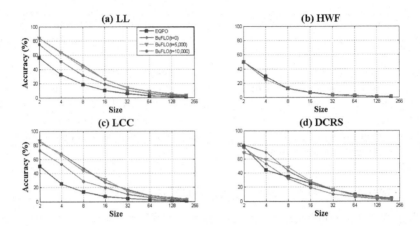

Fig. 3. Comparisons for $case_{[40,60]}$ with HTTPS traffic, EQPO *vs* BuFLO with $d = 1000, \rho = 20$.

DCRS classifiers, comparing with the three discussed countermeasures. For the HWF classifier, the comparing results get tangled up in performances.

Figure 3 shows the comparison results with BuFLO for transmissions via HTTPS. We set the parameters of BuFLO as $d = 1000, \rho = 20$, and t as 0, 5,000, 10,000, respectively. As demonstrated in Fig. 3, the performances of the EQPO are different in the 4 addressed classifiers. It is the most effective countermeasure to defend against the LL and LCC classifiers, comparing with the three sets of BuFLO parameters. For the HWF classifier, the EQPO also reach to the best except some smaller ks. For the DCRS classifier, the EQPO method is in average.

5.3 Visiting Webpages with EQPO Method via SSH

HTTPS can protect the content of transmitted packet, but it discloses real communicated peers. In some applications, we often require tunnel-based transmissions to hide real communicating IP address. The tunnel-based transmission means that the entire specified IP packet is encapsulated into a new IP packet, i.e., that specified packet is as the payload of that new packet. Thus the real IP addresses and ports are protected from any MITM attackers if the tunnel is encrypted. A typical encrypted tunnel is the secure shell (SSH) tunnel.

We also conduct some experiments to test our proposed method against the 4 discussed classifiers on EQPOSSH, comparing with the results with other 4 countermeasures on TrSSH.

Figure 4 shows the comparison results with PadMTU, PadRand, and Morphing countermeasures for transmissions via SSH.

For the case of transmission over SSH tunnel is shown in Fig. 4. It demonstrates similar comparing results as in HTTPS case. It is the most effective countermeasure to defend against the LL, LCC, and DCRS classifiers, comparing with the three discussed countermeasures. For the HWF classifier, the EQPO method gets tangled up with other three countermeasures.

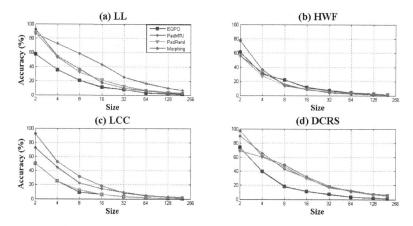

Fig. 4. Comparisons for $case_{[40,60]}$ with SSH traffic, EQPO vs PadMTU, PadRand, and Morphing.

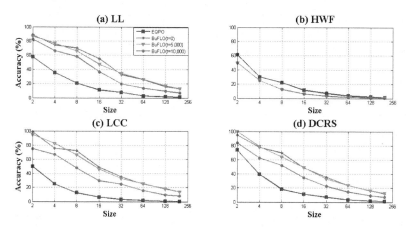

Fig. 5. Comparisons for $case_{[40,60]}$ with SSH traffic, EQPO vs BuFLO with $d = 1000, \rho = 20$.

Figure 5 demonstrates the comparison results with BuFLO for transmissions via SSH. We also set the parameters of BuFLO as $d = 1000, \rho = 20$, and t as 0, 5,000, 10,000, respectively. The case for transmission over SSH tunnel is shown in Fig. 5. It demonstrates that the EQPO is the most effective method against the LL, LCC and DCRS classifiers. For the HWF classifier, the EQPO method is weak in performances comparing with other three BuFLO scenarios.

5.4   Time Cost for the EQPO Method

To evaluate the time costs of the proposed EQPO method, we compare the time costs visiting EQPO-enabled webpages with visiting traditional webpages. We first construct 10 webpages whose object number ranged in 3 cases, i.e., ranged

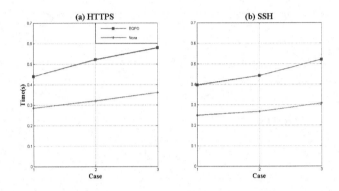

Fig. 6. Loading time: traditional pages and EQPO-enable pages

in [20, 40], [40, 60], and [60, 80], respectively, and transform them into EQPO-enabled pages, respectively. We visit each page 10 times, record the total time for loading objects, and then compute the average of time cost.

Figure 6 demonstrates the comparison results in visiting two types of web-pages. It shows that as the number of objects increasing, the extra time cost is also increased. Comparing to the traditional webpage rendering, the loading time of the EQPO-enabled webpages is proportional to that of the traditional web-pages. For example, in the case of HTTPS transmission, visiting EQPO-enabled pages with 40–60 objects needs 0.52 s in average while visiting traditional pages needs 0.32 s. For the pages with 60–80 objects, it averagely needs 0.58 s to visit EQPO-enabled pages while needs 0.36 s to visit traditional pages.

5.5   Discussions

The idea of our proposed EQPO method is on generating the same web traffic for different webpages. Our conducted experiments demonstrate that the EQPO method can defend against the 4 state of the art TA classifiers more effectively that the other addressed countermeasure. However, this method may introduce extra computation costs in both server side and client side. Since the pseudo-object is composed in base64-encode, it may increase at least 15 % of the network traffic volumes.

Our POC implementation is immature. The object retrieving process is AJAX-based and is over a single connection. Additionally, we do not consider the possible negative effects introduced by cookies, scripts, or caches. And also, we do not consider the case that the objects in page are from different web servers.

The proposed method requires that the original webpages translating into EQPO-enabled webpages. Actually, it is difficult to require different website masters translate their managed webpages into EQPO-enabled webpages. However, our proposed method is suitable for an individual website whose webpages are maintained by a single website master. For example, the clinic websites. Even if the attacker can infer the website the user visited, he cannot infer which the specified webpage the user visited.

5.6   Related Work

Encrypting web traffic is a common strategy to preserve users' privacy when surfing the Web. However, the current encryption suites are focused on protecting contents in flight and some other side traffic features cannot be effectively protected. A traffic analysis attack could use these features to infer the users' web browsing habits and their network connections. Identifying webpage on encrypted traffic is an important class of traffic analysis attacks.

Sun et al. [10] proposed a classifier based on the Jaccard coefficient similarity metric, and reliably identified a large number of webpages in 100,000 webpages. They also proposed some countermeasures against TA attacks but our proposed method is not addressed. Bissias et al. [1] used cross-correlation to determine webpage similarity with features of packet length and timing. Liberatore et al. [5] showed that it is possible to infer webpages with naïve Bayes classifier by observing only the lengths and the directions of packets. Herrmann et al. [4] suggested a multinomial naïve Bayes classifier for page identification that examines normalized packet counts.

Panchenko et al. [9] developed a Support Vector Machine(SVM) based classifier to identify webpages transmitted on onion routing anonymity networks (such as Tor). They used a variety of features, include some totaling data, based on volume, time, and direction of the traffic. Dyer et al. [3] provided a comprehensive analysis of general-purpose TA countermeasures. Their research showed that it is the choosing features, not the analysis tools, that mainly influence the accuracy of webpage identification.

Some other attacks are not only depended on network packets. Wang et al. [14] proposed a new webpage identifying technique on Tor tunnel. They interpreted the data by using the structure of Tor elements as a unit of data rather than network packets. Miller et al. [8] proposed an attack on clustering techniques for pages in a website. They used Gaussian distribution to determine the distance between clusters and identify specified webpages.

Padding extra bytes to packets is a standard countermeasure. Various padding strategies have been proposed to change encrypted web traffic [3]. However, this kind of countermeasures is on a single non-MTU packet, it is vulnerable when using coarse-grain traffic features [3,9]. Traffic morphing [15] tries to make a webpage traffic similar to another given webpage. This method is also focused on the fine-grain packets and is limited in changing coarse-grain features. Sending specified packets at fixed intervals [3] can reduce the correlation between the observed traffic and the hidden information and demonstrate more capabilities against the coarse-grain feature based analysis. However, it also introduces traffic overhead or delay in communication.

Some countermeasure proposals are on application-level. The browser-based obfuscation method, such as the HTTPOS method [7], takes the existing HTTP and TCP implementations to generate randomized requests with different object data requirements at client. It changes the number of requests from clients and the distribution of response packet volumes from servers. The HTTPOS method is on splitting the response packets by introducing special HTTP requests or TCP packets.

The authors also proposed some other methods based on artificial pseudo-objects and the data URI schema [12,13]. The sizes and numbers of pseudo-objects are different in different visits, and hence makes the traffic demonstrate different features in different visits.

6   Conclusion

We have proposed a countermeasure method, EQPO, to defend against webpage identification based traffic analysis by introducing equal-sized pseudo-objects in a page set. By forcing requests and responses on the same number of pseudo-objects, the traffic for a set of webpages could exhibit similar traffic patterns and is difficult to identify a specified webpage. Possible future work may include reducing the computation costs in client side and server side and make it more compatible in current web applications.

Acknowledgments. This paper was partially supported by the National Natural Science Foundation of China under Grant 61472091.

References

1. Bissias, G.D., Liberatore, M., Jensen, D., Levine, B.N.: Privacy vulnerabilities in encrypted HTTP streams. In: Danezis, G., Martin, D. (eds.) PET 2005. LNCS, vol. 3856, pp. 1–11. Springer, Heidelberg (2006)
2. Chen, S., Wang, R., Wang, X., Zhang, K.: Side-channel leaks in web applications: a reality today, a challenge tomorrow. In: Proceedings of IEEE S&P 2010, pp. 191–206 (2010)
3. Dyer, K., Coull, S., Ristenpart, T., Shrimpton, T.: Peek-a-Boo, I still see you: why traffic analysis countermeasures fail. In: Proceedings of IEEE S&P 2012, pp. 332–346 (2012)
4. Herrmann, D., Wendolsky, R., Federrath, H.: Website fingerprinting: attacking popular privacy enhancing technologies with the multinomial Naïve-bayes classifier. In: Proceedings of CCSW 2009, pp. 31–42 (2009)
5. Liberatore, M., Levine, B.: Inferring the source of encrypted HTTP connections. In: Proceedings of ACM CCS 2006, pp. 255–263 (2006)
6. Lu, L., Chang, E.-C., Chan, M.C.: Website fingerprinting and identification using ordered feature sequences. In: Gritzalis, D., Preneel, B., Theoharidou, M. (eds.) ESORICS 2010. LNCS, vol. 6345, pp. 199–214. Springer, Heidelberg (2010)
7. Luo, X., Zhou, P., Chan, E., Lee, W., Chang, R.: HTTPOS: sealing information leaks with browserside obfuscation of encrypted flows. In: Proceedings of NDSS 2011 (2011)
8. Miller, B., Huang, L., Joseph, A.D., Tygar, J.D.: I know why you went to the clinic: risks and realization of HTTPS traffic analysis. In: De Cristofaro, E., Murdoch, S.J. (eds.) PETS 2014. LNCS, vol. 8555, pp. 143–163. Springer, Heidelberg (2014)

9. Panchenko, A., Niessen, L., Zinnen, A., Engel, T.: Website fingerprinting in onion routing based anonymization networks. In: Proceedings of ACM WPES 2011, pp. 103–114 (2011)

10. Sun, Q., Simon, D., Wang, Y., Russell, W., Padmanabhan, V., Qiu, L.: Statistical identification of encrypted web browsing traffic. In: Proceedings of IEEE S&P 2002, pp. 19–30 (2002)

11. Sweeney, L.: k-anonymity: a model for protecting privacy. Int. J. Uncertainty Fuzziness Knowl. Based Syst. **10**(5), 557–570 (2002)

12. Tang, Y., Lin, P., Luo, Z.: Obfuscating encrypted web traffic with combined objects. In: Huang, X., Zhou, J. (eds.) ISPEC 2014. LNCS, vol. 8434, pp. 90–104. Springer, Heidelberg (2014)

13. Tang, Y., Lin, P., Luo, Z.: psOBJ: Defending against traffic analysis with pseudo-objects. In: Au, M.H., Carminati, B., Kuo, C.-C.J. (eds.) NSS 2014. LNCS, vol. 8792, pp. 96–109. Springer, Heidelberg (2014)

14. Wang, T., Goldberg, I.: Improved website fingerprinting on tor. In: Proceedings of WPES 2013, pp. 201–212 (2013)

15. Wright, C., Coull, S., Monrose, F.: Traffic morphing: an efficient defense against statistical traffic analysis. In: Proceedings of NDSS 2009, pp. 237–250 (2009)

16. Masinter, L.: The "data" URL scheme. http://www.ietf.org/rfc/rfc2397.txt

17. https://github.com/kpdyer/traffic-analysis-framework

18. http://masaka.cs.ohiou.edu/eblanton/tcpurify/

A Blind Dual Color Images Watermarking Method via SVD and DNA Sequences

Xiangjun Wu[1,2(✉)] and Haibin Kan[2]

[1] College of Software, Henan University, Kaifeng 475004, China
wuhsiang@yeah.net
[2] Shanghai Key Lab of Intelligent Information Processing,
School of Computer Science, Fudan University, Shanghai 200433, China

Abstract. This paper proposes a new blind color images watermarking scheme based on SVD and DNA sequences, in which a color watermark is embedded into a color host image for copyright protection. Firstly, the color watermark is encrypted by using DNA encoding and CML. Secondly, the color host image is partitioned into 4×4 non-overlapping pixel blocks and then the SVD transform is performed on each selected pixel block. Finally, the DNA sequence watermark (encrypted watermark) is embedded into the host image by modifying the matrix V. The watermark can be extracted from the watermarked image without resorting to the original host image and the original watermark. The experimental results show that the proposed watermarking scheme has not only good transparency, but strong robustness against the common image processing attacks and geometric attacks.

Keywords: Digital watermarking · Color image watermark · SVD · DNA sequences · Blind extraction · Robustness

1 Introduction

Due to the rapid development of Internet and multimedia technology, copyright protection of the multimedia information has received growing attention. Among various solutions for this issue, the digital watermarking is regarded as a powerful one. Digital watermarking is a technique that inserts a watermark into the multimedia host data [1]. The watermark can later be detected or extracted from the watermarked data for identifying the copyright owner. An efficient watermarking method should satisfy some essential requirements including imperceptibility, robustness, security etc. [2, 3].

According to the application of the host data during watermark extraction, the watermarking techniques can be divided into three categories: non-blind, semi-blind and blind [2]. The non-blind watermarking methods need the host data when watermark is extracted. Unfortunately, they are weak against the ambiguity attack [4]. The semi-blind watermarking schemes also require the host data or additional information during watermark detection, while the blind watermarking approaches can recover the watermark without referring to the host data. In real life, the blind watermarking schemes are preferred in view of the portability and availability of the host data.

© Springer International Publishing Switzerland 2016
D. Lin et al. (Eds.): Inscrypt 2015, LNCS 9589, pp. 246–259, 2016.
DOI: 10.1007/978-3-319-38898-4_15

As considering the domain in which the watermark is embedded, the image watermarking techniques can also be broadly classified into two categories: the spatial-domain and frequency-domain (transform domain) methods [5]. The spatial-domain schemes are simple but generally fragile to common image processing operations or other attacks [6, 7]. The watermarking methods based on transform domains such as discrete Fourier transform (DFT), discrete cosine transform (DCT), discrete wavelet transform (DWT), the singular value decomposition (SVD) and so forth, have more robustness and invisibility than the spatial-domain ones [8–13]. Although DCT and DWT are frequently used to design the digital watermarking methods in the past decades, the SVD transform has attracted more and more consideration due to the excellent properties of the SVD from the perspective of image processing, i.e., (i) the size of the matrix is not fixed; (ii) it has a good stability. In other words, when a small perturbation is exerted to an image, the fluctuation of its singular values is very small; (iii) singular values represent intrinsic algebraic image properties [11].

In the past few years, many SVD-based image watermarking methods have been proposed [11, 14–18]. In [11], a digital image watermarking method is proposed based on SVD. However, Zhang and Li [19] argued that the extracted watermark is not the embedded watermark but relate to the reference watermark. Fan et al. [14] modified the elements in the first column of U/V component for watermarking and used the V/U component to compensate visible distortion when embedding watermark into the component of SVD. But some image pixels will be modified incorrectly. Lai and Tsai [15] developed a hybrid image-watermarking scheme using DWT and SVD, in which the watermark is divided into two parts and then they were separately embedded in the singular values of the LH and HL sub-bands. Unfortunately, this scheme is impractical when the singular values generated from the original image are required to extract the watermark [20]. The SVD-based image watermarking schemes [16–18] are robust against the common image manipulations and geometric attacks. However, these methods usually lead to the false positive problem [21]. In addition, it is noted that the SVD-based watermarking methods [14–18] mainly embedded the gray-scale or binary watermarks into the gray-scale host images. It is known that the color images contain more information and are more prevalent in real applications compared with the gray-scale ones. So it is essential to develop the color image watermarking algorithms.

Most recently, some dual color images watermarking schemes are presented based on SVD [22–24]. Goléa et al. [22] introduced a block-SVD based blind dual color images watermarking scheme, in which one or more singular values must be modified to keep the order of singular values. However, modifying the singular values in this method will deteriorate the quality of the watermarked image. Su et al. [23] proposed a dual color images watermarking based on the improved compensation of SVD. In this method, the watermark bits are embedded into 4×4 blocks by modifying the second row first column and the third row first column elements of U component after SVD. In [24], a blind dual color images watermarking scheme was presented by analyzing the orthogonal matrix U via SVD. But the robustness of the dual color images watermarking methods [23, 24] is unsatisfactory for some severe attacks. To our best knowledge, there are few papers on the dual color images watermarking schemes based on SVD and DNA sequences.

Motivated by the above discussions, in this paper, we propose a secure dual color images watermarking scheme based on SVD and DNA sequences. First, the color watermark is encrypted by DNA sequences and CML for enhancing the security of the watermark. Then the host image is decomposed into 4×4 non-overlapping pixel blocks and the SVD is performed on each selected pixel block. The DNA sequence watermark is finally inserted into the host image by modifying the matrix V from the SVD transformation. Without the original watermark and host image information, the watermark can be extracted from the watermarked image by the watermark extraction algorithm. The experimental results show that the proposed watermarking scheme is effective. The main contributions of this paper are listed as follows: (1) encrypt the color watermark using DNA encoding and CML; (2) divide the color host image into 4×4 non-overlapping pixel blocks and perform the SVD transform on each selected pixel block; (3) embed the DNA sequence watermark (encrypted watermark) into the host image by modifying the elements in the first column of the matrix V; (4) our proposed method is a blind watermarking scheme, i.e., the host image and the original watermark are not required to extract the embedded watermark.

The rest of this paper is organized as follows. Section 2 gives a brief description on the SVD, DNA sequences and CML. In Sect. 3, the proposed method including the watermark embedding and extraction are described in detail. Section 4 presents the experimental results. Finally, the conclusions are drawn in Sect. 5.

2   Some Basics

2.1   SVD

SVD plays an important role in image processing, signal analysis and data compression. From the viewpoint of image processing, a digital image can be considered as a matrix. Let A denote an image with size of $m \times n$ ($m \leq n$), and r be the rank of the matrix A. Then the SVD of the matrix A can be given as follows:

$$A = U \times S \times V^{\mathrm{T}}, \tag{1}$$

where U and V are the $m \times m$ and $n \times n$ orthogonal matrices, respectively. $S = \mathrm{diag}(\alpha_1, \alpha_2, \cdots, \alpha_m)$ is a $m \times n$ diagonal matrix, where α_i are called as the singular values and satisfy $\alpha_1 \geq \alpha_2 \geq \cdots \geq \alpha_r > \alpha_{r+1} = \cdots = \alpha_m = 0$. The columns of U and V, i.e., U_i and V_i, are the left and right singular vectors, respectively. Each singular value specifies the luminance (energy) of the image. The singular vectors U_i and V_i denote the horizontal and vertical details (edges) of the image, respectively.

From various numerical experiments, it is found that the matrix V has an interesting feature, i.e., all of the elements in the first column are negative and their values are very close [23, 24]. So we will make full use of this property in our watermarking algorithm.

2.2   DNA Sequences

In recent years, the researchers have attempted to apply the DNA sequences to design the image encryption algorithms owing to huge potential of parallel computing ability, immense information storage density, and ultra-low energy consumption [25, 26]. The single-strand DNA sequence consists of four bases, i.e., A, C, G and T, where A and T, and C and G are complementary pairs, respectively. In modern electronic computer, the information is represented by the binary system. As is well known, the binary system has only two numbers, i.e., 0 and 1, and they are complementary to each other. Obviously, 00 and 11 are complementary, and 01 and 10 are also complementary. Therefore, we can use them to represent four bases A, C, G and T, respectively. There are total 24 kinds of coding schemes. However, according to the Watson-Crick complementary rule [27], only 8 kinds of them meet this rule, which are listed in Table 1. In our work, the DNA XOR operation is employed. We define the XOR operation for the DNA sequences based on the traditional XOR operation in the binary. There are also eight types of the DNA XOR rules. Table 2 gives one type of the DNA XOR operation, which is used to encrypt the watermark. As can be seen, a base in each row or column is unique and the DNA XOR is a reflexive operation.

2.3 Coupled Map Lattice (CML)

A coupled map lattice (CML) is a dynamical system with discrete time, discrete space and continuous state [28], which can be described by

$$x_{t+1}(n) = (1-\varepsilon)f(x_t(n)) + \frac{\varepsilon}{2}[f(x_t(n-1)) + f(x_t(n+1))], \qquad (2)$$

where $t = 1, 2, 3, \cdots$ is the time index, $n = 1, 2, \cdots, L$ is the lattice site index, $\varepsilon \in (0, 1)$ is the coupling constant, and $f(\cdot)$ is the mapping function. In this paper, we choose the Logistic map $f(x) = 1 - \mu x^2$, where the parameter $\mu \in (0, 2)$, and $x \in (0, 1)$. The periodic boundary condition $x_t(n) = x_t(n + L)$ is used in the CML, where L is the length of CML.

3 The Proposed Watermarking Method

In this section, a new secure dual color images watermarking scheme is proposed based on the SVD, DNA sequences and CML. The core idea is to encrypt the watermark by DNA sequences and CML followed by the embedding of the DNA sequence watermark via modifying the values of the first column of the matrix V. Without loss of generality, let the host image H be a color image with size of $M \times N$, and the watermark W be a color image with size of $m \times n$.

3.1 Watermark Embedding Algorithm

The watermark embedding process can be described as follows:
Step 1. Decompose the host image H and the watermark W into the R, G, B components, respectively. The color components H_θ and W_θ ($\theta = R, G, B$) are obtained. For improving the security of the watermarking method, the color watermark W is

Table 1. The encoding and decoding rules for DNA sequences.

	Rule 1	Rule 2	Rule 3	Rule 4	Rule 5	Rule 6	Rule 7	Rule 8
A	00	00	01	01	10	10	11	11
C	01	10	00	11	00	11	01	10
G	10	01	11	00	11	00	10	01
T	11	11	10	10	01	01	00	00

Table 2. The XOR operation for DNA sequences.

XOR	A	C	G	T
A	T	G	C	A
C	G	T	A	C
G	C	A	T	G
T	A	C	G	T

encrypted using DNA sequences and CML (2). Iterate CML (2) with the parameters ε_0, μ_0 and the initial conditions $x_0(1)$, $x_0(2)$, $x_0(3)$ to generate the key streams. Then transform the image matrices and the key streams into the DNA sequence matrices according to the encoding rules (given in Table 1). The DNA sequence watermark $\tilde{W}$ can be obtained by using the DNA XOR operation (given in Table 2). Here the parameters ε_0, μ_0 and the initial conditions $x_0(1)$, $x_0(2)$, $x_0(3)$ are used as the secret keys.

Step 2. Each component of the host image H is equally divided into non-overlapping 4×4 pixel blocks. Randomly select the embedding blocks in H_θ for embedding the corresponding component watermark $\tilde{W}_\theta$.

Step 3. Perform the SVD transform on each selected pixel block and obtain the matrix V.

Step 4. The DNA sequence watermark $\tilde{W}$ is embedded by changing the relations between v_1 and v_3, and v_2 and v_4 of the matrix V, where $v_1 = V(1,1)$, $v_2 = V(2,1)$, $v_3 = V(3,1)$ and $v_4 = V(4,1)$. If the embedded DNA watermark base is 'A', $v_1 - v_3 > 0$, $v_2 - v_4 > 0$, $|v_1 - v_3| > \delta$ and $|v_2 - v_4| > \delta$. If the embedded DNA watermark base is 'T', $v_1 - v_3 > 0$, $v_2 - v_4 < 0$, $|v_1 - v_3| > \delta$ and $|v_2 - v_4| > \delta$. If the embedded DNA watermark base is 'C', $v_1 - v_3 < 0$, $v_2 - v_4 > 0$, $|v_1 - v_3| > \delta$ and $|v_2 - v_4| > \delta$. If the embedded DNA watermark base is 'G', $v_1 - v_3 < 0$, $v_2 - v_4 < 0$, $|v_1 - v_3| > \delta$ and $|v_2 - v_4| > \delta$. Here δ represents an embedding threshold. When the aforementioned conditions are violated, the elements v_1, v_2, v_3 and v_4 should be modified by the following formulae given in Eqs. 3, 4, 5 and 6.

$$\text{if } w = \text{'A'}, \begin{cases} v_1 = -(\bar{V}_1 - \delta/2) \\ v_2 = -(\bar{V}_2 - \delta/2) \\ v_3 = -(\bar{V}_1 + \delta/2) \\ v_4 = -(\bar{V}_2 + \delta/2) \end{cases} \tag{3}$$

$$\text{if } w = \text{'T'}, \quad \begin{cases} v_1 = -(\bar{V}_1 - \delta/2) \\ v_2 = -(\bar{V}_2 + \delta/2) \\ v_3 = -(\bar{V}_1 + \delta/2) \\ v_4 = -(\bar{V}_2 - \delta/2) \end{cases}, \tag{4}$$

$$\text{if } w = \text{'C'}, \quad \begin{cases} v_1 = -(\bar{V}_1 + \delta/2) \\ v_2 = -(\bar{V}_2 - \delta/2) \\ v_3 = -(\bar{V}_1 - \delta/2) \\ v_4 = -(\bar{V}_2 + \delta/2) \end{cases}, \tag{5}$$

$$\text{if } w = \text{'G'}, \quad \begin{cases} v_1 = -(\bar{V}_1 + \delta/2) \\ v_2 = -(\bar{V}_2 + \delta/2) \\ v_3 = -(\bar{V}_1 - \delta/2) \\ v_4 = -(\bar{V}_2 - \delta/2) \end{cases}, \tag{6}$$

where w denotes the DNA watermark base, $\bar{V}_1 = (|v_1| + |v_3|)/2$ and $\bar{V}_2 = (|v_2| + |v_4|)/2$.

Step 5. The inverse SVD is performed to all selected blocks. One can obtain the R, G, B components of the watermarked image.

Step 6. Recombine the watermarked R, G, B components and the resulting watermarked image P is obtained.

3.2 Watermark Extraction Algorithm

The detailed extraction process can be formulated as follows:

Step 1. Divide the watermarked image P into the R, G, B components, and then decompose equally each watermarked component into the non-overlapping 4×4 pixel blocks.

Step 2. Choose the pixel blocks using the same random numbers as those in Sect. 3.1. Then perform SVD on each selected pixel blocks to get the matrix V.

Step 3. Extract the DNA sequence watermark $\hat{W}$ according to the following formula:

$$w = \begin{cases} \text{'A'}, & \text{if } (v_1 > v_3) \text{ and } (v_2 > v_4) \\ \text{'T'}, & \text{if } (v_1 > v_3) \text{ and } (v_2 < v_4) \\ \text{'C'}, & \text{if } (v_1 < v_3) \text{ and } (v_2 > v_4) \\ \text{'G'}, & \text{if } (v_1 < v_3) \text{ and } (v_2 < v_4) \end{cases}. \tag{7}$$

Step 4. By CML and the same secret keys, each watermark component can be extracted through executing the inverse process of DNA encryption presented in Sect. 3.1.

Step 5. Reconstruct the final watermark W' from the three recovered components.

4   Experimental Results and Discussions

In this section, to investigate the performance of the proposed method, six different color images of size 512×512 shown in Fig. 1 are used as the host images, and three different color images of size 32×32 displayed in Fig. 2 are regarded as the watermarks. In the following simulations, Fudan badge is embedded into Lena and Panda images, Multi-color image is embedded into Airplane and Lake images, and IEEE logo is embedded into House and Terrace images. We arbitrarily choose the embedding threshold as $\delta = 0.05$. In our work, peak signal-to-noise ratio (*PSNR*) [8, 9] and the normalized cross-correlation (*NC*) [23, 24] are employed to evaluate the imperceptibility and robustness of the proposed scheme, respectively.

(a) Lena   (b) Panda   (c) Airplane

(d) Lake   (e) House   (f) Terrace

Fig. 1. The original color host images.

(a) Fudan badge   (b) Multi-color image   (c) IEEE logo

Fig. 2. The color watermarks.

4.1 Imperceptibility Analysis

Figure 3 shows the watermarked images and corresponding *PSNR* values. As can be seen, the proposed method has a better watermark transparency. It is known that if the value of *PSNR* is more than 30 dB, the difference between the original and processed images is unnoticeable. Figure 4 displays the extracted watermarks and the *NC* values under no attacks. As can be seen, all the *NC* values are extremely close to 1, which indicates that the extracted watermarks are very similar to the original ones. It is virtually impossible to distinguish the differences between the original watermarks and the original ones.

(a) Lena (*PSNR* = 34.3407)   (b) Panda (*PSNR* = 33.0422)   (c) Airplane (*PSNR* = 32.2057)

(d) Lake (*PSNR* = 34.2142)   (e) House (*PSNR* = 31.9434)   (f) Terrace (*PSNR* = 34.7388)

Fig. 3. The watermarked images.

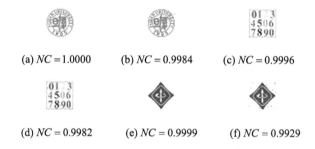

(a) *NC* = 1.0000   (b) *NC* = 0.9984   (c) *NC* = 0.9996

(d) *NC* = 0.9982   (e) *NC* = 0.9999   (f) *NC* = 0.9929

Fig. 4. The extracted watermarks without attacks.

4.2 Robustness Analysis

JPEG compression attack is one of the common attacks. In this experiment, the watermarked images are lossy-compressed with different quality factors. As shown in Fig. 5, the proposed scheme can effectively extract the watermark, which illustrates the proposed scheme is robust against the JPEG compression.

To measure the robustness of the proposed method, many other attacks are also tested with the Lena and House images. Table 3 provides the comparison of the NC and visual perception results with other watermarking algorithms [22, 24]. As can be seen, the proposed scheme is more robust against median filtering, sharpening, Gamma correction, histogram equalization and cropping attacks compared with those methods in [22, 24]. The watermarking method in [24] has a better robustness against salt & pepper noise and resizing attacks, while the presented scheme in [22] has a more satisfying robustness against contrast adjustment.

(a) Attacked Lena image

(b) Extracted watermark

(c) Attacked House image

(d) Extracted watermark

Fig. 5. Results for JPEG compression with quality factor 30.

4.3 Security Analysis

Apart from the robustness considerations, the security is also important for an efficient watermarking method. In order to guarantee the security of the watermark, a watermarking scheme should be highly sensitive to the secret keys. In other words, a slight change in the secret keys will result in the failure of the perfect extraction. In the proposed technique, the parameters ε, μ and the initial conditions $x_0(1), x_0(2), x_0(3)$ are used as the secret keys. All of these keys are private keys, and the keys in the extraction

Table 3. The extracted watermarks and *NC* values under various attacks.

Attacks	Images	Ref. [22]	Ref. [24]	Proposed method
		Extract watermarks / *NC*	Extract watermarks / *NC*	Extract watermarks / *NC*
Salt & Pepper Noise (5%)	Lena	0.9387	0.9524	0.9047
	House	0.9496	0.9562	0.8985
Contrast adjustment	Lena	0.9970	0.9729	0.9383
	House	0.9828	0.9703	0.8733
median filtering	Lena	0.8976	0.8677	0.9042
	House	0.8943	0.8585	0.8981
Sharpening	Lena	0.2120	0.7813	0.9934
	House	0.9014	0.7147	0.9893
Gamma correction	Lena	0.9652	0.8643	0.9996

	House			
		0.9705	0.8867	0.9991
Histogram equalization	Lena			
		0.9484	0.7989	0.9976
	House			
		0.9589	0.7090	0.9954
	Lena			
Resizing (512⊠ 256⊠ 512)		0.7953	0.9193	0.8701
	House			
		0.8052	0.9160	0.8625
	Lena			
Resizing (512⊠ 1024⊠ 512)		0.9863	0.9996	0.9951
	House			
		0.9550	0.9994	0.9898
	Lena			
Cropping (25%)		0.8693	0.9444	0.9581
	House			
		0.8566	0.9475	0.9528

stage must be identical to those in the embedding stage. If the computational precision is 10^{-15}, the key space of the proposed technique is larger than 2^{172}, which makes the brute-force attack infeasible.

Let the initial key set be $K_0 = (\varepsilon_0, \mu_0, x_0(1), x_0(2), x_0(3))$. We slightly modify the secret keys as follows: (i) $K_1 = (\varepsilon'_0, \mu_0, x_0(1), x_0(2), x_0(3))$ where $\varepsilon'_0 = \varepsilon_0 + 10^{-15}$; (ii) $K_2 = (\varepsilon_0, \mu'_0, x_0(1), x_0(2), x_0(3))$ where $\mu'_0 = \mu_0 - 10^{-15}$; (iii) $K_3 = (\varepsilon_0, \mu_0, x'_0(1), x_0(2), x_0(3))$ where $x'_0(1) = x_0(1) + 10^{-15}$; (iv) $K_4 = (\varepsilon_0, \mu_0, x_0(1), x'_0(2), x_0(3))$ where $x'_0(2) = x_0(2) - 10^{-15}$; (v) $K_5 = (\varepsilon_0, \mu_0, x_0(1), x_0(2), x'_0(3))$ where $x'_0(3) = x_0(3) + 10^{-15}$. Figure 6 shows the sensitivity test of the proposed method by extracting the watermark using different secret keys. As can be seen, the correct watermark can be recovered only by the right keys. Therefore, our proposed watermarking scheme is highly sensitive to the secret keys, and all correct keys are necessary for perfect extraction.

Secret keys	Correct keys	Wrong keys				
	K_0	K_1	K_2	K_3	K_4	K_5
Extracted watermarks						

Fig. 6. The extracted watermarks using different secret keys.

5 Conclusions

In this paper, a secure dual color images watermarking scheme is presented based on SVD, DNA sequences and CML, in which both the host image and the watermark are the color images. Different from the existing image watermarking schemes, we insert the DNA sequences instead of binary numbers into the host images. The watermark is embedded by modifying the elements in the first column of the matrix V from the SVD transformation. The experimental results have validated the effectiveness and robustness of the proposed watermarking scheme.

Acknowledgements. This research was jointly supported by the National Natural Science Foundation of China (Grant Nos. 61004006 and 61203094), China Postdoctoral Science Foundation (Grant Nos. 2013M530181 and 2015T80396), the Natural Science Foundation of Henan Province, China (Grant No. 13230010254), Program for Science & Technology Innovation Talents in Universities of Henan Province, China (Grant No. 14HASTIT042), the Foundation for University Young Key Teacher Program of Henan Province, China (Grant No. 2011GGJS-025), Shanghai Postdoctoral Scientific Program (Grant No. 13R21410600), the Science & Technology Project Plan of Archives Bureau of Henan Province (Grant No. 2012-X-62) and the Natural Science Foundation of Educational Committee of Henan Province, China (Grant No. 13A520082).

References

1. Swanson, M.D., Kobayashi, M., Tewfik, A.H.: Multimedia data embedding and watermarking technologies. Proc. IEEE **86**(6), 1064–1087 (1998)
2. Hartung, F., Kutter, M.: Multimedia watermarking techniques. Proc. IEEE **87**(7), 1079–1107 (1999)
3. Bianchi, T.: Secure watermarking for multimedia content protection: a review of its benefits and open issues. IEEE Signal Proc. Mag. **30**(2), 87–96 (2013)
4. Cox, I.J., Miller, M.L., Bloom, J.A.: Digital Watermarking. Morgan Kaufmann Publishers, San Francisco (2001)
5. Barni, M., Bartolini, F., De Rosa, A., Piva, A.: Optimal decoding and detection of multiplicative watermarks. IEEE Trans. Signal Process. **51**(4), 1118–1123 (2003)
6. Chan, C.K., Cheng, L.M.: Hiding data in images by simple LSB substitution. Pattern Recogn. **37**(3), 469–474 (2004)
7. Lee, I.-S., Tsai, W.-H.: Data hiding in grayscale images by dynamic programming based on a human visual model. Pattern Recogn. **42**(7), 1604–1611 (2009)
8. Nikolaidis, A., Pitas, I.: Asymptotically optimal detection for additive watermarking in the DCT and DWT domains. IEEE Trans. Image Process. **12**(5), 563–571 (2003)
9. Hu, H.T., Hsu, L.Y.: Robust, transparent and high-capacity audio watermarking in DCT domain. Signal Process. **109**, 226–235 (2015)
10. Keyvanpour, M., Bayat, F.M.: Blind image watermarking method based on chaotic key and dynamic coefficient quantization in the DWT domain. Math. Comput. Model. **58**(1–2), 56–67 (2013)
11. Liu, R., Tan, T.: An SVD-based watermarking scheme for protecting rightful ownership. IEEE Trans. Multimedia **40**(1), 121–128 (2002)
12. Verma, V.S., Jha, R.K., Ojha, A.: Significant region based robust watermarking scheme in lifting wavelet transform domain. Expert Syst. Appl. **42**(21), 8184–8197 (2015)
13. Solachidis, V., Pitas, I.: Circularly symmetric watermark embedding in 2D DFT domain. IEEE Trans. Image Process. **10**(11), 1741–1753 (2001)
14. Fan, M.Q., Wang, H.X., Li, S.K.: Restudy on SVD-based watermarking scheme. Appl. Math. Comput. **203**(2), 926–930 (2008)
15. Lai, C.C., Tsai, C.C.: Digital image watermarking using discrete wavelet transform and singular value decomposition. IEEE Trans. Inst. Measure. **59**(11), 3060–3063 (2010)
16. Bhatnagar, G., Wu, Q.M.J., Raman, B.: A new robust adjustable logo watermarking scheme. Comput. Secur. **31**(1), 40–58 (2012)
17. Song, C., Sudirman, S., Merabti, M.: A robust region-adaptive dual image watermarking technique. J. Vis. Commun. Image Represent. **23**(3), 549–568 (2012)
18. Bhatnagar, G., Wu, Q.M.J.: A new logo watermarking based on redundant fractional wavelet transform. Math. Comput. Model. **58**(1–2), 204–218 (2013)
19. Zhang, X.P., Li, K.: Comments on SVD-based watermarking scheme for protecting rightful ownership. IEEE Trans. Multimedia **7**(3), 593–594 (2005)
20. Asikuzzaman, Md., Alam, Md.J., Pickering, M.R.: A blind and robust video watermarking scheme in the DT CWT and SVD domain. In: 31st Picture Coding Symposium (PCS 2015), pp. 1–5 (2015)
21. Guo, J.M., Prasetyo, H.: False-positive-free SVD-based image watermarking. J. Vis. Commun. Image Represent. **25**(5), 1149–1163 (2014)
22. Goléa, N.E.H., Seghir, R., Benzid, R.: A bind RGB color image watermarking based on singular value decomposition. In: IEEE/ACS International Conference on Computer Systems and Applications (AICCSA 2010), pp. 1–5 (2010)

23. Su, Q., Niu, Y., Zhao, Y., Pang, S., Liu, X.: A dual color images watermarking scheme based on the optimized compensation of singular value decomposition. Int. J. Electron. Commun. (AEU) **67**(8), 652–664 (2013)
24. Su, Q., Niu, Y., Wang, G., Jia, S., Yue, J.: Color image blind watermarking scheme based on QR decomposition. Signal Process. **94**, 219–235 (2014)
25. Zhang, Q., Guo, L., Wei, X.: Image encryption using DNA addition combining with chaotic maps. Math. Comput. Model. **52**(11–12), 2028–2035 (2010)
26. Liu, H., Wang, X., Kadir, A.: Image encryption using DNA complementary rule and chaotic maps. Appl. Soft Comput. **12**(5), 1457–1466 (2012)
27. Watson, J.D., Crick, F.H.C.: A structure for deoxyribose nucleic acid. Nature **171**, 737–738 (1953)
28. Kaneko, K.: Overview of coupled map lattices. Chaos **2**(3), 279 (1992)

On the Application of Clique Problem for Proof-of-Work in Cryptocurrencies

Samiran Bag[1][✉], Sushmita Ruj[2], and Kouichi Sakurai[1]

[1] Kyushu University, Fukuoka, Japan
bag@inf.kyushu-u.ac.jp, sakurai@csce.kyushu-u.ac.jp
[2] Indian Statistical Institute, Kolkata, India
sush@isical.ac.in

Abstract. In this work we propose a scheme that could be used as an alternative to the existing proof of work(PoW) scheme for mining in Bitcoin P2P network. Our scheme ensures that the miner must do at least a non-trivial amount of computation for solving the computational problem put forth in the paper and thus solving a PoW puzzle. Here, we have proposed to use the problem of finding the largest clique in a big graph as a replacement for the existing Bitcoin PoW scheme. In this paper, we have dealt with a graph having $O(2^{30})$ vertices and $O(2^{48})$ edges which is constructed deterministically using the set of transactions executed within a certain time slot. We have discussed some algorithms that can be used by any Bitcoin miner to solve the PoW puzzle. Then we discuss an algorithm that could perform this task by doing $O(2^{80})$ hash calculations. We have also proposed an improvement to this algorithm by which the PoW puzzle can be solved by calculating $O(2^{70.5})$ hashes and using $O(2^{48})$ space. This scheme is better than the existing proof of work schemes that use Hashcash, where a lucky miner could manage to find a solution to the proof of work puzzle by doing smaller amount of computation though it happens with very low probability. Bitcoin incentivizes the computing power of miners and hence, it is desirable that miners with more computing power always wins. Also, the Bitcoin PoW scheme only incentivizes computing power of miners but our PoW scheme incentivizes both computing power and memory of a miner. In our proposed scheme only the miner cannot randomly find a largest clique without knowing the clique number of the graph.

Keywords: Random graph · Clique problem · Proof of work · Bitcoin

1 Introduction

Bitcoin is a cryptocurrency that does not rely on any centralized trusted server such as a bank. Here, users maintain a publicly auditable ledger called the Bitcoin block chain. This block chain contains all the valid blocks of valid transactions conducted by users and "accepted" by the Bitcoin network. These blocks are generated by 'Bitcoin miners' by solving a proof of work puzzle [13]. This puzzle

© Springer International Publishing Switzerland 2016
D. Lin et al. (Eds.): Inscrypt 2015, LNCS 9589, pp. 260–279, 2016.
DOI: 10.1007/978-3-319-38898-4_16

is adopted from Back's Hashcash [2]. In the classical model the miners would need to construct a block by doing a brute force search with the SHA-256 hash function in order to find a nonce such that the hash of all the transactions and the nonce along with the header of the previous block happens to fall below some predefined target value. If a Bitcoin miner can find such a nonce, then she broadcasts it along with the transactions she considers to be valid. A miner who solves the proof of work puzzle, that is a miner who finds the appropriate nonce is given some amount of Bitcoins as incentive and the block that she broadcasted is mostly accepted by all the users in the entire Bitcoin network. The consistency of the Bitcoin network largely depends upon the honesty of Bitcoin miners who win the proof of work puzzle. This incentive is given to lure honest Bitcoin users into investing their computational resources for keeping Bitcoin secure and to keep dishonest users at bay. In the existing Bitcoin mining scheme a miner needs to compute $O(2^{70})$ hash computations on an average for finding a nonce such that the hash of the nonce and the set of valid transactions is less than a predefined limit under SHA-256 hash scheme. There are other altcoins like Permacoin [11] that uses proof of storage rather than proof of work. Another paper [18] by Sengupta et al. proposed Retriecoin that uses proof of retrievability scheme to improve the communication and storage requirement of Permacoin. Details about Bitcoin are provided in Sect. 2.

Motivation: The existing Bitcoin mining scheme is simple to implement and has got sufficient mean computational complexity for being a computationally expensive mining scheme. The main downside of this scheme is that a Bitcoin miner may solve the puzzle in a single attempt that is the nonce she chooses at first instance might produce a hash which is less than the desired limit. The probability of getting such a nonce in polynomial time is very low, though it is not equal to zero. In fact the probability of solving the PoW puzzle in the i'th attempt is same as solving it in the first attempt. The work of Tromp [21] is proposed as a substitute for the current Bitcoin mining Algorithm. In this scheme the author proposed to use graph problems for Bitcoin PoW. The paper proposes to construct a big graph depending upon a hash based computation and then examines the presence of subgraph (usually an L cycle) in the original graph. Tromp did not discuss any method to apply his scheme in Bitcoin network. Also, the solution of the PoW puzzle could be stolen while in transmission and then the miner who has stolen it could re-transmit is as her solved PoW puzzle.

Contribution: In this paper we introduce another proof-of-work scheme that could be used as an alternative to the existing Bitcoin mining puzzle. Our scheme proposes to employ clique finding problem in a giant random graph for the proof of work puzzle. Finding the maximum clique in a graph is NP-hard. Hence, we choose clique finding problem for the PoW scheme for Bitcoin mining. In our scheme a miner needs to find the largest possible clique in a graph containing exponential number of vertices. The construction of the random graph itself requires computation of exponentially many hashes. Thus, finding the largest clique in the graph becomes a hard problem. We also have proposed an Algorithm that could find the largest cliques in $O(2^{80})$ time which is not too higher than the capacity of Bitcoin network as of April 2015 according to the publicly available information on WWW.BITCOIN.INFO. We further improve the run-time to $O(2^{70.5})$. Our work is motivated by Tromp's paper [21], but is better than [21]. In our scheme, however the graph is constructed deterministically from the set of transactions conducted during a certain time slot. So, any peer of the Bitcoin network can construct the graph deterministically from the set of transactions she wants to verify. Tromp's work does not discuss how to relate his proposed PoW scheme to Bitcoin mining and hence is incomplete. Our scheme however is proposed exclusively for the Bitcoin network. Thus, our scheme is more applicable to the decentralized Bitcoin networking scenario. In our scheme, the miners try to find the largest possible clique within a bounded time what we call an epoch. A miner who can find the largest possible clique within this time-bound is the one who wins the mining game. As we shall see in later parts of the paper that a miner's probability of finding such a clique in the first few attempts is much lower than that of the current PoW schemes that use Hashcash. Chan et al. [4] showed that there can be no algorithm running time $O(n^{o(k)})$ to test whether a given graph has a clique of size k. So, to find a clique, one needs to do a lot of computation and the difficulty of solving our PoW puzzle is based on this result.

Organization: The rest of the paper is organized as follows: In Sect. 2 we do a background study of Bitcoin cryptocurrency and clique problem. In Sect. 3, we have discussed some previous results on random graphs that will be used in our paper. In Sect. 4.1 we discuss our PoW schemes for Bitcoin mining. We provide two mining algorithms namely Figs. 1 and 2 in Sect. 4.1, the second algorithm being a modification of the first one. The mining algorithms use a variant of clique finding problem as the proof of work puzzle. In Sect. 5, we give a detailed theoretic study of our PoW scheme. We also provide two algorithms for solving the proposed PoW scheme. These are Figs. 3 and 4. Figure 3 does random search over the set of vertices of the graph to find the maximum clique whereas Fig. 4 does an efficient search for the maximum clique. We have discussed some improvement of the Fig. 4 that will allow any miner to efficiently find the largest clique and thus to solve the PoW puzzle. In Sect. 6, we give a comparison between our scheme and other PoW schemes proposed for Bitcoin. We conclude the paper in Sect. 7.

2   Related Work

We discuss these topics for our background study.

(1) Bitcoin: Bitcoin is a decentralized cash system that does not depend on a centralized server such as a Bank. Its users make online payments by digitally signing every transaction with their secret key. The corresponding public key can be used to publicly verify the authenticity of the transaction. Bitcoin network maintains a publicly auditable ledger called Bitcoin block chain that is aimed at preventing double spending of Bitcoins. In every 10 min a new block is added to

the block chain that consists of valid transactions. To create a new block Bitcoin users need to solve a PoW puzzle that requires nontrivial amount of computation. Bitcoin network uses Back's Hashcash [2] for PoW. A Bitcoin block is constructed by users called miners and it requires one to execute a nontrivial amount of computation. For mining Bitcoin, a miner needs to find a nonce such that the hash of the nonce, the merkle root of all valid transactions as well as some other parameter starts with some predefined number of zeros. Since the hash function is secure under random oracle model, the only way of finding such a nonce is to do a brute force over all the possible values of the nonce until a nonce is found that gives the hash the requires number of zeros. This is called the Bitcoin mining puzzle and it needs to be solved by at least one miner every time a block is constructed. Typically it takes about ten minutes to construct a block in the Bitcoin network. The number of zeros which are required for the hash to be valid solution defines the difficulty level of the Bitcoin mining puzzle and is updated regularly to ensure that the time to construct a block remains nearly equal to ten minutes. A Bitcoin miner is given a reward of 25 Bitcoins when she can succesfully create a valid block. This incentive attracts Bitcoin miners into expending their computational resources for solving the mining puzzle. At present the Bitcoin network computes 400 million Gigahashes per second. Since, mining takes a nontrivial amount of computation, typically equivalent to the processing power of millions of desktop machines it is not possible for an individual miner to solve the mining puzzle on her own. Hence, many individual miners form a big computational force by combining their processing power to crack down the mining puzzle together. The group of miners is called a mining pool and are administrated by a miner called pool operator. In every pool, the pool operator is solely responsible for distribution of jobs to individual miners as well as for sharing the reward if the pool can solve the mining puzzle.

(2) Clique problem: A clique is a complete subgraph of a graph. Clique problem involves finding two types of cliques viz. maximal clique and maximum clique. A maximal clique is one that cannot be extended to form a clique of bigger size i.e. there is no bigger clique in the graph that contains the former as a subgraph. Again, a maximum clique of graph is a clique that has the size equal to that of the largest clique in the same graph. Clique problem is defined as the problem of finding the largest clique in a graph or listing all maximal cliques in the graph. Moon et al. [12] showed that the number of maximal cliques of a graph could be $O(3^{n/3})$. The simplest Algorithm that can list all maximum clique of a graph is the Bron-Kerbosch Algorithm [3] that has a mean run time of $O(3^{n/3})$. A variant of this was proposed by Tomita et al. [20] and it has a worst case complexity of $O(3^{n/3})$. The Algorithm finds all the maximal cliques in the graph and returns the largest one. Tarjan et al. [19] proposed an Algorithm that improved the complexity to $O(2^{n/3})$ using a recursive backtracking scheme that eliminates some recursive calls intelligently. Further improvements were proposed by Jian [8]. This Algorithm can list all largest cliques in $O(2^{0.304n})$ time. Robson [16] again used dynamic programming technique to reduce it to $O(2^{0.276n})$. But his

Algorithm uses higher amount of space. The best Algorithm known till today is by Robson [17] that has a run time of $O(2^{0.249n})$.

Cliques of a fixed size k can be found in polynomial time by examining all subsets of size k and checking whether each of them forms a clique or not. This would take $O(\binom{n}{k}k^2)$ time. There has been a lot of research done on finding all triangles in a graph. Chiba et al. [5] proposed an Algorithm that finds all triangles in $O(e^{\frac{3}{2}})$ time, e being the number of edges in the graph. The Algorithm can enumerate all triangles in a sparse graph faster. Alon et al. [1] improved the $O(e^{\frac{3}{2}})$ Algorithm for finding triangles to $O(e^{1.41})$ by using fast matrix multiplication. The idea of faster matrix multiplication has also been used for finding bigger cliques in [6,9,14,22,23].

3 Preliminaries

3.1 Random Graph

A random graph $\mathcal{G}_{n,p}$ is a graph (V, E) on a set of vertices V, $|V| = n$ such that

$$\forall v_1, v_2 \in V, Pr[\{v_1, v_2\} \in E] = p.$$

That is in a random graph with n vertices, the probability of occurrence of an edge between a pair of vertices is given by p. This model of construction of graph was first proposed by Erdős and Rényi and is named after them.

Theorem 1. [7] *Let $Z(n, p)$ denote the size of largest complete subgraph of a graph $\mathcal{G}_{n,p}$. The sequence $\{Z(n,p)\}$ of random variables satisfies*

$$\lim_{n \longrightarrow \infty} Pr[Z(n, p) \to \frac{2 \log n}{\log \frac{1}{p}} + O(\log n)] = 1$$

Theorem 2. [10] *For $n \geq 1$, $0 < p < 1$, for $1 \leq k \leq n$,*

$$\left\{ \sum_{j=\max(0,2k-n)}^{k} \frac{\binom{n-k}{k-j}\binom{k}{j}}{\binom{n}{k}} p^{-j(j-1)/2} \right\}^{-1} \leq Prob[Z(n,p) \geq k] \leq \binom{n}{k} p^{k(k-1)/2}$$

The above Theorems show that for any random graph having high number of vertices the size of the largest subgraph approaches a certain value. In Theorem 2, Matula proved using computation that the density of $Z(n, p)$ is very spiked for a big random graph. There is a single spike where more than 99 % of the observed data accumulate. He computed that $Pr[Z(10^{10}, 1/4) = 30] > 0.9997$ [10]. If we have a random graph $\mathcal{G}_{n,p}$, then if n is very high, typically having exponential size, then the size of the largest complete subgraph will be $\approx \frac{2 \log n}{\log 1/p} + O(\log n)$ with an very high probability. Due to this behavior of the density of $Z(n, p)$, it is possible to apply clique finding problem for solving Bitcoin proof of work.

3.2   Bitcoin Transactions

Bitcoin users, as mentioned above, make online payments by making a transaction. A Bitcoin transaction is hashed with the payee's public key and is signed by the payer to validate the payment. A transaction can have multiple input fields as well as output fields and the number of all of them determines the actual size of the transaction. Each transaction input requires at least 41 bytes for references to the previous transaction and other headers and each transaction output requires an additional 9 bytes of headers. Besides, each transaction has an at least 10 bytes long header. Thus, 166 bytes is the minimum size of a Bitcoin transaction.

3.3   Notations

We have used the following notations in this paper:

- K_i is a clique of size i.
- C_i is the set of all K_is in the graph.
- For any integer $m > 0, [m] = \{1, 2, \ldots, m\}$.
- $R(\eta, q)$ is the random graph having η vertices, where the probability of occurrence of an edge between any pair of vertices is q.

4   Our Scheme

4.1   The First Scheme

Let us define 'epoch' to be the period of time between construction of two consecutive blocks. The time of construction of a block in the Bitcoin network happens to be 10 to 15 min or may be less. Therefore an epoch may last for a time between 10 to 15 min so that all transactions executed after the commencement of an epoch and before the expiry of that epoch are included in the Bitcoin block for that epoch. Transactions which are executed sharply before the expiry of an epoch may not be included in the block for the current epoch due to the delay of the transaction in traveling across the Bitcoin network and therefore it could be safely accommodated in the next block if their validity is acknowledged by the miner. Note that the transaction set for which a miner attempts to find a PoW solution may not be the same for all miners. Without loss of generality we may assume that the number of transactions occurring in a particular block is a power of 2. It can be forcibly done either by accommodating some extra transactions to the next block or by using some predefined dummy transactions.

As we have discussed we are going to propose the problem of finding the largest clique in a big graph. Here we shall be dealing with a graph having 2^{30} vertices. We also fix the probability of occurrence of an edge between any two vertices of the graph equal to $\frac{1}{2^{12}}$. The reason behind fixing the values of these two crucial parameters is to have the difficulty level of solving the PoW in the close range of the amount of computation needed to solve the Bitcoin PoW puzzle as of June 2015. In future, as the computational power of the Bitcoin network will increase, the difficulty level could be adjusted by increasing the graph size as it is done in the case of the current Hashcash based Bitcoin mining strategy. Another way of adjusting the difficulty is to increase the number of edges of the graph. As we will see below, the probability of occurrence of any edge of the graph can be adjusted. A higher probability of occurrence of edges will create a denser graphs. The size of maximum clique will shoot up(as the size of the maximum clique depends on the probability of occurrence of edge). The difficulty of finding the maximum clique will also increase. The detailed study of the effect of graph size and the density is beyond the scope of this paper and we will study them in future. The clique finding problem takes $O(n^k)$ time for fixed clique of size k, where n is the number of vertices. Chen et al. [4] Showed that there can be no algorithm that runs in $O(n^{o(k)})$ to test that a given graph has a clique of size k or not. This result provides a surety that finding a clique of size k gets hard as the number of vertices n increases. So, by increasing the number of vertices n we can increase the difficulty of solving the PoW puzzle details of which will be incorporated in a future work.

Let, $\tau = \{T_i : 1 \leq i \leq 2^\nu\}$ be the set of 2^ν transactions that the miner wants to validate in a certain epoch ε_0 where ν is a fixed integer. The set τ may not be the same for all miners and depends upon the choice of a particular miner. This set τ is the set of transactions that this particular miner wants to include in the block created by her. Like the Bitcoin network in our scheme the miners have some degree of freedom in choosing the transactions for any epoch. Let pk be the public key of the miner. We use the set τ to build the set of vertices of the graph that will be ultimately used by this miner for solving the PoW puzzle. Let, $\mathcal{T}$ be a set such that $\mathcal{T} = \{L_{(i-1)*2^{30-\nu}+j} : L_{(i-1)*2^{30-\nu}+j} = T_i || pk || j, 1 \leq i \leq 2^\nu, 0 \leq j \leq 2^{30-\nu} - 1\}$. Hence, $|\mathcal{T}| = 2^\nu * 2^{30-\nu} = 2^{30} = n$(say). The set $\mathcal{T}$ contains some elements which are the concatenation of a transaction from the set τ, the public key pk of the miner and a nonce that takes the value from 0 to $2^{30-\nu} - 1$. So, $|\mathcal{T}| = |\tau| \times 2^{30-\nu} = 2^{30}$. The miner constructs a graph $\mathcal{G}_{n,p} = (V, E)$ such that $V = \mathcal{T}, n = |\mathcal{T}|$. The set of vertices is the set $\mathcal{T}$ itself. We define the set of edges as $E = \{(u, v) : u, v \in V, H(u||v) = 0^{12} || \{0, 1\}^*\}$, where $H(\cdot)$ is a hash function. Hence, a single hash needs to be calculated to check the existence of an edge between a pair of vertices. Thus in the graph any two vertices will be connected by an edge if the hash of the concatenation of the two vertices starts with 12 zeros. So, the probability of occurrence of an edge between any two vertices is given by $p = \frac{1}{2^{12}}$. We choose this to be the probability of occurrence of an edge for having the expected size of the maximum clique as low as 5 for this graph. Since, we use hash function for computing the edges of the graph, the final graph will be random for every miner and will have 2^{30} vertices. So, the difficulty of finding the largest possible clique will be same in all the graphs. Thus, this scheme is fair and the public key of the miner has negligible effect on the graph other than making it different from all other graphs on which other miners are mining. It is also easy to see that the graph need not be stored anywhere as the miner can generate it using the above technique during runtime. It can be easily seen that one can increase the size of the graph changing the number of nonces used to compute the vertices of this graph. Here, we chose $2^{30-\nu}$ to be the number of nonces that are concatenated with the transaction and the public key to compute the vertices. If we choose $2^{\lambda-\nu}$ to be the number of nonces, we will get a graph of 2^λ vertices. Also, if we tune the probability of occurrence of any edge we can get sparser or denser graph. So, thus we can adjust the size of the largest clique in the graph as it follows Theorem 1 making the clique number very close to $\frac{2\lambda}{\log \frac{1}{p}}$. From [4], we can say that as the clique number increases the difficulty to find it also increases. The details of this will be covered in a future paper.

Our Proof-of-Work scheme is to find the largest complete subgraph in $\mathcal{G}_{n,p}$ constructed as above. So, if the miner can find the solution of this PoW that is if she can find the largest clique and his PoW is accepted all transactions in τ will be validated. We provide our PoW scheme in Fig. 1. In this Algorithm, a miner tries to find the maximum clique by searching the adjacency matrix of the graph $\mathcal{G}_{n,p}$. So, the problem of finding the maximum complete subgraph of $\mathcal{G}_{n,p}$ is reduced to the problem of finding the maximum submatrix of the adjacency matrix of $\mathcal{G}_{n,p}$ with all entries equal to '1' except the diagonal entries. Let, B be the adjacency matrix of the graph $\mathcal{G}_{n,p}$. The dimension of B is $|V| \times |V|$. Let B' be an arbitrary submatrix of B. We define the index set $I_{B'}$ of B as $I_{B'}$ contains the indices of each element of B' in the original matrix B.

- **Setup** Let pk be the public key of the miner. $\tau = \{T_1, T_2, \ldots, T_{2^\nu}\}$ are the 2^ν transactions occurring in a particular epoch that the miner wishes to include in the block. $\mathcal{T} = \{L_{(i-1)*2^{30-\nu}+j} : L_{(i-1)*2^{30-\nu}+j} = T_i \| pk \| j, 1 \le i \le 2^\nu, 0 \le j \le 2^{30-\nu} - 1\}$. $V = \mathcal{T}$.
- **Proof of Work Computation** In a scratch off attempt a miner computes a $V \times V$ matrix B as follows: For $i \le j$

$$B[i][j] = B[j][i] = \begin{cases} * & \text{if } i = j \\ 1 & \text{if } H(v_i \| v_j) \in \{0^{12} \| \{0,1\}^*\} \\ 0 & \text{elsewhere} \end{cases}$$

  where $v_i, v_j \in V, i \le j$. The user finds the largest square submatrix B' of B such that each and every entry of B' is a 1 except the diagonal elements. If she finds such a submatrix then she broadcasts it along with the index set $I_{B'}$.
- **Verification** The verifier checks whether $\forall i, j \in I, i \ne j$

$$H(v_i \| v_j) \in \{0^{12} \| \{0,1\}^*\}$$

  If the Verification is successful then the user caches the proof of work and the block generated by the prover. After a predefined time has elapsed the verifier checks the cache and selects a proof with the highest size of B'. Ties are broken by selecting the proof that came first and the corresponding block is accepted and added to the Bitcoin block chain.

Fig. 1. Algorithm 1: new proof of work scheme for mining in Bitcoin network

Remark 1. Every miner should send all the transactions, her public key and her solution(largest clique) across the Bitcoin network. The peers will require to check that the solution indeed forms a clique. For example if the largest clique has six vertices, $v_1, v_2, \ldots, v_5$, where $v_i = \{T_i\|pk\|j_i\}, 1 \leq i \leq 5$. So, the miner should send $(T_1, T_2, \ldots, T_5)$ and $\{j_1, j_2, \ldots, j_5\}$. She also needs to send the set of all transactions τ. From these sets, a peer can easily check whether the set of 5 vertices indeed form a clique or not. Later we will see that a peer needs to calculate 10–15 hashes to check that the vertices indeed form a clique.

It could be observed that in our scheme in Fig. 1, a Bitcoin miner does not necessarily require to find the biggest clique in the graph. It is sufficient if she finds a clique which is larger than any other clique found by other miners. The Bitcoin network, according to our Fig. 1, selects the largest clique from all the cliques it received from all the miners as potential solutions. Hence, a miner may send multiple cliques of different sizes as they are found if the miner uses a bottom up approach to find them. A network peer will store every clique it receives during a particular epoch and when the epoch ends, it will open the cache and choose the largest clique it has received. The peer can alternately store a clique as long as it does not receive a bigger clique from some miner. Thus, the peer will only store one clique as a potential solution at any time and if a clique of bigger size is received it would drop the earlier one and store the new one. If there are multiple cliques of largest size from different miners, the network may break ties by selecting the one that came first. So, a miner may send a clique of size κ even before she is able to confirm that there is no clique of size $\kappa + 1$ which could take a longer time. This would allow miners to play safely, eliminating the chance of not being able to send a single solution before the epoch ends even if she had indeed found a solution before the epoch ended. Since, the epoch is fixed and the miners must submit their solution before the epoch ends, there will be no instability of the block chain due to delayed arrival of PoW solutions. All solutions appearing late will be rejected. Someone can argue that multiple miners may find the solution before the epoch ends and this could cause branching in the block chain. However, branching does happen in Bitcoin block chain and ultimately only one branch wins the race. The same can be true in our scheme. Besides, we have fixed a window of time within which the solution is to be submitted and among all the solutions only the one must be accepted that came first. This would lower the probability of branching in the block chain. Even if a fork happens, only a single branch will grow longer as the original Bitcoin block chain does.

It can be noted that no other miner can steal a miner's solution and relay that as her own solution, because of the inclusion of the public key(pk) of the miner in the calculation of the giant graph. This is incorporated in the construction of the graph to ensure that every graph constructed by every miner is separate but the difficulty in finding the maximum clique will be same in all the graphs because of the result of Lemma 1 which says that all such graphs that different miners are working on are statistically indistinguishable from a random graph $R(2^{30}, \frac{1}{2^{12}})$ and hence the difficulty of finding the maximum clique will be same in all of them. So, the amount of computation needed to find the maximum clique will be equivalent for all miners. Also, the size of the maximum clique will be same in all of them because of Theorem 1.

In our scheme there is no way by which a verifier can determine the exact size of the largest clique of the random graph $\mathcal{G}_{n,p}$ without repeating the same procedure used by miners. However, there are some sharp bounds that allow a verifier to guess the value of the largest clique with a high degree of accuracy as explained in Sect. 3.1. With the help of this bound, the Bitcoin peers can guess the size of the maximum clique with a great degree of accuracy. Since, the graphs on which the different miners are working are statistically indistinguishable, the size of maximum cliques will be same in all of them with a very high probability.

4.2   A Modified Mining Algorithm

According to Theorem 1, the size of the largest clique in a random graph is of the order of $\frac{2\log n}{\log(1/p)} + O(\log n)$, n being the number of vertices(2^{30} in our case) and p being the probability of occurrence of an edge between a pair of vertices($\frac{1}{2^{12}}$ in our case). The probability that the size of the largest clique will be anything beyond this tends to zero as n tends to infinity.

- **Setup** Same as Algorithm 1.
- **Proof of Work Computation** In a scratch off attempt a miner computes a $V \times V$ matrix B as follows: For $i \leq j$

$$B[i][j] = B[j][i] = \begin{cases} * & \text{if } i = j \\ 1 & \text{if } H(v_i || v_j) \in \{0^{12} || \{0,1\}^*\} \\ 0 & \text{elsewhere} \end{cases}$$

  where $v_i, v_j \in V, i \leq j$. The user finds the largest square submatrix B'_0 of B such that each and every entry of B'_0 is a 1 except the diagonal elements. If she finds such a submatrix then she broadcasts it along with the index set I_0.
  Step $r = 1, \ldots$ do {
  if the epoch has ended then exit
  else find another square submatrix from B such that $[B'_r] = [B'_0]$. If such a matrix could be found then broadcast it along with the index set I_r.
  }
- **Verification** The verifier holds separate caches for storing the solutions from different miners. The verifier checks every submatrix for being a valid clique as shown in Algorithm 1. If the Verification is successful then the user caches the proof of work along with the index set which corresponds to the set of vertices The verifier rejects all PoW whose index sets have a non-null intersection with a PoW present in its cache. After the epoch has expired the verifier checks its cache and selects a miner who has so far sent the highest number of proofs(largest cliques). Ties are broken arbitrarily and the corresponding block is accepted and added to the Bitcoin block chain.

Fig. 2. Algorithm 2: yet another proof of work scheme for mining in Bitcoin network

Note that there could be many largest cliques in the graph $\mathcal{G}_{n,p}$. So, many miners may find more than one largest cliques if the 'epoch' is sufficiently long. Therefore, it could be possible to decide the winning miner not by selecting the first solution that was submitted to the Bitcoin network but by the number of cliques a miner has found. That is we can choose a winner by selecting a miner who has found the highest number of distinct cliques in the graph. So, we modify the Fig. 1 as follows:

(1) The miner needs to find as many cliques as she can from the graph.
(2) All those cliques will be on an exclusive subset of vertices of the graph.

Thus, instead of finding a single largest clique, the miners are encouraged to compute as many cliques as they can within a bounded time. The miner who computes maximum number of cliques is the winner of the Bitcoin mining game. Note, that the expected number of largest cliques of size k in our graph is given by $\binom{2^{30}}{k}(\frac{1}{2})^{6k(k-1)} \approx 2^{6k(6-k)}$, where k is given by $5+30c, c \ll 1$. Note that putting $k = 5$, we find that the number of cliques of size 5 is 2^{30}. Similarly, putting $k = 6$, the expected number of cliques of size 6 comes to be only 1. Again the expected number of cliques of size 7 happens to be a very small fraction $\approx \frac{1}{2^{42}}$. This shows that c must be a small constant so that $k = 5$ or 6 with a very high probability conforming to the results of [7].

In Fig. 2, we give a modified version of Fig. 1. Here, as discussed before, we allow the miners to find multiple largest cliques in the graph. In Fig. 1, just finding a single clique would have been sufficient. But in Fig. 2 the miner who sends maximum number of cliques wins. So, a miner who uses incremental approach to construct cliques in Fig. 1, sending them as soon as they are found, will need to find the size of the maximum clique first, then only she can start computing those cliques one by one using her own strategy. Therefore, a Bitcoin miner will be more encouraged to co-operate to constitute a big pool of miners to add up their computational resources. Note that since finding a clique is very costly operation, so only few mining pools having sufficient resources can afford to do mining for our PoW scheme. In Sect. 5, we provide some algorithms that can be used for finding maximum cliques for our PoW schemes i.e. they can be used for finding maximum cliques in our graph $\mathcal{G}_{n,p}$. To verify the PoW solutions one peer does not need to compute the entire graph. Instead she only needs to check the solutions found by the miners indeed form cliques. In our model the size of the largest clique will be 5 or 6 with a very high probability according to Theorem 1. So in order to check one maximum clique, the verifiers will require to compute 10 or 15 hashes. Also the peers don't need to be sure that the clique found by a miner is the largest clique in the graph. She can only choose the biggest one she has received and can accept the corresponding block. This may seem to be slightly inefficient compared to the classical Bitcoin mining scheme where only a single hash needs to be calculated. But in this scheme we ensure that the miner who invests maximum amount of computation can only win the mining game unlike the existing scheme *where a lucky solver can get a solution much earlier than others who possess much higher amount of computational resources.* In our scheme, a solver needs to calculate a large number of hashes to ensure that a significantly large part of the graph is constructed to find the size of the largest clique that fall within a narrow range of 5 or 6 with a very high probability.

5 Analysis of Our Scheme

Lemma 1. *Let C be the incidence matrix of a random graph $R(2^{30}, \frac{1}{2^{12}})$ and B be as defined in Fig. 1. For every probabilistic polynomial time distinguisher Δ,*

$$|Pr[\Delta(B) = 1] - Pr[\Delta(C) = 1]| \leq negl$$

provided $H()$ is a secure hash function.

Proof. If the above lemma is not true, then the hash function will be such that for any two vertices, $v_1, v_2, Pr[e(v_1, v_2)] \neq \frac{1}{2^{12}}$. So, for a pair of any $T_1, T_2 \in \tau, 0 \leq m, n \leq 2^{30-\nu} - 1, Pr\left[H((T_1||pk||m))||(T_2||pk||n)) = 0^{12}||\{0, 1\}^*\right] \neq \frac{1}{2^{12}}$. So, there exists x, y such that $Pr[H(x) = a] \neq Pr[H(y) = a]$, for some a belonging to the domain of the output of the hash function. This shows that $H()$ is not a secure hash function. So, our assumption is incorrect.   ∎

Thus, it is apparent that finding a maximum clique in the giant graph is as difficult as finding the same in a random graph $R(2^{30}, \frac{1}{2^{12}})$. Finding a maximum clique in any graph is known to be NP-hard problem [15]. In later sections of this paper we attempt to construct feasible Algorithms for miners to compute the solution of the graph based proof of work puzzle proposed in Figs. 1 and 2. Before we propose feasible algorithms for finding maximum cliques in the graph, we show that the miner cannot construct a clique by manipulating the transactions i.e. the miner cannot compute transactions that forms a clique in the graph and add it to the graph. The only way a miner can find a maximum clique is by searching the entire graph.

Lemma 2. *The miner cannot compute transactions $T_1, T_2, \ldots, T_5$, (not all necessarily distinct), such that $H((T_i||pk||j)||(T_k||pk||l)) = 0^{12}||\{0, 1\}^*, 1 \leq i, k \leq 5, i < k, 0 \leq j, l \leq 2^{30-\nu}, (T_i, j) \neq (T_k, l)$.*

Proof. The miner will need to find transactions $T_1, T_2, \ldots, T_5$(not all distinct) such that there exists $\{j_1, j_2, \ldots, j_5\}$, $0 \leq j_i \leq 2^{30-\nu}$, such that $H((T_i||pk||j_i)||(T_k||pk||j_k)) = 0^{12}||\{0, 1\}^*, \forall i, k \in \{1, 2, 3, 4, 5\}, i < k$. For any $i, k \in \{1, 2, \ldots, 5\}$ and $0 \leq j_i, j_k \leq 2^{30-\nu}, Pr[H((T_i||pk||j_i)||(T_k||pk||j_k)) = 0^{12}||\{0, 1\}^*] = \frac{1}{2^{12}}$. So, finding two transactions T_i, T_k such that $H((T_i||pk||j_i)||(T_k||pk||j_k)) = 0^{12}||\{0, 1\}^*, 1 \leq i, k \leq 5, i < k, 0 \leq j_i, j_k \leq 2^{30-\nu}, (T_i, j_i) \neq (T_k, j_k)$ takes 2^{12} hash computations. So, finding a set of transactions $\{T_1, T_2, \ldots, T_5\}$(not all necessarily distinct) and indices $\{j_1, j_2, \ldots, j_5\}$ (not all necessarily distinct) such that $H((T_i||pk||j_i)||(T_k||pk||j_k)) = 0^{12}||\{0, 1\}^*, 1 \leq i, k \leq 5, i < k, j_i, j_k \in \{j_1, j_2, \ldots, j_5\}, (T_i, j_i) \neq (T_k, j_k)$ takes on an average $(2^{12})^{\binom{5}{2}} = 2^{120}$ hash computations. As we shall see later, this is higher than the computation required to find a maximum clique in the graph.   ∎

Since our graph contains 2^{30} vertices, all the maximum cliques could be listed by doing an exhaustive search, checking all possible $\binom{2^{30}}{k} \approx 2^{30k}$ subgraphs of k vertices. Lemma 3 shows that by doing a brute force search a maximum clique could be found in time exponential in k when the size of the maximum clique k is known beforehand.

Lemma 3. *The expected amount of computation needed to find the first largest clique in brute force search method is $O\left(2^{6k(k-1)}\right)$, where k is the size of the largest clique.*

Proof. Let N^k be the number of iteration required to find the first clique of size k. The miner's strategy is as follows:

Step 1: $\Omega = \emptyset$.
For each Step i, The miner chooses a set Γ_i of k vertices randomly from the set of vertices V such that $\Gamma_i \notin \Omega$.

If the vertices in Γ_i form a clique then she outputs it. Else she updates Ω as $\Omega = \Omega \cup \Gamma_i$. Now, the expected number of iterations needed to find the first clique is

$$E(N^k) = \sum_{i=1}^{\infty} i(1-r)^{i-1}r,$$

where $r = p^{\frac{1}{2}k(k-1)}$. So, $E(N^k) = r\sum_{i=1}^{\infty} i(1-r)^{i-1} = \frac{1}{r}$. Since, in our scheme $p = \frac{1}{2^{12}}$, the lemma follows.   ∎

The method of Lemma 3 finds a clique of size k by randomly choosing k new vertices each time and checking them whether they form a clique or not. The method incurs an additional space complexity of $O\left(2^{6k(k-1)}k\right)$ to store the set Ω before the first clique is found. Now, in our model, the value of k roughly equals to 5 or 6, so, the average amount of computation required is of the order of 2^{120} or 2^{180} if brute force search is applied. In each of such searches the miner needs to search that a particular $k \times k$ submatrix of the big adjacency matrix B of the random graph $\mathcal{G}_{n,p}$ has all entries equal to '1' except the diagonal entries. Note that our verification Algorithm does not verify whether the clique which is given by the miner is the biggest one or not. It just selects the largest clique from all solutions given by the miners.

5.1   Finding Cliques in the Graph: Algorithms

Here, we give a naive Algorithm to find a clique of arbitrary size using random search. This Fig. 3 supports parallel computation of candidate cliques of the random graph. We know that the clique number of the random graph takes the value of $k + O(\log n)$ with a very high probability, where $k = \frac{2\log n}{\log \frac{1}{p}}$, n being the number of vertices of the graph(2^{30} in our case) and p being the probability of occurrence of an edge between any two vertices($\frac{1}{2^{12}}$ in our case). Thus if the clique number k is known a clique could be found using Fig. 3. In each of the iterations of Fig. 3, a miner only needs to check whether a particular $k \times k$ submatrix of the adjacency matrix of the graph contains all entries equal to '1' excepting the diagonal entries. Since the adjacency matrix is very high in size, it will not be possible to store it at a particular place. Neither will it be possible to share it by different miners spread across the entire globe. Hence, the miner who executes Fig. 3 needs to compute the hashes on the fly and check for existence of a clique. From Lemma 4, we can see that the number of hashes required to be calculated is $O(2^{6k(k-1)})$. In our case $k = 5$ or 6. So, for $k = 5$ the amount of hash computations needed will be $O(2^{120})$. For $k = 6$, however it is much higher. Hence, the Fig. 3 is impractical for use in mining as it would increase the computation vigorously. We need to find an Algorithm that would allow the miners to find a solution in reasonable time. We attempt to construct a feasible Algorithm in later part of this paper.

Input: Let, $P_m, 1 \leq m \leq c$ be the id of a miner of some pool having c members. $LH_T : \{0,1\}^{T+x} \rightarrow \{0,1\}^T$ is a function that outputs the most significant T bits of any binary bit string of length at least T. n is the number of vertices in the graph $\mathcal{G}_{n,p}$. Each vertex is represented by $\lceil \log n \rceil$ bits. $\mathcal{H}(\cdot)$ is SHA-256 hash function.
Output: Find a clique of size k in the graph $\mathcal{G}_{n,p}$.
    P_m chooses a random nonce s_m.
    while epoch has not expired **do**
        Calculate $\Lambda = LH_{k \log n}(\mathcal{H}(m||s_m))$ as the left most $k\lceil \log n \rceil$ bits of $\mathcal{H}(m||s_m)$.
        Choose k vertices $\langle v_1, v_2, \ldots, v_k \rangle \in V$ using the bitstring Λ.
        if $V_1, V_2, \ldots, V_k$ form a clique **then**
            output it and exit.
        end if
    end while

Fig. 3. Algorithm 3: a parallel algorithm to compute a candidate maximum clique of size k

Lemma 4. *The Fig. 3 takes $O(2^{6k(k-1)})$ time to find the first clique of size k if $k < 6$.*

Proof. In every iteration, Fig. 3 selects k vertices randomly and checks whether they form a clique of size k or not. If they do, the Algorithm outputs it. Now the difference between the Fig. 3 and the method described in Lemma 3 is that in Lemma 3 every time a distinct set of vertices is chosen. But in Fig. 3, since the set of vertices are chosen depending upon the output of the hash function, it may so happen that some set of vertices that have previously been verified for formation of clique is again selected for the same verification. This is because Fig. 3 does not store the set of vertices that are selected and verified for formation of clique. Thus we would be doing the same computation that we did previously. This may entail some extra computation. Now, we can try to quantify the amount of computation needed. This has been done below:

There are $\binom{2^{30}}{k} \approx 2^{30k}$ different sets of vertices that needed to be checked for existence of clique. Figure 3 randomly chooses one of these 2^{30k} cliques in every iteration and examines them for existence of a clique. Let us enumerate all 2^{30k} subsets of vertices arbitrarily. Let X_i denote the random variable such that

$$X_i = \begin{cases} 1 & \text{if Algorithm 3 never selects the } i^{\text{th}} \text{subset} \\ 0 & \text{elsewhere} \end{cases}$$

Let, Fig. 3 needs to do L iteration to find a clique of size k. Also, let $Y = \sum_{i=1}^{2^{30k}} X_i$ Now, $E(Y) = E(\sum_{i=1}^{2^{30k}} X_i) = \sum_{i=1}^{2^{30k}} E(X_i) = \sum_{i=1}^{2^{30k}} P[X_i = 1] = \sum_{i=1}^{2^{30k}} \frac{(2^{30k}-1)^L}{2^{30kL}}$. Thus, $E(Y) = 2^{30k} \frac{(2^{30k}-1)^L}{2^{30kL}}$.

Hence, the number of sets of vertices that are not selected by Fig. 3 is $2^{30k} \frac{(2^{30k}-1)^L}{2^{30kL}}$. So, the number of distinct sets of k vertices that are checked by Fig. 3 after L iterations are completed is $2^{30k} \left(1 - \left(\frac{2^{30k}-1}{2^{30k}}\right)^L\right)$. By Lemma 3 this must be equal to $2^{6k(k-1)}$. Equating them we get $1 - \left(\frac{2^{30k}-1}{2^{30k}}\right)^L = \frac{1}{2^{6k(6-k)}}$. Now if $L << 2^{30k}$, we get $\frac{L}{2^{30k}} \approx \frac{1}{2^{6k(6-k)}}$. Whence we get $L \approx 2^{6k(k-1)}$. Hence, the time complexity of Fig. 3 is $O(2^{6k(k-1)})$. ∎

The complexity of Fig. 3 is equal to the average number of iteration needed to find a clique in Lemma 3. So, for $k < 6$, the Fig. 3 has same computational complexity as the method of Lemma 3. The space complexity of Fig. 3 is $O(1)$ for $k < 6$. From Fig. 3 and Lemma 4, we can state the following Theorem:

Theorem 3. *If the size of the maximum clique k is less than 6, then there exists an Algorithm that finds it in $O(2^{6k(k-1)})$ time and with $O(1)$ space.*

Another method of finding the maximum clique is to take a bottom up approach, starting from a set of all κ-cliques of the graph and trying to grow them in size. Once, we find some candidate cliques of size κ using some Algorithm, we can try to find some larger cliques, if there is any. The method of finding a clique of size $\kappa + 1$ is as follows:

(1) Let C be the set of cliques of size κ output by some arbitrary Algorithm. A mining pool may store all candidates cliques at a shared location so that other miners could access them whenever necessary.

(2) Now the miners can try to grow every clique $c \in C$ by checking whether any of the $2^{30} - \kappa$ forms a clique of size $\kappa + 1$ with the vertices of c. This would require $O(|C|2^{30})$ computation. We have calculated the amount of computation needed to find a clique of size $\kappa + 1$ from the set C of all cliques of size κ in Lemma 5.

Lemma 5. *The computation needed to find all cliques of size $\kappa + 1$, if it exists given the set of all cliques of size κ is $O(2^{30+6k(6-\kappa)}\kappa)$.*

Proof. The expected number of cliques of size κ is $\omega = \binom{n}{\kappa} p^{\frac{1}{2}\kappa(\kappa-1)}$. In our case, $n = 2^{30}, p = \frac{1}{2^{12}}$. So, $\omega = \binom{2^{30}}{\kappa}/2^{6\kappa(\kappa-1)} \approx \frac{2^{30\kappa}}{2^{6\kappa(\kappa-1)}} = 2^{6\kappa(5-\kappa+1)} = 2^{6\kappa(6-\kappa)}$. Now, from this set of candidate cliques of size κ, a clique of size $\kappa + 1$ can be found using the above method. So, the total computation needed is $\omega * 2^{30} * \kappa$. Hence, proved. ∎

We shall now show how the result of Lemma 5 could be used to construct an Algorithm that finds all maximum cliques in reasonable time. Let us define $C_i = \{\langle v_1, v_2, \ldots, v_i \rangle\}$ be the set of all i-cliques of the graph $\mathcal{G}_{n,p} = (V, E)$.

Input: C_3.
Output: List all largest clique of the random graph.
    for $i := 3; C_i \neq \emptyset; i + +$ **do**
        while C_i is not empty **do**
            Choose a clique $\langle v_1, v_2, \ldots, v_i \rangle \in C_i$
            for all $v \in V \setminus \{v_1, v_2, \ldots, v_i\}$ **do**
                if $\bigcup_{j=1}^{i}\{v_i, v\} \subset E$ **then**
                    $C_{i+1} = C_{i+1} \cup \langle v_1, v_2, \ldots, v_i, v \rangle$.
                end if
            end for
            $C_i = C_i \setminus \langle v_1, v_2, \ldots, v_i \rangle$.
        end while
    end for

Fig. 4. Algorithm 4: an algorithm to list all largest cliques

Lemma 6. *The computational complexity of Fig. 4 is $O(2^{85.58})$*

Proof. The Fig. 4 takes a bottom up approach in computing the maximum clique. It first takes as input the set of all cliques of size 3. Then it tries to grow the size of the cliques by checking if any of the other vertices of the graph creates a bigger clique with it. Thus in every iteration of the outer for loop of Fig. 4, it uses the set of all cliques of size i to construct a set of all cliques of size $i + 1$. Now, according to Lemma 5, the complexity in each iteration is given by $O(2^{30+6i(6-i)}i)$. Without loss of generality we may assume that we run the Algorithm until C_6 is computed. So, the complexity in each iteration should be as follows;

For $i = 3$, the run time is $O(2^{84} * 3)$
For $i = 4$, the run time is $O(2^{78} * 4)$
For $i = 5$, the run time is $O(2^{60} * 5)$.

It is apparent that the complexity of the above Algorithm is $O(3 * 2^{84})$. Now, we can pre-compute C_3 using the Chiba & Nishizeki Algorithm [5] in $O(2^{72})$ time. So, the total complexity of finding all cliques of size 6(if at least one exists) is $O(3 * 2^{84}) \approx O(2^{85.58})$. ∎

This is a reasonable amount of computation for Bitcoin miners. The present hash rate of Bitcoin network is $10^{22} \approx 2^{73}$ and soon it is expected to reach the order of 2^{85} when our Fig. 4 will become feasible to be applied for minting Bitcoin.

The space complexity of Fig. 4 is determined as $M = \max_{i=3}^{6}(|\mathcal{C}_i|) = O(\max_{i=3}^{6} 2^{6\kappa(6-\kappa)})$. It is easy to see that $M = O(2^{54})$. Hence, the space complexity is of the order of a petabyte. Hence, the mining pool requires to store the set $\mathcal{C}_i$ in a distributed fashion. Since, a mining pool may contain tens of thousands of miners it is possible to afford this much space for executing Fig. 4.

Further reduction of the computational complexity of finding $\mathcal{C}_6$ can be done by computing all K_4s of the graph as follows:

(1) Find all quadrangles of the graph using the Chiba & Nishizeki Algorithm [5].
(2) Find all K_4s using the set of quadrangles.
(3) Then using the set of all K_4s ($\mathcal{C}_4$) try to find $\mathcal{C}_6$ using Fig. 4.

Lemma 7. *The method stated above improves the computational complexity of Fig. 4 to $O(2^{80})$.*

Proof. The set of all quadrangles could be found using Chiba & Nishizeki Algorithm in $2^{70.5}$ time [5]. We calculate the expected number of quadrangles as follows; the number of ways 4 vertices could be chosen out of 2^{30} vertices is $\binom{2^{30}}{4}$. Now, we can create 3 quadrangles from every set of 4 vertices, each of them with a probability of $(\frac{1}{2^{12}})^4$. So the expected number of quadrangles will be $\binom{2^{30}}{4}(\frac{1}{2^{12}})^4 * 3 < 2^{69.6}$. So, in order to check whether each of them is a subgraph of a clique, $O(2^{69.6})$ computations would be required. Thus, $\mathcal{C}_4$ could be generated. Thereafter we could follow Fig. 4 and Lemma 5 to find $\mathcal{C}_6$. It can be seen from Lemma 6, that in order to compute $\mathcal{C}_5$ from $\mathcal{C}_4$ it takes $O(2^{80})$ computation which is the dominating part of the entire complexity evaluation, because computing all K_6 from the set of K_5s takes $O(2^{60} * 5)$ time. Thus, the overall run time is dominated by $O(2^{80})$. ∎

Theorem 4. *There exists an Algorithm that outputs the set of maximum cliques in $O(2^{70.5})$ time and $O(2^{48})$ space.*

Proof. The Algorithm works as follows:

(1) It finds all quadrangles using Chiba & Nishizeki Algorithm. Whenever a quadrangle is found the Algorithm checks whether it could be a part of a K_4 or not. This could be done by computing only two hashes per quadrangle. If a quadrangle could be extended to a K_4 store it in $\mathcal{C}_4$. So, the total amount of computation needed is bounded by the total number of quadrangles in the graph which is $O(2^{69.6})$.
(2) Without loss of generality we may assume $\mathcal{C}_4$ stores all K_4s such that $\forall (u_1, u_2, u_3, u_4) \in \mathcal{C}_4$, $u_1 < u_2 < u_3 < u_4$. Store the set of K_4s $\mathcal{C}_4$ in a sorted array $A[]$ such that for every $i, j \in |\mathcal{C}_4|$, $i < j$ if $A[i] = \{v_1, v_2, v_3, v_4\}$ and $A[j] = \{v_1', v_2', v_3', v_4'\}$, then $\exists l \in \{1, 2, 3, 4\}$ such that $v_m = v_m'$, $\forall 1 \leq m < l$ and $v_l < v_l'$. This would require $|\mathcal{C}_4| \log |\mathcal{C}_4|$ time. Since, $|\mathcal{C}_4| = 2^{48}$, the amount of computation needed to do this is $(48 * 2^{48}) \approx O(2^{54})$.

(3) We can now use the sorted array $A[]$ to find all K_5s. For all $i, 1 \leq i < |C_4|$, if $A[i] = \{u_1, u_2, u_3, v\}$ and $A[i+t] = \{u_1, u_2, u_3, v'\}$, where $t > 1$, check whether $\{u_1, u_2, u_3, v, v'\}$ is a clique or not. This could be checked by computing a single hash. If $\{u_1, u_2, u_3, v, v'\}$ is a K_5 add it to C_5. This would take $O(|C_4|)$ or $O(2^{48})$ computations (see Lemma 8). The space needed to store C_5 is $O(2^{30})$.

(4) Now, the set C_5 could be used to compute the set C_6 if it is nonempty. This could be done using the same method stated in step 2 and 3 or in Fig. 4.

It could be checked that the time complexity of this Algorithm is determined by the time needed to find all quadrangles using Chiba & Nishizeki Algorithm which is $O(2^{70.5})$. Similarly, the space complexity is $O(|C_4|)$ or $O(2^{48})$. ∎

The time complexity of the Algorithm described in Theorem 4 is same as the average computation needed to solve the existing proof of work puzzle which is equivalent to $O(2^{73})$ hash computations.

Lemma 8. *Let $A[]$ be a sorted array of all K_4s such that for every $i \in [|C_4|]$, if $A[i] = \{v_1, v_2, v_3, v_4\}$, then $v_1 < v_2 < v_3 < v_4$ and for every $i, j \in |C_4|, i < j$ if $A[i] = \{v_1, v_2, v_3, v_4\}$ and $A[j] = \{v'_1, v'_2, v'_3, v'_4\}$, then $\exists l \in \{1, 2, 3, 4\}$ such that $v_m = v'_m, \forall 1 \leq m < l$ and $v_l < v'_l$. Then finding every pairs $(i, j), i \neq j$ such that $A[i] = \{v_1, v_2, v_3, v_4\}$ and $A[j] = \{v_1, v_2, v_3, v'_4\}$ takes $O(|A|)$ time.*

Proof. For any arbitrary $i \in [|C_4|]$, Let $A[i] = \{v_1, v_2, v_3, v_4\}$. Therefore, all other K_4s like $\{v_1, v_2, v_3, z\}$ if there are any will be adjacent to $A[i]$ in the array. If there are Δ such cliques $A[j]$ whose first three vertices are same as that of $A[i]$, then all such pairs (i, j) could be found in $O(\Delta^2)$ time. Whether any of these pairs could be extended to form a K_5 can be ascertained by computing a single hash per such pair. Now, for any three vertices $\{u, v, w \in V\}$, $\Delta = \{i : i \in [|C_4|], \{u, v, w\} \subset A[i]\}$. So, $E(\Delta) = 2^{30} * (\frac{1}{2^{12}})^3 = \frac{1}{2^6}$. Now, the total time needed to find every pairs $(i, j), i \neq j$ such that $A[i] = \{v_1, v_2, v_3, v_4\}$ and $A[j] = \{v_1, v_2, v_3, v'_4\}$ is $O\left(|A|\left\{\max(1, E^2(\Delta))\right\}\right)$ or $O(|A|)$. ∎

6   Comparison with Other Schemes

As we have stated earlier, the existing Hashcash based proof of work scheme has high variance where, a lucky miner may find the solution of a proof of work scheme sooner than others. Similarly, the cycle detection scheme of Tromp [21] does not ensure that the miners will need to do at least a fixed amount of computation to find the challenged subgraph in the large graph. In our scheme, however the miner needs to do an $O(2^{70.5})$ hash computations in order to find the most possible solution that is a clique of size 5 or 6. Here, a miner would need to figure out the size of the largest clique before doing any random search similar to that of Fig. 3. So in our scheme the miner cannot find a maximum clique without doing the minimum amount of computation needed. Even if an adversary makes a wild guess of the size of the maximum clique to be k, she will have to find one of the $2^{6k(6-k)}$ expected number of cliques of size k present in the graph. If she follows the method described in Lemma 3, it would take $O(2^{6k(k-1)})$ computation on an average to find the first one which is $O(2^{120})$ for $k = 5$ and $O(2^{180})$ for $k = 6$. The probability that she could find it at the first iteration is $\frac{1}{2^{6k(k-1)}}$. For, $k = 5$ this would be $\frac{1}{2^{120}}$ and for $k = 6$, this would be $\frac{1}{2^{180}}$. So her best option is to execute the sequential steps of described in Theorem 4 and do the $O(2^{70.5})$ computations (that includes $O(2^{70.5})$ hash calculations) to find the maximum clique or the set of all maximum cliques. The method of Theorem 4 has low variance as the steps entail constant number of operation to be performed by every miner.

The main downside of Tromp's work [21] is that it does not discuss how the scheme could be applied to the decentralized Bitcoin network. On the other hand, we have discussed how the graph can be build using the set of transactions that a miner wants to verify. Thus, our scheme is compatible to the Bitcoin network.

7   Conclusion

In this paper we propose a new proof of work scheme for cryptocurrencies such as Bitcoin. This proof of scheme makes use of the set of transactions to construct a giant graph deterministically. The miners are required to find the largest clique of in this graph as a solution to the proof of work puzzle. We also have proposed an Algorithm that can be used to find a solution of this puzzle by performing $O(2^{70.5})$ hash calculations which is commensurate to the hashpower of the current Bitcoin network. In this paper we used fixed parameter for our system model. However as the computational power of the Bitcoin network would increase in future, one may use our generic model to manipulate the difficulty level of the puzzle. The study of the effect of every parameter of our model on the difficulty of the puzzle will open a scope of future work.

Acknowledgements. The authors were partially supported by JSPS and DST under the Japan-India Science Cooperative Program of research project named: "Computational Aspects of Mathematical Design and Analysis of Secure Communication Systems Based on Cryptographic Primitives." The third author is partially supported by JSPS Grants-in-Aid for Scientific Research named "KAKEN-15H02711".

References

1. Alon, N., Yuster, R., Zwick, U.: Finding and counting given length cycles (extended abstract). In: van Leeuwen, J. (ed.) ESA 1994. LNCS, vol. 855, pp. 354–364. Springer, Heidelberg (1994)
2. Back, A.: Hashcash - a denial of service counter-measure. Technical report, August 2002. (implementation released in March 1997)
3. Bron, C., Kerbosch, J.: Algorithm 457: finding all cliques of an undirected graph. Commun. ACM **16**(9), 575–577 (1973)
4. Chen, J., Huang, X., Kanj, I.A., Xia, G.: Strong computational lower bounds via parameterized complexity. J. Comput. Syst. Sci. **72**(8), 1346–1367 (2006)

5. Chiba, N., Nishizeki, T.: Arboricity and subgraph listing algorithms. SIAM J. Comput. **14**(1), 210–223 (1985)
6. Eisenbrand, F., Grandoni, F.: On the complexity of fixed parameter clique and dominating set. Theor. Comput. Sci. **326**(13), 57–67 (2004)
7. Grimmett, G.R., McDiarmid, C.J.H.: On colouring random graphs. Math. Proc. Cambridge Philos. Soc. **77**, 313–324 (1975)
8. Jian, T.: An o(20.304n) algorithm for solving maximum independent set problem. IEEE Trans. Comput. **C−35**(9), 847–851 (1986)
9. Kloks, T., Kratsch, D., Mller, H.: Finding and counting small induced subgraphs efficiently. Inf. Process. Lett. **74**(34), 115–121 (2000)
10. Matula, D.W.: On the complete subgraph of random graph. In: Combinatory Mathematics and Its Applications, pp. 356–369, Chappel Hill, N.C (1970)
11. Miller, A., Juels, A., Shi, E., Parno, B., Katz, J.: Permacoin: repurposing bitcoin work for data preservation (2014)
12. Moon, J.W., Moser, L.: On cliques in graphs. Isr. Jo. Math. **3**(1), 23–28 (1965)
13. Nakamoto, S.: Bitcoin: a peer-to-peer electronic cash system. Consulted **1**(2012), 28 (2008)
14. Nešetřil, J., Poljak, S.: Poljak.: on the complexity of the subgraph problem. Commentationes Mathematicae Universitatis Carolinae **26**(2), 415–419 (1985)
15. Pattabiraman, B., Patwary, M.M.A., Gebremedhin, A.H., Liao, W., Choudhary, A.: Fast algorithms for the maximum clique problem on massive sparse graphs. In: Bonato, A., Mitzenmacher, M., Prałat, P. (eds.) WAW 2013. LNCS, vol. 8305, pp. 156–169. Springer, Heidelberg (2013)
16. Robson, J.M.: Algorithms for maximum independent sets. J. Algorithms **7**(3), 425–440 (1986)
17. Robson, J.M.: Finding a maximum independent set in time o (2n/4). Technical report 1251–01, LaBRI, Université de Bordeaux I (2001)
18. Sengupta, B., Bag, S., Ruj, S., Sakurai, K.: Bitcoin based on compact proofs of retrievability. (to appear in International Conference on Distributed Computing and Networking, 2015)
19. Tarjan, R.E., Trojanowski, A.E.: Finding a maximum independent set. Technical report, Stanford University, Stanford, CA, USA (1976)
20. Tomita, E., Tanaka, A., Takahashi, H.: The worst-case time complexity for generating all maximal cliques and computational experiments. Theor. Comput. Sci. **363**(1), 28–42 (2006)
21. Tromp, J.: Cuckoo cycle: a memory bound graph-theoretic proof-of-work. Cryptology ePrint Archive, Report 2014/059 (2014). http://eprint.iacr.org/
22. Vassilevska, V., Williams, R.: Finding, minimizing, and counting weighted subgraphs. In: Proceedings of the Forty-first Annual ACM Symposium on Theory of Computing, STOC 2009, pp. 455–464. ACM, New York (2009)
23. Yuster, R.: Finding and counting cliques and independent sets in r-uniform hypergraphs. Inf. Process. Lett. **99**(4), 130–134 (2006)

Cloud Security

Proxy Provable Data Possession with General Access Structure in Public Clouds

Huaqun Wang[1,2(✉)] and Debiao He[3]

[1] Dalian Ocean University, Dalian, China
wanghuaqun@aliyun.com
[2] State Key Laboratory of Cryptology, Beijing, China
[3] State Key Lab of Software Engineering Computer School,
Wuhan University, Wuhan, China

Abstract. Since public clouds are untrusted by many consumers, it is important to check whether their remote data keeps intact. Sometimes, it is necessary for many clients to cooperate to store their data in the public clouds. For example, a file needs many clients' approval before it is stored in the public clouds. Specially, different files need different client subsets' approval. After that, these stored remote data will be proved possession by the verifier. In some cases, the verifier has no ability to perform remote data possession proof, for example, the verifier is in the battlefield because of the war. It will delegate this task to its proxy. In this paper, we propose the concept of proxy provable data possession (PPDP) which supports a general access structure. We propose the corresponding system model, security model and a concrete PPDP protocol from n-multilinear map. Our concrete PPDP protocol is provably secure and efficient by security analysis and performance analysis. Since our proposed PPDP protocol supports the general access structure, only the clients of an authorized subset can cooperate to store the massive data to PCS (Public Cloud Servers), and it is impossible for those of an unauthorized subset to store the data to PCS.

Keywords: Cloud computing · Provable data possession · Proxy cryptography · Access control

1 Introduction

Cloud computing is an emerging technology where the client can rent the storage and computing resource of cloud computing servers. The client only needs a terminal device, such as smart phone, tablet, *etc.* Cloud computing servers have huge storage space and strong computation capability. In order to apply for

H. Wang was partly supported by the Natural Science Foundation of China through projects (61272522, 61572379,61501333), by the Program for Liaoning Excellent Talents in University through project (LR2014021), and by the Natural Science Foundation of Liaoning Province (2014020147).

D. Lin et al. (Eds.): Inscrypt 2015, LNCS 9589, pp. 283–300, 2016.
DOI: 10.1007/978-3-319-38898-4_17

data storing or remote computing, the end clients can access cloud computing servers via a web browser or a light weight desktop or mobile application, *etc.* In cloud computing, cloud servers can provide three types service: Infrastructure as a Service, Platform as a Service and Application as a Service. The end nodes are some capacity-limited electronic facilities, for example, personal computer, tablet, remote desktop, mini-note, mobile. These end nodes can access the cloud computing networking to get computing service by via a web browser, *etc.*

Generally, cloud computing can be divided into three different types: public cloud, private cloud and hybrid cloud. Public cloud service may be free or offered on a pay-per-usage model. The main benefits of public cloud service can be listed as follows: easy and inexpensive set-up due to the reason that the corresponding costs are covered by the provider; better scalability; cheaper due to pay-per-usage model; *etc.* Public clouds are external or publicly available cloud environments that are accessible to any client, whereas private clouds are internal or private cloud environments for particular organizations. Hybrid clouds are composed of public clouds and private clouds. More security responsibilities for the clients are indispensable to cloud service providers. It is more critical in public clouds for their own properties.

Public clouds' infrastructure and computational resources are owned and operated by outside public cloud service providers which deliver services to the general clients via a multi-tenant platform. Thus, the clients can not look into the public cloud servers' management, operation, technical infrastructure and procedures. This property incurs some security problems due to the reason that the clients can not control their remote data. For the clients, one of the main concerns about moving data to a public cloud infrastructure is security. Specially, the clients need to ensure their remote data is kept intact in public clouds. It is important to study remote data integrity checking since the public cloud servers (PCS) may modify the clients' data to save the storage space or other aims. Or, some inevitable faults make some data lost. Thus, it is necessary to design provable data possession protocol in public clouds.

1.1 Motivation

We consider the application scenario below.

In a big supermarket, the different managers will move the massive data to the public clouds. The data has to do with sale, capital, staff member, *etc.* These different data needs to get different approvals before they are moved to the public clouds. Such as, before sale data is moved, these data must be approved by salesman and sales manager; before staff member data is moved, these data must be approved by human resource manager and the chairman; capital data will have to be approved by the salesman, the chief financial officer and the chairman before they are moved to public clouds, *etc.*

There exist many application scenarios that the data must be approved by multi clients before they are moved to the public clouds. Since different data needs different client subset's approval, it is necessary to study provable data possession protocol which supports a general access structure. In order to ensure their data security, the verifier has to check their remote data possession at regular intervals. In some situations, the verifier is restricted to access the network, e.g., in prison because of comitting crime, in the battlefield because of the war, *etc.* Thus, the verifier has to delegate its remote data possession proof task to the proxy. After that, the proxy will perform the remote data possession proof protocol based on their warrant. This real social requirement motivates us to study proxy provable data possession with general access structure in public clouds.

1.2 Related Work

It is important to ensure the clients' remote data integrity since the clients do not control their own data. In 2007, a provable data possession (PDP) model was proposed by *G. Ateniese et al.* [1]. PDP is a lightweight probable remote data integrity checking model. After that, they proposed dynamic PDP model and designed the concrete dynamic PDP scheme based on symmetric cryptography algorithm [2]. In order to support data insert operation, *Erway et al.* proposed a full-dynamic PDP scheme from authenticated skip table [3]. *F. Sebe et al.* designed a provable data possession scheme by using factoring large numbers difficult problem [4]. *Wang* proposed the concept of proxy provable data possession [5]. After that, identity-based provable data possession were proposed [6,7]. In order to ensure critical data secure, some clients copy them and get their replications. Then, they move these original data and replicated data to multi PCS. In this case, client must ensure its remote data intact on multi PCS, *i.e.*, multi-replica provable data possession [8–11]. At the same time, as a stronger remote data integrity checking model, proofs of retrievability (PORs) was also proposed [12]. After that, *H. Shacham* gave the first PORs protocol with full security proofs in the strong security model [12,13]. It can be also applied into the fields, pay-TV [14], medical/health data [15], etc. Some research results have been gotten in the field of PORs [16–19]. Provable data possession is an important model which gives the solution of remote data integrity checking. At the same time, it is also very meaningful to study special PDP models according to different application requirements.

1.3 Private PDP and PPDP

From the role of the PDP verifier, it can be divided into two categories: private PDP and public PDP. In the *CheckProof* phase of private PDP, some private information is needed. On the contrary, private information is not needed in the *CheckProof* phase of public PDP. Public PDP provides no guarantee of privacy and can easily leak information. Private PDP is necessary in some cases.

A supermarket sells goods every day and stores the sale records in the public clouds. The supermarket can check these sale records integrity periodically by using PDP model. It would not like other entities to perform the checking task. If the competitors can perform the integrity checking, they can get the sale information by performing many times integrity queries. Without loss of generality, we assume that the queried block sequence is $\{m_{s_1}, m_{s_2}, \cdots, m_{s_c}\}$. The symbols $s_1, s_2, \cdots, s_c$ denote the queried block indices where $s_1 \leq s_2 \leq \cdots \leq s_c$. By making s_c bigger gradually until the PCS can not reply valid response, the competitors can get the biggest number $\hat{s}_c$. Making use of block size and $\hat{s}_c$, the competitors can get the supermarket's sale record data size. Then, they can evaluate its sale volume for every day. It is dangerous for the supermarket. In this case, private PDP is necessary.

In private PDP, when the verifier has no ability to perform PDP protocol, it will delegate the PDP task to the proxy according to the warrant. Thus, it is important and meaningful to study PPDP with the general access structure.

Table 1. Notations and descriptions

Notations	Descriptions
$\mathcal{A}$	General access structure
$\mathcal{A}_i$	Valid subset to move the file to PCS
U_{j_l}	the l-th member in the subset $\mathcal{A}_j$
(x_{j_l}, X_{j_l})	Private/public key pair of U_{j_l}
(y, Y)	Private/public key pair of PCS
(z, Z)	Private/public key pair of dealer
(m_i, T_i)	Block-tag pair
Σ	ordered collection of tags
$F = \{m_1, \cdots, m_n\}$	Stored file
$\mathcal{G}_1, \mathcal{G}_2$	two multiplicative groups
$\hat{e}$	the bilinear map from $\mathcal{G}_1$ to $\mathcal{G}_2$
q	the order of $\mathcal{G}_1$ and $\mathcal{G}_2$
π	pseudo-random permutation
H, h	cryptographic hash function
f, Ω	two pseudo-random functions
$chal = (c, k_1, k_2)$	the challenge, *i.e.*, c denotes the size of the challenged block set, k_1, k_2 are two different random numbers
$(\omega, cert)$	warrant-certificate pair
PCS	public cloud server
PPDP	proxy provable data possession

1.4   Our Contribution

In this paper, we propose the concept, system model and security model of PPDP protocol with general access structure. Then, by making use of the n-multiinear pairings and some difficult problems, we design a concrete and provably secure PPDP protocol which supports general access structure. Finally, we give the formal security proof and performance analysis. Through security analysis and performance analysis, our protocol is shown secure and efficient.

1.5   Organization

The rest of the paper is organized as follows. Section 2 introduces the preliminaries. Section 3 describes our PPDP protocol with general access structure, the formal security analysis and performance analysis. Finally, Sect. 4 gives a conclusion.

The notations throughout this paper are listed in Table 1.

2   Preliminaries

In this section, we propose the system model and security model of PPDP with general access structure. Then, the bilinear pairing, multilinear map and some corresponding difficult problems are reviewed in this section.

2.1   System Model and Security Model

The system consists of four different network entities: *Client, PCS, Dealer, Proxy*. They can be shown as the following.

1. *Client*, who has massive data to be stored on PCS for maintenance and computation, can be either individual consumer or organization, such as desktop computers, laptops, tablets, smart phones, *etc.*;
2. *PCS*, which is managed by public cloud service provider, has significant storage space and computation resource to maintain *client'* massive data;
3. *Dealer* is delegated to store multi-clients' data to PCS where the multi-client subset belongs to the concrete general access structure. It is trusted by all the clients.
4. *Proxy*, which is delegated to check *Client*'s data possession, has the ability to check *Client*'s data possession according to the warrant ω.

In the system model, there exists a general access structure $\mathcal{A} = \{\mathcal{A}_1, \mathcal{A}_2, \cdots, \mathcal{A}_{n'}\}$. In order to store some special files, all the clients in some subset $\mathcal{A}_j$ cooperate to approve and move the special files to PCS via the entity *Dealer*. The clients no longer store the special files locally. The clients can perform the remote data possession proof or delegate it to the proxy in special cases.

We start with the precise definition of PPDP with general access structure, followed by the formal security definition. Before that, we define the general access structure in our PPDP protocol.

Definition 1 (General Access Structure). *For the client set* $\mathcal{U} =$ $\{U_1, U_2, \cdots, U_n\}$, *the clients in* $\mathcal{U}$'s *subset* $\mathcal{A}_j = \{U_{j_1}, U_{j_2}, \cdots, U_{j_{n_j}}\}$ *can cooperate to approve and store the file* F *to PCS where* $j = 1, 2, \cdots, n'$ *and* $\mathcal{A}_j \subseteq \mathcal{U}$. *Denote* $\mathcal{A} = \{\mathcal{A}_1, \mathcal{A}_2, \cdots, \mathcal{A}_{n'}\}$. *Then,* $\mathcal{A}$ *is regarded as the general access structure.*

Without loss of generality, suppose the stored file F is divided into n blocks, i.e., $F = \{m_1, m_2, \cdots, m_n\}$.

Definition 2 (PPDP with General Access Structure). *For general access structure, PPDP is a collection of six polynomial time algorithms (*SetUp, TagGen, CertVry, CheckTag, GenProof, CheckProof*) among* PCS, Client, Dealer *and* Proxy *such that:*

1. $SetUp(1^k) \rightarrow (sk, pk)$ *is a probabilistic polynomial time key generation algorithm. Input a security parameter* k, *it returns a private/public key pair for every running. Every client* $U_{j_l} \in \mathcal{A}_j$ *can get its private/public key pair* (x_{j_l}, X_{j_l}). *PCS can also get its private/public key pair* (y, Y). *On the other hand, the client set* $\mathcal{A}_j$ *also prepares the warrant* ω_j *and the corresponding certificate* $cert_j$, *where* ω_j *points out the restriction conditions that the Proxy can perform the remote data possession checking task. The warrant-certificate pair* $(\omega_j, cert_j)$ *is sent to the Proxy.*

2. $TagGen(x_{j_l}, X_{j_l}, Y, m_i, U_{j_l} \in \mathcal{A}_j) \rightarrow T_i$ *is a probabilistic polynomial time algorithm that is run by all members of* $\mathcal{A}_j$ *and* Dealer *to generate the block tag* T_i. *Input the private/public key pair* (x_{j_l}, X_{j_l}) *for all the* $U_{j_l} \in \mathcal{A}_j$, *PCS's public key* Y *and a file block* m_i, *this algorithm returns the block tag* T_i.

3. $CertVry(\omega_j, cert_j) \rightarrow \{\text{"success"}, \text{"failure"}\}$ *is run by the proxy in order to validate the warrant-certificate pair. If the pair is valid, it outputs "Success" and accepts the pair ; otherwise, it outputs "failure" and rejects the pair.*

4. $CheckTag(m_i, T_i, y, X_{j_l}, Y, U_{j_l} \in \mathcal{A}_j) \rightarrow \{\text{"success"}, \text{"failure"}\}$ *is a determined polynomial time algorithm that is run by the PCS to check whether the block-tag pair* (m_i, T_i) *is valid or not. Input the block-tag pair* (m_i, T_i), *PCS's private/public key pair* (y, Y) *and the clients' public key* X_{j_l} *for all* $U_{j_l} \in \mathcal{A}_j$, *the algorithm returns "success" or "failure" denoting the pair is valid or not respectively.*

   Notes: *CheckTag phase is important in order to prevent the malicious clients. If the malicious clients store invalid block-tag pairs to PCS, PCS will accept them if CheckTag phase does not exist. When the malicious clients check these data's integrity, PCS's response will not pass the verification. The malicious clients will require PCS to pay compensation. Thus, PCS's benefits will be harmed.*

5. $GenProof(X_{j_l}, y, Y, F, \Sigma, chal, U_{j_l} \in \mathcal{A}_j) \rightarrow V$ *is a polynomial time algorithm that is run by the PCS in order to generate a proof of data integrity, where* $\Sigma = \{T_1, T_2, \cdots, T_n\}$ *is the ordered collection of tags. Input the public keys* $(X_{j_l}, Y, U_{j_l} \in \mathcal{A}_j)$, *an ordered collection* F *of blocks, an ordered collection of tags* Σ *and a challenge chal. Upon receiving the challenge from the proxy,*

it returns a data integrity proof V for some blocks in F that are determined by the challenge chal.

6. *$CheckProof(X_{j_l}, Y, chal, V, auxiliary\ data, U_{j_l} \in \mathcal{A}_j) \rightarrow \{\text{"success"},$ "failure"} is a polynomial time algorithm that is run by the proxy in order to check the PCS's response V. Input the public keys X_{j_l}, Y for $U_{j_l} \in \mathcal{A}_j$, a challenge chal, PCS's response V and some auxiliary data, this algorithm returns "success" or "failure" denoting whether V is valid or not for the data integrity checking of the blocks determined by chal.*

For the general access structure, in order to ensure that PPDP protocol is secure and efficient, the following requirements must be satisfied:

1. For the general access structure, the PPDP protocol only be performed by the clients or the delegated proxy.
2. *Dealer* should not be required to keep an entire copy of the files and tags.
3. The protocol should keep secure even if the PCS is malicious. If the PCS has modified some block tag pairs that are challenged, the response V can only pass the *CheckProof* phase with negligible probability. In other words, PCS has no ability to forge the response V in polynomial time.

According to the above security requirements, for general access structure, we define what is a secure PPDP protocol against malicious PCS (security property (3)) below. Without loss of generality, suppose the stored file is F and it is grouped into n blocks, *i.e.*, $F = \{m_1, m_2, \cdots, m_n\}$. Let the general access structure be $\mathcal{A} = \{\mathcal{A}_1, \mathcal{A}_2, \cdots, \mathcal{A}_{n'}\}$. Suppose the subset $\mathcal{A}_j = \{U_{j_1}, U_{j_2}, \cdots, U_{j_{n_j}}\} \in \mathcal{A}$ has the right to approve to store the file F to PCS.

Definition 3 (Unforgeability).*For general access structure, PPDP protocol is unforgeable if for any (probabilistic polynomial time) adversary $\mathbb{A}$ the probability that $\mathbb{A}$ wins the following PPDP game is negligible. For the general access structure, the PPDP game between the challenger $\mathcal{C}$ and the adversary $\mathbb{A}$ can be shown below:*

1. *SetUp: $\mathcal{C}$ generates system parameters params, clients' private/public key pairs (x_{j_l}, X_{j_l}) for all $U_{j_l} \in \mathcal{A}_j$, the proxy's private/public key pair (z, Z) and PCS's private/public key pair (y, Y). Then, it sends $(params, X_{j_l}, Y, y, Z, z, U_{j_l} \in \mathcal{A}_j)$ to the adversary $\mathbb{A}$. $\mathcal{C}$ keeps $(x_{j_l}, U_{j_l} \in \mathcal{A}_j)$ confidential and sends y, z to $\mathbb{A}$, i.e., y, z are known to $\mathbb{A}$. It is consistent with the real environment since the adversary $\mathbb{A}$ simulates PCS or the collusion of PCS and the proxy.*
2. *First-Phase Queries: $\mathbb{A}$ adaptively makes a number of different queries to $\mathcal{C}$. Each query can be one of the following.*
   - Hash *queries. $\mathbb{A}$ makes Hash function queries adaptively. $\mathcal{C}$ responds the Hash values to $\mathbb{A}$.*
   - Tag *queries. $\mathbb{A}$ makes block tag queries adaptively. For a query m_{1_1} queried by $\mathbb{A}$, $\mathcal{C}$ computes the tag $T_{1_1} \leftarrow$ TagGen$(x_{j_l}, y, z, X_{j_l}, Y, Z, m_{1_1}, U_{j_l} \in \mathcal{A}_j)$ and sends it back to $\mathbb{A}$. Without loss of generality, let $\{m_{1_1}, m_{1_2}, \cdots, m_{1_i}, \cdots, m_{1_{|\mathbb{I}_1|}}\}$ be the blocks which have been submitted for tag queries. Denote the index set as $\mathbb{I}_1$, i.e., $1_i \in \mathbb{I}_1$.*

3. *Challenge:* $\mathcal{C}$ *generates a challenge chal which defines a ordered collection* $\{j_1, j_2, \cdots, j_c\}$, *where* $\{j_1, j_2, \cdots, j_c\} \not\subseteq \mathbb{I}_1$ *is a set of indexes and* c *is a positive integer.* $\mathcal{C}$ *is required to provide a data integrity proof for the blocks* $m_{j_1}, \cdots, m_{j_c}$.

4. *Second-Phase Queries: Similar to the First-Phase Queries. Without loss of generality, let* $\{m_{2_1}, m_{2_2}, \cdots, m_{2_i}, \cdots, m_{2_{|\mathbb{I}_2|}}\}$ *be the blocks which have been submitted for tag queries. Denote the index set as* $\mathbb{I}_2$, *i,e.,* $2_i \in \mathbb{I}_2$. *The restriction is that* $\{j_1, j_2, \cdots, j_c\} \not\subseteq \mathbb{I}_1 \cup \mathbb{I}_2$.

5. *Forge:* $\mathbb{A}$ *returns a data integrity checking response* V *for the blocks indicated by chal.*

We say that $\mathbb{A}$ *wins the above game if* $CheckProof(X_{j_l}, Y, chal, V, auxiliary$ $data, U_{j_l} \in \mathcal{A}_j) \to$ "*success*" *with nonnegligible probability.*

Definition 3 states that, for the challenged blocks, a malicious PCS cannot produce a valid remote data integrity checking response if some challenged block tag pairs have been modified. It is a very important security property. On the other hand, if the response can pass the proxy's verification, what is the probability of all the data keeps intact ? The following definition states clearly the status of the blocks that are not challenged. In practice, a secure PPDP protocol also needs to guarantee that after validating the PCS's response, the proxy can be convinced that all of his outsourced data have been kept intact with a high probability. This observation gives the following security definition.

Definition 4 ((ρ, δ) Security). *For general access structure, a PPDP protocol is (ρ, δ) secure if PCS corrupted ρ fraction of the whole blocks, the probability that the corrupted blocks are detected is at least δ.*

In order to explain the definition 4, we give a concrete example. Suppose PCS stored $\ddot{n}$ block-tag pairs. The instrument troubles or malicious operations make $\ddot{l}$ block-tag pairs lost for PCS. Then, the corrupted fraction of the whole blocks is $\rho = \frac{\ddot{l}}{\ddot{n}}$. Suppose the clients query $\ddot{c}$ block-tag pairs' integrity checking. If the probability that the corrupted blocks can detected is at least δ, we call this scheme satisfies the (ρ, δ) security.

2.2 Bilinear Pairings, Multilinear Map and Difficult Problem

Let $\mathcal{G}_1$ and $\mathcal{G}_2$ be two cyclic multiplicative groups with the same prime order q. Let $\hat{e} : \mathcal{G}_1 \times \mathcal{G}_1 \to \mathcal{G}_2$ be a bilinear map. The bilinear map $\hat{e}$ can be constructed by the modified Weil or Tate pairings [20,21] on elliptic curves. The group $\mathcal{G}_1$ with such a map $\hat{e}$ is called a bilinear group. The Computational Diffie-Hellman (CDH) problem is assumed hard while the Decisional Diffie-Hellman (DDH) problem is assumed easy on the group $\mathcal{G}_1$ [22]. We give their expression below.

Gap Diffie-Hellman Problem (GDH). Let g is the generator of $\mathcal{G}_1$. For instance, given unknown $a, b, c \in \mathcal{Z}_q^*$ and $g, g^a, g^b, g^c \in \mathcal{G}_1$, it is recognized that there exists an efficient algorithm to determine whether $ab = c \bmod q$ by verifying $\hat{e}(g^a, g^b) = \hat{e}(g, g)^c$ in polynomial time (DDH problem), while no efficient algorithm can compute $g^{ab} \in \mathcal{G}_1$ with non-negligible probability within polynomial time (CDH problem). An algorithm $\mathbb{A}$ is said to (t, ϵ)-break the CDH problem on $\mathcal{G}_1$ if $\mathbb{A}$'s advantage is at least ϵ in time t, , *i.e.*,

$$Adv_{\mathcal{A}}^{CDH} = \Pr[\mathcal{A}(g, g^a, g^b) = g^{ab} : \forall a, b \in \mathcal{Z}_q^*] \geq \epsilon$$

The probability is taken over the choice of a, b and $\mathbb{A}$'s coin tosses.

A group $\mathcal{G}_1$ is a (t, ϵ)-GDH group if the DDH problem on $\mathcal{G}_1$ is efficiently computable and there exists no algorithm can (t, ϵ)-break the CDH problem on $\mathcal{G}_1$.

We say that $\mathcal{G}_1$ satisfies the CDH assumption if for any randomized polynomial time (in k) algorithm $\mathbb{A}$ we have that $Adv_{\mathcal{A}}^{CDH}(k)$ is a negligible function. In this paper, our multi-client PDP protocol come from the GDH group $\mathcal{G}_1$.

Next, we give the definition of an n-multilinear map. Multilinear map was proposed in the Ref. [23]. Many experts have proposed the concrete implementation [24,25]. We view the groups $\mathcal{G}_1$ and $\mathcal{G}_n$ as multiplicative groups.

Definition 5. *A map $\hat{e}_n : \mathcal{G}_1^n \to \mathcal{G}_n$ is an n-multilinear map if it satisfies the following properties:*

1. $\mathcal{G}_1$ *and* $\mathcal{G}_n$ *are groups of the same prime order* q;
2. *If* $a_1, \cdots, a_n \in \mathcal{Z}_q^*$ *and* $g_1, \cdots, g_n \in \mathcal{G}_1$ *then*

$$\hat{e}_n(g_1^{a_1}, \cdots, g_n^{a_n}) = \hat{e}_n(g_1, \cdots, g_n)^{a_1 a_2 \cdots a_n}$$

3. *The map $\hat{e}_n$ is non-degenerate in the following sense: if $g \in \mathcal{G}_1$ is a generator of $\mathcal{G}_1$ then $\hat{e}_n(g, \cdots, g)$ is a generator of $\mathcal{G}_n$.*

Multilinear Diffie-Hellman Problem. Given $g, g^{a_1}, \cdots, g^{a_{n+1}}$ in $\mathcal{G}_1$, it is hard to compute $\hat{e}_n(g, \cdots, g)^{a_1 \cdots a_{n+1}}$ in $\mathcal{G}_n$.

n-multilinear map has been used in the encryption, key management, hash function *etc.* [26–28].

3 Our Proposed PPDP Protocol with General Access Structure

3.1 Construction of PPDP Protocol with General Access Structure

First, we introduce some additional notations which will be used in the construction of our PPDP protocol with general access structure. Let g be a generator of $\mathcal{G}_1$. Suppose the stored file F (maybe encoded by using error-correcting code, such as, Reed-Solomon code) is divided into n blocks $(m_1, m_2, \cdots, m_n)$ where $m_i \in \mathcal{Z}_q^*$, *i.e.*, $F = (m_1, m_2, \cdots, m_n)$. The following functions are given below:

$$\begin{aligned}
f &: \mathcal{Z}_q^* \times \{1, 2, \cdots, n\} \to \mathcal{Z}_q^* \\
\Omega &: \mathcal{Z}_q^* \times \{1, 2, \cdots, n\} \to \mathcal{Z}_q^* \\
\pi &: \mathcal{Z}_q^* \times \{1, 2, \cdots, n\} \to \{1, 2, \cdots, n\} \\
H &: \{0, 1\}^* \to \mathcal{Z}_q^* \\
h &: \mathcal{Z}_q^* \times \mathcal{Z}_q^* \to \mathcal{G}_1^*
\end{aligned}$$

where f and Ω are two pseudo-random functions, and π is a pseudo-random permutation, H and h are cryptographic hash functions. For general access structure, PPDP protocol construction consists of six phases: *SetUp, TagGen, CertVry, CheckTag, GenProof, CheckProof.*

SetUp: PCS picks a random number $y \in \mathcal{Z}_q^*$ as its private key and computes $Y = g^y$ as its public key. The proxy picks a random number $z \in \mathcal{Z}_q^*$ as its private key and computes $Z = g^z$ as its public key. Suppose there are $\bar{n}$ clients $\mathcal{U} = \{U_1, U_2, \cdots, U_{\bar{n}}\}$. Let the general access structure be $\mathcal{A} = \{\mathcal{A}_1, \mathcal{A}_2, \cdots, \mathcal{A}_s\}$, where $\mathcal{A}_j = \{U_{j_1}, U_{j_2}, \cdots, U_{j_{n_j}}\} \subseteq \mathcal{U}, 1 \leq j \leq s$. For every $\mathcal{A}_j$, the dealer picks a random $u_j \in \mathcal{G}_1$ as $\mathcal{A}_j$'s public key. For any client $U_i \in \mathcal{U}$, it picks a random $x_i \in \mathcal{Z}_q^*$ as its private key and computes $X_i = g^{x_i}$ as its public key. $\mathcal{A}_j$'s warrant consists of the description ω_j of the constraints for which remote data possession proof is delegated together with a certificate $cert_j$. $cert_j$ is the multi-signature on ω_j of all the clients in $\mathcal{A}_j$ by using the concrete algorithms [29, 30]. Once delegated, the proxy can perform the data possession proof by using its private key z and warrant-certification pair $(\omega_j, cert_j)$. The clients send $(\omega_j, cert_j)$ to the proxy. The system parameter set is $params = \{\mathcal{G}_1, \mathcal{G}_2, \mathcal{G}_{n_j+1}, \hat{e}_{n_j+1}, \hat{e}, f, \Omega, \pi, H, h, q, u_j, X_i, \mathcal{A}_j \in \mathcal{A}, U_i \in \mathcal{U}\}$.

TagGen$(x_{j_l}, F, i, U_{j_l} \in \mathcal{A}_j)$: Suppose the valid client subset $\mathcal{A}_j$ generates the corresponding tags for the file $F = (m_1, m_2, \cdots, m_n)$. Denote the set $\bar{\mathcal{A}}_{j_l} = \mathcal{A}_j / U_{j_l}$. For every block m_i, the clients $\{U_{j_1}, U_{j_2}, \cdots, U_{j_{n_j}}\}$ in $\mathcal{A}_j$ and the dealer generate the tag T_i. In $\mathcal{A}_j$, all the clients cooperate to generate the multi-signature $cert_j$ on the warrant ω_j. The warrant-certification pair $(\omega_j, cert_j)$ are sent to the proxy. For $U_{j_l} \in \{U_{j_1}, U_{j_2}, \cdots, U_{j_{n_j}}\}$, it performs the following procedures:

1. U_{j_l} computes

$$t_j = H(\hat{e}_{n_j+1}(X_{j_1}, \cdots, X_{j_{l-1}}, X_{j_{l+1}}, \cdots, X_{j_{n_j}}, Y, Z)^{x_{j_l}}, \omega_j)$$

$$W_{i,j} = \Omega_{t_j}(i), \quad T_{i,j_l} = (h(t_j, W_{i,j})u_j^{m_i})^{x_{j_l}};$$

2. U_{j_l} sends the block-tag pair (m_i, T_{i,j_l}) and the corresponding warrant ω_j to dealer.

After receiving all the block-tag pairs (m_i, T_{i,j_l}), where $m_i \in F$, $U_{j_l} \in \mathcal{A}_j$, the dealer computes $T_i = \prod_{U_{j_l} \in \mathcal{A}_j} T_{i,j_l}$. Then it uploads the block-tag pair (m_i, T_i) and the corresponding warrant ω_j to PCS. When the above procedures are performed n times, all the block-tag pairs (m_i, T_i) are generated and uploaded to PCS for $1 \leq i \leq n$.

CertVry$(\{(\omega_j, cert_j), X_{j_i}, U_{j_i} \in \mathcal{A}_j\})$: Upon receiving the clients' warrant-certification pair $(\omega_j, cert_j)$, the proxy verifies its validity. If it is valid, the proxy accepts ω_j; otherwise, the proxy rejects it and queries the *Clients* for new warrant-certification pair.

$CheckTag((m_i, T_i), 1 \leq i \leq n)$: Given $\{(m_i, T_i), 1 \leq i \leq n\}$, for every i and $A_j \in \mathcal{A}$, PCS computes

$$\hat{t}_j = H(\hat{e}_{n_j+1}(X_{j_1}, \cdots, X_{j_{n_j}}, Z)^y, \omega_j), \quad \hat{W}_{i,j} = \Omega_{\hat{t}_j}(i)$$

Then, it verifies whether the following formula holds.

$$\hat{e}(T_i, g) \stackrel{?}{=} \hat{e}(h(\hat{t}_j, \hat{W}_{i,j})u_j^{m_i}, \prod_{U_{j_l} \in A_j} X_{j_l})$$

If it holds, PCS accepts it; otherwise, it is rejected.

$GenProof(pk, F, chal, \Sigma)$: Let $F, chal, \Sigma$ denote $F = (m_1, m_2, \cdots, m_n)$, $chal = (c, k_1, k_2)$, $\Sigma = (T_1, \cdots, T_n)$ where $chal$ is the proxy's challenge. In this phase, the dealer asks the PCS for remote data integrity checking of c file blocks whose indices are randomly chosen for each challenge. It can prevent the PCS from anticipating which blocks will be queried in each challenge. The number $k_1 \in Z_q^*$ is the random key of the pseudo-random permutation π. The number $k_2 \in Z_q^*$ is the random key of the pseudo-random function f. On the other hand, the proxy sends $(\omega_j, cert_j)$ to PCS. PCS verifies whether the signature $cert_j$ is valid. If it is valid, PCS compares this ω_j with its stored warrant ω'_j. When $\omega_j = \omega'_j$ and the proxys query complys with the warrant ω_j, PCS performs the procedures as follows. Otherwise, PCS rejects the proxys query.

1. For $1 \leq r \leq c$, PCS computes $i_r = \pi_{k_1}(r), a_r = f_{k_2}(r)$ as the indexes and coefficients of the blocks for which the proof is generated.
2. PCS computes $T = \prod_{r=1}^{c} T_{i_r}^{a_r}$, $\hat{m} = \sum_{r=1}^{c} a_r m_{i_r}$.
3. PCS outputs $V = (T, \hat{m})$ and sends V to the proxy as the response to the $chal$ query.

$CheckProof(chal, X_{j_l}, V, U_{j_l} \in A_j)$: Upon receiving the response V from PCS, the proxy performs the procedures below:

1. For $1 \leq r \leq c$, the proxy computes

$$t_j = H(\hat{e}_{n_j+1}(X_{j_1}, \cdots, X_{j_{n_j}}, Y)^z, \omega_j)$$
$$v_r = \pi_{k_1}(r), \quad a_r = f_{k_2}(r), \quad W_{v_r, j} = \Omega_{t_j}(v_r)$$

2. The proxy checks whether the following formula holds.

$$\hat{e}(T, g) \stackrel{?}{=} \hat{e}(\prod_{r=1}^{c} h(t_j, W_{v_r, j})^{a_r} u_j^{\hat{m}}, \prod_{U_{j_l} \in A_j} X_{j_l})$$

If the above formula holds, then the proxy outputs "success". Otherwise the proxy outputs "failure".

Notes: In the subset A_j, any client U_{j_l} can also perform the phase $CheckProof$ since the following formula holds:

$$\hat{e}_{n_j+1}(X_{j_1}, \cdots, X_{j_{l-1}}, X_{j_{l+1}}, \cdots, X_{j_{n_j}}, Y, Z)^{x_{j_l}}$$
$$= \hat{e}_{n_j+1}(X_{j_1}, \cdots, X_{j_{n_j}}, Y)^z$$

Thus, U_{j_l} can also calculate t_j and finish $CheckProof$.

3.2   Correctness Analysis and Security Analysis

The correctness analysis and security analysis of our proposed PPDP protocol can be given by the following lemmas and theorems:

Theorem 1. *If Client, Dealer, and PCS are honest and follow the proposed procedures, then the uploaded block-tag pairs can pass PCS's tag checking.*

Proof. In the phases *TagGen* and *CheckTag*, for all $U_{j_l} \in \mathcal{A}_j$,

$$
\begin{aligned}
\bar{t}_j &= H(\hat{e}_{n_j+1}(X_{j_1}, \cdots, X_{j_{n_j}}, Z)^y, \omega_j) \\
&= H(\hat{e}_{n_j+1}(g, \cdots, g, g)^{yz \prod_{U_{j_l} \in \mathcal{A}_j} x_{j_l}}, \omega_j) \\
&= t_j \\
&= \hat{t}_j
\end{aligned}
$$

Then, we can get $W_{i,j} = \bar{W}_{i,j} = \hat{W}_{i,j}$. By using *TagGen*, we know that

$$
\begin{aligned}
\hat{e}(T_i, g) &= \hat{e}(\prod_{U_{j_l} \in \mathcal{A}_j} (h(t_j, W_{i,j}) u_j^{m_i})^{x_{j_l}}, g) \\
&= \hat{e}(h(t_j, W_{i,j}) u_j^{m_i}, g^{\sum_{U_{j_l} \in \mathcal{A}_j} x_{j_l}}) \\
&= \hat{e}(h(t_j, W_{i,j}) u_j^{m_i}, \prod_{U_{j_l} \in \mathcal{A}_j} X_{j_l})
\end{aligned}
$$

Theorem 2. *If the proxy and PCS are honest and follow the proposed procedures, the response V can pass the proxy's data integrity checking, i.e., our PPDP protocol satisfies the correctness.*

Proof. Based on *TagGen* and *GenProof*, we know that $T = \prod_{r=1}^{c} T_{i_r}^{a_r}$. Thus,

$$
\begin{aligned}
\hat{e}(T, g) &= \hat{e}(\prod_{r=1}^{c} T_{i_r}^{a_r}, g) \\
&= \hat{e}(\prod_{r=1}^{c} (h(t_j, W_{i_r,j}) u_j^{m_{i_r}})^{a_r}, \prod_{U_{j_l} \in \mathcal{A}_j} X_{j_l}) \\
&= \hat{e}(\prod_{r=1}^{c} h(t_j, W_{i_r,j})^{a_r} u_j^{\sum_{r=1}^{c} a_r m_{i_r}}, \prod_{U_{j_l} \in \mathcal{A}_j} X_{j_l}) \\
&= \hat{e}(\prod_{r=1}^{c} h(t_j, W_{i_r,j})^{a_r} u_j^{\hat{m}}, \prod_{U_{j_l} \in \mathcal{A}_j} X_{j_l})
\end{aligned}
$$

Lemma 1. *Let $(\mathcal{G}_1, \mathcal{G}_2)$ be a (T', ϵ')-GDH group pair of order q. Let $\mathcal{A}_j$ be the tag generating subset. Then the tag generation scheme TagGen is $(T, q_S, q_H, q_h, \epsilon)$-existentially unforgeable under the adaptive chosen-message attack for all T and ϵ satisfying $\epsilon' \geq \frac{\epsilon}{(q_s+1)e}$ and $T' \leq T + c_{\mathcal{G}_1}(q_h + 2q_S) + c_{\hat{e}_{n_j}} q_H$, where $c_{\mathcal{G}_1}$ is the time cost of exponentiation on $\mathcal{G}_1$, $c_{\hat{e}_{n_j}}$ is the time cost of n_j-linear map. Here e is the base of the natural logarithm, and q_S, q_H, q_h are the times of Tag query, H-query and h-query respectively. n_j is the cardinal number of the tag generating subset $\mathcal{A}_j$.*

Proof. It is similar with Ref. [5]. We omit the proof procedures due to the page limits.

Lemma 2. *Let the challenge be chal $= (c, k_1, k_2)$. Then, the queried block-tag pair set is $\{(m_{\pi_{k_1}(i)}, T_{\pi_{k_1}(i)}), 1 \le i \le c\}$. If some block tag pairs are modified, the grouped block tag pair $(\hat{m}, T)$ can pass the proxy's verification only with negligible probability.*

Proof. We will prove this lemma by contradiction. It is assumed that the forged block tag pair $(\hat{m}, \hat{T})$ can pass the dealer's integrity checking, *i.e.*,

$$\hat{e}(\hat{T}, g) = \hat{e}(\prod_{r=1}^{c} h(t_j, W_{i_r,j})^{a_r} u_j^{\hat{m}}, \sum_{U_{j_l} \in \mathcal{A}_j} X_{j_l})$$

We prove this lemma from two cases.

Case 1, PCS makes use of the modified block tag pair to generate the grouped block tag pair and the block indexes satisfy the challenge requirements:

$$\hat{e}(\prod_{r=1}^{c} \hat{T}_{i_r}^{a_r}, g) = \hat{e}(\prod_{r=1}^{c} h(t_j, W_{i_r,j})^{a_r} u_j^{\sum_{r=1}^{c} a_r \hat{m}_{i_r}}, \prod_{U_{j_l} \in \mathcal{A}_j} X_{j_l})$$

where $a_r = f_{k_2}(r)$ and $i_r = \pi_{k_1}(r)$ are pseudo random, $1 \le r \le c$. Then,

$$\prod_{r=1}^{c} \hat{e}(\hat{T}_{\hat{m}_{i_r}}^{a_r}, g) = \prod_{r=1}^{c} \hat{e}(h(t_j, W_{i_r,j}) u_j^{\hat{m}_{i_r}}, \prod_{U_{j_l} \in \mathcal{A}_j} X_{j_l})^{a_r}$$

Let the generator of $\mathcal{G}_2$ be d, and

$$\hat{e}(\hat{T}_{i_r}, g) = d^{\hat{y}_r}$$

$$\hat{e}(h(t_j, W_{i_r,j}) u_j^{\hat{m}_{i_r}}, \prod_{U_{j_l} \in \mathcal{A}_j} X_{j_l}) = d^{y_r}$$

Then we can get

$$d^{\sum_{r=1}^{c} a_r \hat{y}_r} = d^{\sum_{j=1}^{c} a_r y_r}$$

$$\sum_{r=1}^{c} a_r \hat{y}_r = \sum_{r=1}^{c} a_r y_r$$

$$\sum_{r=1}^{c} a_j (\hat{y}_r - y_r) = 0 \bmod (q-1) \tag{1}$$

According to *Lemma 1*, a single block *Tag* is existential unforgeable. So, there exist at least two different indexes r such that $\hat{y}_r \ne y_r$. Suppose there are $s \le c$ pairs $(\hat{y}_r, y_r)$ such that $\hat{y}_r \ne y_r$. Then, there exist q^{s-1} tuples $(a_1, a_2, \cdots, a_c)$ satisfying the above Eq. (1). Since $(a_1, a_2, \cdots, a_c)$ is a random vector, the probability that the tuple satisfies the Eq. (1) is not greater than $q^{s-1}/q^c \le q^{c-1}/q^c = q^{-1}$. This probability is negligible.

Case 2, the PCS substitutes the other valid block-*Tag* pairs for modified block-*Tag* pairs:

To the challenge $chal = (c, k_1, k_2)$, PCS can get queried block tag pairs index set $\{i_1, i_2, \cdots, i_c\}$. Without loss of generality, we assume s block tag pairs are modified and their index set is $\{i_1, i_2, \cdots, i_s\}$ where $s \leq c$. PCS substitutes s valid block tag pairs for the s modified pairs. Without loss of generality, suppose the s valid block tag pairs indexes are $\mathcal{V} = \{v_1, v_2, \cdots, v_s\}$. PCS computes the grouped block tag pair as follows:

$$T = \prod_{r=s+1}^{c} T_{i_r}^{a_r} \prod_{v \in \mathcal{V}} T_v^{a_v}, \quad \hat{m} = \sum_{r=s+1}^{c} a_r m_{i_r} + \sum_{v \in \mathcal{V}} a_v m_v$$

where $a_r = f_{k_2}(r)$ for all $1 \leq r \leq c$ and $a_{v_i} = a_i$ for $1 \leq i \leq s$.

Assume the forged group block tag pair can pass the dealer's checking, $i.e.$,

$$\hat{e}(T, g) = \hat{e}(\prod_{r-1}^{c} h(t_j, W_{i_r,j})^{a_r} u_j^{\hat{m}}, \prod_{U_{j_l} \in \mathcal{A}_j} X_{j_l})$$

Since some block tag pairs are valid, $i.e.$, for $s + 1 \leq r \leq c$,

$$\hat{e}(T_{i_r}, g) = \hat{e}(h(t_j, W_{i_r,j}) u_j^{m_{i_r}}, \prod_{U_{j_l} \in \mathcal{A}_j} X_{j_l})$$

We can get the following formula:

$$\hat{e}(\prod_{v \in \mathcal{V}} T_v^{a_v}, g) = \hat{e}(\prod_{r=1}^{s} h(t_j, W_{i_r,j})^{a_r} u_j^{\sum_{v \in \mathcal{V}} a_v m_v}, \prod_{U_{j_l} \in \mathcal{A}_j} X_{j_l}) \text{ On the other hand,}$$

$$\hat{e}(\prod_{v \in \mathcal{V}} T_v^{a_v}, g) = \hat{e}(\prod_{v \in \mathcal{V}} h(t_j, W_{v,j})^{a_v} u_j^{\sum_{v \in \mathcal{V}} a_v m_v}, \prod_{U_{j_l} \in \mathcal{A}_j} X_{j_l})$$

Thus,

$$\hat{e}(\prod_{r=1}^{s} h(t_j, W_{i_r,j})^{a_r} u_j^{\sum_{v \in \mathcal{V}} a_v m_v}, \prod_{U_{j_l} \in \mathcal{A}_j} X_{j_l})$$
$$= \hat{e}(\prod_{v \in \mathcal{V}} h(t_j, W_{v,j})^{a_v} u_j^{\sum_{v \in \mathcal{V}} a_v m_v}, \prod_{U_{j_l} \in \mathcal{A}_j} X_{j_l})$$

We can get $\prod_{r=1}^{s} h(t_j, W_{i_r,j})^{a_r} = \prod_{v \in \mathcal{V}} h(t_j, W_{v,j})^{a_v}$. The probability that the above formula holds is q^{-1} because of h is hash oracle. It is negligible.

Based on Case 1 and Case 2, the forged group block tag pair can pass the dealer's checking with the probability no more than q^{-1}. It is negligible.

Lemma 1 states that an untrusted PCS cannot forge individual tag to cheat the proxy. Lemma 2 implies that the untrusted PCS cannot aggregate fake tags to cheat the dealer.

Theorem 3. *According to our proposed PPDP protocol with general access structure, if some queried block tag pairs are modified, PCS's response can only pass the proxy's CheckProof phase with negligible probability based on the assumption that the CDH problem is hard on $\mathcal{G}_1$.*

Proof. Suppose the stored blocks set is $\{m_1, m_2, \cdots, m_n\}$. We denote the challenger as $\mathcal{C}$ and the adversary as $\mathbb{A}$. Let the public parameters be $params = \{\mathcal{G}_1, \mathcal{G}_2, \hat{e}, f, \Omega, \pi, H, h, q\}$. Input (g, g^a, g^b), the goal of $\mathcal{C}$ is to compute the value g^{ab}. Let the client subset that can generate tag is $\mathcal{A}_j$. First, $\mathcal{C}$ picks random $z_{j_l} \in \mathcal{Z}_q^*, u_j \in \mathcal{G}_1$ and calculates $X_{j_l} = (g^a)^{z_{j_l}}$ for all $U_{j_l} \in \mathcal{A}_j$. u_j can be regarded as the public parameter of the access subset $\mathcal{A}_j$. Let X_{j_l} be the client U_{j_l}'s public key. The corresponding private key is unknown to $\mathcal{C}$. The challenger maintains three tables T_H, T_h, T which are initialized empty. PCS picks a random $y \in \mathcal{Z}_q^*$ and computes $Y = g^y$. Let (y, Y) be the PCS's private/public key pair. PCS picks a random $z \in \mathcal{Z}_q^*$ and computes $Z = g^z$. Let (z, Z) be the proxy's private/public key pair. Then, $\mathcal{C}$ answers all the queries that $\mathbb{A}$ makes.

H-Oracle, h-Oracle, Tag-Oracle are the same as the corresponding procedures in the Lemma 1.

We consider the challenge $chal = (c, k_1, k_2)$. Assume the forged aggregated block-tag pair $(\hat{m}, T)$ can pass the dealer's data integrity checking, *i.e.*,

$$\hat{e}(\hat{T}, g) = \hat{e}(\prod_{r=1}^{c} h(t_j, W_{i_r, j})^{a_r} u_j^{\hat{m}}, \prod_{U_{j_l} \in \mathcal{A}_j} X_{j_l}) \tag{2}$$

where $a_j = f_{k_2}(j)$ are random, $1 \leq j \leq c$.

According to Lemmas 1 and 2, we know that if some queried block-tag pairs are corrupted, the verification formula (2) holds with negligible probability. Thus, our propose multi-client PDP protocol is provably unforgeable in the random oracle model.

Theorem 4. *For the general access structure, the proposed PPDP protocol is $(\frac{d}{n}, 1 - (\frac{n-d}{n})^c)$-secure. The probability P_R of detecting the modification satisfies:*

$$1 - (\frac{n-d}{n})^c \leq P_R \leq 1 - (\frac{n-c+1-d}{n-c+1})^c$$

where n denotes the stored block-tag pair number, d denotes the modified block-tag pair number, and the challenge is $chal = (c, k_1, k_2)$.

Proof. It is similar with the Ref. [5]. We omit it due to the page limits.

3.3   Performance Analysis

In this section, we analyze the performance of our proposed PPDP protocol in terms of computation and communication overheads.

Computation: In our proposed PPDP protocol, suppose there exist n message blocks and the tag generating client subset is $\mathcal{A}_j$ which comprises n_j clients. In the *TagGen* phase, the clients need to perform n_j n_j-linear map, n_j exponentiations on the group G_{n_j+1} and $2nn_j$ exponentiations on the group G_1. On the other hand, the proxy needs to perform 1 n_j-linear map, 1 exponentiations on the group G_{n_j+1}, $2nn_j$ bilinear pairings. In the *CheckTag* phase, PCS has to compute 1 n_{j_1}-linear map, 1 exponentiations on the group G_{n_j+1}, $2n$ bilinear pairings and n exponentiations on the group G_1. In the *GenProof* phase, PCS needs to perform c exponentiations on the group G_2. In the *CheckProof* phase, the proxy can perform 1 n_j-linear map (it can be pre-computed and stored in the *TagGen* phase), 2 bilinear pairings, and $c+1$ exponentiations on G_1. Compared to the pairings and exponentiation, other operations, such as hashing, permutation, multiplication, *etc.*, are omitted since their costs are negligible.

Communication: The communication overhead mostly comes from the PPDP queries and responses. In PPDP query, the proxy needs to send $\log_2 c$ bits and 2 elements in $\mathcal{Z}_q^*$ to PCS. In the response, the PCS responds 1 element in G_1 and 1 element in $\mathcal{Z}_q^*$ to the proxy. Thus, our PPDP protocol has low communication cost.

Notes: Our proposed PPDP protocol is a general remote data integrity checking method with the general access structure. The idea is motivated by the application requirements which has been given in the subsection 1.1. The existing PDP protocols can only be applied for single client. It is not enough because the multi-client PDP and proxy PDP are also indispensable in some application fields. Of course, single client PDP is only the special case of our protocol when the size of the valid subset is 1 and the proxy is omitted. In general access structure, the PPDP protocol is proposed for the first time. It can be used in many application fields.

4 Conclusion

In this paper, we proposes a PPDP protocol with general access structure. We give its concept, security model, formal security proof and performance analysis. It is shown that our PPDP protocol is provably secure and efficient. It can be used in the public clouds to ensure remote data integrity.

References

1. Ateniese, G., Burns, R., Curtmola, R., Herring, J., Kissner, L., Peterson, Z., Song, D.: Provable data possession at untrusted stores. In: Capitani, D., di Vimercati, S., Syverson, P. (eds.) CCS 2007, pp. 598–609. ACM, New York (2007)
2. Ateniese, G., Di Pietro, R., Mancini, L.V., Tsudik, G.: Scalable and efficient provable data possession. In: Liu, P., Molva, R. (eds.) SecureComm 2008, pp. 9:1–9:10. ACM, New York (2008)
3. Erway, C.C., Küpçü, A., Papamanthou, C., Tamassia, R.: Dynamic provable data possession. ACM Trans. Inf. Syst. Secur. **17**(4), 1–29 (2015). 15
4. Sebé, F., Domingo-Ferrer, J., Martinez-Balleste, A., Deswarte, Y., Quisquater, J.J.: Efficient remote data possession checking in critical information infrastructures. IEEE Trans. Knowl. Data Eng. **20**(8), 1034–1038 (2008)
5. Wang, H.: Proxy provable data possession in public clouds. IEEE Trans. Serv. Comput. **6**(4), 551–559 (2013)

6. Wang, H., Wu, Q., Qin, B., Domingo-Ferrer, J.: Identity-based remote data possession checking in public clouds. IET Inf. Secur. **8**(2), 114–121 (2014)

7. Wang, H.: Identity-based distributed provable data possession in multicloud storage. IEEE Trans. Serv. Comput. **8**(2), 328–340 (2015)

8. Curtmola, R., Khan, O., Burns, R., Ateniese, G.: MR-PDP: multiple-replica provable data possession. In: ICDCS 2008, pp. 411–420. IEEE Press (2008)

9. Barsoum, A.F., Hasan, M.A.: Provable possession and replication of data over cloud servers (2010). http://www.cacr.math.uwaterloo.ca/techreports/2010/cacr2010-32.pdf

10. Hao, Z., Yu, N.: A multiple-replica remote data possession checking protocol with public verifiability. In: ISDPE 2010, pp. 84–89. IEEE Press (2010)

11. Barsoum, A.F., Hasan, M.A.: On Verifying Dynamic Multiple Data Copies over Cloud Servers(2011). http://eprint.iacr.org/2011/447.pdf

12. Juels, A., Kaliski Jr., B.S.: PORs: Proofs of retrievability for large files. In: ACM CCS 2007, pp. 584–597. ACM, New York (2007)

13. Shacham, H., Waters, B.: Compact proofs of retrievability. In: Pieprzyk, J. (ed.) ASIACRYPT 2008. LNCS, vol. 5350, pp. 90–107. Springer, Heidelberg (2008)

14. Wang, H.: Anonymous multi-receiver remote data retrieval for pay-TV in public clouds. IET Inf. Secur. **9**(2), 108–118 (2014)

15. Wang, H., Wu, Q., Qin, B., Domingo-Ferrer, J.: FRR: fair remote retrieval of outsourced private medical records in electronic health networks. J. Biomed. Inform. **50**, 226–233 (2014)

16. Bowers, K.D., Juels, A., Oprea, A.: Proofs of retrievability: theory and implementation. In: ACM CCSW 2009, pp. 43–54. ACM, New York (2009)

17. Zheng, Q., Xu, S.: Fair and dynamic proofs of retrievability. In: CODASPY 2011, pp. 237–248. ACM, New York (2011)

18. Dodis, Y., Vadhan, S., Wichs, D.: Proofs of retrievability via hardness amplification. In: Reingold, O. (ed.) TCC 2009. LNCS, vol. 5444, pp. 109–127. Springer, Heidelberg (2009)

19. Zhu, Y., Wang, H., Hu, Z., Ahn, G.J., Hu, H.: Zero-knowledge proofs of retrievability. Sci. China Inf. Sci. **54**(8), 1608–1617 (2011)

20. Boneh, D., Franklin, M.: Identity-based encryption from the weil pairing. In: Kilian, J. (ed.) CRYPTO 2001. LNCS, vol. 2139, pp. 213–229. Springer, Heidelberg (2001)

21. Miyaji, A., Nakabayashi, M., Takano, S.: New explicit conditions of elliptic curve traces for FR-reduction. IEICE Trans. Fundam. Electron. commun. comput. sci. **84**(5), 1234–1243 (2001)

22. Boneh, D., Lynn, B., Shacham, H.: Short signatures from the weil pairing. In: Boyd, C. (ed.) ASIACRYPT 2001. LNCS, vol. 2248, pp. 514–532. Springer, Heidelberg (2001)

23. Boneh, D., Silverberg, A.: Applications of multilinear forms to cryptography. Contemp. Math. **324**(1), 71–90 (2003)

24. Huang, M.D., Raskind, W.: A multilinear generalization of the tate pairing. Contemp. Math. **518**, 255–263 (2010)

25. Garg, S., Gentry, C., Halevi, S.: Candidate multilinear maps from ideal lattices. In: Johansson, T., Nguyen, P.Q. (eds.) EUROCRYPT 2013. LNCS, vol. 7881, pp. 1–17. Springer, Heidelberg (2013)

26. Hohenberger, S., Sahai, A., Waters, B.: Full domain hash from (leveled) multilinear maps and identity-based aggregate signatures. In: Canetti, R., Garay, J.A. (eds.) CRYPTO 2013, Part I. LNCS, vol. 8042, pp. 494–512. Springer, Heidelberg (2013)

27. Freire, E.S.V., Hofheinz, D., Paterson, K.G., Striecks, C.: Programmable hash functions in the multilinear setting. In: Canetti, R., Garay, J.A. (eds.) CRYPTO 2013, Part I. LNCS, vol. 8042, pp. 513–530. Springer, Heidelberg (2013)

28. Coron, J.-S., Lepoint, T., Tibouchi, M.: Practical multilinear maps over the integers. In: Canetti, R., Garay, J.A. (eds.) CRYPTO 2013, Part I. LNCS, vol. 8042, pp. 476–493. Springer, Heidelberg (2013)

29. Bagherzandi, A., Jarecki, S.: Identity-based aggregate and multi-signature schemes based on RSA. In: Nguyen, P.Q., Pointcheval, D. (eds.) PKC 2010. LNCS, vol. 6056, pp. 480–498. Springer, Heidelberg (2010)

30. Kawauchi, K., Minato, H., Miyaji, A., Tada, M.: A multi-signature scheme with signers' intentions secure against active attacks. In: Kim, K. (ed.) ICISC 2001. LNCS, vol. 2288, pp. 175–196. Springer, Heidelberg (2002)

31. Kumanduri, R., Romero, C.: Number Theory with Computer Applications, pp. 479–508. Prentice Hall, New Jersey (1998)

32. Miller, V.S.: Use of elliptic curves in cryptography. In: Williams, H.C. (ed.) CRYPTO 1985. LNCS, vol. 218, pp. 417–426. Springer, Heidelberg (1986)

33. Rivest, R.L., Hellman, M.E., Anderson, J.C., Lyons, J.W.: Responses to NIST's proposal. Commun. ACM **35**(7), 41–54 (1992)

A Provable Data Possession Scheme with Data Hierarchy in Cloud

Changlu Lin[1($\boxtimes$)], Fucai Luo[1], Huaxiong Wang[1,2], and Yan Zhu[3]

[1] College of Mathematics and Computer Science, Fujian Normal University,
Fuzhou 350117, China
cllin@fjnu.edu.cn
[2] Division of Mathematical Sciences, School of Physical and Mathematical Sciences,
Nanyang Technological University, Singapore, Singapore
hxwang@ntu.edu.sg
[3] School of Computer and Communication Engineering,
University of Science and Technology Beijing, Beijing 100083, China
zhuyan@ustb.edu.cn

Abstract. In recent years, numerous provable data possession (PDP) schemes have been proposed for checking the availability and integrity of data stored on cloud storage server (CSS) which is not fully trusted. However, these schemes do not work with specific subsets of verifiers, and they do not efficiently support dynamic enrollment and revocation of verifiers. In this paper, we propose a novel provable data possession scheme under hierarchical data framework in cloud environment. Our scheme can be considered as a generalization of privately verifiable PDP schemes. Specifically, data of different values are integrated into a data hierarchy, and clients are classified and authorized different access permissions according to their amounts of payment. Furthermore, our scheme allows the data owner to efficiently enroll and revoke clients. The scheme satisfies existential unforgeability against malicious CSS based on the hardness of the computational Diffie-Hellman problem.

Keywords: Cloud storage server · Provable data possession · Data hierarchy · Computational Diffie-Hellman assumption

1 Introduction

Outsourcing data to cloud storage servers (CSS), which alleviates the burden of local data storage and maintenance, has been receiving remarkable attentions thanks to its comparably low cost, high scalability, location-independent platform. However, CSS, in most of the cases, is not fully trusted. For instance, for its own benefits, it might delete or modify some data while concealing these behaviors from the data owner, or it might not store data in a fast storage specified by the contract with the owner. Thus, security issues such as availability and integrity of outsourced data are taken into consideration. It is a crucial

© Springer International Publishing Switzerland 2016
D. Lin et al. (Eds.): Inscrypt 2015, LNCS 9589, pp. 301–321, 2016.
DOI: 10.1007/978-3-319-38898-4_18

demand that data owners and data users (clients) are able to get strong evidences (proofs) that CSS still possesses the outsourced data, and that they have not been tampered or partially deleted over time. Meanwhile, it is impractical for the data owner and clients to download all outsourced data on cloud in order to validate its integrity, because of the immense transmission overheads across the network. Therefore, an efficient mechanism to validate the integrity of outsourced data with inexpensive computation, transmission, communication and storage overheads is highly desirable.

Scenario. *Data outsourced in CSS are frequently downloaded by clients (e.g., the employees of the company that owns the data). The clients must be able to validate the data integrity before downloading, to ensure that they can be safely used. Furthermore, in practice, according to the different payments for the data from the clients, the data owner may determine which data the clients are allowed to use through controlling their ability to validate the corresponding integrity proof.*

The existing Proofs of Retrievability (PORs) [1, 20] and Provable Data Possession (PDP) [3, 7, 9, 10, 12] schemes do not address this issue in the above scenario, since they are either publicly verifiable or privately verifiable. Furthermore, if one client wants to use multiple data, the client needs to keep the corresponding multiple secret keys when we implement the aforementioned PDP schemes. This will aggravate the burden of clients and the key managements and distributions become more complicated. In brief, the existing cannot solve these issues effectively, and they do not efficiently support dynamic enrollment and revocation of verifiers. These observations inspire us to explore a new solution in which not only specific subsets of verifiers are allowed to validate the proofs of data integrity, but also the verifiers can be efficiently added or revoked.

Related Works. A number of PORs [1, 4, 17, 20] and PDPs [2, 3, 5–10, 12–16] in various models have been proposed to enable the verification of data availability and integrity. Most of the existing PDP schemes rely on probabilistic proof techniques. In these challenge-response protocols, upon receiving a challenge from a verifier, CSS calculates a PDP (proof) as the response, so that the .verifier can check the validity of the PDP. Few PORs are more efficient than their PDP counterparts, based on an observation made by Barsoum *et al.* [3]. Sookhak *et al.* [11] surveyed the techniques of the existing remote data auditing schemes (i.e., they are the generalization of PDP schemes) and classified them into three different classes: replication-based model, erasure-coding-based model, and network-coding-based model. They also investigated the similarities and differences of these models. From the view of the security requirements, the challenges and the issues are described to offer a lightweight and efficient security mechanism.

All existing PDP schemes can be divided into two categories based on the role of the verifiers in the model: private verification and public verification. Most of the PDP schemes are publicly verifiable, while some others [2, 16] provide both private and public verification. However, public verification is undesirable in many circumstances, and private verification is necessary in some applications to prevent the disclosure of some confidential information. Shen and Tzeng [10] proposed a delegable PDP model, in which data owner generates verifiable tags for CSS and delegation key for the delegated verifier. The delegated verifier then checks the PDP obtained from CSS, but he cannot re-delegate this verification ability to others. Wang [14] also proposed a proxy PDP model, in which the authorized proxy verifier sends a warrant obtained from data owner and a challenge to CSS. CSS then validates the warrant and generates a PDP as the response. Based on elliptic curves, Ren et al. [8] proposed a designated-verifier PDP model in which data owner designates a verifier to check the integrity of outsourced data, while the verifier is stateless and independent from CSS. Wang et al. [16] proposed identity-based remote data possession checking in public clouds. In their PDP scheme, a trusted private key generator (PKG) generates system public key, master secret key and private key, the client accepts the private key after validating its correctness by using the public key. The client then challenges CSS, and CSS computes a PDP as the response. It appoints some people to validate the PDP without certificate management and verification.

Nevertheless, aforementioned PDP schemes only considered a small number of clients, and they do not address specific subsets of clients. Ren et al. [9] then proposed an attributed-based PDP model, in which attribute-based signature is utilized to construct the homomorphic authenticator. In their scheme, only the clients who satisfied the strategy can validate the PDP, since the homomorphic authenticator contains an attribute strategy. The data owner can adaptively authorize clients to use data through controlling their ability to validate the PDP by the attribute strategy in [9]. Nevertheless, their PDP scheme is inefficient in case of the number of clients is changing overtime or is large. Additionally, in their scheme, the enrollment and revocation of clients are inefficient.

When a large number of users access the data stored in cloud, these data are not only files but also are considered as file system. Role-based access control (RBAC) is one of the most important access systems to control the permissions. The hierarchical structure is used to manage the user permission in RBAC. The data belonging to the data owner will be treated as a special structure, called data hierarchy, to support our new PDP scheme. The idea of data hierarchy is inspired by the role-key hierarchy in the role-based cryptosystem proposed by Zhu et al. [18,19]. They introduced a new RBAC and used them to enforce fine-grained policies for sharing resources as well as to support various security features, including encryption, identification, and signature. They also provided several role-based cryptosystems such as role-based encryption, role-based signature and role-based authentication to achieve the RBAC system based on the rich algebraic structure of elliptic curves. Moreover, the role and user revocation, anonymity and tracing are also implemented.

Our Contributions. In this paper, we propose a provable data possession with data hierarchy on cloud, in which the data owner constructs a data hierarchy, and classifies the clients into the data hierarchy according to their different amounts of payment for the data. In brief, the data owner controls clients' ability to access the data by classifying them into the data hierarchy, and the data can be used if clients can access the corresponding data. The capability to validate the PDP is equivalent to the ability to access the corresponding data which is authorized by data owner in the data hierarchy. Our main contributions of the work are summarized as follows:

(1) We propose a new PDP with data hierarchy in which a data hierarchy is constructed. The system model and security model are also presented formally.
(2) In our proposed PDP scheme, CSS is stateless and independent of data owner. CSS sends a commitment value which is obtained from the data owner to the trusted third party to arbitrate between them in case the controversy on the integrity of data happens.
(3) The data owner can add or revoke clients easily and efficiently.
(4) We provide the full security proof for the proposed scheme based on the hardness of the computational Diffie-Hellman problem and provide the full security proof.
(5) We analyze the performance of the proposed PDP scheme. Comparing it with others, we draw the conclusion that our proposed PDP with data hierarchy is more feasible.

Organization. In Sect. 2, the system model, bilinear pairs and hardness assumptions will be given, and we give the partial order relation and a data hierarchy, meanwhile, the role-based cryptosystem (RBC) [18] will be reviewed in brief. We provide the definition of the proposed PDP scheme with data hierarchy and its security definition in Sect. 3. In Sect. 4, we present the concrete PDP scheme with data hierarchy. We analyze the security of the proposed PDP scheme with data hierarchy including the full security proof in Sect. 5. The performance analysis is presented in Sect. 6. Finally, Sect. 7 concludes the paper.

2   Preliminaries

2.1   System Model

Our proposed PDP scheme with data hierarchy involves three different entities as illustrated in Fig. 1.

- The *data owner*, who has a huge number of data files stored in clouds and is entitled to access and manipulate outsourced (stored) data, he/she can be a company or an organization who has a large number of clients (data users), and the data owner constructs a data hierarchy to manage the clients efficiently according to their different amount of payment for the data.
- The *clients*, who are data users and are required to pay for using (or downloading) some data, but they are not entitled to access and manipulate stored data. They have to validate the availability and integrity of the data before using them.

– The *cloud storage server* (CSS), which provides data storage services and has enough storage space and significant computation resources but is manipulated by manpower such as the CSS providers.

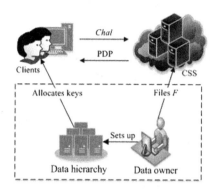

Fig. 1. PDP model with data hierarchy

The paper assumes that CSS is untrusted. It implies that CSS maybe tamper or partially delete the outsourced data for their own benefits, while we assume that the data is secure when it is transmitted to CSS through a secure channel. We do not consider the security issues when the data has been sent to CSS and CSS has confirmed it since CSS must take responsibility for this security issue. Once such security problem has occurred, CSS either compensates for it or conceals it which is also included in untrusted situation. We also assume that the data owner first constructs a data hierarchy, then authorizes or revokes the permissions of the clients to validate the PDP. The data owner is also able to validate the PDP if it is necessary.

2.2 Bilinear Pairings

Let $\mathbb{S} = (p, \mathbb{G}_1, \mathbb{G}_2, \mathbb{G}_T, e)$ be a bilinear map group system, where $\mathbb{G}_1$, $\mathbb{G}_2$ and $\mathbb{G}_T$ are three multiplicative cyclic groups of prime order p. Moreover, $e : \mathbb{G}_1 \times \mathbb{G}_2 \longrightarrow \mathbb{G}_T$ is a bilinear map which has the following properties : for any $G \in \mathbb{G}_1$, $H \in \mathbb{G}_2$ and all $a, b \in \mathbb{Z}_p$, we have

– bilinearity: $e(G^a, H^b) = e(G, H)^{ab}$;
– non-degeneracy: $e(G, H) \neq 1_{\mathbb{G}_T}$ unless G or $H = 1$;
– computability: $e(G, H)$ is efficiently computable.

Definition 1 (Computation Diffie-Hellman (CDH) Problem). *Given* g, g^x, $h \in \mathbb{G}$ *for some group* $\mathbb{G}$ *and* $x \in \mathbb{Z}_p$, *to compute* h^x.

Definition 2 (q-Strong Diffie-Hellman (SDH) Problem [18]**).** *Given* $\langle G, [x]G, [x^2]G, \cdots, [x^q]G \rangle$, *to compute* $\langle c, [\frac{1}{x+c}]G \rangle$ *where* $c \in \mathbb{Z}_p^*$ *and* G *be a generator chosen from* $\mathbb{G}_1$ *(or* $\mathbb{G}_2$*).*

2.3   Partial Order Relation and Data Hierarchy

We describe briefly the partial order relation, more details about it are referred to Zhu *et al.*'s work [18]. Let $\Psi = \langle P, \preceq \rangle$ be a partially ordered set with partial order relation "$\preceq$" on a set P. A partial order is a transitive, reflexive and antisymmetric binary relation. Two distinct elements x and y in Ψ are said to be comparable if $x \preceq y$ or $y \preceq x$. Otherwise, they are incomparable, denoted by $x \parallel y$. An order relation "$\preceq$" on P gives rise to a relation "$\prec$" of strict partial order: $x \prec y$ in P iff $x \preceq y$ and $x \neq y$. We define the predecessors and successors of elements in $\Psi = \langle P, \preceq \rangle$ as follows: for an element x in P, $\uparrow x = \{y \in P : x \preceq y\}$ denotes the set of predecessors of x. $\downarrow x = \{y \in P : y \preceq x\}$ denotes the set of successors.

In the paper, the data owner first constructs a data hierarchy. An example of the data hierarchy, $\langle D, \preceq \rangle$, which is transformed from the role-key hierarchy in RBAC [18], is shown in Fig. 2, in which the circle denotes data and the triangle denotes user, and more powerful (senior) data (the data allocated to senior roles) are in the higher level and less powerful (junior) data (the data allocated to junior roles) toward lower level in the data hierarchy. Specifically, data have been authorized different permissions with the partial order relation illuminated in the data hierarchy, and users are classified into different data, thus obtain the corresponding permissions of data (means they can validate its PDP in the paper) based on the different amount of payment for the data. It is essentially a cryptographic order relation for the set of data denoted by $\Omega = \langle U, K, D, P, \preceq \rangle$, where U denotes set of users, K denotes the set of keys including the data-key set PK and the user-key set SK, D denotes the set of data and P denotes the set of permissions. The six basic conditions for the data hierarchy and its security goals are referred to Zhu *et al.* [18], we will not present it here due to the limited space.

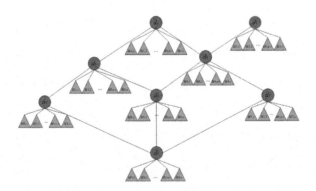

Fig. 2. The example of the data hierarchy

2.4   Review of the Role-Based Cryptosystem

We here highlight Zhu *et al.*'s role-based cryptosystem (RBC) [18], more details are referred to [18]. Let $\Omega = \langle U, K, D, P, \preceq \rangle$ be a data key hierarchy with partial-order "$\preceq$". We assume that the total number of the data and users are m and N in Ω respectively, i.e., $D = \{d_1, d_2, \cdots, d_m\}$, $\mid D \mid = m$, and $\mid U \mid = N$, where D and U denote the set of data and users, respectively. The RBC scheme is constructed as follows.

– *Setup*(κ, ψ): Let $S = (P, \mathbb{G}_1, \mathbb{G}_2, \mathbb{G}_T, e)$ be a bilinear map group system with the random elements $G \in \mathbb{G}_1$ and $H \in \mathbb{G}_2$, where $\mathbb{G}_1$, $\mathbb{G}_2$ and $\mathbb{G}_T$ are three multiplicative cyclic groups of prime order p. The algorithm chooses a random integer $\tau_i \in \mathbb{Z}_p^*$ for each data d_i in the data hierarchy, where d_i denotes the i-th data in D. Then, defines

$$\begin{cases} U_i = G^{\tau_i} \in \mathbb{G}_1 & \forall d_i \in R, \\ V = e(G, H) & \in \mathbb{G}_T. \end{cases}$$

Each τ_i is called as the secret of a data and U_i is the identity of a data. Defines $U_0 = G^{\tau_0}$ by computing from a random $\tau_0 \in \mathbb{Z}_p^*$. Thus, the public parameter is

$$params = \langle H, V, U_0, U_1, \cdots, U_m \rangle,$$

and $mk = \langle G, \tau_0, \tau_1, \cdots, \tau_m \rangle$ is kept privately.
– *GenRKey*$(params, d_i)$: This is an assignment algorithm from the setup parameter *params*. According to the data hierarchy, for any data d_i, it computes the data key pk_i as follows:

$$\begin{cases} pk_i = \langle H, V, W_i, \{U_k\}_{d_k \in \uparrow d_i} \rangle, \\ W_i = U_0 \cdot \prod_{d_i \not\preceq d_k} U_k, \end{cases}$$

where pk_i denotes the public data key of data $d_i \in D$ and $\{U_k\}_{d_k \in \uparrow d_i}$ is the set of all data in $\uparrow d_i$ which denotes the control domain for the data d_i. It is clear that $W_i = G^{\tau_0 + \sum_{d_i \not\preceq d_k} \tau_k}$. For the sake of simplicity, lets $\zeta_i = \tau_0 + \sum_{d_i \not\preceq d_k} \tau_k$, so that it has $W_i = G^{\zeta_i}$.
– *AddUser*$(mk, ID, u_{i,j})$: Given $mk = \langle G, \{\tau_i\}_{i=0}^m \rangle$ and a user index $u_{i,j}$ in the data d_i (where $u_{i,j}$ denotes the j-th user in data d_i), the data owner generates a unique private key by randomly selecting a fresh $x_{i,j} = Hash(ID, u_{i,j}) \in \mathbb{Z}_p$ as a $lab_{i,j}$ ($lab_{i,j} = x_{i,j}$) of $u_{i,j}$ which is public and defines

$$\begin{cases} A_{i,j} = G^{\frac{x_{i,j}}{\zeta_i + x_{i,j}}} \in \mathbb{G}_1, \\ B_{i,j} = H^{\frac{1}{\zeta_i + x_{i,j}}} \in \mathbb{G}_2. \end{cases}$$

Therefore, the user $u_{i,j}$ takes $sk_{i,j} = \langle A_{i,j}, B_{i,j} \rangle$ as the corresponding secret key.

Finally, the above procedure outputs the set of public data keys $\{pk_i\}$ and the set of user secret keys $\{sk_{i,j}\}$.

Since the users can been added into the RBC dynamically, collusion attack, which implies that two or more users belonged to different roles may collaborate to reveal the user key that they are unknown before, is possible in RBC. Zhu et al. [18] proved that the proposed RBC is secure against this collusion attack and got the following theorem.

Theorem 1. *Given a role-key hierarchy $\Omega = \langle U, K, D, P, \preceq \rangle$, the RBC scheme is (m, N, t)-collusion secure against collusion under Strong Diffie-Hellman (SDH) problem.*

3   Formal Definitions

In this section, we present the formal definition of the PDP scheme with data hierarchy and the corresponding security definition.

3.1   Definition of the PDP Scheme with Data Hierarchy

Definition 3 (PDP with Data Hierarchy). *A PDP protocol with data hierarchy is a collection of two algorithms (KeyGen, TagGen) and two interactive proof systems (GenProof, CheckProof), which are described in details below.*

KeyGen *(1^κ). The key generation algorithm is run by the data owner to setup the scheme. Input a bilinear map group system $\mathbb{S} = (p, \mathbb{G}_1, \mathbb{G}_2, \mathbb{G}_T, e)$, according to the role-based cryptosystem in Sect. 2, it outputs the system public parameters params, the public keys pk and the secret keys sk.*

TagGen *(sk, pk, F). The block-tag generation algorithm is also run by the data owner. Input the private keys sk and the public keys pk, the block m_k (the data owner splits file F into n blocks, i.e., $F = (m_1, m_2, \cdots, m_n)$, where $m_k \in \mathbb{Z}_p^*$), it outputs the tuple $\{\Phi_k, (m_k, T_{m_k})\}$, where Φ_k denotes the k-th record of metadata, (m_k, T_{m_k}) denotes the $k - th$ block-tag pair. Denotes all the metadata $\{\Phi_k\}$ as Φ and all the block-tag pairs $\{(m_k, T_{m_k}), k \in [1, n]\}$ as Σ.*

GenProof *($pk, chal, \Sigma$). This proof generation algorithm is run by CSS in response to a query from a client. Input the public keys pk, a chal from the query of the client $u_{j,k}$ and Σ, it outputs a PDP (proof) Γ as the response.*

CheckProof *($pk, sk_{j,k}, chal, \Gamma$). This checking algorithm is run by the client $u_{j,k}$ to validate the PDP. Input the public keys pk, the private key $sk_{j,k}$, the chal and the PDP Γ, the client $u_{j,k}$ outputs "success" or "failure".*

Remark 1. This definition is similar to other PDP schemes [5,8–10,13,14,16], except that the **KeyGen** is more complicated since the more public keys are generated in our scheme. We note that some PDP schemes in [3,5] encrypt outsourced data beforehand, while our PDP focus on the checking the integrity of outsourced data and the outsourced data is known for CSS, but they are secret for the clients except that the clients can validate its integrity after paying for them.

3.2 Security Definition

We know that the security of outsourced data, i.e., the availability and integrity, are the most important issues on outsourcing data to CSS. In this subsection, we follow the security definition in [8,14], and define the security for our proposed PDP scheme. We say that there is no (polynomial-time algorithm) adversary $\mathscr{A}$ (malicious CSS) that can successfully construct a valid proof so as to cheat verifier with non-negligible probability. The security definition of the game model between a challenger and an adversary $\mathscr{A}$ is provided as follows.

Setup. For any data d_i and the corresponding allocated file F_i, the challenger implements **KeyGen** (1^κ) to generates (sk, pk), and keeps sk secretly, while sends pk to the adversary $\mathscr{A}$.

First-Phase Queries. The adversary $\mathscr{A}$ adaptively queries the hash, block-tag pairs and PDP, which are described as follows.

- H-**query**: Any time $\mathscr{A}$ queries the hash, if the challenger has the hash value, then gives it to $\mathscr{A}$ directly. Otherwise, the challenger computes the hash value and sends it to $\mathscr{A}$.
- **Block-tag pairs query**: For the file F_i, the adversary $\mathscr{A}$ selects a block $m_{i,k}$ from F_i and sends it to the challenger, the challenger implements **TagGen**(sk, pk, F_i) to compute the tag $T_{m_{i,k}}$, then sends the block-tag pair $(m_{i,k}, T_{m_{i,k}})$ to $\mathscr{A}$. Let Π_1 be the set of indices of block tags that have been queried.
- **PDP query**: The adversary $\mathscr{A}$ sends a challenge $chal_1 = (c_1, k_{1,1}, k_{1,2})$ which defines an order collection $\{i_1, i_2, \cdots, i_{c_1}\}$, where $c_1 \in [1, n]$, is a positive integrity and denotes the number of block data queried by $\mathscr{A}$, and $k_{1,1}, k_{1,2} \in \mathbb{Z}_p^*$. The challenger runs **TagGen**(sk, pk, F_i) to generate the commitment value V^{t_i}, then runs **GenProof**$(\pi, f, chal_1, \Sigma_i)$ to generate a valid PDP, after that, sends the PDP and commitment value V^{t_i} to $\mathscr{A}$.

Challenge. The challenger generates a challenge $chal = (c, k_1, k_2)$ which defines an order collection $\{i_1, i_2, \cdots, i_c\}$, where $c \in [1, n]$, is a positive integrity and denotes the number of block data queried by the challenger, and $k_1, k_2 \in \mathbb{Z}_p^*$, requests the adversary $\mathscr{A}$ to provide a PDP for the corresponding blocks.

Second-Phase Queries. Similar to the first-phase queries. Let Π_2 be the set of indices of block tags in the **Block-tag pairs query**, $chal_2 = (c_2, k_{2,1}, k_{2,2})$ denotes the challenge in **PDP query**.

Answer. The adversary $\mathscr{A}$ sends a forged PDP $\Gamma_i' = (\mu_i', T_i')$ as the response to the challenger.

We say that the adversary $\mathscr{A}$ wins the game if

(1) The PDP $\Gamma_i' = (\mu_i', T_i')$ can pass the checking.
(2) The challenge $chal = (c, k_1, k_2) \notin \{chal_1, chal_2\}$ and the order collection $\{i_1, i_2, \cdots, i_c\} \nsubseteq (\Pi_1 \cup \Pi_2)$.

We define $AdvSig_{\mathscr{A}}$ to be the probability that $\mathscr{A}$ wins the above game, which taken over the coin tosses of $\mathscr{A}$ and the challenger. Based on the adversary's advantage in definition, we provide a security definition below:

Definition 4 (Unforgeability). *An adversary $\mathscr{A}$ (malicious CSS) is said to be a $(\varepsilon, t, q_h, q_b, q_p)$-forgery against the PDP scheme if $\mathscr{A}$ runs in time at most t, $\mathscr{A}$ makes at most q_h hash queries, at most q_b block-tag pairs queries and at most q_p PDP queries, while $AdvSig_{\mathscr{A}}$ is at least ε. The PDP scheme is $(\varepsilon, t, q_h, q_b, q_p)$-existentially unforgeable under the attack if no $(\varepsilon, t, q_h, q_b, q_p)$-forgery exists.*

4 Our Construction of the PDP Scheme with Data Hierarchy

Our PDP scheme with data hierarchy comprises with algorithms (**KeyGen**, **TagGen**) and algorithm (**GenProof, CheckProof**), where the **KeyGen** is more complicated involving a data hierarchy than the others schemes. The data owner (maybe a company or an organization) constructs a data hierarchy, generates the private keys of clients and outputs public keys by implementing the algorithm **KeyGen**, and then sends the private keys to clients and posts the public parameters on the public cloud. Specifically, the data owner first constructs a role-key hierarchy in RBAC, then allocates different data into different roles according to their different values. The role-key hierarchy is implemented by a data hierarchy (different data is bound with different roles). The clients, who are regarded as different users, are classified into the data hierarchy according to their payment for the data or the contracts. That is, the clients are authorized different permissions to access different values data. Next, the data owner computes the block-tag pairs by operating algorithm **TagGen**, and CSS generates a PDP (proof) by implementing algorithm **GenProof** as a response to the query of the client. Finally, the client operates **CheckProof** to check the PDP. The permission to access the data is achieved by validating the corresponding PDP, i.e., the data cannot be used by clients if the clients cannot validate its PDP. We describe them in detail as follows.

First, we define the following cryptographic functions which are used in our PDP scheme:

$$
\begin{aligned}
h(\cdot) &: \mathbb{G}_T \longrightarrow \mathbb{Z}_p^*, \\
H(\cdot) &: \{0,1\}^* \longrightarrow \mathbb{G}_1, \\
f &: \mathbb{Z}_p^* \times \{1, 2, \cdots, n\} \longrightarrow \mathbb{Z}_p^*, \\
\pi &: \mathbb{Z}_p^* \times \{1, 2, \cdots, n\} \longrightarrow \{1, 2, \cdots, n\},
\end{aligned}
$$

where, f is a pseudo-random function and π is a pseudo-random permutation. The parameters $\{h, H, f, \pi\}$ are made publicly. Our concrete PDP scheme with data hierarchy is constructed as follows.

KeyGen (1^κ). First, with a bilinear map group system $\mathbb{S} = (p, \mathbb{G}_1, \mathbb{G}_2, \mathbb{G}_T, e)$, the data owner constructs a data hierarchy based on the RBC mentioned in Sect. 2, and classifies the clients into the data hierarchy. The data owner generates corresponding public parameters and private keys, sends private keys to clients and posts the public parameters on the public cloud. The processes are provided below.

(1) The data owner constructs a data hierarchy by implementing RBC. For data d_i, the data owner generates public parameters

$$\begin{cases} pk_i = \langle H, V, W_i, \{U_k\}_{d_k \in \uparrow d_i} \rangle, \\ W_i = U_0 \cdot \prod_{d_i \npreceq d_k} U_k, \end{cases}$$

and classifies the clients as users of data into the data hierarchy. The data owner then generates the clients private keys $sk_{i,j} = \langle A_{i,j}, B_{i,j} \rangle$ and sends them to the corresponding clients $u_{i,j}$ in data d_i. The data owner also generates manager private keys $mk = \langle G, \tau_0, \tau_1, \cdots, \tau_m \rangle$, then stores them on local storage. Finally, the above process outputs the set of data keys $\{pk_i\}$ and the set of clients keys $\{sk_{i,j}\}$.

(2) For the data d_i, the data owner randomly chooses $t_i \in \mathbb{Z}_p^*$, and generates the following values,

$$\begin{cases} C_1 = W_i^{t_i} \in \mathbb{G}_1, \\ C_2 = H^{t_i} \in \mathbb{G}_2, \\ C_3 = V^{t_i} \in \mathbb{G}_T, \\ U_k' = U_k^{t_i} \in \mathbb{G}_1 \quad d_k \in \uparrow d_i. \end{cases}$$

$\hat{p}_i = \langle C_1, C_2, \{U_k'\}_{d_k \in \uparrow d_i} \rangle$ is posted as the public parameters of data d_i, while $C_3 = V^{t_i}$ is sent to CSS as a commitment for data d_i.

(3) The data owner selects a random element $u \in \mathbb{G}_1$ and a random value $x \in \mathbb{Z}_p^*$, chooses a generator $g \in \mathbb{G}_2$, then computes $v \leftarrow g^x$.
Finally, the data owner outputs secret keys $sk = (mk, x)$ and the public parameters $pk = (pk_i, \hat{p}_i, h, H, f, \pi, u, v, g, p)$.

TagGen (sk, pk, F_i). For data d_i, the data owner allocates the file F_i to it. The data owner splits file F_i into n blocks, i.e., $F_i = (m_{i,1}, m_{i,2}, \cdots, m_{i,n})$, where $m_{i,\ell} \in \mathbb{Z}_p^*$ of which $\ell \in [1, n]$. For every block $m_{i,\ell}$, the data owner performs the procedures as follows:

(1) Calculates $h(V^{t_i})$, and generates $u\|\ell$, then, computes $H(u\|\ell)$, where $\ell \in [1, n]$.

(2) Calculates

$$T_{m_{i,\ell}} = (H(u\|\ell)u^{h(V^{t_i})}u^{m_{i,\ell}})^x.$$

(3) Outputs $(m_{i,\ell}, T_{m_{i,\ell}})$.

After the procedures are performed at n times, all block tags are generated by the data owner, then, the data owner denotes the collection of all block-tag pairs $\{(m_{i,\ell}, T_{m_{i,\ell}}), \ell \in [1, n]\}$ as

$$\Sigma_i = \{(m_{i,1}, T_{m_{i,1}}), (m_{i,2}, T_{m_{i,2}}), \cdots, (m_{i,n}, T_{m_{i,n}})\},$$

and sends the block-tag pairs collection Σ_i and commitment value V^{t_i} to CSS. Then, CSS stores the block-tag pairs collection Σ_i and the commitment value V^{t_i}. After confirming CSS has stored the outsourced data, the data owner deletes the block-tag pairs collection Σ_i and commitment value V^{t_i} from its local storage but stores the private key sk and the corresponding metadata Φ_i. Next, we assume a client $u_{j,k}$ in data d_j, where $d_i \preceq d_j$, queries CSS.

GenProof $(\pi, f, chal, \Sigma_i)$. Upon receiving a query from the client $u_{j,k}$ which is $chal = (c, k_1, k_2)$, where $c \in [1, n]$ is the number of queried block data, $k_1, k_2 \in \mathbb{Z}_p^*$. CSS performs the procedures as follows:

(1) For $1 \leq j \leq c$, CSS calculates the indexes and coefficients of the blocks for which the PDP is generated: $a_j = \pi_{k_1}(j), v_j = f_{k_2}(j)$.

(2) CSS calculates

$$T_i = \prod_{j=1}^{c} T_{m_{i,a_j}}^{v_j}, \mu_i = \sum_{j=1}^{c} v_j m_{i,a_j}.$$

(3) CSS outputs $\Gamma_i = (\mu_i, T_i)$, and sends Γ_i to the client $u_{j,k}$ as the response to the query.

CheckProof $(pk, sk_{j,k}, chal, \Gamma_i)$. Upon receiving the response Γ_i from CSS, the client $u_{j,k}$ performs the procedures as follows:

(1) For $1 \leq j \leq c$, the client $u_{j,k}$ calculates the indexes and coefficients of the blocks: $a_j = \pi_{k_1}(j), v_j = f_{k_2}(j)$, in the same way with CSS.

(2) With the public parameters $\hat{p}_i$ of pk and private keys $sk_{j,k}$, the client $u_{j,k}$ calculates the value V^{t_i}. The client $u_{j,k}$ can figure the value V^{t_i} out due to $d_i \preceq d_j$.

(3) With the value V^{t_i}, the client $u_{j,k}$ calculates $h(V^{t_i})$, then checks whether the following equation holds:

$$e(T_i, g) \overset{?}{=} e(\prod_{j=1}^{c} H(u\|a_j)^{v_j} u^{h(V^{t_i})v_j}, v) \cdot e(u^{\mu_i}, v). \tag{1}$$

(4) If it holds, the client $u_{j,k}$ outputs "success". Otherwise, the client $u_{j,k}$ declares "failure".

– **Revocation.** The clients have to pay for the data in order to use them, and the clients might have signed contracts with the data owner. If the contract expires, the clients will not extend the contract, or other reasons, and if the whole clients of a data d_j are to be revoked, i.e., the data d_j is to be revoked, what can the date user do? Actually, this two situations can be solved easily. The data owner implements **AddUser** algorithm in role-based encryption scheme with revocation mechanism [18] to generate users' private keys and public labels, and generates new values $\hat{p}_i = \langle C_1, C_2, \{U_k'\}_{d_k \in \uparrow d_i} \rangle$ in step 2 and the commitment value $C_3 = V_{R_u}^{t_i}$ in **KeyGen** to revoke them, where,

$$\begin{cases} C_1 = & W_i^{t_i} & \in \mathbb{G}_1, \\ C_2 = & B_{R_u}^{t_i} & \in \mathbb{G}_2, \\ C_3 = & V_{R_u}^{t_i} & \in \mathbb{G}_T, \\ U_k' = U_k^{t_i} \in \mathbb{G}_1 & d_k \in \uparrow d_i. \end{cases}$$

The values $R_u, C_1, C_2, V_{R_u}^{t_i}, U_k'$ are same with the corresponding values in role-based encryption scheme with revocation mechanism [18], the details are not presented here due to the limited space.

– **Adding.** When some new clients pay for the data and are authorized to use the data. It implies that the clients should be added into the data hierarchy by data owner. The data owner just classifies them into data hierarchy based on their payment or the contracts, and then authorizes them the corresponding permission to access the data by implementing the **AddUser** procedure in Sect. 2.4.

Remark 2. When the controversy occurs between CSS and the data owner, for example, CSS asserts the data F_i is complete but the data owner does not agree with it. In this case, CSS sends the commitment value V^{t_i} to the trusted third party to arbitrate between them. This enables the trusted third party to validate the PDP since anyone can validate the PDP if he/she figures the value V^{t_i} out. In such situation, CSS may reveal the commitment value V^{t_i} to the clients privately. However, CSS cannot know the identities of clients since the information about it is kept secretly, meanwhile, CSS cannot publish the commitment value V^{t_i} naturally (the data owner can give clear indication of this in the contract with CSS). Moreover, in the worse situation, CSS reveals the commitment value V^{t_i} successfully. This will damage the interests of the data owner, but it will not release the secret keys of the data owner. That is, this will not threat to the security in our PDP scheme. In brief, the commitment value V^{t_i} is useless to CSS except for solving the controversy about the integrity of data.

5 Security Analysis

In this section, we describe the security analysis for our proposed PDP scheme with data hierarchy. We first verify the correctness and then we prove that there are no clients in data d_j can validate the PDP of file F_i if $d_i \npreceq d_j$. We finally prove that our scheme is secure against the forgery attack if the CDH problem is hard.

Theorem 2. *If the data owner, CSS and client are honest and they follow our proposed processes, then any challenge-response can pass clients' checking, namely, the PDP scheme satisfies the correctness.*

Proof. According to the proposed PDP scheme with data hierarchy in Sect. 4, any challenge-response can pass checking if Eq. (1) is satisfied. If $d_i \preceq d_j$, any client $u_{j,k}$ in data d_j can verify the PDP. With the public parameters $\hat{p}_i$ of pk and private keys $sk_{j,k}$, we first have

$$V^{t_i} = e\left(C_1 \cdot \prod_{d_\ell \in \Gamma(d_j, d_i)} U'_\ell, B_{j,k} \right) \cdot e(A_{j,k}, C_2) \tag{2}$$

$$= e\left(W_i^{t_i} \cdot \prod_{d_\ell \in \Gamma(d_j, d_i)} U'_\ell, B_{j,k} \right) \cdot e(A_{j,k}, H^{t_i})$$

$$= e\left((U_0 \cdot \prod_{d_i \npreceq d_k} U_k)^{t_i} \cdot (\prod_{d_\ell \in \cup_{d_j \npreceq d_k}\{d_k\} \setminus \cup_{d_i \npreceq d_k}\{d_k\}} U_\ell)^{t_i}, B_{j,k} \right) \cdot e(A_{j,k}, H^{t_i})$$

$$= e\big(G^{\zeta_j \cdot t_i}, H^{\frac{1}{\zeta_j + x_{j,k}}}\big) \cdot e\big(G^{\frac{x_{j,k}}{\zeta_j + x_{j,k}}}, H^{t_i}\big)$$

$$= e(G, H)^{\frac{\zeta_j \cdot t_i}{\zeta_j + x_{j,k}}} \cdot e(G, H)^{\frac{x_{j,k} \cdot t_i}{\zeta_j + x_{j,k}}} = e(G, H)^{t_i} = V^{t_i},$$

where $\Gamma(d_j, d_i)$ denotes $\cup_{d_j \not\preceq d_k}\{d_k\} \setminus \cup_{d_i \not\preceq d_k}\{d_k\}$ and $d_\ell \in \Gamma(d_j, d_i) \subseteq \uparrow d_i$. Then, with the value V^{t_i}, we have

$$e\Big(T_i, g\Big) = e\Big(\prod_{j=1}^{c} T_{m'_{i,a_j}}^{v_j}, g\Big) = e\Big(\prod_{j=1}^{c} \big(H(u\|a_j)u^{h(V^{t_i})}u^{m_{i,a_j}}\big)^{v_j}, v\Big)$$

$$= e\Big(\prod_{j=1}^{c} H(u\|a_j)^{v_j} u^{h(V^{t_i})v_j}, v\Big) \cdot e\Big(\prod_{j=1}^{c} u^{v_j m_{i,a_j}}, v\Big)$$

$$= e\Big(\prod_{j=1}^{c} H(u\|a_j)^{v_j} u^{h(V^{t_i})v_j}, v\Big) \cdot e\Big(u^{\mu_i}, v\Big). \qquad \square$$

Theorem 3. *In our proposed PDP scheme with data hierarchy, there are no clients in data d_j can validate the PDP of file F_i other than validating it by guessing, if $d_i \not\preceq d_j$.*

Proof. According to procedures of the proposed PDP scheme with data hierarchy, the client has to figure the value V^{t_i} out by the Eq. (2). Otherwise, he/she cannot calculate the value $h(V^{t_i})$ that result in failing to verify the Eq. (1). In Zhu et al.'s security proof [18], since $r_i \not\preceq r_j$, clients $\{u_{j,k}\}_{k\in[1,N]}$ in role r_j have no permissions to access the role r_i, namely, the clients $\{u_{j,k}\}_{k\in[1,N]}$ in role r_j cannot recover the corresponding message M due to they cannot calculate V^{t_i}. Above all, the clients $\{u_{j,k}\}_{k\in[1,N]}$ in data d_j cannot calculate V^{t_i} result in failing to validate the PDP of file F_i but validating it by guessing, which completes the proof. $\qquad \square$

Lemma 1. *If all the outsourced block tags are intact, while challenged blocks are modified or lost by the malicious CSS, the malicious CSS can compute a valid PDP with negligible probability.*

Proof. Without loss of generality, we assume the challenge is $chal = (c, k_1, k_2)$, the forged PDP generated by the malicious CSS is $\Gamma'_i = (\mu'_i, T'_i)$, where,$T'_i = \prod_{j=1}^{c} T_{m'_{i,a_j}}^{v_j}$ and $\mu'_i = \sum_{j=1}^{c} v_j m'_{i,a_j}$. We also assume the block tags are intact , so $T'_i = \prod_{j=1}^{c} T_{m'_{i,a_j}}^{v_j}$ is calculated correctly, while $\mu'_i = \sum_{j=1}^{c} v_j m'_{i,a_j}$ is not. If the forged PDP $\Gamma'_i = (\mu'_i, T'_i)$ can pass the checking, we have

$$e(T'_i, g) = e(\prod_{j=1}^{c} H(u\|a_j)^{v_j} u^{h(V^{t_i})v_j}, v) \cdot e(u^{\mu'_i}, v), \qquad (3)$$

where $\prod_{j=1}^{c} H(u\|a_j)^{v_j} u^{h(V^{t_i})v_j}$ is calculated correctly by verifier. On the other hand, if the challenged blocks are intact, we have

$$e(T_i, g) = e(\prod_{j=1}^{c} H(u\|a_j)^{v_j} u^{h(V^{t_i})v_j}, v) \cdot e(u^{\mu_i}, v). \qquad (4)$$

where $\mu_i = \sum_{j=1}^{c} v_j m_{i,a_j}$ is an expected value which is calculated correctly. Dividing the Eq. (4) by Eq. (3), we have

$$e(T_i' \cdot T_i^{-1}, g) = e(u^{\mu_i' - \mu_i}, v),$$

since we assume all the outsourced block tags are intact, so $T_i' = T_i$, then, we have,

$$e(u^{\mu_i' - \mu_i}, v) = e(T_i' \cdot T_i^{-1}, g) = 1,$$

from the above equation, we have $u^{\mu_i' - \mu_i} = 1$, resulting in $\mu_i' = \mu_i$, namely, $\sum_{j=1}^{c} v_j m_{i,a_j}' = \sum_{j=1}^{c} v_j m_{i,a_j}$, then we get

$$\sum_{j=1}^{c} v_j(m_{i,a_j}' - m_{i,a_j}) = 0. \qquad (5)$$

Suppose there exist $s \le c$ indices of blocks have been modified or lost, such that the corresponding index pairs (m_{i,a_j}', m_{i,a_j}) have the property $m_{i,a_j}' \ne m_{i,a_j}$, then there exist at most $(p-1)^{s-1}$ tuples $(v_1, v_2, \cdots, v_c)$ satisfy the Eq. (5). Thus, if there exist $s \le c$ indices of blocks have been modified or lost, the Eq. (5) holds only with the negligible probability less than $(p-1)^{s-1}/(p-1)^c \le (p-1)^{c-1}/(p-1)^c = 1/(p-1)$, which completes the proof. $\square$

Theorem 4. *Suppose the (ε', t')-CDH problem is hard in bilinear groups, then our PDP scheme with data hierarchy is $(\varepsilon, t, q_h, q_b, q_p)$-existentially unforgeable. That is, no adversary $\mathscr{A}$ (malicious CSS) can pass the checking procedure, i.e., the verification Eq. (1) in CheckProof cannot been satisfied. If the original blocks of file F_i and even only one of the blocks are incomplete, except by responding with the correctly computed PDP $\Gamma_i = (\mu_i, T_i)$, where, $T_i = \prod_{j=1}^{c} T_{m_{i,a_j}}^{v_j}$ and $\mu_i = \sum_{j=1}^{c} v_j m_{i,a_j}$. For all ε and t satisfying*

$$\varepsilon' \ge \varepsilon \cdot (p-2)/(p-1) \text{ and } t' \ge t + t_h \cdot q_h + t_b \cdot q_b + t_p \cdot q_p + t_c,$$

where, t_h is the cost of computing hash function, t_b is the cost of generating queried block tag, t_p is the cost of generating queried PDP, and t_c is the computational cost in solving the CDH problem.

Proof. Based on the security definition model, we analyze the possible attacks and countermeasures. First, it is infeasible for the adversary $\mathscr{A}$ to pass the checking procedures using blocks with different indices even if the data owner uses the same secret key x with all tags of blocks. This is because that the hash value $H(u\|a_j)$ and the block indices are embedded into the block tag which specified the tag so as to prevent using the tag to obtain a valid PDP(proof) for the different block. Second, the value $u^{h(V^{t_i})}$ of tags is fixed and independent which is computed by the data owner beforehand and must be computed by verifiers when checking the PDP afterward. It is useful for the verifiers but useless for the adversary $\mathscr{A}$. Above all, the goal of the adversary $\mathscr{A}$ is to generate a forged PDP which is not computed correctly but can pass the verification procedures. We will construct an algorithm $\mathscr{B}$ (which plays as a verifier) that, by interacting with $\mathscr{A}$, to solve the CDH problem if we assume that the adversary $\mathscr{A}$ wins the game.

The CDH problem is that: given g, g^x, $h \in \mathbb{G}$ for some group $\mathbb{G}$, where $x \in \mathbb{Z}_p$ is unknown, compute h^x. We suppose that the algorithm $\mathscr{B}$ expects to compute h^x. We also assume $\mathscr{B}$ can compute all values except for x in our proposed PDP scheme. $\mathscr{B}$ sets $u = g^\alpha h^\beta$ for α, $\beta \in \mathbb{Z}_p^*$ as a public parameter beforehand.

Setup. For data d_i and the corresponding allocated file F_i, $\mathscr{B}$ implements **Key-Gen**(1^κ) to generates (sk, pk), and keeps sk secretly, while sends pk to the adversary $\mathscr{A}$, where, private keys $sk = (mk, x)$ and the public parameters $pk = (pk_i, \hat{p}_i, h, H, f, \pi, u, \upsilon, g, p)$.

First-Phase Queries. It involves three kinds of queries:

- **H-query**: Any time $\mathscr{A}$ queries the hash, if $\mathscr{B}$ has the hash value, then gives it to $\mathscr{A}$ directly. Otherwise, $\mathscr{B}$ computes the hash value and sends it to $\mathscr{A}$.
- **Block-tag pairs query**: For the file F_i, the adversary $\mathscr{A}$ selects a block $m_{i,k}$ from F_i and sends it to $\mathscr{B}$, $\mathscr{B}$ implements **TagGen**(sk, pk, F_i) to compute the tag $T_{m_{i,k}}$, then sends the block-tag pair $(m_{i,k}, T_{m_{i,k}})$ to $\mathscr{A}$, where $T_{m_{i,k}} = (H(u\|k)\ u^{h(V^{t_i})}u^{m_{i,k}})^x$. Let Π_1 be the set of indices of block tags that have been queried.
- **PDP query**: The adversary $\mathscr{A}$ sends a challenge $chal_1 = (c_1, k_{1,1}, k_{1,2})$ which defines an order collection $\{i_1, i_2, \cdots, i_{c_1}\}$, where $c_1 \in [1, n]$, is a positive integrity and denotes the number of block data queried by $\mathscr{A}$, and $k_{1,1}, k_{1,2} \in \mathbb{Z}_p^*$. $\mathscr{B}$ runs **TagGen**(sk, pk, F_i) to generate the commitment value V^{t_i} and runs **GenProof**$(\pi, f, chal_1, \Sigma_i)$ to generate a valid PDP, then sends them to $\mathscr{A}$.

Challenge. $\mathscr{B}$ generates a challenge $chal = (c, k_1, k_2)$ which defines an order collection $\{i_1, i_2, \cdots, i_c\}$, where $c \in [1, n]$, is a positive integrity and denotes the number of block data queried by the challenger, and $k_1, k_2 \in \mathbb{Z}_p^*$, requests the adversary $\mathscr{A}$ to provide a PDP for the corresponding blocks.

Second-Phase Queries. Similar to the first-phase queries. Let Π_2 be the set of indices of block tags in the **Block-tag pairs query**, $chal_2 = (c_2, k_{2,1}, k_{2,2})$ denotes the challenge in **PDP query**.

Answer. The adversary $\mathscr{A}$ sends a forged PDP $\Gamma_i' = (\mu_i', T_i')$ as the response to $\mathscr{B}$.

Output. If the PDP $\Gamma_i' = (\mu_i', T_i')$, where $T_i' = \prod_{j=1}^{c} T_{m_{i,a_j}'}^{v_j}$ and $\mu_i' = \sum_{j=1}^{c} v_j m_{i,a_j}'$,
is valid, i.e., $\mathscr{A}$ wins the game, we have

(1) The PDP $\Gamma_i' = (\mu_i', T_i')$ can pass the checking, namely, the Eq. (1) holds.

(2) The challenge $chal = (c, k_1, k_2) \notin \{chal_1, chal_2\}$ and the order collection $\{i_1, i_2, \cdots, i_c\} \nsubseteq (\Pi_1 \cup \Pi_2)$.

We assume $\Gamma_i = (\mu_i, T_i)$ is an expected response with correctly values μ_i and T_i, along with $T_i = \prod_{j=1}^{c} T_{m_{i,a_j}}^{v_j}$ and $\mu_i = \sum_{j=1}^{c} v_j m_{i,a_j}$. According to the above 2), if the adversary $\mathscr{A}$ wins the game, we know that $\mathscr{A}$ cannot send a valid PDP to $\mathscr{B}$ obtained from the queries, thus, the corresponding block-tag pairs of the challenged blocks also cannot be obtained from the queries. According to the Lemma 1, since the challenged blocks are incomplete, we have $\mu_i' = \mu_i$ holds only with the negligible probability less than $1/(p-1)$. Then $\Gamma_i' = \Gamma_i$ holds with the negligible probability less than $1/(p-1)$. In other word, $\Gamma_i' \neq \Gamma_i$ holds with the overwhelming probability $(p-2)/(p-1)$. Then, $\mathscr{B}$ begins to solve the CDH problem as follows.

The response $\Gamma_i' = (\mu_i', T_i')$ of the adversary $\mathscr{A}$ is valid, i.e. $\Gamma_i' = (\mu_i', T_i')$ satisfies the Eq. (1), thus, $\mathscr{B}$ computes

$$e(T_i', g) = e(\prod_{j=1}^{c} H(u\|a_j)^{v_j} u^{h(V^{t_i})v_j}, v) \cdot e(u^{\mu_i'}, v). \tag{6}$$

At the same time, the expected proof $\Gamma_i = (\mu_i, T_i)$ also satisfies the Eq. (1), then, $\mathscr{B}$ also computes

$$e(T_i, g) = e(\prod_{j=1}^{c} H(u\|a_j)^{v_j} u^{h(V^{t_i})v_j}, v) \cdot e(u^{\mu_i}, v). \tag{7}$$

Since $\Gamma_i' \neq \Gamma_i$, then, there are two cases:

(a) $T_i' = T_i$, $\mu_i' \neq \mu_i$. $\mathscr{B}$ gets $e(T_i', g) = e(T_i, g)$ and $T_i' \cdot T_i^{-1} = 1$, then according to the Eqs. (6) and (7), $\mathscr{B}$ defines $\Delta\mu = \mu_i' - \mu_i \neq 0$, dividing the Eq. (7) by Eq. (6) to obtain

$$e(T_i' \cdot T_i^{-1}, g) = e(u^{\Delta\mu_i}, v). \tag{8}$$

Then, $\mathscr{B}$ transforms Eq. (8) aboved to obtain

$$e(T_i' \cdot T_i^{-1}, g) = e(u^{x\Delta\mu_i}, g). \tag{9}$$

Since $u = g^\alpha h^\beta$, $\mathscr{B}$ computes h^x by the Eq. (9) as follows:

$$(g^\alpha h^\beta)^{x\Delta\mu_i} = T_i' \cdot T_i^{-1}$$
$$h^{x\beta\Delta\mu_i} = 1 \cdot v^{-\alpha\Delta\mu_i}$$
$$h^x = v^{-\frac{\alpha}{\beta}}. \tag{10}$$

Thus, the given CDH problem is solved by $\mathscr{B}$.

(b) $T_i' \neq T_i$, $\mu_i' \neq \mu_i$. The same to the above processes, $\mathscr{B}$ divides the Eq. (4) by Eq. (3) result in $e(T_i' \cdot T_i^{-1}, g) = e(u^{\Delta\mu_i}, v)$, then, computes h^x as follows:

$$(g^\alpha h^\beta)^{x\Delta\mu_i} = T_i' \cdot T_i^{-1}$$
$$h^{x\beta\Delta\mu_i} = T_i' \cdot T_i^{-1} \cdot v^{-\alpha\Delta\mu_i}$$
$$h^x = (T_i' \cdot T_i^{-1} \cdot v^{-\alpha\Delta\mu_i})^{\frac{1}{\beta\Delta\mu_i}}. \tag{11}$$

Thus, the given CDH problem is also solved by $\mathscr{B}$.

Hence, if the adversary $\mathscr{A}$ breaks the proposed PDP scheme at a polynomial time t in non-negligible advantage ε, the algorithm $\mathscr{B}$ solves CDH problem at the polynomial time t' in non-negligible advantage $\varepsilon' \geq \varepsilon \cdot (p-2)/(p-1)$. Let t_h be the cost of computing hash function, t_b be the cost of generating queried block tag, t_p be the cost of generating queried PDP, and t_c be the computational cost in solving the CDH problem, then the total running time t' of $\mathscr{B}$ are no less than the total time of the running time t of $\mathscr{A}$, the time $t_h \cdot q_h$ for responding to q_h hash queries, the time $t_b \cdot q_b$ for q_b block tag pairs queries. the time $t_p \cdot q_p$ for q_p PDP queries, and the time t_c in computing the CDH problem. Therefore, the advantage $\varepsilon' \geq \varepsilon \cdot (p-2)/(p-1)$ and the total running time is $t' \geq t + t_h \cdot q_h + t_b \cdot q_b + t_p \cdot q_p + t_c$ as required. This completes the proof. □

6   Performance Analysis

We analyze performance of our proposed PDP scheme with data hierarchy mainly in four aspects: the computational cost of each algorithm, the communication cost of each phase, the storage cost of the data owner, the clients and CSS, and comparison with some other private verifiability PDP schemes.

Table 1. Computational cost of each algorithm

Algorithm	Addition	Scalar exponentiation	Hash	Pairing
KeyGen	$O(m)$	$m + 4 + 2m \cdot N$	$m \cdot N$	1
TagGen	0	$2n + 1$	$n + 1$	0
GenProof	$3c$	0	0	0
CheckProof	0	$4c + 1$	$c + 1$	4

- m denotes the number of data and N denotes the number of clients in each data.
- $O(m)$ denotes a polynomial of m.
- c and n denote the number of data blocks queried by the clients, and the number of data blocks, respectively.

Table 1 shows the computational cost of each algorithm. We calculate the numbers of additions in $\mathbb{Z}_p^*$, scalar exponentiations in $\mathbb{G}_1$, $\mathbb{G}_2$ and $\mathbb{G}_T$, hashes, and pairings.

Table 2. Communication cost

Phase	Communication cost
KeyGen	$mN \cdot (q_1 + q_2)$
TagGen	$n \cdot (p + q_1 - 1) + q_t$
GenProof	$c + 3p + q_1 - 3$

- m denotes the number of data and N denotes the number of clients in each data.
- q_1, q_2 and q_t denote the size of the element in $\mathbb{G}_1$, $\mathbb{G}_2$ and $\mathbb{G}_T$.
- c, n denote the number of data blocks queried by the clients, and the number of data blocks, respectively.

Table 3. Storage cost

Entity	Storage cost
Data owner	$(p - 1) \cdot (m + 2) + q_1$
Clients	$mN \cdot (q_1 + q_2)$
CSS	$mn \cdot (p + q_1 - 1) + q_t \cdot m$

- m denotes the number of data and N denotes the number of clients in each data.
- q_1, q_2 and q_t denote the size of the element in $\mathbb{G}_1$, $\mathbb{G}_2$ and $\mathbb{G}_T$.
- n denotes the number of data blocks.

Table 2 presents the communication cost of each phase in our PDP scheme. In the **KenGen** phase, the data owner generates private keys and delivers them to the corresponding clients of data in the data hierarchy through a secure channel. In the **TagGen** phase, the data owner computes all block-tag pairs and sends them to CSS via a secure channel. In the **GenProof** phase, the verifier queries CSS and CSS makes a respond.

Table 3 presents the storage cost of the data owner, the clients and CSS. The data owner needs to store the private keys sk, the clients receive the private keys from data owner and have to store them, and CSS stores the block-tag pairs collection Σ_i and the commitment value V^{t_i}. We compare the proposed PDP with data hierarchy with some other private verifiability PDP scheme on the number of the verifiers and whether they support the adding/revoking the users or not in Table 4.

From the Tables 1, 2 and 3, we find that the total cost of computational, communication and storage are lesser, except that it costs more in **KeyGen** process. In Table 4, we note that only our proposed PDP with data hierarchy satisfies both multi-verifiers, and to add or revoke clients easily and efficiently.

Remark 3. We note that our proposed PDP scheme is similar with [5], except for our **KeyGen** algorithm is more complicated since we introduce the hierarchical data framework. Comparing with the privately verifiable PDP [8–10,14,16], the total computational cost of our PDP is the nearly same with them. In our PDP scheme, we introduce the multiple clients and a data hierarchy in which clients' private keys and public parameters of the data hierarchy are computed, and a commitment value has to be figured out by the verifier in **CheckProof** algorithm. They involve two bilinear pairs computation and a few multiplication in $\mathbb{G}_1$. Moreover, though the **KeyGen** algorithm consumes more, it is no great

Table 4. Comparison with other PDP schemes

Protocols	Architecture of verifying	Add/Revoke or not
PDP in [10]	Single Verifier	*Yes*
PDP in [14]	Single Verifier	*Yes*
PDP in [8]	Single Verifier	*No*
PDP in [16]	Single Verifier	*Yes*
PDP in [9]	Multi-verifiers	*No*
Our PDP	Multi-verifiers	*Yes*

impact on the entire system performance since the **KeyGen** algorithm is only implemented at a time during the files life.

7   Conclusions

In this paper, we propose a provable data possession scheme which is the first provably secure PDP scheme that operates with a data hierarchy.

The capability to validate the PDP is authorized by the data hierarchy, while the clients are added or revoked efficiently by data owner. Moreover, CSS in our PDP scheme is stateless and independent of data owner. The controversy on the integrity of data can be solved efficiently without revealing the secret keys of the data owner. A small shortcoming is that the setup procedure consumes more time since it involves constructing a data hierarchy. A subject of our future work is to design a less time-consuming setup procedure.

Acknowledgements. The first author would like to thank Dr. Khoa Nguyen for many helpful and valuable comments on the write-up. We are also very grateful to four anonymous referees, who pointed out several inaccuracies and suggested improvements in the presentation of the paper. This work is supported by the National 973 Program (Grant No. 2013CB329605), the National Natural Science Foundation of China (Grant Nos. 61572132, 61170264, 61472032, and U1405255, and the Natural Science Foundation of Fujian Province (Grant No. 2015J01239).

References

1. Armknecht, F., Bohli, J.M., Karame, G.O., Liu, Z., Reuter, C.A.: Outsourced proofs of retrievability. In: Proceedings of the 21st ACM Conference on Computer and Communications Security, CCS 2014, pp. 831–843 (2014)
2. Ateniese, G., Burns, R., Curtmola, R., Herring, J., Kissner, L., Peterson, Z., Song, D.: Provable data possession at untrusted stores. In: Proceedings of the 14th ACM Conference on Computer and communications security, CCS 2007, pp. 598–609 (2007)
3. Barsoum, A.F., Hasan, M.A.: Provable possession and replication of data over cloud servers. Centre For Applied Cryptographic Research (CACR), University of Waterloo, Report 2010/32 (2010). http://www.cacr.math.uwaterloo.ca/techreports/2010/cacr2010-32.pdf

4. Bowers, K.D., Juels, A., Oprea, A.: Proofs of retrievability: Theory and implementation. In: Proceedings of the ACM Workshop on Cloud Computing Security, CCSW 2009, pp. 43–54 (2009)
5. Curtmola, R., Khan, O., Burns, R., Ateniese, G.: MR-PDP: Multiple-replica provable data possession. In: Proceedings of the 28th International Conference on Distributed Computing Systems, ICDCS 2008, pp. 411–420 (2008)
6. Erway, C.C., Küpü, A., Papamanthou, C., Tamassia, R.: Dynamic provable data possession. ACM Trans. Inf. Syst. Secur. (TISSEC) **17**(4), 1–29 (2015). article 15
7. Gritti, C., Susilo, W., Plantard, T.: Efficient dynamic provable data possession with public verifiability and data privacy. In: Foo, E., Stebila, D. (eds.) ACISP 2015. LNCS, vol. 9144, pp. 395–412. Springer, Heidelberg (2015)
8. Ren, Y., Xu, J., Wang, J., Kim, J.-U.: Designated-verifier provable data possession in public cloud storage. Int. J. Secur. Its Appl. **7**(6), 11–20 (2013)
9. Ren, Y., Yang, Z., Wang, J., Fang, L.: Attributed based provable data possession in public cloud storage. In: Proceedings of the Tenth International Conference on Intelligent Information Hiding and Multimedia Signal Processing, IIH-MSP 2014, pp. 710–713 (2014)
10. Shen, S.-T., Tzeng, W.-G.: Delegable provable data possession for remote data in the clouds. In: Qing, S., Susilo, W., Wang, G., Liu, D. (eds.) ICICS 2011. LNCS, vol. 7043, pp. 93–111. Springer, Heidelberg (2011)
11. Sookhak, M., Gani, A., Talebian, H., Akhunzada, A., Khan, S.U., Buyya, R., Zomaya, A.Y.: Remote data auditing in cloud computing environments: A survey, taxonomy, and open issues. ACM Comput. Surv. **47**(4), Article No. 65, p. 34 (2015)
12. Sun, X., Chen, L., Xia, Z., Zhu, Y.: Cooperative provable data possession with stateless verification in multicloud storage. J. Comput. Inf. Syst. **10**(8), 3403–3411 (2014)
13. Tan, S., Tan, L., Lin, X., Jia, Y.: An efficient method for checking the integrity of data in the Cloud. China Commun. **11**(9), 68–81 (2014)
14. Wang, H.: Proxy provable data possession in public clouds. IEEE Trans. Serv. Comput. **6**(4), 551–559 (2012)
15. Wang, Q., Wang, C., Li, J., Ren, K., Lou, W.: Enabling public verifiability and data dynamics for storage security in cloud computing. IEEE Trans. Parallel Distrib. Syst. **22**(5), 847–859 (2011)
16. Wang, H., Wu, Q., Qin, B., Domingo-Ferrer, J.: Identity-based remote data possession checking in public clouds. IET Inf. Secur. **8**(2), 114–121 (2014)
17. Yuan, J., Yu, S.: Proofs of retrievability with public verifiability and constant communication cost in cloud. In: Proceedings of the International Workshop on Security in Cloud Computing - Cloud Computing 2013, pp. 19–26 (2013)
18. Zhu, Y., Ahn, G.-J., Hu, H.X., Ma, D., Wang, H.X.: Role-based cryptosystem: A new cryptographic RBAC system based on role-key hierarchy. IEEE Trans. Inf. Forensics Secur. **8**(12), 2138–2153 (2013)
19. Zhu, Y., Ahn, G.-J., Hu, H.X., Wang, H.X.: Cryptographic role-based security mechanisms based on role-key hierarchy. In: Proceedings of the 5th ACM Symposium on Information, Computer and Communications Security, ASIACCS 2010, pp. 314–319 (2010)
20. Zhu, Y., Wang, H.X., Hu, Z.X., Ahn, G.-J., Hu, H.X.: Zero-knowledge proofs of retrievability. Sci. China Inf. Sci. **54**(8), 1608–1617 (2011)

Threshold Broadcast Encryption with Keyword Search

Shiwei Zhang$^{(\boxtimes)}$, Yi Mu, and Guomin Yang

School of Computing and Information Technology,
Centre for Computer and Information Security Research,
University of Wollongong, Wollongong, Australia
{sz653,ymu,gyang}@uow.edu.au

Abstract. Many users store their data in a cloud, which might not be fully trusted, for the purpose of convenient data access and sharing. For efficiently accessing the stored data, keyword search can be performed by the cloud server remotely with a single query from the user. However, the cloud server cannot directly search the data if it is encrypted. One of solutions could be to allow the user to download the encrypted data, in order to carry out a search; however, it might consume huge network bandwidth. To solve this problem, the notion of keyword search on encrypted data (searchable encryption) has been proposed. In this paper, a special variant of searchable encryption with threshold access is studied. Unlike some previous proposals which have fixed group and fixed threshold value, we define a new notion named *Threshold Broadcast Encryption with Keyword Search* (TBEKS) for dynamic groups and flexible threshold values. We formalize the security of a TBEKS scheme via a new security model named IND-T-CKA which captures indistinguishability against chosen keyword attacks in the threshold setting. We also propose the first practical TBEKS scheme with provable security in our IND-T-CKA security model, assuming the hardness of the Decisional Bilinear Diffie-Hellman problem.

Keywords: Searchable encryption · Keyword search · Cloud security

1 Introduction

Cloud computing [8] provides flexible computing resources, including data storage, to end users. Users are able to upload their data to the cloud for later access by themselves or by other users (i.e., data sharing) via the Internet. In other words, on-demand data access is available via the Internet where users can search and then download what they need. To prevent a huge amount of network bandwidth consumption, the search operations are usually done by the cloud instead of letting users download all the data and search locally.

Meanwhile, to ensure the privacy of the users, some sensitive data should be protected against the cloud server while the keyword search functionality is maintained. Specifically, the data to be searched and the keyword used in the search

© Springer International Publishing Switzerland 2016
D. Lin et al. (Eds.): Inscrypt 2015, LNCS 9589, pp. 322–337, 2016.
DOI: 10.1007/978-3-319-38898-4_19

operation should be inaccessible by any non-authorised parties, including the cloud. With such a demand, various searchable encryption schemes [1,3,4,7,11] have been proposed to enable secure searching over encrypted data. In a public key searchable encryption scheme, Bob encrypts both the data and the keywords under Alice's public key and uploads the ciphertexts to the cloud. As both the data and the keywords are protected, it is hard for the cloud server to gain any information about the data. To perform search operations, Alice generates a trapdoor for a keyword [1,3,4] or multiple keywords [7] and transfers it to the cloud via a secure communication channel. Upon receiving Alice's trapdoor of the keyword, the cloud server searches the whole database and returns the search results back to Alice. Finally, Alice downloads the ciphertexts from the cloud based on the search results, and decrypts them to get the original data.

In the normal searchable encryption schemes, the accessibility to the data and its search operation is authorised to a user [1,4] or a set of users [12,15] where any *single* user in the authorised user set can perform the search and the decryption operations. However, a single identity may not be trustful in some scenarios. For instance, a research team of a company is developing a new product and needs to access the company database. The head of the research department does not trust any single member of the research team to access the database, since an individual member may leak the secrets of the company for monetary purposes. To reduce the risk of a single point failure, a threshold searchable encryption scheme is more suitable where the accessibility to the data is decentralised from a single member to n members of the team where searching the database and decrypting a ciphertext both require at least t members to work together. To be more precise, in order to perform a search operation successfully, the cloud needs to obtain for a keyword at least t trapdoors from the n authorised users. If such a threshold searchable encryption also supports dynamic groups and flexible threshold values, the company can specify different classifications for different data by changing the authorised user set and the threshold value t. This paper aims to provide a practical solution for this problem.

1.1   Related Work

Boneh et al. [4] introduced the searchable encryption, namely public encryption with keyword search (PEKS), and defined the security model that the adversary cannot identify the keyword from the ciphertext without a trapdoor. Xu et al. [14] argued that PEKS is insecure under key guessing attack (KGA) since the remote server can always create a ciphertext of a keyword and test it with the target trapdoor. If the keyword space is in polynomial size, the adversary can get the keyword from the target trapdoor in polynomial time. Xu et al. [14] also proposed a method to enhance the security under KGA by encrypting and searching for the fuzzy keyword instead of the exact keyword.

Boneh et al. [4] showed that PEKS implies identity-based encryption (IBE) but not vice versa. Nevertheless, Abdalla et al. [1] proposed a PEKS scheme generically constructed using anonymous identity-based encryption (AnonIBE).

They also proposed identity-based searchable encryption (IBKS) from a 2-leveled anonymous hierarchical identity-based encryption (AnonHIBE). Similarly, searchable broadcasting encryption, namely broadcast encryption with keyword search (BEKS), can be constructed using 2-leveled anonymous hierarchical identity-coupling broadcast encryption (AnonHICBE) [2].

Searchable encryption can be divided into the single user setting (e.g. PEKS) and the multi-user setting. Broadcast encryption with keyword search [2] and attribute-based encryption with keyword search (ABKS) [12,15] are in the multi-user setting. In BEKS, the keyword is encrypted for a set of users. If a user is in the target set, the user can generate the trapdoor for testing the ciphertext. In ABKS, the keyword is encrypted under a policy or with attributes. Only the user who has a match of the policy and the attributes can generate the trapdoor for testing the ciphertext.

However, in both BEKS and ABKS, the individual target user has the full ability to generate the trapdoor. Wang et al. [13] decentralised the ability of trapdoor generation to multi-user in a threshold manner, which requires at least k of n users to generate the trapdoor. Siad [10] gave a formal definition of threshold public key encryption with keyword search (TPEKS), and generically constructed a TPEKS scheme with threshold (n, t)-IBE but no concrete scheme is provided. In Wang et al.'s scheme [13], a trusted centralised manager is required to generate the private keys for all users. To enhance the security, Siad's scheme [10] leverages a distributed protocol in private keys generation instead of a trusted third party.

We find that both schemes [10,13] are limited to a fixed number of users and fixed threshold value at the key generation stage. It makes adding or removing a user impossible, and changing the threshold value for individual ciphertext impossible. To encrypt a keyword for different set of users or with different threshold value, we have to generate the private keys for all the users in the target set. If it is an (n, t)-TPEKS scheme where t is the threshold value such that $0 < t \leq n$ and n is the maximum number of users, the users have to store $O(n \cdot 2^n)$ private-public key pairs for the possible ciphertexts, although they may share the same global public parameters.

1.2   Our Contribution

In this paper, we introduce a new notion named *Threshold Broadcast Encryption with Keyword Search* (TBEKS). We provide a formal definition of TBEKS and a formal security model, named indistinguishability in the threshold setting against chosen keyword attack (IND-T-CKA), to capture its security. Moreover, we construct a practical TBEKS scheme and prove that it is IND-T-CKA secure under the Decisional Bilinear Diffle-Hellman (DBDH) assumption in random oracle model.

In our TBEKS definition and scheme, users are ad hoc, i.e., they can generate their own private-public key pair individually. The data owner selects a target set of users and threshold value t to encrypt a keyword, and then uploads the full ciphertext to the remote server. To search the files containing a certain keyword, at least t users of the target user set need to generate their trapdoor shares for that keyword, and transfer those trapdoor shares to the remote server, in order to enable the remote server to perform the search operation. Our scheme does not fix the user group and the threshold value at the system setup, and only one private-public key pair is required for each user. Thus we solve Wang et al.'s open problem for dynamic group [13].

1.3   Paper Organisation

The rest of this paper is organised as follows. In Sect. 2, we review some essential tools and assumptions, including threshold secret sharing schemes and bilinear maps. We define TBEKS and its security model in Sect. 3. Then we propose our TBEKS scheme in Sect. 4 and prove that it is secure under the security model defined in Sect. 3.2. Finally, conclusion is addressed in Sect. 5.

2   Preliminaries

2.1   Threshold Secret Sharing Scheme

Shamir's secret sharing scheme [9] divides a secret s into n pieces $s_1, \ldots, s_n$ using a $k - 1$ degree polynomial and distributes to n users. If and only if k users or more come together, they can recover s by polynomial interpolation. Knowing $k - 1$ pieces of s does not reveal any information about s. This scheme is also called (k, n) *Threshold Secret Sharing Scheme*. Details are shown as follows.

Let $GF(q)$ be a finite field with order q where $q > n$. Each user U_i is associated with a public unique number $u_i \in GF(q)$. We also represent randomly choosing r from a space $\mathbb{S}$ by $r \in_R \mathbb{S}$. To share a secret $s \in GF(q)$ among a user set $S = \{U_1, \ldots, U_n\}$, a random $k - 1$ degree polynomial is picked as $p(x) = s + \sum_{j=1}^{k-1} a_j x^j$ where $a_j \in_R GF(q)$. Each user in the user set S gets a share $s_i = p(u_i)$. When k users come together and form a user set $A \subset S$, we can recover $p(x) = \sum_{U_i \in A} \Delta_i^A s_i$ where $\Delta_i^A = \prod_{U_\ell \in A \wedge i \neq \ell} \frac{x - u_\ell}{u_i - u_\ell}$. Then we can recover $s = p(0)$. Obviously, we can recover any point by

$$s_j = p(u_j) = \sum_{U_i \in A} \Delta_{ij}^A s_i \quad \text{where } \Delta_{ij}^A = \prod_{U_\ell \in A \wedge i \neq \ell} \frac{u_j - u_\ell}{u_i - u_\ell}.$$

By defining $u_0 = 0$, we have $s = \sum_{U_i \in A} \Delta_{i0}^A s_i$.

2.2   Bilinear Maps

A bilinear map is a function that maps two group spaces to a third space. For simplicity, we exploit the same bilinear map used in [5] where the first two group spaces are the same. Let $\mathbb{G}_1$ be a additive group, $\mathbb{G}_2$ be a multiplicative group, and both are cyclic groups of prime order q. Let P be a generator of $\mathbb{G}_1$. A bilinear map $e : \mathbb{G}_1 \times \mathbb{G}_1 \to \mathbb{G}_2$ has the following properties:

- Bilinearity: $\forall a, b \in \mathbb{Z}_q, e(aP, bP) = e(P, P)^{ab}$.
- Non-Degeneracy: $e(P, P) \neq 1$.
- Efficiency: It can be computed for any possible input efficiently.

There is a computational hard problem named *Decisional Bilinear Diffie-Hellman problem* (DBDH) coming with the bilinear map that we can rely on to construct our cryptographic scheme. The definition of DBDH problem is shown as follows.

Definition 1 (Decisional Bilinear Diffie-Hellman Problem). *Let a, b, c be uniformly and independently chosen from $\mathbb{Z}_q$, and T be uniformly and independently chosen from $\mathbb{G}_2$. Giving two probability distributions $\mathcal{D}_{BDH} = (P, aP, bP, cP, e(P, P)^{abc})$ and $\mathcal{D}_{rand} = (P, aP, bP, cP, T)$, there is an algorithm $\mathcal{A}$ can distinguish $\mathcal{D}_{BDH}$ and $\mathcal{D}_{rand}$ with advantage:*

$$Adv_{\mathcal{A}}^{DBDH} = \frac{1}{2} \left| \Pr[1 \leftarrow \mathcal{A}(D \xleftarrow{U} \mathcal{D}_{BDH})] - \Pr[1 \leftarrow \mathcal{A}(D \xleftarrow{U} \mathcal{D}_{rand})] \right|$$

where $D \xleftarrow{U} \mathcal{D}_{BDH}$ means that D is uniformly and independently chosen from $\mathcal{D}_{BDH}$. The advantage can be represented alternatively as

$$D_0 \xleftarrow{U} \mathcal{D}_{BDH}, \quad D_1 \xleftarrow{U} \mathcal{D}_{rand}, \quad b \xleftarrow{U} \{0, 1\},$$

$$Adv_{\mathcal{A}}^{DBDH} = \left| \Pr[b = b' \leftarrow \mathcal{A}(D_b)] - \frac{1}{2} \right|.$$

The DBDH problem is computational hard if and only if the advantage $Adv_{\mathcal{A}}^{DBDH}$ is negligible. In other words, it is hard to distinguish whether a vector is chosen from $\mathcal{D}_{BDH}$ or $\mathcal{D}_{rand}$ other than a random guess. Our scheme is secure based on the assumption of the hardness of the DBDH problem.

3   Threshold Broadcast Encryption with Keyword Search

3.1   Definition

Generally speaking, a *Threshold Broadcast Encryption with Keyword Search* (TBEKS)[1] scheme is used along with a *Threshold Broadcast Encryption* (TBE) scheme [6], where the former encrypts the keywords and the latter encrypts the message[2]. Independent private-public key pairs are suggested for the combination of the above mentioned system. In TBKES, there are three roles involved, including the **data owner** who encrypts the message and the keywords, the **server** who stores the ciphertexts and performs the requested search, and the **user** who has the access to the decryption of the message and generates search queries. TBEKS works as follows. The **data owner** chooses a set of **users** and a threshold value t, and encrypts the message under TBE and the keyword

---

[1] We choose the name TBEKS in order to separate it from TPEKS.

[2] In a storage system, messages are actually files.

under TBEKS. Then the **data owner** combines the ciphertexts and uploads them to the **server**. To perform a search operation, at least t **users** generate their individual trapdoors for the same target keyword W and upload the trapdoors to the **server** via a secure communication channel. After that, the **server** searches the whole database of ciphertexts with the given trapdoors and returns the result message indices back. Upon receiving the indices, the **users** retrieve the corresponding ciphertexts and decrypt them with at least t **users** working together. Note that only the trapdoors are required to be transferred via a secure communication channel.

Formally, we present the definition of *Threshold Broadcast Encryption with Keyword Search* as follows.

Definition 2 (Threshold Broadcast Encryption with Keyword Search). *A threshold broadcast encryption with keyword search scheme, involving the data users, the servers and the users U_i, consists of the following five possibly probabilistic polynomial time algorithms:*

- *params $\leftarrow Setup(1^k)$: The randomised system setup algorithm takes a security parameter 1^k, and outputs a set of parameters used in the system widely. This algorithm can be run by anyone whereas all users are required to agree on the same parameters.*
- *$(PK_i, SK_i) \leftarrow KeyGen(params)$: The randomised user key generation algorithm takes a system parameter params, and outputs a pair of secret key SK_i and public key PK_i of a user U_i. This algorithm is run by the users individually.*
- *$C \leftarrow TBEKS(\{PK_1, \ldots, PK_n\}, t, W)$: The randomised keyword encryption algorithm takes a set of public keys $\{PK_1, \ldots, PK_n\}$ of n target users, a threshold value t and a keyword W, and outputs a ciphertext C of the keyword W. This algorithm is run by the data owner.*
- *$T \leftarrow Trapdoor(SK_i, W)$: The possibly randomised trapdoor generation algorithm takes the secret key SK_i of a user U_i and a keyword W, and outputs a user trapdoor T of the keyword W. This algorithm is run by the users individually.*
- *$1/0 \leftarrow Test(\{T_1, \ldots, T_t\}, C)$: The deterministic test algorithm takes t trapdoors $T_i \leftarrow Trapdoor(SK_i, W)$ and a keyword ciphertext $C \leftarrow TBEKS(\{PK_1', \ldots, PK_n'\}, t', W')$, and outputs*

$$\begin{cases} 1 & \text{if } W = W' \wedge t \geq t' \wedge \{PK_1, \ldots, PK_t\} \subset \{PK_1', \ldots, PK_n'\}, \\ 0 & \text{otherwise.} \end{cases}$$

where $(PK_i, SK_i) \leftarrow KeyGen(params)$. This algorithm is run by the servers and the results will be sent back to each user involved.

In addition, we require the scheme to be correct.

Definition 3 (Correctness). *A threshold broadcast encryption with keyword search scheme is correct if the following statement is always true:*

$\forall params \leftarrow Setup(1^k), \quad \forall(SK, PK) \leftarrow KeyGen(params),$

$\forall n, t \in \mathbb{Z}^+ \wedge t \leq n, \quad \forall W \in \{0,1\}^*, \quad \forall C \leftarrow TBEKS(\{PK_1, \ldots, PK_n\}, t, W),$

$\forall S \subset \{1, \ldots, n\} \wedge t \leq |S| \leq n,$

$Test(\{T \mid T \leftarrow Trapdoor(SK_i, W) \wedge i \in S\}, C) = 1.$

3.2 Security Model

In Definition 2, we implicitly allow the server to combine the trapdoors for a keyword freely without any interaction with related users. For instance, the data owner creates $C_1 \leftarrow TBEKS(\{PK_1, PK_2\}, 2, W)$. To search for C_1, the users U_1, U_2 generate $T_i \leftarrow Trapdoor(SK_i, W)$ for $i = 1, 2$. Later, the data owner creates $C_2 \leftarrow TBEKS(\{PK_2, PK_3\}, 2, W)$ for the same keyword W. Similarly, to search for C_2, the users U_2, U_3 generate $T'_i \leftarrow Trapdoor(SK_i, W)$ for $i = 2, 3$. If $Trapdoor$ is a deterministic algorithm, the server can easily link T_1, T_2 and T'_3 together that they are created for the same keyword since $T_2 = T'_2$. As a result, if the data owner creates $C_3 \leftarrow TBEKS(\{PK_1, PK_2, PK_3\}, 3, W)$, the server can search C_3 by $Test(\{T_1, T_2, T'_3\}, C_3) = 1$. However, the server gains no information, especially the keyword encrypted in the trapdoors, other than the test result. Instead, this provides a feature that the server can cache the uploaded trapdoors from the users.

Because of this feature, we consider that the server is *honest but curious.* Importantly, we do not allow the server to collude with any users. Otherwise, the user U_1 and the server can learn the keyword in the ciphertext. For example, the server gets T_1, T_2 and T_3 from the users to test C_3. Then the server can use T_2 and T_3 to test C_2, and return the result to the user U_1. Now, the user U_1 knows the keyword of a ciphertext while U_1 is not in the target user set.

We also do not consider the keyword guessing attack (KGA) [14], since the server can always create a ciphertext $C = TBEKS(\{PK\}, 1, W)$ for all the keywords W with the user's trapdoor T. Commonly, if the keyword space is polynomial sized, the server can get the corresponding keyword W of C in polynomial time. However, this kind of attack can be prevented by Xu et al.'s method [14]. The BDOP scheme [4] also does not consider this attack.

In threshold broadcast encryption with keyword schemes, many users are involved. We consider that all users are registered before creating the ciphertexts, as the adversary may be able to register the private-public key pair of the target user. Now we define the *indistinguishability in the threshold setting against chosen keyword attack* (IND-T-CKA) game (Game 1) where an active adversary $\mathcal{A}$ tries to distinguish two encryptions of keywords W_0 and W_1 with the security parameter k:

1. The challenger runs the $Setup(1^k)$ algorithm to generate a set of system-wide parameters and passes them to the adversary $\mathcal{A}$.

$Game_{IND-T-CKA}^{k}$:
$$\mathcal{U}, \mathcal{C}, \mathcal{W} \leftarrow \emptyset$$
$$params \leftarrow Setup(1^k)$$
$$(\mathcal{S}, t, W_0, W_1) \leftarrow \mathcal{A}^{\mathcal{O}_{KeyGen}, \mathcal{O}_{Corrupt}, \mathcal{O}_{Trapdoor}}(params)$$
$$b \in_R \{0, 1\}$$
$$C \leftarrow TBEKS(\{PK_i\}_{i \in \mathcal{S}}, t, W_b)$$
$$b' \leftarrow \mathcal{A}^{\mathcal{O}_{KeyGen}, \mathcal{O}_{Corrupt}, \mathcal{O}_{Trapdoor}}(C)$$

$\mathcal{O}_{KeyGen}$:
$$(PK_i, SK_i) \leftarrow KeyGen(params)$$
$$\mathcal{U} \leftarrow \mathcal{U} \cup \{U_i\}$$
return PK_i

$\mathcal{O}_{Corrupt}$:
$$\mathcal{C} \leftarrow \mathcal{C} \cup \{U_i\} \subset \mathcal{U}$$
return SK_i

$\mathcal{O}_{Trapdoor}$:
$$T \leftarrow Trapdoor(SK_i, W)$$
$$\mathcal{W} \leftarrow \mathcal{W} \cup \{W\}$$
return T

$$Adv_{\mathcal{A}}^{IND-T-CKA} = \left| \Pr\left[b = b' \wedge |\mathcal{S} \cap \mathcal{C}| < t \wedge W_0, W_1 \notin \mathcal{W}\right] - \frac{1}{2} \right|$$

Game 1: IND-T-CKA

2. The adversary can adaptively ask the challenger to register a user and obtain the public key of that user by querying the key generation oracle $\mathcal{O}_{KeyGen}$. At the same time, the challenger records the requested user U_i in the user list $\mathcal{U}$.

3. The adversary can adaptively ask the challenger to obtain the secret key of a registered user $U_i \in \mathcal{U}$ by querying the collusion oracle $\mathcal{O}_{Corrupt}$. At the same time, the challenger records the requested user U_i in the collusion list $\mathcal{C}$.

4. The adversary can adaptively ask the challenger to obtain the user U_i's trapdoor of a keyword W by querying the trapdoor generation oracle $\mathcal{O}_{Trapdoor}$. At the same time, the challenger records the requested keyword W in the keyword list $\mathcal{W}$. For the corrupted users $U_i \in \mathcal{C}$, the adversary can compute the trapdoor by itself using the users U_i's secret key SK_i. Hence, the keyword list $\mathcal{W}$ only contains the requested keywords of uncorrupted users.

5. At some point, the adversary $\mathcal{A}$ outputs a set $\mathcal{S}$ of users, a threshold value t and two keywords W_0 and W_1 to be challenged. The adversary is restricted that W_0 and W_1 are not in the list $\mathcal{W}$ as they are not queried to $\mathcal{O}_{Trapdoor}$.

The adversary is also restricted that it cannot corrupt t users or more in the user set $\mathcal{S}$.

6. The challenger randomly selects b to be 0 or 1, and gives a ciphertext $C = TBEKS(\{PK_i\}_{i \in \mathcal{S}}, t, W_b)$ to the adversary $\mathcal{A}$.
7. The adversary can continue to query all three oracles $\mathcal{O}_{KeyGen}$, $\mathcal{O}_{Corrupt}$, $\mathcal{O}_{Trapdoor}$ with the same restrictions.
8. Eventually, the adversary $\mathcal{A}$ outputs a bit b'. If $b = b'$, the adversary wins the game.

We define the advantage of winning Game 1 as

$$Adv_{\mathcal{A}}^{IND-T-CKA} = \left| \Pr \left[b = b' \wedge |\mathcal{S} \cap \mathcal{C}| < t \wedge W_0, W_1 \notin \mathcal{W} \right] - \frac{1}{2} \right|.$$

Definition 4 (IND-T-CKA Security). *A threshold broadcast encryption with keyword search (TBEKS) scheme is indistinguishable in the threshold setting against chosen keyword attack (IND-T-CKA) if $Adv_{\mathcal{A}}^{IND-T-CKA}$ is a negligible function for all adversary $\mathcal{A}$ winning the Game 1 in polynomial time.*

4   Construction

4.1   The Scheme

We build our TBEKS scheme based on Daza et al.'s TBE scheme [6] using the idea similar to Boneh et al.'s PEKS scheme [4]. The main idea of the construction is to use the secret keys of users as the shares of a shared secret in an $(n, 2n - t)$ threshold secret sharing scheme. The shared secret works as the secret key of a dummy user in [4]. Since all computations are done with points on the elliptic curve and due to the hardness of discrete logarithm problem (DLP), the secret shares are computationally secure.

Our TBEKS scheme works as follows.

- $params \leftarrow Setup(1^k)$: Given a security parameter 1^k, this algorithm generates a prime number of q bits and groups $\mathbb{G}_1$ and $\mathbb{G}_2$ of order q where there is a bilinear map $e : \mathbb{G}_1 \times \mathbb{G}_1 \to \mathbb{G}_2$. This algorithm also picks a random generator P of $\mathbb{G}_1$. After that, the algorithm picks two hash functions $H : \{0,1\}^* \to \mathbb{G}_1$ and $H' : \mathbb{G}_1 \to \mathbb{Z}_q$. Note that the hash function H is used to hash a keyword W into a point on the elliptic curve and the hash function H' is used to hash the public key PK_i of a user U_i to an domain input $u_i = H'(PK_i)$ used in the threshold secret sharing scheme. Hence, it is required H' to be a collision resistant hash function. Alternatively, instead of using a hash function H', a user U_i can select its own unique u_i and then register along with its public key PK_i to a certification authority. Thus each user U_i has a unique public key PK_i associated with a unique public value u_i. For the system simplicity, we use the hash function H'.

$$\mathbb{G}_1 = \langle P \rangle, \quad e : \mathbb{G}_1 \times \mathbb{G}_1 \to \mathbb{G}_2, \quad H : \{0,1\}^* \to \mathbb{G}_1, \quad H' : \mathbb{G}_1 \to \mathbb{Z}_q.$$

return $params = (q, \mathbb{G}_1, \mathbb{G}_2, P, e, H, H')$.

- $(PK_i, SK_i) \leftarrow KeyGen(params)$: With the system wide parameters $param$, each user U_i randomly chooses a secret key $SK_i = x_i \in_R \mathbb{Z}_q^+$. Then its public key can be computed as $PK_i = x_i P$.

$$SK_i = x_i \in \mathbb{Z}_q^+, \quad PK_i = x_i P.$$

return (PK_i, SK_i).
- $C \leftarrow TBEKS(\{PK_1, \ldots, PK_n\}, t, W)$: To encrypt a keyword W, the data owner first obtains all the public keys of the target users $S = \{U_1, \ldots, U_n\}$. Then the data owner obtains the associated input values $\mathcal{U} = \{u_i = H'(PK_i) \mid U_i \in S\}$. Having n input values u_i and $PK_i = x_i P$, we can recover/construct a polynomial $p(u_j)P = \sum_{U_i \in S} \Delta_{ij}^S PK_i$. To form an $(n, 2n - t)$ threshold secret sharing scheme, the data owner chooses $n - t$ unique domain input values $\mathcal{D} \leftarrow \mathbb{Z}_q^{n-t}$, where $\mathcal{U} \cap \mathcal{D} = \emptyset$, for $n - t$ dummy users. Given the $n - t$ dummy users as a dummy user set D, the data owner can compute the public keys of the dummy users by computing $PK_j = p(u_j)P$ for all $u_j \in \mathcal{D}$. The detail of this algorithm works as follows.

$$Q = p(0)P = \sum_{U_i \in S} \Delta_{i0}^S PK_i, \quad s \in_R \mathbb{Z}_q^*, \quad C_1 = sP, \quad C_2 = e(H(W), Q)^s,$$

For each dummy user $U_j \in D$,

$$PK_j = p(u_j)P = \sum_{U_i \in S} \Delta_{ij}^S PK_i, \quad K_j = e(sH(W), PK_j).$$

return $C = (S, t, D, C_1, C_2, \{K_j\}_{U_j \in D})$.

To improve computational efficiency, it is possible to reuse Q and PK_j for the same user set S and different keywords, since these two variables are irrelevant to the keyword W and the randomness s. When calculating K_j, we can calculate $sH(W)$ before the loop so that all we need is a pairing operation. For the ciphertext C, it is not necessary to include all the public keys and the associated domain input values for both real users and dummy users since we only need the domain input values later. For better efficiency, this algorithm can return the ciphertext as $C = (\mathcal{U}, t, \mathcal{D}, C_1, C_2, \{K_j\}_{U_j \in D})$. As the values in $\mathcal{D}$ are not required to be chosen uniformly, the data owner can choose a continuous interval that $\mathcal{D} = \{r, r + 1, r + 2, \ldots, r + n - t - 1\}$ where $r \in_R \mathbb{Z}_q$. Thus $\mathcal{D}$ can be represented in two numbers in $\mathbb{Z}_q$. Hence, for the best result, the ciphertext size is $(n + 3)\mathbb{Z}_q + (n - t + 2)\mathbb{G}_1$.
- $T \leftarrow Trapdoor(SK_i, W)$: The user U_i generates the trapdoor T for the keyword W simply using its secret key. Then the user U_i uploads the trapdoor to the server via a secure communication channel.

$$T = x_i H(W).$$

return T.
- $1/0 \leftarrow Test(\{T_1, \ldots, T_t\}, C)$: Upon receiving t trapdoors from the users $A = \{U_1, \ldots, U_t\}$ where $|A \subset S| = t$, the server can run the following algorithm.

If more than t target trapdoors are uploaded, the server only picks the first t trapdoors.

$$\text{For each user } U_i \in A, \quad K_i = e(T_i, C_1),$$

$$B = A \cup D, \quad K = \prod_{U_i \in B} K_i^{\Delta_{i0}^B}.$$

return $K \overset{?}{=} C_2$.

Theorem 1. *The proposed threshold broadcast encryption with keyword search scheme is correct.*

Proof. Correctness is verified as following. First, K_i can be calculated as

$$K_i = e(sH(W), PK_i) = e(sH(W), x_iP) = e(x_iH(W), sP) = e(T_i, C_1).$$

Then we continue to verify the correctness of K,

$$K = \prod_{U_i \in B} K_i^{\Delta_{i0}^B} = \prod_{U_i \in B} e(sH(W), PK_i)^{\Delta_{i0}^B} = e(H(W), \sum_{U_i \in B} \Delta_{i0}^B PK_i)^s.$$

Since the $(n, 2n - t)$ threshold secret sharing scheme is constructed by n real users S, distributing shares to $n - t$ dummy users D, any n users in $S \cup D$ can recover the polynomial p used in the $TBEKS$ algorithm. Having $|D| = n - t$ and $S \cap D = \emptyset$ and $|A \subset S| = t$, we conclude that $A \cap D = \emptyset$ and further $|B = (A \cup D) \subset (S \cup D)| = n$. Thus the algorithm can recover the polynomial p with the users B. Then we have,

$$\sum_{U_i \in B} \Delta_{i0}^B PK_i = p(0)P = Q.$$

Finally,

$$K = e(H(W), Q)^s = C_2.$$

4.2   Security Proof

Theorem 2. *The proposed threshold broadcast encryption with keyword search scheme is IND-T-CKA secure. If an adversary $\mathcal{A}$ can win Game 1 with the advantage ε, an algorithm $\mathcal{S}$ can be constructed to solve DBDH problem in polynomial time with the advantage $\varepsilon' \geq \frac{\varepsilon}{2e^2(q_C+1)(q_T+1)}$, querying $\mathcal{O}_{Corrupt}$ for at most q_C times and $\mathcal{O}_{Trapdoor}$ for at most q_T times.*

Proof. Let $\delta = (P, aP, bP, cP, T)$ be an instance of DBDH problem (recall Definition 1) that a simulator $\mathcal{S}$ is challenged to distinguish that $\delta \in \mathcal{D}_{BDH}$ or $\delta \in \mathcal{D}_{rand}$. From the DBDH instance $\mathbb{D}$, the simulator $\mathcal{S}$ is also given two groups $\mathbb{G}_1$ and $\mathbb{G}_2$ of the same order q, a generator P of $\mathbb{G}_1$ and a bilinear map $e : \mathbb{G}_1 \times \mathbb{G}_1 \to \mathbb{G}_2$. The simulator $\mathcal{S}$ further chooses two hash functions $H : \{0, 1\}^* \to \mathbb{G}_1$ and $H' : G_1 \to \mathbb{Z}_q$. Then the simulator $\mathcal{S}$ packs those parameters as $params = (q, \mathbb{G}_1, \mathbb{G}_2.P, e, H, H')$ and passes $param$ to the adversary $\mathcal{A}$.

At the same time, the simulator $\mathcal{S}$ simulates the three oracles as follows.

- $\mathcal{O}_H$: The hash function H is viewed as a random oracle for the adversary $\mathcal{A}$ simulated by the simulator $\mathcal{S}$. Upon requesting the hash value of the keyword W_i, the simulator $\mathcal{S}$ randomly tosses a coin $c_i \in \{0, 1\}$ such that $\Pr[c_i = 0] = \alpha$ where α is determined later. The simulator $\mathcal{S}$ also chooses a random value $a_i \in_R \mathbb{Z}_q^+$. Then the simulator $\mathcal{S}$ computes the hash value h_i as

$$h_i = \begin{cases} a_i aP & \text{if } c_i = 0, \\ a_i P & \text{if } c_i = 1. \end{cases}$$

The distribution of $\{h_i\}$ is indistinguishable with a random distribution of $\mathbb{G}_1$. After that, the simulator $\mathcal{S}$ returns h_i to the adversary $\mathcal{A}$. In addition, the simulator $\mathcal{S}$ maintains a hash list $\mathcal{H} = \{W_i, c_i, a_i.h_i\}$. If the requested keyword W is on the list $\mathcal{H}$, the simulator $\mathcal{S}$ returns h_i directly.

- $\mathcal{O}_{KeyGen}$: To create a user U_i, the simulator $\mathcal{S}$ randomly tosses a coin $d_i \in \{0, 1\}$ such that $\Pr[d_i = 0] = \beta$ where β is determined later. The simulator also chooses a random value $x_i \in \mathbb{Z}_q^+$. Then the simulator $\mathcal{S}$ computes the secret key SK_i and the public key PK_i as follows.

$$SK_i = \begin{cases} \text{unknown} & \text{if } d_i = 0, \\ x_i & \text{if } d_i = 1. \end{cases} \quad PK_i = \begin{cases} x_i bP & \text{if } d_i = 0, \\ x_i P & \text{if } d_i = 1. \end{cases}$$

In the case of $d_i = 0$, the secret key $SK_i = x_i b$ cannot be computed by and is unknown to the simulator $\mathcal{S}$ since it is computational hard to compute b from bP. The distribution of $\{PK_i\}$ is indistinguishable with a random distribution of $\mathbb{G}_1$. After that, the simulator $\mathcal{S}$ returns PK_i to the adversary $\mathcal{A}$. In addition, the simulator $\mathcal{S}$ maintains a user key list $\mathcal{K} = \{U_i, d_i, SK_i, PK_i\}$.

- $\mathcal{O}_{Corrupt}$: Upon requesting the secret key SK_i of a created user U_i, the simulator $\mathcal{S}$ searches the user key list $\mathcal{K}$ and checks the corresponding d_i value. If $d_i = 0$, the simulator $\mathcal{S}$ **aborts** since the secret key SK_i is unknown to $\mathcal{S}$. Otherwise, the simulator $\mathcal{S}$ returns SK_i to the adversary $\mathcal{A}$.

- $\mathcal{O}_{Trapdoor}$: To create a created user U_i's trapdoor of a keyword W_j, the simulator $\mathcal{S}$ first looks up the hash list $\mathcal{H}$ for W_j. If $c_j = 0$, the simulator $\mathcal{S}$ simply **aborts**. Otherwise, the simulator $\mathcal{S}$ computes the trapdoor $T = a_j PK_i$ and returns it to the adversary $\mathcal{A}$. The correctness is verified as follows.

$$T = a_j PK_i = \begin{cases} a_j x_i bP = x_i b a_j P = x_i b h_i = SK_i H(W_j) & \text{if } d_i = 0, \\ a_j x_i P = x_i a_j P = x_i h_i = SK_i H(W_j) & \text{if } d_i = 1. \end{cases}$$

Although the trapdoor is still able to be simulated in the case of $c_j = 1 \wedge d_i = 0$ as $T = SK_i H(W_j) = x_i h_j = x_i a_j aP$, the simulator $\mathcal{S}$ still aborts for the simplicity of this proof. In other words, the probability ε' of solving the DBDH problem is greater if the simulator $\mathcal{S}$ does not abort in the above case but makes the proof harder. As long as ε' is not negligible, it is still acceptable.

At some point, the adversary $\mathcal{A}$ outputs a target user set S, a target threshold value t and two target keyword W_0 and W_1. The simulator looks up the hash list $\mathcal{H}$ for W_0 and W_1. If the keyword is not on the list $\mathcal{H}$, the simulator asks the $\mathcal{O}_H$ oracle for its hash value and then the keyword is on the list. If the corresponding values c_0 and c_1 of the keywords W_0 and W_1 is equals to 1, the simulator $\mathcal{S}$ **aborts**. Otherwise, the simulator $\mathcal{S}$ randomly picks $b \in \{0, 1\}$ such that $c_b = 0$. If there is only one $c = 0$, the simulator $\mathcal{S}$ has no choice and the value b is fixed. Then we have $H(W_b) = a_b a P$. Due to the restrictions to the adversary $\mathcal{A}$, at least one user U_σ in S has not been corrupted. The simulator $\mathcal{S}$ looks up the user key list $\mathcal{K}$ for that user. If the corresponding value $d_\sigma = 1$, the simulator $\mathcal{S}$ **aborts**. Otherwise, the simulator $\mathcal{S}$ divides the user set S into two sets $S_0 = \{U_i \in S \mid d_i = 0\}$ and $S_1 = \{U_i \in S \mid d_i = 1\}$. Intuitively, $S_0 \neq \emptyset$ because of the existence of U_σ. After that, the simulator $\mathcal{S}$ sets $C_1 = cP$. Before simulating C_2, we first seek how the genuine ciphertext is computed:

$$
\begin{aligned}
C_2 &= e(H(W), Q)^c = e(H(W), \sum_{U_i \in S} \Delta_{i0}^S PK_i)^c \\
&= e(H(W), \sum_{U_i \in S_0} \Delta_{i0}^S PK_i)^c \cdot e(H(W), \sum_{U_i \in S_1} \Delta_{i0}^S PK_i)^c \\
&= e(a_b a P, \sum_{U_i \in S_0} \Delta_{i0}^S x_i b P)^c \cdot e(a_b a P, \sum_{U_i \in S_1} \Delta_{i0}^S x_i P)^c \\
&= e(P, P)^{abc \cdot a_b \sum_{U_i \in S_0} \Delta_{i0}^S x_i} \cdot e(aP, cP)^{a_b \sum_{U_i \in S_1} \Delta_{i0}^S x_i}.
\end{aligned}
$$

The simulator $\mathcal{S}$ replaces the $e(P, P)^{abc}$ part with T and sets C_2 as

$$
C_2 = T^{a_b \sum_{U_i \in S_0} \Delta_{i0}^S x_i} \cdot e(aP, cP)^{a_b \sum_{U_i \in S_1} \Delta_{i0}^S x_i}.
$$

After that, the simulator $\mathcal{S}$ selects a set D of $n - t$ dummy users with the same restrictions in the normal construction (i.e. $\mathcal{U} \cap \mathcal{D} = \emptyset$). Similar to C_2, the simulator $\mathcal{S}$ computes K_j as

$$
K_j = T^{a_b \sum_{U_i \in S_0} \Delta_{ij}^S x_i} \cdot e(aP, cP)^{a_b \sum_{U_i \in S_1} \Delta_{ij}^S x_i}.
$$

Finally, the simulator $\mathcal{S}$ packs the ciphertext $C = (S, t, D, C_1, C_2, \{K_j\}_{U_j \in D})$ and sends to the adversary $\mathcal{A}$. Note that the resulted ciphertext C is consistent only if $T = e(P, P)^{abc}$.

Eventually, the adversary $\mathcal{A}$ outputs a guess b'. If $b = b'$, it means the ciphertext is consistent and it is believed that $T = e(P, P)^{abc}$. Hence, the simulator $\mathcal{S}$ outputs $\delta \in \mathcal{D}_{BDH}$. Otherwise, the simulator $\mathcal{S}$ outputs $\delta \in \mathcal{D}_{rand}$.

Lemma 1. *Let ρ be the probability of the simulator $\mathcal{S}$ not aborting. The advantage ε' of the simulator $\mathcal{S}$ solving the DBDH problem is at least $\frac{\rho \varepsilon}{2}$, assuming the probability of $\delta \in \mathcal{D}_{BDH}$ is $\frac{1}{2}$ and the adversary $\mathcal{A}$ wins Game 1 with the advantage ε.*

Proof. We prove this lemma by calculating the probability of the simulator $\mathcal{S}$ succeeding. If $\delta \in \mathcal{D}_{rand}$, the behaviour of the adversary $\mathcal{A}$ is unpredictable.

Thus, the simulator $\mathcal{A}$ succeeds at least better than a random guess with succeeding probability of $\frac{1}{2}$. Similarly, if the simulator S aborts, we just have a random guess. Otherwise, we take the result of the adversary $\mathcal{A}$ with the correct probability of $\frac{1}{2} + \varepsilon$. Let R be the event that the simulator S succeeds with a random guess. We have,

$\Pr[S \text{ succeeds}]$

$$= \frac{1}{2} \Pr[S \text{ succeeds} \mid \delta \in \mathcal{D}_{BDH}] + \frac{1}{2} \Pr[S \text{ succeeds} \mid \delta \in \mathcal{D}_{rand}]$$

$$\geq \frac{1}{2} (\Pr[S \text{ does not abort}] \cdot \Pr[\mathcal{A} \text{ succeeds}] + \Pr[S \text{ aborts}] \cdot \Pr[R]) + \frac{1}{2} \Pr[R]$$

$$= \frac{1}{2} \left(\Pr[S \text{ does not abort}] \cdot (\frac{1}{2} + \varepsilon) + \Pr[S \text{ aborts}] \cdot \frac{1}{2} \right) + \frac{1}{2} \cdot \frac{1}{2}$$

$$= \frac{1}{2} \left(\rho \cdot (\frac{1}{2} + \varepsilon) + (1 - \rho) \cdot \frac{1}{2} \right) + \frac{1}{2} \cdot \frac{1}{2}$$

$$= \frac{1}{2} \rho \varepsilon + \frac{1}{2}.$$

Since $\Pr[S \text{ succeeds}] \geq \frac{1}{2} \rho \varepsilon + \frac{1}{2}$ and $Adv_S^{DBDH} = \left| \Pr[S \text{ succeeds}] - \frac{1}{2} \right|$, we have

$$Adv_S^{DBDH} \geq \frac{\rho \varepsilon}{2}.$$

Lemma 2. *The simulator S does not abort with the probability ρ at least*

$$\frac{1}{e^2 (q_C + 1)(q_T + 1)},$$

querying $\mathcal{O}_{Corrupt}$ for at most q_C times and $\mathcal{O}_{Trapdoor}$ for at most q_T times.

Proof. There are 4 possible points that the simulator S may abort.

1. The simulator S aborts in answering $\mathcal{O}_{Corrupt}$ queries if $d_i = 0$. The single abort probability is β. The probability of not aborting for all $\mathcal{O}_{Corrupt}$ queries is $\Pr[E_1] = (1 - \beta)^{q_C}$.
2. The simulator S aborts in answering $\mathcal{O}_{Trapdoor}$ queries if $c_i = 0$. The single abort probability is α. The probability of not aborting for all $\mathcal{O}_{Trapdoor}$ queries is $\Pr[E_2] = (1 - \alpha)^{q_T}$.
3. The simulator S aborts in the challenge phase if $c_0 = c_1 = 1$. The probability of not aborting for this event is $\Pr[E_3] = 1 - (1 - \alpha)^2 = 2\alpha - \alpha^2$. Since $\alpha \in [0, 1]$, we have

$$\alpha \leq 1 \implies \alpha^2 \leq \alpha \implies 0 \leq \alpha - \alpha^2 \implies \alpha \leq 2\alpha - \alpha^2 = \Pr[E_3].$$

4. The simulator S aborts in the challenge phase if $d_\delta = 1$. The probability of not aborting for this event is $\Pr[E_4] = \beta$.

Since all the events are independent, we have

$$\rho = \Pr[E_1] \cdot \Pr[E_2] \cdot \Pr[E_3] \cdot \Pr[E_4] \geq \alpha(1-\alpha)^{q_T}\beta(1-\beta)^{q_C}.$$

The function $\alpha(1-\alpha)^{q_T}\beta(1-\beta)^{q_C}$ is maximised when $\alpha = \frac{1}{q_T+1}$ and $\beta = \frac{1}{q_C+1}$. Now we have

$$\rho \geq \frac{1}{q_T+1}(1-\frac{1}{q_T+1})^{q_T} \cdot \frac{1}{q_C+1}(1-\frac{1}{q_C+1})^{q_C}$$

$$\geq \frac{1}{q_T+1}\left(\lim_{q_T\to\infty}(1-\frac{1}{q_T+1})^{q_T}\right) \cdot \frac{1}{q_C+1}\left(\lim_{q_C\to\infty}(1-\frac{1}{q_C+1})^{q_C}\right)$$

$$= \frac{1}{e^2(q_C+1)(q_T+1)}.$$

Note that the statement $(1-\frac{1}{q_T+1})^{q_T} \geq \frac{1}{e} = \lim_{q_T\to\infty}(1-\frac{1}{q_T+1})^{q_T}$ is always true for any q_T. Similar statement for q_C also applies.
Combining $Adv_{\mathcal{S}}^{DBDH} \geq \frac{\rho\varepsilon}{2}$ from Lemma 1 and $\rho \geq \frac{1}{e^2(q_C+1)(q_T+1)}$ from Lemma 2, we have

$$\varepsilon' = Adv_{\mathcal{S}}^{DBDH} \geq \frac{\varepsilon}{2e^2(q_C+1)(q_T+1)}.$$

5   Conclusion

In this paper, we defined Threshold Broadcast Encryption with Keyword Search (TBEKS) scheme and its IND-T-CKA security model. We proposed the first TBEKS scheme and proved it is IND-T-CKA secure, assuming the hardness of the Decisional Bilinear Diffie-Hellman problem. In our TBEKS scheme, we consider the server to be honest but curious and we do not allow the server to collude with the users. It is an open problem to build a scheme that is secure against a malicious server that may collude with other users.

References

1. Abdalla, M., Bellare, M., Catalano, D., Kiltz, E., Kohno, T., Lange, T., Malone-Lee, J., Neven, G., Paillier, P., Shi, H.: Searchable encryption revisited: consistency properties, relation to anonymous ibe, and extensions. J. Cryptology **21**(3), 350–391 (2008)
2. Attrapadung, N., Furukawa, J., Imai, H.: Forward-secure and searchable broadcast encryption with short ciphertexts and private keys. In: Lai, X., Chen, K. (eds.) ASIACRYPT 2006. LNCS, vol. 4284, pp. 161–177. Springer, Heidelberg (2006)
3. Baek, J., Safavi-Naini, R., Susilo, W.: On the integration of public key data encryption and public key encryption with keyword search. In: Katsikas, S.K., López, J., Backes, M., Gritzalis, S., Preneel, B. (eds.) ISC 2006. LNCS, vol. 4176, pp. 217–232. Springer, Heidelberg (2006)
4. Boneh, D., Di Crescenzo, G., Ostrovsky, R., Persiano, G.: Public key encryption with keyword search. In: Cachin, C., Camenisch, J.L. (eds.) EUROCRYPT 2004. LNCS, vol. 3027, pp. 506–522. Springer, Heidelberg (2004)

5. Boneh, D., Franklin, M.: Identity-based encryption from the weil pairing. In: Kilian, J. (ed.) CRYPTO 2001. LNCS, vol. 2139, pp. 213–229. Springer, Heidelberg (2001)
6. Daza, V., Herranz, J., Morillo, P., Ràfols, C.: CCA2-secure threshold broadcast encryption with shorter ciphertexts. In: Susilo, W., Liu, J.K., Mu, Y. (eds.) ProvSec 2007. LNCS, vol. 4784, pp. 35–50. Springer, Heidelberg (2007)
7. Hwang, Y.H., Lee, P.J.: Public key encryption with conjunctive keyword search and its extension to a multi-user system. In: Takagi, T., Okamoto, T., Okamoto, E., Okamoto, T. (eds.) Pairing 2007. LNCS, vol. 4575, pp. 2–22. Springer, Heidelberg (2007)
8. Mell, P., Grance, T.: The nist definition of cloud computiing. Technical report, National Institue of Standards and Technology (2011)
9. Shamir, A.: How to share a secret. Commun. ACM 22(11), 612–613 (1979)
10. Siad, A.: Anonymous identity-based encryption with distributed private-key generator and searchable encryption. In: 2012 5th International Conference on New Technologies, Mobility and Security (NTMS), pp. 1–8, May 2012
11. Song, D.X., Wagner, D., Perrig, A.: Practical techniques for searches on encrypted data. In: 2000 IEEE Symposium on Security and Privacy, S P 2000, Proceedings, pp. 44–55 (2000)
12. Sun, W., Yu, S., Lou, W., Hou, T., Li, H.: Protecting your right: verifiable attribute-based keyword search with fine-grained owner-enforced search authorization in the cloud. IEEE Trans. Parallel Distrib. Syst. PP(99), 1 (2014)
13. Wang, P., Wang, H., Pieprzyk, J.: Threshold privacy preserving keyword searches. In: Geffert, V., Karhumäki, J., Bertoni, A., Preneel, B., Návrat, P., Bieliková, M. (eds.) SOFSEM 2008. LNCS, vol. 4910, pp. 646–658. Springer, Heidelberg (2008)
14. Xu, P., Jin, H., Wu, Q., Wang, W.: Public-key encryption with fuzzy keyword search: a provably secure scheme under keyword guessing attack. IEEE Trans. Comput. 62(11), 2266–2277 (2013)
15. Zheng, Q., Xu, S., Ateniese, G.: Vabks: verifiable attribute-based keyword search over outsourced encrypted data. In: INFOCOM, 2014 Proceedings IEEE, pp. 522–530, April 2014

Key Management and Public Key Encryption

Secret Sharing Schemes with General Access Structures

Jian Liu[1,2], Sihem Mesnager[3(✉)], and Lusheng Chen[4]

[1] School of Computer Software, Tianjin University,
Tianjin 300072, People's Republic of China
jianliu.nk@gmail.com
[2] CNRS, UMR 7539 LAGA, Paris, France
[3] Department of Mathematics, University of Paris VIII, University of Paris XIII,
CNRS, UMR 7539 LAGA and Telecom ParisTech, Paris, France
smesnager@univ-paris8.fr
[4] School of Mathematical Sciences, Nankai University,
Tianjin 300071, People's Republic of China
lschen@nankai.edu.cn

Abstract. Secret sharing schemes with general monotone access structures have been widely discussed in the literature. But in some scenarios, non-monotone access structures may have more practical significance. In this paper, we shed a new light on secret sharing schemes realizing general (not necessarily monotone) access structures. Based on an attack model for secret sharing schemes with general access structures, we redefine perfect secret sharing schemes, which is a generalization of the known concept of perfect secret sharing schemes with monotone access structures. Then, we provide for the first time two constructions of perfect secret sharing schemes with general access structures. The first construction can be seen as a democratic scheme in the sense that the shares are generated by the players themselves. Our second construction significantly enhances the efficiency of the system, where the shares are distributed by the trusted center (TC).

Keywords: Secret sharing schemes · General access structures · Information rate · Orthogonal arrays · Resilient functions

1 Introduction

Secret sharing schemes were first introduced by Blakley [5] and Shamir [29] independently in 1979. Besides secure information storage, secret sharing schemes have numerous other applications in cryptography such as secure multiparty computations [4,14,15], key-distribution problems [23], multi-receiver authentication schemes [33] etc. Note that minimal codes introduced and studied in the literature have applications in secret sharing (see for instance [12,16,18]).

This work is supported by the National Key Basic Research Program of China under Grant 2013CB834204. Due to the limited pages, a full version of this paper is available in [24].

© Springer International Publishing Switzerland 2016
D. Lin et al. (Eds.): Inscrypt 2015, LNCS 9589, pp. 341–360, 2016.
DOI: 10.1007/978-3-319-38898-4_20

The secret sharing schemes given in [5,29] are for the threshold case, i.e., the qualified groups that can reconstruct the secret key are all the subsets with cardinality no smaller than a threshold. A (t,n) threshold scheme is a method where n pieces of information of the secret key K, called *shares* are distributed to n players so that the secret key can be reconstructed from the knowledge of any t or more shares and the secret key can not be reconstructed from the knowledge of fewer than t shares. But in reality, there are many situations in which it is desirable to have a more flexible arrangement for reconstructing the secret key. Given some n players, one may want to designate certain authorized groups of players who can use their shares to recover the key. This kind of scheme is called secret sharing scheme for general access structure, which generalizes the threshold case. Formally, a secret sharing scheme for general access structure is a method of sharing a secret K among a finite set of players $\mathcal{P} = \{P_1, \ldots, P_n\}$ in such a way that

1. if the players in $A \subseteq \mathcal{P}$ are qualified to know the secret, then by pooling together their partial information, they can reconstruct the secret K,
2. any set $B \subset \mathcal{P}$ which is not qualified to know K, cannot reconstruct the secret K.

The threshold secret sharing schemes have received considerably attention, see e.g. [13,17,26,27]. Secret sharing schemes for general monotone access structures were first studied by Ito et al. [21]. The access structure defined in [21] is a set of qualified groups Γ which satisfies the monotone property that if $A \in \Gamma$ and $A \subseteq B$, then $B \in \Gamma$. Secret sharing schemes for general monotone access structures have got a lot of attention, and there exist a wide range of general methods of constructing monotone secret sharing schemes [2,3,7,22]. The approaches to the construction of monotone secret sharing schemes based on linear codes can be found in [6,25]. To our best knowledge, all the known secret sharing schemes are designed for realizing monotone access structures. We refer to [15] for a survey on monotone secret sharing schemes.

A secret sharing scheme can be represented by a set of recovery algorithms which realizes an access structure such that only qualified groups can reconstruct the secret key by pooling their shares. For example, in the bank teller problem described in Chap. 13 of [30], any two out of three tellers are authorized to reconstruct the secret key. It is quite natural to assume that three tellers are permitted to make a requirement on two of them to execute the recovery algorithm and reconstruct the secret key, then any group with two or more tellers is a qualified group. Hence, the access structure considered in this scenario has monotone property. However, for some scenarios, the requirement on fewer players of a group to recover the secret key is not available, and secret sharing schemes with non-monotone access structures may be more preferable. For a secret sharing scheme, it is reasonable to assume that the access structure is public and in the reconstruction phase, the players are anonymous, that is to say, the players will not disclose which group they belong to.

Scenario 1. Suppose that on the network, there are several groups of users who share a large amount of information resources stored by the network center (e.g., a secure cloud storage server) with a secret key. Once the secret key is recovered, only the users who pool their shares will get the access to download the information. For some reasons, the users of the same group are not willing to download their information together with an outsider who does not belong to this group. So, only when all the users of the same group pool their shares, the secret key can be reconstructed, and if an outsider joins, the reconstruction reveals nothing about the secret key.

The access structure in the above scenario is non-monotone, since there exist $A \in \Gamma$ and $B \notin \Gamma$ such that $A \subseteq B$. A secret key can always be recovered by all the users in a qualified group A, but if an outsider, say P, intrudes, the reconstruction by the users in the unqualified (i.e., forbidden) group $B = A \bigcup P$ reveals nothing about the secret key. Consider that in Scenario 1, different groups of users have independently purchased the access to the database from the network center, and the payment of each group is afforded by every user of this group, thus the costs of the users from different groups may be different. Of course, the users belonging to one group do not hope to download their data together with an outsider. We consider data mining as another example for Scenario 1. Suppose different groups of market investigators are employed by different companies respectively to gather some information from the network center. Because of the market competition, the companies do not hope to disclose what they are gathering to each other, i.e., the market investigators of one company are not willing to reconstruct the secret key and download their information together with an outsider. Thus, the access structures here should be non-monotone.

Secret sharing scheme is also a key tool for secure multiparty computation (MPC) (see [1,14,15]). Secure MPCs solve the problem that n players want to compute some agreed function with their inputs private. For instance, two millionaires want to know who is richer without disclosing their wealths to each other. This millionaire problem, first introduced by Yao [32], is a secure MPC problem which can be solved by monotone secret sharing schemes. A secure MPC protocol can be described as that every player shares his input with all the players by employing some secret sharing scheme, then the players in a qualified group can compute the result of the agreed function, and the players in a forbidden group cannot learn anything about the result and the inputs of the other players, where the qualified and the forbidden groups are determined by the access structure of the employed secret sharing scheme (see [1,14] for more details). In the following scenario, a secure MPC with non-monotone access structure is preferable.

Scenario 2. Suppose that the employees of several different companies are interested in their salary level by comparing their incomes, i.e., they want to know the ranking of the average income of each company by sharing their incomes privately. To avoid the risk of embarrassment, the employees of one company are not willing to compute the ranking result together with an outsider who does not belong to this company. So, only all the employees of the same company can compute the ranking result, and if an outsider joins, the computation reveals nothing about the result.

In the above scenarios, we must guarantee that if B is a qualified group but $A \supseteq B$ is not, then the players in A cannot make a requirement on the players in B to reconstruct the secret key or to compute the result of the agreed function. When all the players in a forbidden group follow the protocol accordingly, they can determine nothing about the secret key or the result of the agreed function.

Perhaps one can find some other means to solve the problems presented in the above scenarios, and Scenario 2 may have less practical significance, but all these are intended primarily as examples to provide us a direction for possible applications of non-monotone secret sharing schemes. Similar practical scenarios could be found.

In this paper, we mainly discuss secret sharing schemes realizing general (not necessarily monotone) access structures. We first describe a general attack model for secret sharing schemes. Afterwards, a formal definition of unconditional security (called perfect) for secret sharing schemes with general access structures is given, which is a generalization of the known perfect monotone secret sharing schemes. Moreover, we propose two constructions for secret sharing schemes realizing general access structures. To the best of our knowledge, this is the first time when constructions of non-monotone secret sharing schemes are proposed. Our first construction is democratic in the sense that the shares are generated by the players themselves instead of distributed by the trusted center (TC). In this construction, TC has to recompute an updated function for every time the secret key changes. The second construction is presented for the sake of efficiency, where the shares are computed and distributed by TC. We also show that the well designed secret sharing schemes presented in this paper are perfect.

This paper is organized as follows. Formal definitions and necessary preliminaries are introduced in Sect. 2. In Sect. 3, we discuss the attack model and the security of secret sharing schemes with general access structures. Perfect democratic secret sharing schemes are constructed in Sect. 4, and perfect secret sharing schemes with distributed shares are constructed in Sect. 5. In the last section, we summarize this paper and indicate some future research directions.

2   Preliminaries

For a secret sharing scheme, we denote a *player* by P_i, where $i = 1, 2, \ldots$, the set of all the players by $\mathcal{P}$, the set of all the subsets of $\mathcal{P}$ by $2^{\mathcal{P}}$, and the *trusted center* of the scheme by TC. The groups authorized to reconstruct the secret key are called *qualified*, and the groups unauthorized to reconstruct the secret key are called *forbidden*. The sets of qualified and forbidden groups are denoted by Γ and Δ respectively, where $\Gamma \subseteq 2^{\mathcal{P}}$ and $\Delta \subseteq 2^{\mathcal{P}}$. If $\Gamma \bigcap \Delta = \emptyset$, then the tuple (Γ, Δ) is called an *access structure*. Moreover, an access structure is called *complete* if $\Gamma \bigcup \Delta = 2^{\mathcal{P}}$. In this paper, we focus on secret sharing schemes with complete access structures. The set of qualified groups Γ is called *monotone increasing* if for each set $A \in \Gamma$, the superset of A is also in Γ. An access structure (Γ, Δ) is called *monotone* if Γ is monotone increasing.

Let $\mathscr{S}$ be a secret sharing scheme. We denote the set of all possible secret keys by $\mathbf{K}$, the set of all possible shares of group $A = \{P_{i_1}, \ldots, P_{i_m}\} \in 2^{\mathcal{P}}$ by $\mathbf{S}(A)$, i.e., $\mathbf{S}(A) = \mathbf{S}(P_{i_1}) \times \cdots \times \mathbf{S}(P_{i_m})$, where $\mathbf{S}(P_{i_j})$ is the set of all possible shares of P_{i_j} and "$\times$" denotes the Cartesian product. For a qualified group $A \in \Gamma$, there exists a recovery algorithm f_A defined on $\mathbf{S}(A)$ which satisfies $f_A(s(A)) = k$, where $k \in \mathbf{K}$ is the secret key that TC wants to share and $s(A) \in \mathbf{S}(A)$ is the shares of the players in A. Then, a secret sharing scheme $\mathscr{S}$ realizing access structure (Γ, Δ) can be viewed as a set of recovery algorithms $\mathcal{F} = \{f_A \mid A \in \Gamma\}$ such that only qualified groups can reconstruct the secret key by pooling their shares.

Definition 1 [8]. *Let $\mathscr{S}$ be a secret sharing scheme, $\mathbf{K}$ be the set of all possible secret keys, and for $1 \leqslant i \leqslant n$, $\mathbf{S}(P_i)$ be the set of all possible shares that P_i might have. Then, the* information rate *of P_i is defined as $\rho_i = \frac{\log_2 |\mathbf{K}|}{\log_2 |\mathbf{S}(P_i)|}$, and the* information rate *of $\mathscr{S}$ is defined as*

$$\rho = \min\{\rho_i \mid 1 \leqslant i \leqslant n\}. \tag{1}$$

In the following, we introduce some definitions and properties of q-ary functions, which will be useful in constructing secret sharing schemes.

Let $\mathbb{F}_q$ be a finite field, where q is a power of a prime, then $\mathbb{F}_q^n$ denotes the n-dimensional vector space over the finite field $\mathbb{F}_q$. In this paper, we always assume $q > 2$. Let $\mathbb{F}_q^* = \mathbb{F}_q \setminus \{0\}$, then $(\mathbb{F}_q^*)^n$ denotes the Cartesian product that $\overbrace{\mathbb{F}_q^* \times \cdots \times \mathbb{F}_q^*}^{n}$. The mappings from the vector space $\mathbb{F}_q^n$ to $\mathbb{F}_q$ are called n-variable q-*ary functions*, which can be uniquely represented in the *algebraic normal form* (ANF), see [28]: $F(x) = \sum_{u \in \mathbb{Z}_q^n} a_u x_1^{u_1} x_2^{u_2} \cdots x_n^{u_n}$, where $\mathbb{Z}_q = \{0, \ldots, q-1\}$, $x = (x_1, \ldots, x_n) \in \mathbb{F}_q^n$, $u = (u_1, \ldots, u_n) \in \mathbb{Z}_q^n$, and $a_u \in \mathbb{F}_q$. In fact, given the values of $F(w)$, $w = (w_1, \ldots, w_n) \in \mathbb{F}_q^n$, the ANF of F can be determined as

$$F(x) = \sum_{w \in \mathbb{F}_q^n} F(w) \prod_{i=1}^{n} \left(1 - (x_i - w_i)^{q-1}\right). \tag{2}$$

For an n-variable q-ary function F, the set $\mathbb{F}_q^n$ is called the *domain set* of F and the vector $(F(v_0), \ldots, F(v_{q^n-1}))$ is called the *value table* of F, where $v_0, \ldots, v_{q^n-1}$ are all the vectors in $\mathbb{F}_q^n$ which have some prescribed order, e.g., the lexicographical order. F is called *balanced* if for any element $a \in \mathbb{F}_q$, the size of the pre-image set satisfies $|F^{-1}(a)| = q^{n-1}$.

More generally, if F is a mapping from $E_1 \subseteq \mathbb{F}_q^n$ to $E_2 \subseteq \mathbb{F}_q$, then E_1 is called the domain set of F, and $(F(v_0), \ldots, F(v_{|E_1|-1}))$ is called the value table of F, where $v_0, \ldots, v_{|E_1|-1}$ are all the vectors in E_1 with some prescribed order. In addition, F called balanced onto E_2 if for any element $a \in E_2$, the size of the pre-image set satisfies $|F^{-1}(a)| = |E_1|/|E_2|$.

For $i = 1, \ldots, m$, let F_i be a mapping from $E_i \subseteq \mathbb{F}_q^n$ to $\mathbb{F}_q$, where $E_1, \ldots, E_m$ are disjoint sets, then the *concatenation function* F of $F_1, \ldots, F_m$ is the mapping from $\bigcup_{i=1}^m E_i$ to $\mathbb{F}_q$ which satisfies $F(x) = F_i(x)$ for $x \in E_i$, where $i = 1, \ldots, m$.

3 The Security of Secret Sharing Schemes with General Access Structures

For a secret sharing scheme with general access structure (Γ, Δ), we assume that the players in $A \in \Delta$ are passively collaborating to pool their shares and try to reconstruct the secret key. Note that the collaborating players are assumed to execute the protocol correctly and every player will keep his share private, i.e., the attack is *passive*. We also assume that the collaborating players are *static*, which means that the set of collaborating players is fixed during the protocol.

Attack Model. The players in $A \in \Delta$ are passively collaborating to find some efficient recovery algorithms to reconstruct the secret key.

For a general access structure, the players in group $A \in \Delta$ will try to guess the secret key by collaborating, and in this case, even if $A \supseteq B \in \Gamma$ (this case only appears in the non-monotone case), the players in B are passively collaborating, and cannot be required to execute their recovery algorithm and reconstruct the secret key independently from A. In Sect. 1, we present two scenarios to show that this assumption is reasonable for the non-monotone case. Note that a secret sharing scheme can be viewed as a set of recovery algorithms $\mathcal{F} = \{f_A \mid A \in \Gamma\}$. Hence, if the players in $A \in \Delta$ are passively collaborating, they can only try to guess the secret key by employing some known reconstruction algorithms that f_B, $B \in \Gamma$, where $B \subseteq A$.

Particularly, for monotone access structures, one can just assume that all the players belonging to a forbidden group (which is a proper subset of a qualified group) are passively collaborating to reconstruct the secret key.

Let A be any subset of players, $B \in \Gamma$, and $k \in \mathbf{K}$, then given $s(A) \in \mathbf{S}(A)$, the conditional probability determined by algorithm f_B is denoted by $\mathrm{Pr}_B(K = k \mid S(A) = s(A))$, which means that by using algorithm f_B, the players in A can guess the secret key correctly with probability $\mathrm{Pr}_B(K = k \mid S(A) = s(A))$. We use $\mathrm{Pr}(K = k \mid S(A) = s(A))$ for short if there is no risk of confusion, and use $\mathrm{Pr}(K = k)$ to denote the *a prior* probability distribution on the secret key set $\mathbf{K}$. Considering the above attack model, we present a formal definition of unconditional security for secret sharing schemes with general access structures.

Definition 2. *A secret sharing scheme $\mathscr{S}$ with access structure (Γ, Δ) and secret key set $\mathbf{K}$ is perfect if $\mathscr{S}$ satisfies the following two properties.*

(i) For any $A \in \Gamma$, the secret key can be reconstructed correctly.

(ii) For any $A \in \Delta$ and any $B \in \Gamma$, where $B \subseteq A$, the conditional probability determined by algorithm f_B satisfies $\mathrm{Pr}(K = k \mid S(A) = s(A)) = \mathrm{Pr}(K = k)$ for every $k \in \mathbf{K}$. In other words, by using algorithm f_B, the players in A can learn nothing about the secret key.

Remark 1. For secret sharing schemes with monotone access structures, the concept of perfect system has been introduced in [8] and widely studied (see [1,30] for a survey). If a secret sharing scheme $\mathscr{S}$ with monotone access structure (Γ, Δ) satisfies (i) for any $A \in \Gamma$, the secret key can be reconstructed correctly, (ii) for any $A \in \Delta$, the players in A can learn nothing about the secret key, then $\mathscr{S}$ is called perfect. From the above discussion, it is easy to see that the concept of perfect system given in Definition 2 is more general, and for the monotone case, Definition 2 coincides with the standard perfect monotone secret sharing schemes.

For perfect secret sharing schemes with monotone access structures, it is proved that the information rate $\rho \leqslant 1$, see [8,30]. We now show that this result still holds for perfect secret sharing schemes with general access structures.

Theorem 1. *For any perfect secret sharing scheme with general access structure, the information rate satisfies $\rho \leqslant 1$.*

Proof. Suppose that $\mathscr{S}$ is a perfect secret sharing scheme with general access structure (Γ, Δ), then there must exist a set $A \in \Gamma$ such that $B = A \setminus \{P_i\} \in \Delta$ for some $P_i \in A$. In fact, if for any $A \in \Gamma$ and any $P_i \in A$, $A' = A \setminus \{P_i\} \in \Gamma$, then $\emptyset \neq A' \in \Gamma$ and $A' \setminus \{P_j\} = A \setminus \{P_i, P_j\} \in \Gamma$. This process can be continued until we get the contradiction that $\emptyset \in \Gamma$.

Without loss of generality, we assume that $A = \{P_1, \ldots, P_m\} \in \Gamma$ and $B = A \setminus \{P_m\} \in \Delta$. Since $B \in \Delta$ and $\mathscr{S}$ is perfect, when the players in B are collaborating to reconstruct the secret key by using the recovery algorithm f_A, then the conditional probability of the secret key is

$$\Pr(K = k \mid S(B) = s(B)) = \Pr(K = k), \tag{3}$$

where $k \in \mathbf{K}$. Since $A \in \Gamma$, then Eq. (3) implies that for any two distinct secret keys $k_1, k_2 \in \mathbf{K}$, there exist two distinct shares $s_1(P_m), s_2(P_m) \in \mathbf{S}(P_m)$ such that

$$f_A(s(B), s_1(P_m)) = k_1, \quad f_A(s(B), s_2(P_m)) = k_2. \tag{4}$$

Therefore, $|\mathbf{S}(P_m)| \geqslant |\mathbf{K}|$, and thus $\rho_m \leqslant 1$. Hence, from (1), we have $\rho \leqslant 1$. $\qquad \square$

4   Democratic Secret Sharing Schemes

In this section, we present a construction of democratic secret sharing schemes with general access structures, where the shares are generated on the set of all possible shares independently by the players themselves. Moreover, we provide a perfect secret sharing scheme with information rate $\rho = 1$.

4.1   A General Description

Let e_i be the *identity vector* in $\mathbb{F}_q^n$ with 1 in the i-th position and zeros elsewhere, where q is a power of a prime. By abuse of notation, we write a set $A = \{i_1, \ldots, i_m\}$ for $A = \{P_{i_1}, \ldots, P_{i_m}\} \subseteq \mathcal{P}$.

In Table 1, we present a construction of democratic secret sharing schemes realizing general access structures.

4.2   Perfect Democratic Secret Sharing Schemes Realizing General Access Structures

As shown in Table 1, given a set of n players $\mathcal{P} = \{P_1, \ldots, P_n\}$ with general access structures (Γ, Δ) and secret key set $\mathbf{K} = \mathbb{F}_q^*$, one can always construct a democratic secret sharing scheme. Clearly, the security of Secret Sharing

Table 1. Secret Sharing Scheme I

Initialization Phase:

1. For a player $P_i \in \mathcal{P}$, where $1 \leqslant i \leqslant n$, P_i randomly chooses $\alpha_i \in \mathbb{F}_q^*$ as his share, and then transmits α_i secretly to the trusted center TC.
2. The access structure (Γ, Δ) is public.

Sharing Phase:

1. Suppose TC wants to share a secret key $k \in \mathbb{F}_q^*$, then TC chooses a q-ary function $F : \mathbb{F}_q^n \to \mathbb{F}_q$ which satisfies

$$F(x) = k, \quad \text{if } x = \sum_{i \in A} \alpha_i e_i \text{ for some } A \in \Gamma, \tag{5}$$

where $e_1, \ldots, e_n$ are the identity vectors in $\mathbb{F}_q^n$, and for other $x \in \mathbb{F}_q^n$, $F(x)$ are carefully chosen.
2. TC computes the algebraic normal form of F by using Eq.(2),

$$F(x) = \sum_{u=(u_1,\ldots,u_n)\in\mathbb{Z}_q^n} a_u x_1^{u_1} \cdots x_n^{u_n}.$$

3. TC publishes the algebraic normal form of F.

Reconstruction Phase:

For any $A = \{P_{i_1}, \ldots, P_{i_m}\} \in 2^{\mathcal{P}}$, the players in A do as follows.

1. **Determine the recovery algorithm.**
   The players in A get

$$f_A(x_{i_1}, \ldots, x_{i_m}) = F(0, \ldots, 0, x_{i_1}, 0, \ldots, 0, x_{i_m}, 0, \ldots, 0)$$

   as their recovery algorithm.
2. **Compute the secret key.**
   The players in A pool their shares and compute $f_A(\alpha_{i_1}, \ldots, \alpha_{i_m})$.

Scheme I depends heavily on the choice of the q-ary function F. In Table 2, an explicit construction of q-ary functions is presented. Employing such functions in Secret Sharing Scheme I, one can get perfect democratic secret sharing schemes. We first introduce some useful notations below.

For $x = (x_1, \ldots, x_n) \in \mathbb{F}_q^n$, let $\mathrm{supp}(x) = \{i \mid x_i \neq 0\}$ denote the support set of x. Then, for an n-variable q-ary function F and a set $A \subseteq \mathcal{P}$, $F|_A$ denotes the restriction of F to the set

$$E_A = \{x \in \mathbb{F}_q^n \mid \mathrm{supp}(x) = A\}, \tag{6}$$

i.e., $F|_A : E_A \to \mathbb{F}_q$ satisfies $F|_A(x) = F(x)$ for $x \in E_A$. For a set $E \subseteq \mathbb{F}_q^n$ and an element $a \in \mathbb{F}_q^n$, $a + E = \{a + e \mid e \in E\}$. For $s(\mathrm{P}_i) \in \mathbf{S}(\mathrm{P}_i)$, $i \in A$, which are the shares of players in A, define $\boldsymbol{s}(A) = \sum_{i \in A} s(\mathrm{P}_i)e_i \in \mathbb{F}_q^n$, then for $B \subseteq A$, we denote by $F|_{\boldsymbol{s}(A\backslash B) \times B}$ the restriction of F to the set

$$E_{\boldsymbol{s}(A\backslash B) \times B} = \boldsymbol{s}(A \setminus B) + E_B. \tag{7}$$

Moreover, for $B \subseteq A$, we denote by $F|_{\boldsymbol{s}(A\backslash B) \times \overline{B}}$ the restriction of F to the set

$$E_{\boldsymbol{s}(A\backslash B) \times \overline{B}} = \{(x_1, \ldots, x_n) \in E_{\boldsymbol{s}(A\backslash B) \times B} \mid x_i \neq s(\mathrm{P}_i), i \in B\}. \tag{8}$$

Clearly, if $B = \emptyset$, then $F|_{\boldsymbol{s}(A\backslash B) \times B} = F|_{\boldsymbol{s}(A\backslash B) \times \overline{B}} = F(\boldsymbol{s}(A))$.

Proposition 1. *For $A \subseteq \mathcal{P}$, the set E_A can be partitioned into disjoint subsets $E_{\boldsymbol{s}(A\backslash B) \times \overline{B}}$, where $B \subseteq A$, i.e.,*

$$E_A = \bigcup_{B \subseteq A} E_{\boldsymbol{s}(A\backslash B) \times \overline{B}}, \tag{9}$$

where for two distinct subsets $B_1, B_2 \subseteq A$,

$$E_{\boldsymbol{s}(A\backslash B_1) \times \overline{B_1}} \bigcap E_{\boldsymbol{s}(A\backslash B_2) \times \overline{B_2}} = \emptyset. \tag{10}$$

Proof. We first prove (10). Since $B_1 \neq B_2$, then without loss of generality, we assume that $i \in B_1$ but $i \notin B_2$. Suppose that there exists $x = (x_1, \ldots, x_n) \in E_{\boldsymbol{s}(A\backslash B_1) \times \overline{B_1}} \bigcap E_{\boldsymbol{s}(A\backslash B_2) \times \overline{B_2}}$, then we have $x_i \neq s(\mathrm{P}_i)$ since $i \in B_1$. However, $i \notin B_2$ implies that $x_i = s(\mathrm{P}_i)$, a contradiction. Hence, (10) holds.

It is obvious that $\bigcup_{B \subseteq A} E_{\boldsymbol{s}(A\backslash B) \times \overline{B}} \subseteq E_A$. Suppose that $|A| = m$, then since the sets $E_{\boldsymbol{s}(A\backslash B) \times \overline{B}}$, where $B \subseteq A$, are disjoint, we have that

$$\left| \bigcup_{B \subseteq A} E_{\boldsymbol{s}(A\backslash B) \times \overline{B}} \right| = \sum_{B \subseteq A} |E_{\boldsymbol{s}(A\backslash B) \times \overline{B}}| = \sum_{i=0}^{m} \binom{m}{i} (q-2)^i = (q-1)^m = |E_A|. \tag{11}$$

Therefore, (9) holds, and we get the desired result. $\square$

By using Proposition 1, we can prove that Construction I in Table 2 outputs a q-ary function.

Table 2. Construction I

Input:	Secret shares $s(\mathrm{P}_i) \in \mathbb{F}_q^*$, $i = 1, \ldots, n$; Secret key $k \in \mathbb{F}_q^*$; Access structure (Γ, Δ).
Output:	A function $F : \mathbb{F}_q^n \to \mathbb{F}_q$.

Step 1: For every subset of players $A \in \Delta$, set $F|_A = 0$, i.e., $F|_A$ is a zero function.

Step 2: For every subset of players $A = \{\mathrm{P}_{i_1}, \ldots, \mathrm{P}_{i_m}\} \in \Gamma$, execute the following two steps.

1. Set $F(s(A)) = k$.

2. Define $n_0 = 1$ and $N_0 = 0$. From $l = 1$ to $l = m$, do as follows. For every $B = \{\mathrm{P}_{j_1}, \ldots, \mathrm{P}_{j_l}\} \subseteq A$, arrange the value table of $F|_{s(A \backslash B) \times \overline{B}}$ as follows.
   (i) Let k appear n_l times, where

$$n_l = (q-1)^{l-1} - \sum_{i=0}^{l-1} \binom{l}{i} n_i. \tag{12}$$

   (ii) Let every element in $\mathbb{F}_q^* \backslash \{k\}$ appear N_l times, where

$$N_l = (q-1)^{l-1} - \sum_{i=0}^{l-1} \binom{l}{i} N_i. \tag{13}$$

Lemma 1. [1] *Construction I outputs an n-variable q-ary function F.*

Lemma 2. *Let F be constructed by Construction I. Then, for any subset of players $A \in \Gamma$ and any non-empty set $B \subseteq A$, the restriction function $F|_{s(A \backslash B) \times B}$ is balanced onto $\mathbb{F}_q^*$.*

Proof. Let $A \in \Gamma$ and $\emptyset \neq B \subseteq A$ with $|A| = m$ and $|B| = l \leqslant m$. According to Step 2 of Construction I, we have that for any subset $C \subseteq B$ with $|C| = s$, where $0 \leqslant s \leqslant l$, the secret key k appears $n_s = (q-1)^{s-1} - \sum_{i=0}^{s-1} \binom{s}{i} n_i$ times in the value table of $F|_{s(A \backslash C) \times \overline{C}}$, where $n_0 = 1$, and every element in $\mathbb{F}_q^* \backslash \{k\}$ appears $N_s = (q-1)^{s-1} - \sum_{i=0}^{s-1} \binom{s}{i} N_i$ times in the value table of $F|_{s(A \backslash C) \times \overline{C}}$, where $N_0 = 0$. Clearly, n_s and N_s depend only on s. Similar to Proposition 1, we can prove that $E_{s(A \backslash B) \times B} = \bigcup_{C \subseteq B} E_{s(A \backslash C) \times \overline{C}}$, where for two distinct subsets $C_1, C_2 \subseteq B$, $E_{s(A \backslash C_1) \times \overline{C_1}} \bigcap E_{s(A \backslash C_2) \times \overline{C_2}} = \emptyset$. Therefore, $F|_{s(A \backslash B) \times B}$ is the concatenation function of $F|_{s(A \backslash C) \times \overline{C}}$, $C \subseteq B$. Note that the number of different $C \subseteq B$ with $|C| = s$ is $\binom{l}{s}$. Thus, in the value table of $F|_{s(A \backslash B) \times B}$, the secret key k appears $\sum_{i=0}^{l} \binom{l}{i} n_i$ times and every element in

---

[1] A formal proof of this lemma is provided in a full version of this paper [24].

$\mathbb{F}_q^* \setminus \{k\}$ appears $\sum_{i=0}^{l} \binom{l}{i} N_i$ times. According to Eqs. (12) and (13), we have $\sum_{i=0}^{l} \binom{l}{i} n_i = \sum_{i=0}^{l} \binom{l}{i} N_i = (q-1)^{l-1}$, which implies that $F|_{s(A \setminus B) \times B}$ is balanced onto $\mathbb{F}_q^*$.   □

Theorem 2. *Let F be constructed by Construction I. Then, Secret Sharing Scheme I is perfect with information rate $\rho = 1$.*

Proof. It is clear that Secret Sharing Scheme I has $\rho = 1$. We now prove that if F is constructed by Construction I, then Secret Sharing Scheme I is perfect.

For any $A \in \Gamma$, the recovery algorithm f_A satisfies $f_A(s(A)) = F(s(A)) = k$, thus the qualified group A can reconstruct the secret key.

For any forbidden group $A \in \Delta$, we assume that the players in A are collaborating to reconstruct the secret key by using some recovery algorithm, say f_B, where $B \in \Gamma$. As discussed in Sect. 3, we must have $B \nsubseteq A$, i.e., $C = A \cap B \subsetneq B$, where $C \subsetneq B$ means that C is a proper subset of B (i.e., $C \subseteq B$ but $C \neq B$). Since $C \subsetneq B$, then $B \setminus C \neq \emptyset$. Due to Lemma 2, the restriction function $F|_{s(C) \times (B \setminus C)}$ is balanced onto $\mathbb{F}_q^*$. Hence, the conditional probability determined by f_B satisfies

$$\Pr(K = \gamma \mid S(C) = s(C)) = \frac{1}{q-1} = \frac{1}{|\mathbf{K}|} = \Pr(K = \gamma), \tag{12}$$

where $\gamma \in \mathbb{F}_q^*$, which implies that for every $k \in \mathbf{K}$, the secret key k can be guessed correctly with probability $\Pr(K = k) = 1/|\mathbf{K}|$. Therefore, according to Definition 2, we get that Secret Sharing Scheme I is perfect.   □

5   Secret Sharing Schemes with Distributed Shares

In Sect. 4, we have shown a construction of perfect democratic secret sharing schemes with information rate $\rho = 1$. Note that in Secret Sharing Scheme I, the shares are generated by the players themselves, but when the secret key that TC wants to share is changed, the function F published by TC should be updated accordingly. This may cause the problem of low efficiency if n is large, because TC has to recompute the ANF of the new function F, and this process needs approximately $\mathcal{O}(nq^n)$ operations over the finite field $\mathbb{F}_q$.

To avoid the drawback of updating the function F, we propose Secret Sharing Scheme II, where the shares of the players are computed and distributed secretly by TC. In Secret Sharing Scheme II, the public q-ary function F is fixed, and when the secret key is changed, the shares distributed to the players by TC will be updated accordingly. Comparing the two constructions, one can see that Secret Sharing Scheme I realizes the democracy, while Secret Sharing Scheme II is designed towards enhancing the efficiency.

5.1   A General Description

Recall that for an n-variable q-ary function F, $F|_A$ denotes the restriction of F to the set $E_A = \{x \in \mathbb{F}_q^n \mid \mathrm{supp}(x) = A\}$, where $A = \{i_1, \ldots, i_m\}$, and we use

$A = \{P_{i_1}, \ldots, P_{i_m}\} \subseteq \mathcal{P}$ to denote a subset of players. In Table 3, we present Secret Sharing Scheme II, in which the shares of the players are computed and distributed secretly by TC. It can be seen that, different from Secret Sharing Scheme I, in Secret Sharing Scheme II, the q-ary function determined by TC does not depend on the choice of the secret key.

Table 3. Secret Sharing Scheme II

Initialization Phase:

1. The set of all the players is $\mathcal{P} = \{P_1, \ldots, P_n\}$.

2. The access structure (Γ, Δ) is public.

Sharing Phase:

1. The trusted center TC chooses a q-ary function $F : \mathbb{F}_q^n \to \mathbb{F}_q$ which satisfies
   (i) For $A \in \Gamma$, $F|_A$ is balanced onto $\mathbb{F}_q^*$, which is carefully chosen.
   (ii) For $A \in \Delta$, $F|_A = 0$, i.e., $F|_A$ is a zero function.
   Then, TC computes the algebraic normal form of F by using Eq.(2),

   $$F(x) = \sum_{u=(u_1,\ldots,u_n)\in\mathbb{Z}_q^n} a_u x_1^{u_1} \cdots x_n^{u_n},$$

   and the algebraic normal form of F is public.

2. Suppose TC wants to share a secret key $k \in \mathbb{F}_q^*$. Then, for every $A = \{P_{i_1}, \ldots, P_{i_m}\} \in \Gamma$, TC randomly chooses

   $$x^{(A)} = \left(0, \ldots, 0, x_{i_1}^{(A)}, 0, \ldots, 0, x_{i_m}^{(A)}, 0, \ldots, 0\right)$$

   $$\in (F|_A)^{-1}(k) = \{x \in \mathbb{F}_q^n \mid F|_A(x) = k\},$$

   and TC transmits the values $x_{i_1}^{(A)}, \ldots, x_{i_m}^{(A)}$ secretly to $P_{i_1}, \ldots, P_{i_m}$ respectively. Finally, for $i = 1, \ldots, n$, the player P_i receives $s(P_i) = \{s^{(A)}(P_i) = x_i^{(A)} \mid A \in \Gamma \text{ and } P_i \in A\}$ as his share.

Reconstruction Phase:

For any $A = \{P_{i_1}, \ldots, P_{i_m}\} \in 2^{\mathcal{P}}$, the players in A do as follows.

1. **Determine the recovery algorithm.**
   The players in A get

   $$f_A(x_{i_1}, \ldots, x_{i_m}) = F(0, \ldots, 0, x_{i_1}, 0, \ldots, 0, x_{i_m}, 0, \ldots, 0) \tag{15}$$

   as their recovery algorithm.

2. **Compute the secret key.**
   The players in A pool their shares and compute $f_A\left(x_{i_1}^{(A)}, \ldots, x_{i_m}^{(A)}\right)$.

5.2   Perfect Secret Sharing Schemes from Orthogonal Arrays

An *orthogonal array*, denoted by $OA_\lambda(t, m, v)$, is a $\lambda v^t \times m$ array of v symbols, such that in any t columns of the array, every possible t-tuple of the symbols appears exactly λ times. An orthogonal array is called *simple* if and only if no two rows are identical. A large set of orthogonal arrays $LOA_\lambda(t, m, v)$ is a set of v^{m-t}/λ simple arrays $OA_\lambda(t, m, v)$ which satisfies that every possible m-tuple of the symbols appears in exactly one of the orthogonal arrays in the set. We refer to [31] for background on orthogonal arrays.

In Table 4, we propose a method to construct q-ary functions by using orthogonal arrays. By employing such functions in Secret Sharing Scheme II, we can get perfect secret sharing schemes.

Theorem 3. *Let F be constructed by Construction II. Then, Secret Sharing Scheme II is perfect.*

Proof. For any subset of players $A = \{P_{i_1}, \dots, P_{i_m}\} \in \Gamma$, the recovery algorithm f_A satisfies $f_A\left(x_{i_1}^{(A)}, \dots, x_{i_m}^{(A)}\right) = F|_A\left(x^{(A)}\right) = k$, where

$$x^{(A)} = \left(0, \dots, 0, x_{i_1}^{(A)}, 0, \dots, 0, x_{i_m}^{(A)}, 0, \dots, 0\right),$$

thus the qualified group A can reconstruct the secret key.

For any forbidden group $A \in \Delta$, we assume that the players in A are collaborating to reconstruct the secret key by using some recovery algorithm, say f_B, where $B \in \Gamma$. As discussed in Sect. 3, we must have $B \nsubseteq A$, i.e., $C = A \cap B \subsetneq B$. Suppose that $|C| = t$, $|B| = m$, then $t < m$. From Step 2.1 of Construction II, we have that for $B \in \Gamma$, there exists $\left\{\mathcal{OA}_\gamma^{(B)} \mid \gamma \in \mathbb{F}_q^*\right\}$, which is a set of $q - 1$ disjoint simple arrays $OA_1(m-1, m, q-1)$. Given $\gamma \in \mathbb{F}_q^*$, every possible $(m-1)$-tuple of $\mathbb{F}_q^*$ occurs exactly one time in $\mathcal{OA}_\gamma^{(B)}$, which implies that every possible

Table 4. Construction II

Input:	A set of players $\mathcal{P} = \{P_1, \dots, P_n\}$ with access structure (Γ, Δ)
Output:	A function $F : \mathbb{F}_q^n \to \mathbb{F}_q$
Step 1:	For every subset of players $A \in \Delta$, set $F|_A = 0$, i.e., $F|_A$ is a zero function
Step 2:	For every subset of players $A = \{P_{i_1}, \dots, P_{i_m}\} \in \Gamma$, execute the following two steps
	1. Choose a large set of orthogonal arrays $LOA_1(m-1, m, q-1)$, i.e., a set of $q-1$ disjoint simple arrays $OA_1(m-1, m, q-1)$, which is denoted by $\left\{\mathcal{OA}_\gamma^{(A)} \mid \gamma \in \mathbb{F}_q^*\right\}$
	2. For $x \in E_A$, denote by $\tilde{x}$ the vector obtained by deleting all the zero coordinates of x. Then, set $F|_A(x) = \gamma$ if and only if $\tilde{x}$ is a row vector of $\mathcal{OA}_\gamma^{(A)}$

t-tuple of $\mathbb{F}_q^*$ occurs exactly $(q-1)^{m-1-t}$ times in $\mathcal{OA}_\gamma^{(B)}$. Hence, from Step 2.2 of Construction II, we have that given $\gamma \in \mathbb{F}_q^*$, for any shares $s^{(B)}(C) \in (\mathbb{F}_q^*)^t$, the conditional probability determined by f_B satisfies

$$\Pr\left(S^{(B)}(C) = s^{(B)}(C) \mid K = \gamma\right)$$
$$= \frac{(q-1)^{m-1-t}}{(q-1)^{m-1}} = \frac{1}{(q-1)^t} = \Pr\left(S^{(B)}(C) = s^{(B)}(C)\right),$$

which implies from Bayes' theorem that

$$\Pr\left(K = \gamma \mid S^{(B)}(C) = s^{(B)}(C)\right)$$
$$= \frac{\Pr\left(S^{(B)}(C) = s^{(B)}(C) \mid K = \gamma\right)\Pr(K = \gamma)}{\Pr\left(S^{(B)}(C) = s^{(B)}(C)\right)} = \Pr(K = \gamma) = \frac{1}{|\mathbf{K}|}. \tag{16}$$

Due to Eq. (16), we have that for every $k \in \mathbf{K}$, the secret key k can be guessed correctly with probability $\Pr(K = k) = 1/|\mathbf{K}|$. Therefore, from Definition 2, Secret Sharing Scheme II is perfect. $\qquad\square$

Remark 2. We can prove that by employing q-ary function F constructed in Construction II, Secret Sharing Scheme I is perfect. In fact, let k be the secret key and $\alpha_1, \ldots, \alpha_n$ be the generated shares of $P_1, \ldots, P_n$ respectively. When we add the constraint that $\tilde{x} = (\alpha_{i_1}, \ldots, \alpha_{i_m})$ is a row vector of $\mathcal{OA}_k^{(A)}$ to Step 2.2 of Construction II, then the function F satisfies $F(s(A)) = k$ for $A \in \Gamma$. Thus, it can be proved similarly as in Theorem 3 that Secret Sharing Scheme I is perfect. By adding this constraint, the output functions in Construction II form a proper subset of all the output functions in Construction I.

5.3   Perfect Secret Sharing Schemes from Resilient Functions

For two integers n and m, the function $F : \mathbb{F}_q^n \to \mathbb{F}_q^m$ is called t-*resilient* if the output value of F satisfies for any $\{i_1, \ldots, i_t\} \subseteq \{1, 2, \ldots, n\}$, any $z_j \in \mathbb{F}_q$, $j = 1, \ldots, t$, and any $\gamma \in \mathbb{F}_q$,

$$\Pr(F(x_1, \ldots, x_n) = \gamma \mid x_{i_1} = z_1, \ldots, x_{i_t} = z_t) = \Pr(F(x_1, \ldots, x_n) = \gamma) = \frac{1}{q^m}. \tag{17}$$

In [19], t-resilient functions from $\mathbb{F}_q^n$ to $\mathbb{F}_q^m$ are characterized in terms of orthogonal arrays. Furthermore, Camion et al. [9] claimed that this characterization holds for t-resilient functions from $\mathcal{F}^n$ to $\mathcal{F}^m$, where $\mathcal{F}$ is a finite alphabet. Inspired by Construction II in Table 4 and the close relationship between orthogonal arrays and t-resilient functions, we find a way to construct perfect secret sharing schemes with general access structures by employing t-resilient functions. The idea of constructing perfect secret sharing schemes by using resilient functions can be found in simplified (n, n)-threshold scheme that all the n players

Table 5. Construction III

Input:	A set of players $\mathcal{P} = \{P_1, \ldots, P_n\}$ with access structure (Γ, Δ).
Output:	A function $F : \mathbb{F}_q^n \to \mathbb{F}_q$, where $q - 1$ is a power of a prime.

Step 1: Choose a one-to-one mapping $\phi : \mathbb{F}_q^* \to \mathbb{F}_{q'}$, where $q' = q - 1$.

Step 2: For every subset of players $A \in \Delta$, set $F|_A = 0$, i.e., $F|_A$ is a zero function.

Step 3: For every subset of players $A = \{P_{i_1}, \ldots, P_{i_m}\} \in \Gamma$, execute the following two steps.

1. Choose an $(m, m-1, q')$ resilient function G_A.

2. For $x \in E_A$, denote by $\tilde{x} = (x_{i_1}, \ldots, x_{i_m})$ the vector obtained by deleting all the zero coordinates of x. By abuse of notation, we use $\phi(\tilde{x})$ to denote $(\phi(x_{i_1}), \ldots, \phi(x_{i_m}))$. Then, define

$$F|_A(x) = \phi^{-1} \circ G_A \circ \phi(\tilde{x}). \tag{18}$$

pool their shares and compute the secret key $k \in \mathbb{Z}_m$ ($\mathbb{Z}_m$ is the residue class ring with m elements) by the formula $k = \sum_{i=1}^n x_i \mod m$, where for $i = 1, \ldots, n$, $x_i \in \mathbb{Z}_m$ is the share of player P_i (see [30, Chap. 13] for more details). Note that the recovery algorithm $F(x) = \sum_{i=1}^n x_i \mod m$, $x = (x_1, \ldots, x_n) \in \mathbb{Z}_m^n$ is indeed an $(n-1)$-resilient function from $\mathbb{Z}_m^n$ to $\mathbb{Z}_m$. When $m = 2$, this idea appears in [17, Chap. 7] for the construction of binary (n, n)-threshold schemes. Resilient functions can also be employed as building blocks of perfect monotone secret sharing schemes from the description of monotone circuit, see [2,3].

For convenience, we denote t-resilient functions from $\mathbb{F}_q^n$ to $\mathbb{F}_q$ by (n, t, q) resilient functions. Clearly, an (n, t, q) resilient function must be (n, t', q) resilient when $t' \leqslant t$. In Table 5, we use (n, t, q) resilient functions to construct q-ary functions. By employing such functions in Secret Sharing Scheme II, we can get perfect secret sharing schemes.

Remark 3. In Construction III, the finite field $\mathbb{F}_q$ should satisfy that $q - 1$ is a power of a prime, i.e., $q = p^s + 1$ for some prime p and positive integer s. If $q = 2^t$ for some $t \geqslant 2$, then p is odd and $p^s = 2^t - 1$. If q is odd, then p is even and $q = 2^t + 1$ for some $t \geqslant 1$. In Table 6, we give some examples of q such that $q - 1$ is a power of a prime.

Theorem 4. *Let F be constructed by Construction III. Then, Secret Sharing Scheme II is perfect.*

Proof. For any subset of players $A = \{P_{i_1}, \ldots, P_{i_m}\} \in \Gamma$, the recovery algorithm f_A satisfies $f_A\left(x_{i_1}^{(A)}, \ldots, x_{i_m}^{(A)}\right) = F|_A\left(x^{(A)}\right) = k$, where

$$x^{(A)} = \left(0, \ldots, 0, x_{i_1}^{(A)}, 0, \ldots, 0, x_{i_m}^{(A)}, 0, \ldots, 0\right),$$

Table 6. All numbers $2 < q < 2^{64}$ satisfying q is a power of a prime and $q-1$ is a power of a prime

q	$q-1$	q	$q-1$
$3 = 3^1$	$2 = 2^1$	$257 = 257^1$	$256 = 2^8$
$4 = 2^2$	$3 = 3^1$	$8192 = 2^{13}$	$8191 = 8191^1$
$5 = 5^1$	$4 = 2^2$	$65537 = 65537^1$	$65536 = 2^{16}$
$8 = 2^3$	$7 = 7^1$	$131072 = 2^{17}$	$131071 = 131071^1$
$9 = 3^2$	$8 = 2^3$	$524288 = 2^{19}$	$524287 = 524287^1$
$17 = 17^1$	$16 = 2^4$	$2147483648 = 2^{31}$	$2147483647 = 2147483647^1$
$32 = 2^5$	$31 = 31^1$	$2305843009213693952 = 2^{61}$	2305843009213693951 $= 2305843009213693951^1$
$128 = 2^7$	$127 = 127^1$		

Note: a^1 means that a is a prime.

thus the qualified group A can reconstruct the secret key.

For any forbidden group $A \in \Delta$, we assume that the players in A are collaborating to reconstruct the secret key by using some recovery algorithm, say f_B, where $B \in \Gamma$. As discussed in Sect. 3, we must have $B \nsubseteq A$, i.e., $C = A \cap B \subsetneq B$. Suppose that $|C| = t$, $|B| = m$, then $t < m$. Let $x \in E_B$ with $\tilde{x} = (x_{i_1}, \ldots, x_{i_m}) \in (\mathbb{F}_q^*)^m$, where $\tilde{x}$ is the vector obtained by deleting all the zero coordinates of x. From Step 3.1 of Construction III, we have that G_B is an $(m, m-1, q')$ resilient function, where $q' = q-1$. Let $\{j_1, \ldots, j_t\} \subseteq \{i_1, \ldots, i_m\}$, then from Eq. (17), we have that for any $z_s \in \mathbb{F}_{q'}$, $s = 1, \ldots, t$, and any $\beta \in \mathbb{F}_{q'}$,

$$\Pr(G_B \circ \phi(\tilde{x}) = \beta \mid \phi(x_{j_1}) = z_1, \ldots, \phi(x_{j_t}) = z_t)$$
$$= \Pr(G_B \circ \phi(\tilde{x}) = \beta) = \frac{1}{q'} = \frac{1}{q-1}. \tag{19}$$

According to Eq. (18), we get that Eq. (19) is equivalent to

$$\Pr(F|_B(x) = \phi^{-1}(\beta) \mid x_{j_1} = \phi^{-1}(z_1), \ldots, x_{j_t} = \phi^{-1}(z_t))$$
$$= \Pr(F|_B(x) = \phi^{-1}(\beta)) = \frac{1}{q-1}. \tag{20}$$

Since ϕ is a one-to-one mapping from $\mathbb{F}_q^*$ to $\mathbb{F}_{q'}$, then given the shares $s^{(B)}(C) \in (\mathbb{F}_q^*)^t$, for any $\gamma \in \mathbb{F}_q^*$, the conditional probability determined by f_B satisfies

$$\Pr\left(K = \gamma \mid S^{(B)}(C) = s^{(B)}(C)\right) = \Pr(K = \gamma) = \frac{1}{q-1} = \frac{1}{|\mathbf{K}|}, \tag{21}$$

which implies that for every $k \in \mathbf{K}$, the secret key k can be guessed correctly with probability $\Pr(K = k) = 1/|\mathbf{K}|$. Therefore, from Definition 2, we have that Secret Sharing Scheme II is perfect. $\qquad \square$

Remark 4. It is proved in [19] that an (m, t, q) resilient function is equivalent to a large set of orthogonal arrays $LOA_{q^{m-1-t}}(t, m, q)$. In fact, in Step 3.2 of Construction III, the sets $\{\tilde{x} \in (\mathbb{F}_q^*)^m \mid F|_A(x) = \gamma, x \in E_A\}$, $\gamma \in \mathbb{F}_q^*$, consist of a large set of orthogonal arrays $LOA_1(m - 1, m, q - 1)$. Hence, Construction III can be seen as a special case of Construction II.

There exist a large amount of constructions of resilient functions over finite fields, e.g. see [10, 11, 20, 34]. We remark that, generally, given the value table of a q-ary function F, it needs approximately $\mathcal{O}(nq^n)$ operations over the finite field $\mathbb{F}_q$ to compute the ANF of F. However, it will be much easier for us to derive the ANF of F by using the known ANFs of resilient functions. We illustrate this process by a simple example in the Appendix, which provides a perfect secret sharing scheme realizing a non-monotone access structure.

For Secret Sharing Scheme II it is clear that the information rate is

$$\rho = \min\left\{\rho_i = \frac{1}{|\{A \in \Gamma \mid P_i \in A\}|} \;\middle|\; 1 \leqslant i \leqslant n\right\}, \tag{22}$$

which depends on the access structure. In the worst case, there may exist a player who joins in 2^{n-1} qualified groups, then according to Eq. (22), the information rate is $\mathcal{O}(2^{-n})$ which is much lower than the upper bound.

We emphasize that in the sharing phase of Secret Sharing Scheme I and Secret Sharing Scheme II, the computational complexity depends on the access structure, which is often exponential in the number of players for practical applications. In general, for non-monotone secret sharing schemes, it is hard to decrease the complexity of the sharing phase (excepting some special access structures).

6   Conclusion

In this paper, we discuss secret sharing schemes realizing general (not necessarily monotone) access structures. For secret sharing schemes with general access structures, the attack model and the definition of unconditional security (called perfect) given in this paper are generalizations of the monotone access structure case. Secret Sharing Scheme I presented in Table 1 is a democratic scheme such that the shares are generated by the players. We prove that if the value table of the q-ary function F is well arranged, Secret Sharing Scheme I is perfect with information rate $\rho = 1$. We propose Secret Sharing Scheme II for the sake of efficiency, which requires the trusted center TC to distribute the shares. By employing orthogonal arrays as well as resilient functions in the construction of q-ary function F, we prove that Secret Sharing Scheme II is perfect.

Appendix: An Example of Secret Sharing Scheme II

We illustrate Secret Sharing Scheme II by the following example, where the q-ary function F is constructed by Construction III in Table 5.

Example 1. Let $\mathcal{P} = \{P_1, P_2, P_3, P_4\}$ and $\Gamma = \{A_1 = \{P_1, P_2, P_3\}, A_2 = \{P_1, P_2, P_4\}, A_3 = \{P_3, P_4\}, A_4 = \{P_1, P_2, P_3, P_4\}\}$. The set of secret keys is $\mathbf{K} = \mathbb{F}_8^* = \{1, \alpha, \alpha^2, \ldots, \alpha^6\}$, where α is a primitive element of $\mathbb{F}_8$. Suppose that TC wants to share $k = \alpha^5$ as the secret key. Following Construction III, TC defines $\phi : \mathbb{F}_8^* \to \mathbb{F}_7$ as $\phi(\gamma) = \log_\alpha \gamma$, which means that if $\gamma = \alpha^a \in \mathbb{F}_8^*$ for some integer a, then $\log_\alpha \gamma = a$. For the access structure Γ, TC chooses

$$\begin{cases} G_{A_1}(z_1, z_2, z_3) = 2z_1 + 3z_2 + z_3, \\ G_{A_2}(z_1, z_2, z_4) = z_1 + 2z_2 + 3z_4, \\ G_{A_3}(z_3, z_4) = 2z_3 + 4z_4, \\ G_{A_4}(z_1, z_2, z_3, z_4) = z_1 + z_2 + z_3 + z_4 + 1 \end{cases} \tag{23}$$

as the 7-ary linear resilient functions (see [9] for more details). After that, TC computes and secretly transmits the shares

$$s(P_1) = \{s_1^{(A_1)} = \alpha, s_1^{(A_2)} = \alpha^2, s_1^{(A_4)} = \alpha\},$$
$$s(P_2) = \{s_2^{(A_1)} = \alpha^2, s_2^{(A_2)} = \alpha^3, s_2^{(A_4)} = \alpha\},$$
$$s(P_3) = \{s_3^{(A_1)} = \alpha^4, s_3^{(A_3)} = \alpha, s_3^{(A_4)} = \alpha\},$$
$$s(P_4) = \{s_4^{(A_2)} = \alpha^6, s_4^{(A_3)} = \alpha^6, s_1^{(A_4)} = \alpha\},$$

to P_1, P_2, P_3, P_4 respectively. From (23), the 8-ary function F is defined as

$$F|_{A_1}(x) = \phi^{-1} \circ G_{A_1} \circ \phi(\tilde{x}) = x_1^2 x_2^3 x_3,$$
$$F|_{A_2}(x) = \phi^{-1} \circ G_{A_2} \circ \phi(\tilde{x}) = x_1 x_2^2 x_4^3,$$
$$F|_{A_3}(x) = \phi^{-1} \circ G_{A_3} \circ \phi(\tilde{x}) = x_3^2 x_4^4,$$
$$F|_{A_4}(x) = \phi^{-1} \circ G_{A_4} \circ \phi(\tilde{x}) = \alpha x_1 x_2 x_3 x_4,$$

where $x \in \mathbb{F}_8^4$, $\tilde{x}$ denotes the vector obtained by deleting all the zero coordinates of x, and for every forbidden group $A \in \Delta = 2^{\mathcal{P}} \setminus \Gamma$, $F|_A = 0$. Finally, TC publishes $F(x) = (1 - x_4^7) x_1^2 x_2^3 x_3 + (1 - x_3^7) x_1 x_2^2 x_4^3 + (1 - x_1^7)(1 - x_2^7) x_3^2 x_4^4 + \alpha x_1 x_2 x_3 x_4 = x_3^2 x_4^4 + x_1^2 x_2^3 x_3 + x_1 x_2^2 x_4^3 - x_1^7 x_3^2 x_4^4 - x_2^7 x_3^2 x_4^4 + \alpha x_1 x_2 x_3 x_4 - x_1^2 x_2^3 x_3 x_4^7 - x_1 x_2^2 x_3^7 x_4^3 + x_1^7 x_2^7 x_3^2 x_4^4$.

Due to Theorem 4, this secret sharing scheme is perfect. In fact, assume that the players in the forbidden group $B = \{P_1, P_3, P_4\} \in \Delta$ are collaborating to reconstruct the secret key. Their recovery algorithm defined in (15) is $f_B(x_1, x_3, x_4) = (1 - x_1^7) x_3^2 x_4^4$, which equals 0 for any $(x_1, x_3, x_4) \in (\mathbb{F}_8^*)^3$. Suppose that they try to use the recovery algorithms

$$f_{A_1}(x_1, x_2, x_3) = F(x_1, x_2, x_3, 0) = x_1^2 x_2^3 x_3,$$
$$f_{A_2}(x_1, x_2, x_4) = F(x_1, x_2, 0, x_4) = x_1 x_2^2 x_4^3,$$
$$f_{A_4}(x_1, x_2, x_3, x_4) = F(x_1, x_2, x_3, x_4) = x_3^2 x_4^4 + x_1^2 x_2^3 x_3 + x_1 x_2^2 x_4^3 - x_1^7 x_3^2 x_4^4$$
$$- x_2^7 x_3^2 x_4^4 + \alpha x_1 x_2 x_3 x_4 - x_1^2 x_2^3 x_3 x_4^7 - x_1 x_2^2 x_3^7 x_4^3 + x_1^7 x_2^7 x_3^2 x_4^4,$$

which are functions defined on $(\mathbb{F}_8^*)^3$, $(\mathbb{F}_8^*)^3$, and $(\mathbb{F}_8^*)^4$ respectively. For the players P_1, P_3, and P_4, the values of $s_2^{(A_1)}$, $s_2^{(A_2)}$, and $s_2^{(A_4)}$ are unknown random values, thus according to (21), the secret key can be guessed correctly with probability $1/|\mathbf{K}|$, i.e., the players in B can learn nothing about the secret key. Similar discussion holds for other forbidden groups.

Moreover, it is clear that the information rate of this scheme is

$$\rho = \min \left\{ \frac{\log_2 |\mathbf{K}|}{\log_2 |\mathbf{S}(P_i)|} \ \middle| \ 1 \leqslant i \leqslant 4 \right\} = \frac{1}{3}.$$

References

1. Beimel, A.: Secret-sharing schemes: a survey. In: Chee, Y.M., Guo, Z., Ling, S., Shao, F., Tang, Y., Wang, H., Xing, C. (eds.) IWCC 2011. LNCS, vol. 6639, pp. 11–46. Springer, Heidelberg (2011)

2. Benaloh, J.C., Leichter, J.: Generalized secret sharing and monotone functions. In: Goldwasser, S. (ed.) CRYPTO 1988. LNCS, vol. 403, pp. 27–35. Springer, Heidelberg (1990)

3. Benaloh, J.: General linear secret sharing (extended abstract). http://research. microsoft.com/pubs/68477/glss.ps

4. Ben-Or, M., Goldwasser, S., Wigderson, A.: Completeness theorems for non-cryptographic fault-tolerant distributed computation. In: Proceedings of the 20th Annual ACM Symposium on Theory of Computing, pp. 1–10. ACM, New York (1988)

5. Blakley, G.R.: Safeguarding cryptographic keys. In: Proceedings of the National Computer Conference, pp. 313–317. AFIPS Press, New York (1979)

6. Blakley, G.R., Kabatianskii, G.A.: Linear algebra aproach to secret sharing schemes. In: Chmora, A., Wicker, S.B. (eds.) Workshop on Information Protection. LNCS, vol. 829, pp. 33–40. Springer, Heidelberg (1994)

7. Brickell, E.F.: Some ideal secret sharing schemes. In: Quisquater, J.-J., Vandewalle, J. (eds.) EUROCRYPT 1989. LNCS, vol. 434, pp. 468–475. Springer, Heidelberg (1990)

8. Brickell, E.F., Stinson, D.R.: Some improved bounds on the information rate of perfect secret sharing schemes. J. Cryptol. **5**(3), 153–166 (1992)

9. Camion, P., Canteaut, A.: Construction of t-resilient functions over a finite alphabet. In: Maurer, U.M. (ed.) EUROCRYPT 1996. LNCS, vol. 1070, pp. 283–293. Springer, Heidelberg (1996)

10. Camion, P., Canteaut, A.: Correlation-immune and resilient functions over a finite alphabet and their applications in cryptography. Des. Codes Crypt. **16**(2), 121–149 (1995)

11. Carlet, C.: More correlation-immune and resilient functions over galois fields and galois rings. In: Fumy, W. (ed.) EUROCRYPT 1997. LNCS, vol. 1233, pp. 422–433. Springer, Heidelberg (1997)

12. Carlet, C., Ding, C., Yuan, J.: Linear codes from perfect nonlinear mappings and their secret sharing schemes. IEEE Trans. Inf. Theory **51**(6), 2089–2102 (2005)

13. Carpentieri, M.: A perfect threshold secret sharing scheme to identify cheaters. Des. Codes Crypt. **5**(3), 183–186 (1995)

14. Cramer, R., Damgård, I.B., Maurer, U.M.: General secure multi-party computation from any linear secret-sharing scheme. In: Preneel, B. (ed.) EUROCRYPT 2000. LNCS, vol. 1807, pp. 316–334. Springer, Heidelberg (2000)

15. Cramer, R., Damgård, I., Nielsen, J.B.: Secure Multiparty Computation and Secret Sharing: An Information Theoretic Approach. https://users-cs.au.dk/jbn/mpc-book.pdf

16. Cohen, G.D., Mesnager, S., Patey, A.: On minimal and quasi-minimal linear codes. In: Stam, M. (ed.) IMACC 2013. LNCS, vol. 8308, pp. 85–98. Springer, Heidelberg (2013)

17. Ding, C., Pei, D., Salomaa, A.: Chinese Remainder Theorem: Applications in Computing, Coding, Cryptography. World Scientific Publishing Co. Pte. Ltd., Singapore (1996)

18. Ding, K., Ding, C.: A class of two-weight and three-weight codes and their applications in secret sharing. IEEE Trans. Inf. Theory 61(11), 5835–5842 (2015)

19. Gopalakrishnan, K., Stinson, D.R.: Three characterizations of non-binary correlation-immune and resilient functions. Des. Codes Crypt. 5(3), 241–251 (1995)

20. Gupta, K.C., Sarkar, P.: Improved construction of nonlinear resilient S-boxes. IEEE Trans. Inf. Theor. 51(1), 339–348 (2005)

21. Ito, M., Saito, A., Nishizeki, T.: Secret sharing schemes realizing general access structure. Electron. Comm. Jpn. Pt. III 72(9), 56–64 (1989)

22. Karchmer, M., Wigderson, A.: On span programs. In: Proceedings of the 8th IEEE Structure in Complexity Theory, pp. 102–111. IEEE (1993)

23. Lee, C.-Y., Wang, Z.-H., Harn, L., Chang, C.-C.: Secure key transfer protocol based on secret sharing for group communications. IEICE Trans. Inf. Syst. E94–D(11), 2069–2076 (2011)

24. Liu, J., Mesnager, S., Chen, L.: Secret sharing schemes with general access structures (full version). Cryptology ePrint Archive, Report 2015/1139 (2015). https://eprint.iacr.org/2015/1139

25. Massey, J.: Minimal codewords and secret sharing. In: Proceedings of the 6th Joint Swedish-Russian International Workshop on Information Theory, pp. 276–279 (1993)

26. McEliece, R.J., Sarwate, D.V.: On sharing secrets and Reed-Solomon codes. Commun. ACM 24(9), 583–584 (1981)

27. Pieprzyk, J., Zhang, X.-M.: Ideal threshold schemes from MDS codes. In: Lee, P.J., Lim, C.H. (eds.) ICISC 2002. LNCS, vol. 2587, pp. 253–263. Springer, Heidelberg (2003)

28. Pless, V., Brualdi, R.A., Huffman, W.C.: Handbook of Coding Theory. Elsevier Science Inc., New York (1998)

29. Shamir, A.: How to share a secret. Commun. ACM 22(11), 612–613 (1979)

30. Stinson, D.R.: Cryptography: Theory and Practice, 3rd edn. CRC Press, Boca Raton (2006)

31. Stinson, D.R.: Combinatorial Designs: Construction and Analysis. Springer, New York (2004)

32. Yao, A.C.: Protocols for secure computations. In: Proceedings of the 23rd Annual Symposium on Foundations of Computer Science, pp. 160–164. IEEE (1982)

33. Zhang, J., Li, X., Fu, F.-W.: Multi-receiver authentication scheme for multiple messages based on linear codes. In: Huang, X., Zhou, J. (eds.) ISPEC 2014. LNCS, vol. 8434, pp. 287–301. Springer, Heidelberg (2014)

34. Zhang, X.-M., Zheng, Y.: Cryptographically resilient functions. IEEE Trans. Inf. Theor. 43(5), 1740–1747 (1997)

CCA Secure Public Key Encryption Scheme Based on LWE Without Gaussian Sampling

Xiaochao Sun[1,2,3], Bao Li[1,2], Xianhui Lu[1,2], and Fuyang Fang[1,2,3(✉)]

[1] State Key Laboratory of Information Security,
Institute of Information Engineering, Chinese Academy of Sciences,
Beijing, China
{xchsun,lb,xhlu,fyfang13}@is.ac.cn
[2] Data Assurance and Communication Security Research Center,
Chinese Academy of Sciences, Beijing 100093, China
[3] University of Chinese Academy of Sciences, Beijing, China

Abstract. We present a CCA secure PKE based on the problem of the LWE with uniform errors. We use one of the instantiations of parameters of LWE with uniform errors suggested by Micciancio and Peikert (CRYPTO 2013). Since the uniform errors do not bear the Fourier-properities as the Gaussian errors, the statistical techniques and tools used by Micciancio and Peikert (EUROCRYPT 2012) to construct CCA secure PKE are not available for LWE with uniform errors. However, we conquer the problem by employing the double-trapdoor mechanism to construct a tag-based encryption with CCA security and transform it to a CCA secure PKE from the generic conversion based on one-time signatures.

Keywords: Public key encryption · Chosen-ciphertext security · LWE · Uniform errors

1 Introduction

The *learning with errors* (LWE) problem, a generalization of the *learning parity with noise* (LPN) problem, has been applied in many cryptographic scenarios in the past decade. Informally, given a random matrix $\mathbf{A} \in \mathbb{Z}_q^{m \times n}$ and $\mathbf{b} = \mathbf{As} + \mathbf{e} \in \mathbb{Z}_q^m$ with a secret vector $\mathbf{s} \in \mathbb{Z}_q^n$ and some error vector $\mathbf{e}$, it asks to recover the secret vector $\mathbf{s} \in \mathbb{Z}_q^n$. Regev set the error distribution to be the discrete Gaussian distribution and the modulus $q = poly(n)$ to be a polynomial in n, and gave a reduction from solving some lattice problems such as GapSVP in the worst-case to this LWE problem in average-case quantumly [Reg05]. Peikert gave a classical reduction with exponential modulus q in [Pei09], and Barkerski et al. showed a classical reduction with polynomial modulus q [BLP+13]. These worst-to-average-case reductions were given according to the Fourier-properties of the Gaussian distribution.

This research is supported by the National Nature Science Foundation of China (No. 61379137 and No. 61272040), the National Basic Research Program of China (973 project) (No. 2013CB338002).

© Springer International Publishing Switzerland 2016
D. Lin et al. (Eds.): Inscrypt 2015, LNCS 9589, pp. 361–378, 2016.
DOI: 10.1007/978-3-319-38898-4_21

The LWE problem with discrete Gaussian errors has been proved to be versatile and useful in cryptographic constructions. Currently, there are many cryptographic applications based on LWE with discrete Gaussian errors, including public key encryption (PKE) under chosen-plaintext attacks (CPA) [Reg05, KTX07, PVW08, LP11] and PKE under chosen-ciphertext secure attacks (CCA) [PW08, Pei09, MP12], identity-based encryption (IBE) [GPV08, CHKP10, ABB10, MP12], oblivious transfer [PVW08], and fully homomorphic encryption (FHE) [Gen09a, Gen09b, BV11, BGV12, Bra12].

CCA FROM LWE. LWE based PKE with CPA security was first proposed by Regev in [Reg05]. Peikert and Waters proposed a new framework to construct CCA secure PKE schemes based on a new primitive named as *lossy trapdoor function* (LTDF), and showed that LTDF can be realized from the LWE assumption [PW08]. Peikert observed that the LWE problem can be used to construct correlated products trapdoor function [Pei09], which can be used to construct CCA secure PKE schemes [RS09]. Using the generic conversion technique in [BCHK07], one can obtain CCA secure PKE schemes from IBE schemes based on the LWE problem [ABB10, CHKP10]. Micciancio and Peikert proposed a new method to generate trapdoors for the LWE problem [MP12]. Based on their new technique, an efficient CCA secure PKE scheme was proposed. In fact, they constructed an *adaptive trapdoor function* (ATDF) based on the LWE problem by using their new technique and achieved CCA security by employing the technique in [KMO10].

LWE with discrete Gaussian errors attracts more attention [Reg05, MR07, Pei09, BLP+13] and has been proved its success in cryptography. However, sampling discrete Gaussian errors usually costs a significant amount of computational resource in key generation and encryption algorithms [Pei10, MP12]. Construction based on LWE without discrete Gaussian errors is not considered until Döttling and Müller-Quade replaced the discrete Gaussian errors with uniform errors [DM13]. They gave the first worst-to-average case reduction for LWE with uniform errors over small interval by using a new tool called *lossy codes*. Micciancio and Peikert showed that the hardness of LWE remains even with small errors, provided that the number of samples is small enough [MP13].

LWE WITH UNIFORM ERRORS. Micciancio and Mol mentioned the importance of understanding LWE problems with various errors [MM11]. Döttling and Müller-Quade used a tool called lossy codes and showed the hardness of the LWE problem with uniform errors in a small interval [DM13]. Lossy codes loss the information of the message on average, but random linear codes can recover the message from a type of errors information theoretically. Moreover, lossy codes are computationally indistinguishable from random linear codes. Concurrently and independently, Micciancio and Peikert gave the new results on the hardness of LWE with uniform errors [MP13]. Particularly, it showed the hardness connection between the LWE problem with linear number of samples, uniform errors and polynomial modulus and standard lattice problems.

Cabarcas, Göpfert and Weiden gave the first CPA secure PKE scheme based LWE with uniform errors [CGW14]. There is no CCA secure PKE scheme from LWE with uniform errors.

1.1 Our Contributions and Technique

In this work we use one of the instantiations of parameters of the LWE with uniform errors which is suggested by Micciancio and Peikert in [MP13] to construct a tag-based encryption, and transform it into a CCA secure PKE. The instantiation includes a linear number of samples $m = cn$ with a constant $c \geq 1$ and an error interval of polynomial magnitude. More precisely,

$$m = 3n, \quad s \geq (6Cn)^5, \quad q \geq \max\{3\sqrt{n}, (4s)^{\frac{3}{2}}, 12n^2 s^2 + 2\},$$

where C is a large enough universal constant and s is the size of the interval from where the errors are sampled uniformly. Our scheme is the first one with CCA security based on the LWE without discrete Gaussian errors.

Since the uniform distribution does not bear the Fourier-properties as the discrete Gaussian distribution which is a powerful tool in standard LWE, we can not give an almost statistically perfect simulation of the challenge ciphertext and provide decryption oracle simultaneously. However, we use the *double-trapdoor mechanism* introduced by Kiltz, Masny and Pietrzak in [KMP14] instantiated with *normal LWE* problem, which gives a computationally indistinguishable simulation. More specifically, the simulator must answer decryption queries without the secret of the instance of Normal LWE (which is one of the trapdoors) during the security reduction. But the simulator knows the other trapdoor thus it can answer decryption queries, and the adversary can not distinguish which trapdoor is used by the simulator.

2 Preliminaries

2.1 Notation

Let $\mathbb{Q}$ be the rational field. Let $\mathbb{Z}_q$ be the q-ary finite field for a prime $q \geq 2$. If $\mathbf{x} = [x_1, \cdots, x_n]^\mathsf{T}$ is a vector over $\mathbb{Z}_q^n$, then $\|\mathbf{x}\|_\infty \triangleq \max_{i=1}^n |x_i|$ denotes its ℓ_∞ norm. If $\mathcal{A}$ is an algorithm, then $y \leftarrow \mathcal{A}(x)$ denotes that $\mathcal{A}$ outputs y with input x, no matter that $\mathcal{A}$ is deterministic or probabilistic. Specially, we denote by $y \leftarrow \mathcal{A}^{\mathcal{O}}(x)$ that when $\mathcal{A}$ has access to an oracle $\mathcal{O}$, $\mathcal{A}$ outputs y with input x. If D is a distribution, then $e \sim D$ denotes that e distributes according to D. Assuming that there is an efficient algorithm that samples from the distribution D, $e \leftarrow D$ denotes that the random variable e is output from the algorithm. If S is a finite set, then $s \leftarrow S$ denotes sampling s from S uniformly at random.

We write $poly(n)$ to denote a function $f(n) = O(n^c)$ for some constant c. A *negligible* function on n, denoted as $negl(n)$ usually, is a function $f(n)$ such that $f(n) = o(n^{-c})$ for every fixed constant c. An event happens with *overwhelming* probability if it happens with probability at least $1 - negl(n)$. Two ensembles of distributions $\{X_n\}$ and $\{Y_n\}$ are computationally indistinguishable, if for any probabilistic polynomial time algorithm $\mathcal{A}$, $|\Pr[\mathcal{A}(X_n) = 1] - \Pr[\mathcal{A}(Y_n) = 1]| \leq negl(n)$.

2.2   Public Key Encryption and Tag-Based Encryption

We recall the definition of the public key encryption scheme and the security definition under adaptive chosen-ciphertext attacks [RS91].

A public key encryption scheme **PKE** = (**KeyGen, Enc, Dec**) consists of three (either probabilistic or deterministic) polynomial time algorithms, where

- **KeyGen**(1^k): a probabilistic algorithm that takes a security parameter k as input, and outputs a pair of public key and secret key (PK, SK).
- **Enc**(PK, μ): a probabilistic algorithm that takes the public key PK and a message μ as input, and outputs a ciphertext CT.
- **Dec**(SK, CT): a deterministic algorithm that takes the secret key SK and a ciphertext CT as input, and outputs a plaintext μ or aborts.

The correctness of the scheme **PKE** requires that for all messages μ, it holds that

$$\Pr[\mathbf{Dec}(\mathsf{SK}, \mathbf{Enc}(\mathsf{PK}, \mu)) \neq \mu : (\mathsf{PK}, \mathsf{SK}) \leftarrow \mathbf{KeyGen}(1^k)] < negl(k).$$

CCA Security. Let $\mathcal{A} = (\mathcal{A}_1, \mathcal{A}_2)$ be a two-stage adversary. The advantage of $\mathcal{A}$ in the CCA experiment of a public key encryption scheme **PKE** = (**KeyGen, Enc, Dec**) is defined as

$$\mathbf{Adv}_{\mathbf{PKE}}^{cca,\mathcal{A}}(k) \triangleq \left| \Pr \left[b = b' : \begin{array}{l} (\mathsf{PK}, \mathsf{SK}) \leftarrow \mathbf{KeyGen}(1^k), \\ (\mu_0, \mu_1, St) \leftarrow \mathcal{A}_1^{\mathbf{Dec}(\mathsf{SK}, \cdot)}(\mathsf{PK}), \\ b \leftarrow \{0, 1\}, \quad \cdot \\ \mathsf{CT}^* \leftarrow \mathbf{Enc}(\mathsf{PK}, \mu_b), \\ b' \leftarrow \mathcal{A}_2^{\mathbf{Dec}(\mathsf{SK}, \cdot)}(\mathsf{PK}, \mathsf{CT}^*, St) \end{array} \right] - \frac{1}{2} \right|,$$

where $\mathcal{A}_1$ and $\mathcal{A}_2$ both have access to a decryption oracle **Dec**(SK, $\cdot$), but $\mathcal{A}_2$ is not allowed to ask the decryption oracle on CT*. And $\mathbf{Adv}_{\mathbf{PKE},t,Q}^{cca}(k) = \max_{\mathcal{A}} \mathbf{Adv}_{\mathbf{PKE}}^{cca,\mathcal{A}}(k)$, where the equation is taken over all adversaries $\mathcal{A}$ that run in time t and make Q decryption queries at most. **PKE** is said to be indistinguishable against adaptive chosen-ciphertext attacks (CCA secure in short) if $\mathbf{Adv}_{\mathbf{PKE},t,Q}^{cca}(k) < negl(k)$ for $t = poly(k)$ and $Q = poly(k)$.

A tag-based encryption (TBE) scheme **TBE** = (**T.KeyGen, T.Enc, T.Dec**) with message space $\mathcal{M}$ and tag space $\mathcal{T}$ consists of three (either probabilistic or deterministic) polynomial time algorithms, where

- **T.KeyGen**(1^k): a probabilistic algorithm that takes a security parameter k as input, and outputs a pair of public key and secret key (PK, SK).
- **T.Enc**(PK, τ, μ): a probabilistic algorithm that takes the public key PK, a tag $\tau \in \mathcal{T}$ and a message $\mu \in \mathcal{M}$ as input, and outputs a ciphertext CT.

- **T.Dec**$(\mathsf{SK}, \tau, \mathsf{CT})$: a deterministic algorithm that takes the secret key SK, a tag $\tau \in \mathcal{T}$ and a ciphertext CT as input, and outputs a message μ or the abort symbol $\perp$.

The correctness of the scheme **TBE** requires that for all messages $\mu \in \mathcal{M}$ and tags $\tau \in \mathcal{T}$, it holds that

$$\Pr[\mathbf{T.Dec}(\mathsf{SK}, \tau, \mathbf{T.Enc}(\mathsf{PK}, \tau, \mu)) \neq \mu : (\mathsf{PK}, \mathsf{SK}) \leftarrow \mathbf{T.KeyGen}(1^k)] < negl(k).$$

Selective-Tag Weak CCA Security. Let $\mathcal{A} = (\mathcal{A}_1, \mathcal{A}_2, \mathcal{A}_3)$ be a three-stage adversary. The advantage of $\mathcal{A}$ in selective-tag weak CCA experiment against a tag-based encryption scheme **TBE** = (**T.KeyGen**, **T.Enc**, **T.Dec**) is defined as

$$\mathbf{Adv}_{\mathbf{TBE}}^{st\text{-}cca,\mathcal{A}}(k) \triangleq \left| \Pr \left[b = b' : \begin{array}{l} (\tau^*, St_1) \leftarrow \mathcal{A}_1(1^k), \\ (\mathsf{PK}, \mathsf{SK}) \leftarrow \mathbf{T.KeyGen}(1^k), \\ (\mu_0, \mu_1, St_2) \leftarrow \mathcal{A}_2^{\mathbf{T.Dec}(\mathsf{SK}, \cdot, \cdot)}(\mathsf{PK}, St_1), \\ b \leftarrow \{0, 1\}, \\ \mathsf{CT}^* \leftarrow \mathbf{T.Enc}(\mathsf{PK}, \tau^*, \mu_b), \\ b' \leftarrow \mathcal{A}_3^{\mathbf{T.Dec}(\mathsf{SK}, \cdot, \cdot)}(\mathsf{PK}, \mathsf{CT}^*, St_2), \end{array} \right] - \frac{1}{2} \right|,$$

where $\mathcal{A}_2$ and $\mathcal{A}_3$ both have access to a decryption oracle $\mathbf{T.Dec}(\mathsf{SK}, \cdot, \cdot)$, but not allowed to ask the decryption oracle on the target tag τ^*. And $\mathbf{Adv}_{\mathbf{TBE},t,Q}^{st\text{-}cca}(k) = \max_{\mathcal{A}} \mathbf{Adv}_{\mathbf{TBE}}^{st\text{-}cca,\mathcal{A}}(k)$, where the equation is taken over all adversaries $\mathcal{A}$ that run in time t and make Q decryption queries at most. **TBE** is said to be selective-tag weakly secure against chosen ciphertext attacks if $\mathbf{Adv}_{\mathbf{TBE},t,Q}^{st\text{-}cca}(k) < negl(k)$ for $t = poly(k)$ and $Q = poly(k)$.

There are mainly two generic transformations from a selective-tag weak CCA secure TBE scheme to a CCA secure PKE. One is to use one-time signatures (OTS) [Kil06, BCHK07]. The transformation sets the verification key to be the tag of the ciphertext of TBE and makes a signature for the ciphertext of TBE. The other one is to replace OTS with message authentication code [BK05].

2.3 One-Time Signatures

A one-time signature scheme **SIG** = (**SIG.Gen**, **SIG.Sign**, **SIG.Verify**) with message space $\mathcal{M}'$ consists of three (either probabilistic or deterministic) polynomial time algorithms, where

- **SIG.Gen**(1^k): a probabilistic algorithm that takes a security parameter k as input, and outputs a pair of verification and signature keys $(\mathsf{VK}, \mathsf{SGK})$.
- **SIG.Sign**$(\mathsf{SGK}, \mathsf{m})$: a probabilistic algorithm that takes the signature key SGK and a message $\mathsf{m} \in \mathcal{M}'$ as input, and outputs a signature σ.
- **SIG.Verify**$(\mathsf{VK}, \mathsf{m}, \sigma)$: an algorithm that takes the verification key VK, the message m and the signature σ as input, and outputs a bit $b \in \{0, 1\}$.

The correctness of the scheme **SIG** requires that for all messages $m \in \mathcal{M}'$, it holds that

$$\Pr[\mathbf{SIG.Verify}(\mathsf{VK}, m, \mathbf{SIG.Sign}(\mathsf{SGK}, m)) = 0 : (\mathsf{VK}, \mathsf{SGK}) \leftarrow \mathbf{SIG.Gen}(1^k)]$$
$$\leq negl(k).$$

Strong One-Time Signature. Let $\mathcal{A}$ be a adversary. The advantage of $\mathcal{A}$ against the signature scheme $\mathbf{SIG} = (\mathbf{SIG.Gen}, \mathbf{SIG.Sign}, \mathbf{SIG.Verify})$ is defined as

$$\mathbf{Adv}_{\mathbf{SIG}}^{ot\text{-}ex\text{-}for,\mathcal{A}}(k) \triangleq$$

$$\Pr\left[\mathbf{SIG.Verify}(\mathsf{VK}, m^*, \sigma^*) = 1 : \begin{array}{l} (\mathsf{VK}, \mathsf{SGK}) \leftarrow \mathbf{SIG.Gen}(1^k), \\ (m^*, \sigma^*) \leftarrow \mathcal{A}^{\mathbf{SIG.Sign}(\mathsf{SGK}, \cdot)}(\mathsf{VK}), \end{array}\right],$$

where the pair (m^*, σ^*) that $\mathcal{A}$ outputs must be different from (m, σ) that $\mathcal{A}$ obtained from the oracle query and $\mathcal{A}$ can only ask the signature oracle $\mathbf{SIG.Sign}(\mathsf{SGK}, \cdot)$ one time. And $\mathbf{Adv}_{\mathbf{SIG}}^{ot\text{-}ex\text{-}for}(k) \triangleq \max_{\mathcal{A}} \mathbf{Adv}_{\mathbf{SIG}}^{ot\text{-}ex\text{-}for,\mathcal{A}}(k)$, where the equation is taken over all the probabilistic time adversary. **SIG** is said be a strong one-time signature scheme if $\mathbf{Adv}_{\mathbf{SIG}}^{ot\text{-}ex\text{-}for,\mathcal{A}}(k) \leq negl(k)$.

2.4 Learning with Errors

In this section we recall the standard decisional learning with errors (LWE) problem and some variants of LWE problem. Let $q \in \mathbb{Z}^*$ be the modulus, $n \in \mathbb{Z}^*$ be the size of the secret vector, m be the number of the samples and χ be the error distribution over $\mathbb{Z}_q$.

Definition 1 ((Decisional) LWE). *The LWE distribution is defined as*

$$D_{\mathrm{LWE}(n,m,\chi)} \triangleq \left((\mathbf{A}, \mathbf{As} + \mathbf{e}) | \mathbf{A} \leftarrow \mathbb{Z}_q^{m \times n}, \mathbf{s} \leftarrow \mathbb{Z}_q^n, \mathbf{e} \leftarrow \chi^m\right).$$

The advantage of a distinguisher $\mathcal{D}$ in distinguishing $D_{\mathrm{LWE}(n,m,\chi)}$ from uniform distribution is defined as

$$\mathbf{Adv}_{\mathrm{LWE}(n,m,\chi)}^{\mathcal{D}} \triangleq \left| \Pr_{\mathbf{A},\mathbf{s},\mathbf{e}} [\mathcal{D}(\mathbf{A}, \mathbf{As} + \mathbf{e}) = 1] - \Pr_{\mathbf{A},\mathbf{u}} [\mathcal{D}(\mathbf{A}, \mathbf{u}) = 1] \right|,$$

where $(\mathbf{A}, \mathbf{As} + \mathbf{e}) \leftarrow D_{\mathrm{LWE}(n,m,\chi)}$, $\mathbf{u} \leftarrow \mathbb{Z}_q^m$. Also define $\mathbf{Adv}_{\mathrm{LWE}(n,m,\chi),t} \triangleq \max_{\mathcal{D}} \mathbf{Adv}_{\mathrm{LWE}(n,m,\chi)}^{\mathcal{D}}$, where the equation is taken over all $\mathcal{D}$'s that run in time t, and say that LWE is hard if $\mathbf{Adv}_{\mathrm{LWE}(n,m,\chi),t} \leq negl(n)$ for $t = poly(n)$.

2.4.1 Normal (Extended) LWE

In [ACPS09], Applebaum et al. gave a version of LWE called normal LWE, where the secret vector hidden in the normal LWE distribution is chosen from the error distribution χ rather than the uniform distribution over $\mathbb{Z}_q^n$ in the standard LWE. Namely,

$$D_{\mathrm{Norm\text{-}LWE}(n,m-n,\chi)} \triangleq \left((\mathbf{A}, \mathbf{At} + \mathbf{x}) | \mathbf{A} \leftarrow \mathbb{Z}_q^{(m-n) \times n}, \mathbf{t} \leftarrow \chi^n, \mathbf{x} \leftarrow \chi^{m-n}\right).$$

Applebaum et al. gave a transformation technique from standard LWE distribution to the normal LWE distribution, and showed that distinguishing normal LWE distribution from an uniform one is as hard as the standard LWE problem at least.

In [AP12], Alperin-Sheriff and Peikert gave a tight reduction from standard LWE to extended-LWE (ELWE) introduced by O'Neil et al. [OPW11], where a hint of the error vector is leaked. Similarly, we use Applebaum et al.'s technique on ELWE samples and obtain normal extended-LWE (Normal ELWE) samples with a hint of the error vector and secret vector.

Definition 2 (Normal ELWE). *The normal ELWE distribution is defined as*

$$D_{\text{Norm-ELWE}(n,m-n,\chi)} \triangleq$$

$$\left(\left(\mathbf{A}, \mathbf{At} + \mathbf{x}, \begin{pmatrix} \mathbf{e} \\ \mathbf{f} \end{pmatrix}, \mathbf{t}^\top \mathbf{e} + \mathbf{x}^\top \mathbf{f} \right) \middle| \mathbf{A} \leftarrow \mathbb{Z}_q^{(m-n) \times n}, \mathbf{t}, \mathbf{e} \leftarrow \chi^n, \mathbf{x}, \mathbf{f} \leftarrow \chi^{m-n} \right).$$

The advantage of a distinguisher $\mathcal{D}$ in distinguishing $D_{\text{Norm-ELWE}(n,m-n,\chi)}$ from uniform distribution is defined as

$$\mathbf{Adv}^{\mathcal{D}}_{\text{Norm-ELWE}(n,m-n,\chi)} \triangleq \left| \Pr \left[\mathcal{D} \left(\mathbf{A}, \mathbf{At} + \mathbf{x}, \begin{pmatrix} \mathbf{e} \\ \mathbf{f} \end{pmatrix}, \mathbf{t}^\top \mathbf{e} + \mathbf{x}^\top \mathbf{f} \right) = 1 \right] \right.$$

$$\left. - \Pr \left[\mathcal{D} \left(\mathbf{A}, \mathbf{u}, \begin{pmatrix} \mathbf{e} \\ \mathbf{f} \end{pmatrix}, \mathbf{t}^\top \mathbf{e} + \mathbf{x}^\top \mathbf{f} \right) = 1 \right] \right|,$$

where $\mathbf{A}, \mathbf{t}, \mathbf{x}, \mathbf{e}, \mathbf{f}$ is as above and $\mathbf{u} \leftarrow \mathbb{Z}_q^{m-n}$. Also define

$$\mathbf{Adv}_{\text{Norm-ELWE}(n,m-n,\chi),t} \triangleq \max_{\mathcal{D}} \mathbf{Adv}^{\mathcal{D}}_{\text{Norm-ELWE}(n,m-n,\chi)},$$

where the equation is taken over all $\mathcal{D}$'s that run in time t, and say that normal ELWE is hard if $\mathbf{Adv}_{\text{Norm-ELWE}(n,m-n,\chi),t} \leq negl(n)$ for $t = poly(n)$.

By a routine hybrid argument we directly have the hardness of the ℓ-fold normal LWE and the ℓ-fold normal ELWE problem (i.e. $\mathbf{T}$ has ℓ rows, and every row is a secret vector in unfolded problem). Namely,

$$\mathbf{Adv}_{\text{Norm-LWE}^\ell(n,m-n,\chi),t} \triangleq \max_{\mathcal{D}} \left| \Pr \left[\mathcal{D} \left(\mathbf{A}, \mathbf{TA} + \mathbf{X} \right) = 1 \right] - \Pr \left[\mathcal{D} \left(\mathbf{A}, \mathbf{U} \right) = 1 \right] \right|$$

$$\leq \ell \cdot \mathbf{Adv}_{\text{Norm-LWE}(n,m-n,\chi),t},$$

and

$$\mathbf{Adv}_{\text{Norm-ELWE}^\ell(n,m-n,\chi),t} \triangleq \max_{\mathcal{D}} \left| \Pr \left[\mathcal{D} \left(\mathbf{A}, \mathbf{TA} + \mathbf{X}, \begin{pmatrix} \mathbf{e} \\ \mathbf{f} \end{pmatrix}, \mathbf{Te} + \mathbf{Xf} \right) = 1 \right] \right.$$

$$\left. - \Pr \left[\mathcal{D} \left(\mathbf{A}, \mathbf{U}, \begin{pmatrix} \mathbf{e} \\ \mathbf{f} \end{pmatrix}, \mathbf{Te} + \mathbf{Xf} \right) = 1 \right] \right|$$

$$\leq \ell \cdot \mathbf{Adv}_{\text{Norm-ELWE}(n,m-n,\chi),t},$$

where $\mathbf{A} \leftarrow \mathbb{Z}_q^{n \times (m-n)}, \mathbf{T} \leftarrow \chi^{\ell \times n}, \mathbf{X} \leftarrow \chi^{\ell \times (m-n)}, \mathbf{e} \leftarrow \chi^n, \mathbf{f} \leftarrow \chi^{m-n}, \mathbf{U} \leftarrow \mathbb{Z}_q^{\ell \times (m-n)}$, and the equation is taken over all $\mathcal{D}$'s that run in time $t = poly(n)$.

2.4.2 LWE with Uniform Errors and Instantiation

In the original paper of LWE problem [Reg05] and the consequent developments of cryptographic constructions based on LWE, the Gaussian errors are employed. In our construction, we use the result given by Micciancio and Peikert about the LWE with small parameters and uniform errors. Namely,

Lemma 1 ([MP13]). *Let $0 < k \leq n \leq m - \omega(\log k) \leq k^{O(1)}$, $\ell = m - n + k$, $s \geq (Cm^{\ell/(n-k)})$ for a large enough universal constant C, and q be a prime such that $\max\{3\sqrt{k}, (4s)^{m/(m-n)}\} \leq q \leq k^{O(1)}$. For any set of $X \subseteq \{-s, \cdots, s\}^m$ of size $|X| \geq s^m$, the $LWE(n, m, \chi)$ function family is pseudorandom with respect to the uniform input distribution $\chi = \mathcal{U}(X)$, under the assumption that $SIVP_\gamma$ is (quantum) hard to approximate, in the worst case, on k-dimensional lattices to within a factor $\gamma = \tilde{O}(\sqrt{k} \cdot q)$.*

We set $k_1 = \frac{n_1}{2}$ and $k_2 = \frac{n_2}{2}$. From Lemma 1, we give two sets of parameters instantiation used in our scheme as follows

$$m_1 = 3n_1, \quad s_1 \geq (3Cn_1)^5, \quad q_1 \geq \max\{3\sqrt{n_1}, (4s_1)^{\frac{3}{2}}\}, \tag{1}$$

and

$$m_2 = 3n_2, \quad s_2 \geq (3Cn_2)^5, \quad q_2 \geq \max\{3\sqrt{n_2}, (4s_2)^{\frac{3}{2}}\}, \tag{2}$$

The $LWE(n_1, m_1, \chi_1)$ function family under the modulus q_1 and $LWE(n_2, m_2, \chi_2)$ function family under the modulus q_2 are pseudorandom, where χ_i is uniform distribution over $X_i \subseteq \{-s_i, \cdots, s_i\}^{m_i}$ of size $|X_i| \geq s_i^{m_i}$, $i = 1, 2$.

To unify the error interval and the modulus in the two parameters instantiations, we set $n_1 = n$, $n_2 = 2n_1 = 2n$ and

$$m = 3n, \quad s \geq (6Cn)^5 \geq \max\{s_1, s_2\}, \quad q \geq (4s)^{\frac{3}{2}} \geq \max\{q_1, q_2, 12n^2s^2 + 2\}. \tag{3}$$

[1]The $LWE(n, m, \chi_1)$ and $LWE(2n, 2m, \chi_2)$ function families under the same prime modulus q are both pseudorandom, where χ_i is uniform distribution over $X_i \subseteq \{-s, \cdots, s\}^{m_i}$ of size $|X_i| \geq s^{m_i}$, $m_1 = 3n$, $m_2 = 6n$, $i = 1, 2$.

3 TBE Scheme from LWE Without Gaussian Error

In this section we present a TBE scheme with selective-tag weak CCA security if the LWE problem is hard under the parameter instantiation defined above. It can be transformed to a standard PKE with CCA security.

3.1 Codes Under ℓ_∞-norm

Before the description of the scheme, we introduce our codes and decoding algorithm. Similarly to [MP12], we use the public "gadget" matrix $\mathbf{G}$ which is viewed

---

[1] $q \geq 12n^2s^2 + 2$ is related to the correctness of the decryption and we will explain it later.

as the generator-matrix of codes. Differently, the codes used in our construction work under ℓ_∞-norm rather than ℓ_2-norm in [MP12]. So we give a new decoding algorithm. We consider the case of 1-dimension. More specifically, let $[q_0, \cdots, q_{k-1}]^\mathsf{T} \in \{0,1\}^k$ be the binary expansion of q. Let $\mathbf{g} = [1, 2^1, \cdots, 2^{k-1}]^\mathsf{T}$ and

$$\mathbf{S}_k(q) = \begin{bmatrix} 2 & -1 & & & & \\ & 2 & -1 & & & \\ & & & \ddots & \ddots & \\ & & & & 2 & -1 \\ q_0 & q_1 & q_2 & \cdots & q_{k-2} & q_{k-1} \end{bmatrix} \in \mathbb{Z}^{k \times k}.$$

Obviously, it holds that $\mathbf{S}_k(q)\mathbf{g} = \mathbf{0} \pmod{q}$. Next, we construct an efficient decoding algorithm.

Lemma 2. *There is an efficient decoding algorithm $\mathcal{D}_1$ to recover $x \in \mathbb{Z}_q$ from $\mathbf{y} = x \cdot \mathbf{g} + \mathbf{e} \in \mathbb{Z}_q^k$, if $\|\mathbf{e}\|_\infty \leq \frac{1}{k} \cdot \lfloor \frac{q}{2} \rfloor$.*

Proof. The decoding algorithm $\mathcal{D}_1$ works as follows:

1. Compute $\mathbf{y}' \leftarrow \mathbf{S}_k(q)\mathbf{y} \pmod{q} \in \mathbb{Z}_q^k$.
2. Compute $\mathbf{e}' \leftarrow \mathbf{S}_k^{-1}(q)\mathbf{y}'$ (not mod q and $\mathbf{S}_k^{-1}(q) \in \mathbb{Q}^{k \times k}$).
3. Compute $\mathbf{y} - \mathbf{e}'$ and obtain $x \cdot \mathbf{g}$ and x finally.

From the decoding, there is

$$\mathbf{y}' = \Big(\mathbf{S}_k(q)\mathbf{y} \pmod{q}\Big) = \Big(\mathbf{S}_k(q)(x \cdot \mathbf{g} + \mathbf{e}) \pmod{q}\Big)$$
$$= \Big(\mathbf{S}_k(q)\mathbf{e} \pmod{q}\Big) = \mathbf{S}_k(q)\mathbf{e} \in \mathbb{Q}^k,$$

where the last equation satisfies

$$\|\mathbf{S}_k(q)\mathbf{e}\|_\infty \leq \max\{\max_{i=1}^{k-1} |2e_{i-1} - e_i|, |\sum_{i=0}^{k-1} e_i q_i|\} \leq \max\{3 \cdot \|\mathbf{e}\|_\infty, k \cdot \|\mathbf{e}\|_\infty\} \leq \lfloor \frac{q}{2} \rfloor.$$

Hence $\mathbf{e}' = \mathbf{e} \in \mathbb{Q}^k$. Finally it is easy to obtain $x \in \mathbb{Z}_q$. $\qquad\square$

In our construction, we need to extend $\mathbf{g}$ and $\mathbf{S}_k(q)$ to $\mathbf{G}$ and $\mathbf{S}$ with higher dimension. More precisely, let $\mathbf{G} = \mathbf{I}_{n \times n} \otimes \mathbf{g} \in \mathbb{Z}_q^{\ell \times n}$ with $k = \lceil \lg q \rceil$ and $\ell = nk$, and $\mathbf{S} = \mathbf{I}_{n \times n} \otimes \mathbf{S}_k(q)$. We can construct the decoding algorithm $\mathcal{D}$ to decode $\mathbf{y} = \mathbf{G}\mathbf{x} + \mathbf{z} \in \mathbb{Z}_q^\ell$ and recover $\mathbf{x} \in \mathbb{Z}_q^n$ by calling $\mathcal{D}_1$ k-wisely in parallel if $\|\mathbf{z}\|_\infty \leq \frac{1}{k} \cdot \lfloor \frac{q}{2} \rfloor$.

3.2 Description of the TBE Scheme

The following parameters and public settings are involved in the scheme.

- The security parameter n.

- m, s, q is defined as in (3) with a prime modulus q, and $\ell = 2n\lceil \log q \rceil$.
- $\chi = \mathcal{U}(X)$ is the uniform distribution over a set $X \subseteq \{-s, \cdots, s\}^m$ with $|X| = s^m$.
- Message space $\mathcal{M} = \{0, 1\}^m$, tag space $\mathcal{T} = \{0, \cdots, q-1\}^n$.
- An encoding function $H : \mathcal{T} \to \mathbb{Z}_q^{(m-n)\times(m-n)}, \tau \mapsto \mathbf{H}_\tau$, such that $\mathbf{H}_0 = \mathbf{0}_{(m-n)\times(m-n)}$, and for all $\tau_1 \neq \tau_2$, $\mathbf{H}_{\tau_1} - \mathbf{H}_{\tau_2} = \mathbf{H}_{\tau_1-\tau_2}$ is full-rank and invertible [ABB10].

Our scheme is described as follows:

T.KeyGen$(1^n) \to$ (PK, SK): The algorithm runs as follows:

$$\mathbf{A} \leftarrow \mathbb{Z}_q^{n\times(m-n)}, \mathbf{T}_0, \mathbf{T}_1 \leftarrow \chi^{\ell\times n}, \mathbf{X}_0, \mathbf{X}_1 \leftarrow \chi^{\ell\times(m-n)},$$

$$\mathbf{B}_0 \leftarrow \mathbf{T}_0\mathbf{A} + \mathbf{X}_0, \mathbf{B}_1 \leftarrow \mathbf{T}_1\mathbf{A} + \mathbf{X}_1,$$

$$\mathbf{C} \leftarrow \mathbb{Z}_q^{m\times(m-n)},$$

$\mathsf{SK} \leftarrow (\mathbf{T}_0, \mathbf{T}_1) \in (\mathbb{Z}_q^{\ell\times n})^2,$

$\mathsf{PK} \leftarrow (\mathbf{A}, \mathbf{B}_0, \mathbf{B}_1, \mathbf{C}) \in \mathbb{Z}_q^{n\times(m-n)} \times \mathbb{Z}_q^{\ell\times(m-n)} \times \mathbb{Z}_q^{\ell\times(m-n)} \times \mathbb{Z}_q^{m\times(m-n)}.$

T.Enc$(\mathsf{PK}, \tau, \mathbf{m}) \to \mathsf{CT}$: The algorithm runs as follows:

$$\mathbf{f} \leftarrow \chi^{m-n}, \mathbf{e} \leftarrow \chi^n, \mathbf{e}_2 \leftarrow \chi^m, \mathbf{T}_0', \mathbf{T}_1' \leftarrow \chi^{\ell\times n}, \mathbf{X}_0', \mathbf{X}_1' \leftarrow \chi^{\ell\times(m-n)}$$

$$
\begin{aligned}
\mathbf{c} &\leftarrow \mathbf{A}\mathbf{f} + \mathbf{e} & \in \mathbb{Z}_q^n, \\
\mathbf{c}_0 &\leftarrow (\mathbf{B}_0 + \mathbf{G}\mathbf{H}_\tau)\mathbf{f} + \mathbf{T}_0'\mathbf{e} - \mathbf{X}_0'\mathbf{f} & \in \mathbb{Z}_q^\ell, \\
\mathbf{c}_1 &\leftarrow (\mathbf{B}_1 + \mathbf{G}\mathbf{H}_\tau)\mathbf{f} + \mathbf{T}_1'\mathbf{e} - \mathbf{X}_1'\mathbf{f} & \in \mathbb{Z}_q^\ell, \\
\mathbf{c}_2 &\leftarrow \mathbf{C}\mathbf{f} + \mathbf{e}_2 + \lfloor \tfrac{q}{2} \rfloor \cdot \mathbf{m} & \in \mathbb{Z}_q^m,
\end{aligned}
$$

$$\mathsf{CT} \leftarrow (\mathbf{c}, \mathbf{c}_0, \mathbf{c}_1, \mathbf{c}_2).$$

T.Dec$(\mathsf{SK}, \tau, \mathsf{CT}) \to (\bot/\mathbf{m})$: The algorithm works as follows:

1. Compute $\mathbf{y}_0 \leftarrow \mathbf{c}_0 - \mathbf{T}_0\mathbf{c}$ and use the decoding algorithm $\mathcal{D}$ of $\mathbf{G}$ to recover $\mathbf{H}_\tau\mathbf{f}$ from $\mathbf{y}_0$. If $\mathcal{D}$ aborts, then the decryption algorithm outputs $\bot$. Otherwise, the decryption algorithm computes $\mathbf{f} = \mathbf{H}_\tau^{-1}\mathbf{H}_\tau\mathbf{f}$.
2. If $\|\mathbf{f}\|_\infty > s$ or $\|\mathbf{c} - \mathbf{A}\mathbf{f}\|_\infty > s$ or $\|\mathbf{c}_0 - (\mathbf{B}_0 + \mathbf{G}\mathbf{H}_\tau)\mathbf{f}\|_\infty > m \cdot s^2$ or $\|\mathbf{c}_1 - (\mathbf{B}_1 + \mathbf{G}\mathbf{H}_\tau)\mathbf{f}\|_\infty > m \cdot s^2$, then output $\bot$. Otherwise, compute $\mathbf{c}_2 - \mathbf{C}\mathbf{f} = \lfloor \tfrac{q}{2} \rfloor \cdot \mathbf{m} + \mathbf{e}_2$ and recover the message $\mathbf{m} \in \{0, 1\}^m$ and output it.

3.2.1   Correctness

We now prove the correctness of the scheme.

Theorem 1 (Correctness). *In the above scheme with the choice of the public settings, for* $(\mathsf{PK}, \mathsf{SK}) \leftarrow$ **T.KeyGen**(1^k), **T.Dec**$(\mathsf{SK}, \tau, \mathsf{CT}) = \mathbf{m}$, *where* $\mathsf{CT} \leftarrow$ **T.Enc**$(\mathsf{PK}, \tau, \mathbf{m})$.

Proof. Since $\mathbf{y}_0 = \mathbf{c}_0 - \mathbf{T}_0\mathbf{c} = \mathbf{GH}_\tau\mathbf{f} + \mathbf{X}_0\mathbf{f} + \mathbf{T}'_0\mathbf{e} - \mathbf{X}'_0\mathbf{f} - \mathbf{T}_0\mathbf{e}$, it needs to bound the error-term $\mathbf{X}_0\mathbf{f} + \mathbf{T}'_0\mathbf{e} - \mathbf{X}'_0\mathbf{f} - \mathbf{T}_0\mathbf{e}$. For any $\mathbf{t}, \mathbf{e} \leftarrow \chi^n$ and $\mathbf{x}, \mathbf{f} \leftarrow \chi^{m-n}$, it has $|\mathbf{t}^\mathsf{T}\mathbf{e}| \leq ns^2$ and $|\mathbf{x}^\mathsf{T}\mathbf{f}| \leq (m-n)s^2$. Since CT is generated from the encryption algorithm,

$$\|\mathbf{X}_0\mathbf{f} + \mathbf{T}'_0\mathbf{e} - \mathbf{X}'_0\mathbf{f} - \mathbf{T}_0\mathbf{e}\|_\infty \leq \|\mathbf{X}_0\mathbf{f}\|_\infty + \|\mathbf{T}'_0\mathbf{e}\|_\infty + \|\mathbf{X}'_0\mathbf{f}\|_\infty + \|\mathbf{T}_0\mathbf{e}\|_\infty$$
$$\leq (m-n)s^2 + ns^2 + (m-n)s^2 + ns^2 = 6ns^2$$
$$\leq \frac{1}{n} \cdot \left(\frac{q}{2} - 1\right) \leq \frac{1}{k} \cdot \lfloor\frac{q}{2}\rfloor.$$

As the construction of $\mathbf{G}$ and the decoding algorithm $\mathcal{D}$, the decryption algorithm recovers $\mathbf{H}_\tau\mathbf{f}$ and further computes $\mathbf{f}$ with $\|\mathbf{f}\|_\infty \leq s$. Furthermore, it has

$$\|\mathbf{c} - \mathbf{Af}\|_\infty = \|\mathbf{e}\|_\infty \leq s,$$
$$\|\mathbf{c}_0 - (\mathbf{B}_0 + \mathbf{GH}_\tau)\mathbf{f}\|_\infty = \|\mathbf{T}'_0\mathbf{e} - \mathbf{X}'_0\mathbf{f}\|_\infty \leq \|\mathbf{T}'_0\mathbf{e}\|_\infty + \|\mathbf{X}'_0\mathbf{f}\|_\infty$$
$$\leq ns^2 + (m-n)s^2 \leq ms^2,$$
$$\text{and } \|\mathbf{c}_1 - (\mathbf{B}_1 + \mathbf{GH}_\tau)\mathbf{f}\|_\infty = \|\mathbf{T}'_1\mathbf{e} - \mathbf{X}'_1\mathbf{f}\|_\infty \leq \|\mathbf{T}'_1\mathbf{e}\|_\infty + \|\mathbf{X}'_1\mathbf{f}\|_\infty$$
$$\leq ns^2 + (m-n)s^2 \leq ms^2.$$

Since $\|\mathbf{e}_2\|_\infty \leq s < \frac{1}{n} \cdot \lfloor\frac{q}{4}\rfloor$, the decryption algorithm computes $\mathbf{m}$ correctly. $\square$

3.2.2   Proof of Security

Theorem 2. *If the normal LWE problem is hard, then the scheme **TBE** is selective-tag weak CCA secure. In particular, for any PPT adversary $\mathcal{A}$ that runs in time t, there is*

$$\mathbf{Adv}_{\mathbf{TBE}}^{st\text{-}cca,\mathcal{A}}(n) \leq 2\ell \cdot \mathbf{Adv}_{\text{Norm-LWE}(n,2n,\chi),t} + 2\ell \cdot \mathbf{Adv}_{\text{Norm-ELWE}(n,2n,\chi),t}$$
$$+ \mathbf{Adv}_{\text{Norm-LWE}(2n,4n,\chi),t} \tag{4}$$

Proof. Before the adversary $\mathcal{A}$ receives the public key from the challenger $\mathcal{C}$, $\mathcal{A}$ commits to the challenge tag τ^*. The challenge ciphertext is $\mathsf{CT}^* = (\mathbf{c}^*, \mathbf{c}_0^*, \mathbf{c}_1^*, \mathbf{c}_2^*)$, and $\mathsf{CT} = (\mathbf{c}, \mathbf{c}_0, \mathbf{c}_1, \mathbf{c}_2)$ denotes one of $Q = poly(n)$ decryption queries with tag τ. $\mathcal{A}(\mathbf{Game}_i) = 1$ denotes the event that $\mathcal{A}$'s output b' equals to the $\mathcal{C}$'s random choice b in **Game i**.

Game 0. This is the original selective-tag weak CCA game between the adversary $\mathcal{A}$ and the challenger $\mathcal{C}$. We have

$$\mathbf{Adv}_{\mathbf{TBE}}^{st\text{-}cca,\mathcal{A}}(k) = \left|\Pr\left[\mathcal{A}(\mathbf{Game}_0) = 1\right] - \frac{1}{2}\right|. \tag{5}$$

Game 1. In this game the challenger $\mathcal{C}$ replaces $\mathbf{B}_1 = \mathbf{T}_1\mathbf{A} + \mathbf{X}_1$ with a uniformly random $\mathbf{B}_1 = \mathbf{U}_1 \leftarrow \mathbb{Z}_q^{\ell \times (m-n)}$. The rest is identical to **Game 0**. The adversary $\mathcal{A}$ can not distinguish **Game 1** from **Game 0** for the hardness of

ℓ-fold normal LWE problem. More specifically, assuming that the adversary $\mathcal{A}$ distinguishes **Game 1** from **Game 0**, we construct an algorithm $\mathcal{A}_1$ to solve ℓ-fold normal LWE problem. Given a tuple $(\mathbf{A}, \mathbf{B}'_1)$, where $\mathbf{B}'_1$ is either $\mathbf{T}_1\mathbf{A} + \mathbf{X}_1$ or uniformly random $\mathbf{U}_1$, $\mathcal{A}_1$ sets $\mathbf{B}_1 \leftarrow \mathbf{B}'_1$. The rest is identical in **Game 0**. $\mathcal{A}_1$ outputs whatever $\mathcal{A}$ outputs.

Since that the decryption oracle works under the secret key $\mathbf{T}_0$ which the challenger $\mathcal{C}$ chooses, it is identical as in **Game 0**. We now analyse the behavior of $\mathcal{A}_1$.

Normal LWE: Since $\mathbf{B}'_1 = \mathbf{T}_1\mathbf{A} + \mathbf{X}_1$, $\mathbf{B}_1 = \mathbf{T}_1\mathbf{A} + \mathbf{X}_1$ is distributed as in **Game 0**. Hence $\mathcal{A}_1$ simulates the challenger in **Game 0** and

$$\Pr\left[\mathcal{A}(\mathbf{Game}_0) = 1\right] = \Pr\left[\mathcal{A}_1\left(\mathbf{A}, \mathbf{B}'_1 = \mathbf{T}_1\mathbf{A} + \mathbf{X}_1\right) = 1\right].$$

Uniform: Since that $\mathbf{B}'_1 = \mathbf{U}_1$ is uniform, $\mathcal{A}_1$ simulates the behavior of the challenger $\mathcal{C}$ in **Game 1** and

$$\Pr\left[\mathcal{A}(\mathbf{Game}_1) = 1\right] = \Pr\left[\mathcal{A}_1\left(\mathbf{A}, \mathbf{B}'_1 = \mathbf{U}_1\right) = 1\right].$$

Therefore,

$$\left|\Pr\left[\mathcal{A}(\mathbf{Game}_1) = 1\right] - \Pr\left[\mathcal{A}(\mathbf{Game}_0) = 1\right]\right| \leq \mathbf{Adv}_{\text{Norm-LWE}^\ell(n, m-n, \chi), t}. \quad (6)$$

Game 2. In this game the challenger $\mathcal{C}$ replaces the uniformly random $\mathbf{B}_1 = \mathbf{U}_1$ with $\mathbf{B}_1 = \mathbf{T}_1\mathbf{A} + \mathbf{X}_1 - \mathbf{GH}_{\tau^*}$ and sets $\mathbf{c}_1^* = (\mathbf{B}_1 + \mathbf{GH}_{\tau^*})\mathbf{f}^* + \mathbf{T}_1\mathbf{e}^* - \mathbf{X}_1\mathbf{f}^*$. The rest is identical to **Game 1**. Note that the decryption oracle works under the secret key $\mathbf{T}_0$ and it answers identically as in **Game 1** for the same ciphertext query. $\mathcal{A}$ can not distinguish **Game 2** from **Game 1** for the hardness of normal extended LWE.

Assuming that $\mathcal{A}$ distinguishes **Game 2** from **Game 1**, we construct an algorithm $\mathcal{A}_1$ to solve the ℓ-fold normal extended LWE problem. Given a quadruple $\left(\mathbf{A}, \mathbf{B}'_1, \begin{pmatrix} \mathbf{e}' \\ \mathbf{f}' \end{pmatrix}, \mathbf{z}_1 = \mathbf{T}_1\mathbf{e}' + \mathbf{X}_1\mathbf{f}'\right)$, where $\mathbf{B}'_1$ is either $\mathbf{T}_1\mathbf{A} + \mathbf{X}_1$ or uniformly random $\mathbf{U}_1$, $\mathcal{A}_1$ sets $\mathbf{B}_1 = \mathbf{B}'_1 - \mathbf{GH}_{\tau^*}$, $\mathbf{f}^* = \mathbf{f}'$, $\mathbf{e}^* = -\mathbf{e}'$, and $\mathbf{c}_1^* = \mathbf{B}'_1\mathbf{f}' - \mathbf{z}_1$. We now analyse the behavior of $\mathcal{A}_1$.

Normal ELWE: Since $\mathbf{B}'_1 = \mathbf{T}_1\mathbf{A} + \mathbf{X}_1$, $\mathbf{B}_1 = \mathbf{T}_1\mathbf{A} + \mathbf{X}_1 - \mathbf{GH}_{\tau^*}$ is distributed as in **Game 2**. $\mathbf{c}^*$, $\mathbf{c}_0^*$ and $\mathbf{c}_2^*$ are distributed as in **Game 2** as well. And

$$\begin{aligned} \mathbf{c}_1^* = \mathbf{B}'_1\mathbf{f}' - \mathbf{z}_1 &= (\mathbf{B}_1 + \mathbf{GH}_{\tau^*})\mathbf{f}' - (\mathbf{T}_1\mathbf{e}' + \mathbf{X}_1\mathbf{f}') \\ &= (\mathbf{B}_1 + \mathbf{GH}_{\tau^*})\mathbf{f}' + \mathbf{T}_1(-\mathbf{e}') - \mathbf{X}_1\mathbf{f}' \\ &= (\mathbf{B}_1 + \mathbf{GH}_{\tau^*})\mathbf{f}^* + \mathbf{T}_1\mathbf{e}^* - \mathbf{X}_1\mathbf{f}^* \end{aligned}$$

has the same distribution as in **Game 2**. Hence $\mathcal{A}_1$ simulates the behavior of the challenger in **Game 2** and

$$\Pr\left[\mathcal{A}(\mathbf{Game}_2) = 1\right]$$
$$= \Pr\left[\mathcal{A}_1\left(\mathbf{A}, \mathbf{B}'_1 = \mathbf{T}_1\mathbf{A} + \mathbf{X}_1, \begin{pmatrix} \mathbf{e}' \\ \mathbf{f}' \end{pmatrix}, \mathbf{z}_1 = \mathbf{T}_1\mathbf{e}' + \mathbf{X}_1\mathbf{f}'\right) = 1\right].$$

Uniform: Since that $\mathbf{B}_1' = \mathbf{U}_1$ is uniform, $\mathbf{B}_1 = \mathbf{B}_1' - \mathbf{GH}_{\tau^*}$ is uniform as well. $\mathbf{c}^*$, $\mathbf{c}_0^*$ and $\mathbf{c}_2^*$ are distributed as in **Game 1**. And

$$\begin{aligned}
\mathbf{c}_1^* = \mathbf{B}_1'\mathbf{f}' - \mathbf{z}_1 &= (\mathbf{B}_1 + \mathbf{GH}_{\tau^*})\mathbf{f}' - (\mathbf{T}_1\mathbf{e}' + \mathbf{X}_1\mathbf{f}') \\
&= (\mathbf{B}_1 + \mathbf{GH}_{\tau^*})\mathbf{f}' + \mathbf{T}_1(-\mathbf{e}') - \mathbf{X}_1\mathbf{f}' \\
&= (\mathbf{B}_1 + \mathbf{GH}_{\tau^*})\mathbf{f}^* + \mathbf{T}_1^*\mathbf{e}^* - \mathbf{X}_1^*\mathbf{f}^*
\end{aligned}$$

has the same distribution as in **Game 1** with $\mathbf{T}_1^* = \mathbf{T}_1$, $\mathbf{X}_1^* = \mathbf{X}_1$, $\mathbf{e}^* = -\mathbf{e}'$ and $\mathbf{f}^* = \mathbf{f}'$. Hence $\mathcal{A}_1$ simulates the behavior of the challenger in **Game 1** and

$$\Pr[\mathcal{A}(\mathbf{Game}_1) = 1] = \Pr\left[\mathcal{A}_1\left(\mathbf{A}, \mathbf{B}_1' = \mathbf{U}_1, \begin{pmatrix}\mathbf{e}'\\\mathbf{f}'\end{pmatrix}, \mathbf{z}_1 = \mathbf{T}_1\mathbf{e}' + \mathbf{X}_1\mathbf{f}'\right) = 1\right],$$

Therefore,

$$|\Pr[\mathcal{A}(\mathbf{Game}_2) = 1] - \Pr[\mathcal{A}(\mathbf{Game}_1) = 1]| \leq \mathbf{Adv}_{\text{Norm-ELWE}^\ell(n,m-n,\chi),t}. \quad (7)$$

Game 3. In this game the challenger $\mathcal{C}$ sets $\mathbf{c}_1^* \leftarrow \mathbf{T}_1\mathbf{c}^*$. The rest is identical to **Game 2**. Note that

$$\begin{aligned}
\mathbf{c}_1^* = \mathbf{T}_1\mathbf{c}^* = \mathbf{T}_1(\mathbf{Af}^* + \mathbf{e}^*) &= (\mathbf{T}_1\mathbf{A} + \mathbf{X}_1)\mathbf{f}^* + \mathbf{T}_1\mathbf{e}^* - \mathbf{X}_1\mathbf{f}^* \\
&= (\mathbf{B}_1 + \mathbf{GH}_{\tau^*})\mathbf{f}^* + \mathbf{T}_1\mathbf{e}^* - \mathbf{X}_1\mathbf{f}^*
\end{aligned}$$

is identical to $\mathbf{c}_1^*$ in **Game 3**. Therefore,

$$\Pr[\mathcal{A}(\mathbf{Game}_3) = 1] = \Pr[\mathcal{A}(\mathbf{Game}_2) = 1]. \quad (8)$$

Game 4. In this game the challenger $\mathcal{C}$ answers the decryption queries with the other "secret key" $\mathbf{T}_1$ (together with the challenge tag τ^*) rather than $\mathbf{T}_0$ in **Game 3**. More precisely, $\mathcal{C}$ computes $\mathbf{y}_1 = \mathbf{c}_1 - \mathbf{T}_1\mathbf{c}$, recovers $(\mathbf{H}_\tau - \mathbf{H}_{\tau^*})\mathbf{f} = \mathbf{H}_{\tau-\tau^*}\mathbf{f}$ from $\mathbf{y}_1$ and computes $\mathbf{f} = \mathbf{H}_{\tau-\tau^*}^{-1}\mathbf{H}_{\tau-\tau^*}\mathbf{f}$. The rest of decryption oracle is the same as the real decryption algorithm. We will show that the modification to the decryption oracle can not change the answers to the decryption queries, since that the two decryption oracles give the same answer to the same decryption query.

Lemma 3. *The decryption oracles in* **Game 3** *and* **Game 4** *have the same output distributions.*

Proof. Let $\mathsf{CT} = (\mathbf{c}, \mathbf{c}_0, \mathbf{c}_1, \mathbf{c}_2)$ be a ciphertext of decryption queries with tag τ. If the decryption oracle in **Game 3** outputs $\mathbf{m}$, then it recovers $\mathbf{f} \in \mathbb{Z}_q^{m-n}$ such that $\|\mathbf{e} \triangleq \mathbf{c} - \mathbf{Af}\|_\infty \leq s$, $\|\mathbf{c}_0 - (\mathbf{B}_0 + \mathbf{GH}_\tau)\mathbf{f}\|_\infty \leq m \cdot s^2$ and $\|\mathbf{r}_1 \triangleq \mathbf{c}_1 - (\mathbf{B}_1 + \mathbf{GH}_\tau)\mathbf{f}\|_\infty \leq m \cdot s^2$. Then $\mathbf{y}_1 = \mathbf{c}_1 - \mathbf{T}_1\mathbf{c} = \mathbf{GH}_{\tau-\tau^*}\mathbf{f} + \mathbf{r}_1 + \mathbf{X}_1\mathbf{f} - \mathbf{T}_1\mathbf{e}$. Since that the error-term

$$\|\mathbf{r}_1 + \mathbf{X}_1\mathbf{f} - \mathbf{T}_1\mathbf{e}\|_\infty \leq \|\mathbf{r}_1\|_\infty + \|\mathbf{X}_1\mathbf{f}\|_\infty + \|\mathbf{T}_1\mathbf{e}\|_\infty \leq 2ms^2,$$

the decryption oracles in **Game 3** and **Game 4** recover the same $\mathbf{f}$. The rest of the decryption oracles are identical in **Game 3** and **Game 4**. Therefore, they have the same output distribution.   □

Therefore,

$$\Pr\left[\mathcal{A}(\mathbf{Game}_4) = 1\right] = \Pr\left[\mathcal{A}(\mathbf{Game}_3) = 1\right]. \tag{9}$$

Game 5. In this game the challenger $\mathcal{C}$ replaces $\mathbf{B}_0 = \mathbf{T}_0\mathbf{A} + \mathbf{X}_0$ with a uniformly random $\mathbf{B}_0 = \mathbf{U}_0 \leftarrow \mathbb{Z}_q^{\ell \times (m-n)}$. The rest is identical to **Game 4**. Note that the decryption oracle works under the other "secret key" $\mathbf{T}_1$ and its answer to the decryption query is identical as in **Game 4**. With the same analysis in **Game 1**, it has

$$|\Pr\left[\mathcal{A}(\mathbf{Game}_5) = 1\right] - \Pr\left[\mathcal{A}(\mathbf{Game}_4) = 1\right]| \leq \mathbf{Adv}_{\text{Norm-LWE}^\ell(n,m-n,\chi),t}. \tag{10}$$

Game 6. In this game the challenger $\mathcal{C}$ replaces the uniformly random $\mathbf{B}_0 = \mathbf{U}_0$ with $\mathbf{B}_0 = \mathbf{T}_0\mathbf{A} + \mathbf{X}_0 - \mathbf{G}\mathbf{H}_{\tau^*}$ and sets $\mathbf{c}_0^* = (\mathbf{B}_0 + \mathbf{G}\mathbf{H}_{\tau^*})\mathbf{f}^* + \mathbf{T}_0\mathbf{e}^* - \mathbf{X}_0\mathbf{f}^*$. The rest is identical to **Game 5**. With the same analysis in **Game 2**, it has

$$|\Pr\left[\mathcal{A}(\mathbf{Game}_6) = 1\right] - \Pr\left[\mathcal{A}(\mathbf{Game}_5) = 1\right]| \leq \mathbf{Adv}_{\text{Norm-ELWE}^\ell(n,m-n,\chi),t}. \tag{11}$$

Game 7. In this game the challenger $\mathcal{C}$ sets $\mathbf{c}_0^* \leftarrow \mathbf{T}_0\mathbf{c}^*$. The rest is identical to **Game 6**. With the same analysis in **Game 3**, it has

$$\Pr\left[\mathcal{A}(\mathbf{Game}_7) = 1\right] = \Pr\left[\mathcal{A}(\mathbf{Game}_6) = 1\right]. \tag{12}$$

Game 8. In this game the challenger $\mathcal{C}$ replaces $\mathbf{c}^* = \mathbf{A}\mathbf{f}^* + \mathbf{e}^*$ with uniformly random $\mathbf{u} \leftarrow \mathbb{Z}_q^m$ and $\mathbf{c}_2^* = \mathbf{C}\mathbf{f}^* + \mathbf{e}_2^* + \lfloor \frac{q}{2} \rfloor \cdot \mathbf{m}_b$ with uniformly random $\mathbf{u}_2 \leftarrow \mathbb{Z}_q^m$ respectively. The hardness of the normal LWE problem ensures that the adversary $\mathcal{A}$ can not distinguish **Game 8** from **Game 7**.

Assuming that $\mathcal{A}$ distinguishes **Game 8** from **Game 7**, we construct an algorithm $\mathcal{A}_1$ to solve normal LWE problem. $\mathcal{A}_1$ is given a tuple $\left(\begin{pmatrix} \mathbf{A}' \\ \mathbf{C}' \end{pmatrix}, \begin{pmatrix} \mathbf{b}' \\ \mathbf{b}'_2 \end{pmatrix} \right)$, where $\mathbf{A}' \leftarrow \mathbb{Z}_q^{n \times (m-n)}$, $\mathbf{C}' \leftarrow \mathbb{Z}_q^{m \times (m-n)}$, $(\mathbf{b}', \mathbf{b}'_2) \in \mathbb{Z}_q^n \times \mathbb{Z}_q^m$ is either $\mathbf{b}' = \mathbf{A}'\mathbf{f}' + \mathbf{e}'$, $\mathbf{b}'_2 = \mathbf{C}'\mathbf{f}' + \mathbf{e}'_2$ for some $\mathbf{f}' \leftarrow \chi^{m-n}$, $\mathbf{e}' \leftarrow \chi^n$, $\mathbf{e}'_2 \leftarrow \chi^m$ or $(\mathbf{b}', \mathbf{b}'_2) = (\mathbf{u}', \mathbf{u}'_2)$ for the uniform $\mathbf{u}'$ and $\mathbf{u}'_2$. It sets $\mathbf{A} \leftarrow \mathbf{A}'$, $\mathbf{C} \leftarrow \mathbf{C}'$ and generates $\mathbf{B}_0$ and $\mathbf{B}_1$ as in **Game 7**. It also sets

$$\mathbf{c}^* \leftarrow \mathbf{b}', \quad \mathbf{c}_2^* \leftarrow \mathbf{b}'_2 + \lfloor \frac{q}{2} \rfloor \cdot \mathbf{m}_b.$$

$\mathbf{c}_0^*$ and $\mathbf{c}_1^*$ are generated as in **Game 7** and the decryption oracle works as in **Game 7** as well. We now analyze the behavior of $\mathcal{A}_1$.

Normal LWE: Since $\mathbf{b}' = \mathbf{A}'\mathbf{f}' + \mathbf{e}'$ and $\mathbf{b}'_2 = \mathbf{C}'\mathbf{f}' + \mathbf{e}'_2$, $\mathbf{c}^* = \mathbf{A}'\mathbf{f}' + \mathbf{e}'$ and $\mathbf{c}_2^* = \mathbf{C}'\mathbf{f}' + \mathbf{e}'_2 + \lfloor \frac{q}{2} \rfloor \cdot \mathbf{m}_b$ are distributed as in **Game 7**. Hence $\mathcal{A}_1$ simulates the behavior of $\mathcal{C}$ in **Game 7** and

$$\Pr\left[\mathcal{A}(\mathbf{Game}_7) = 1\right] = \Pr\left[\mathcal{A}_1\left(\begin{pmatrix} \mathbf{A}' \\ \mathbf{C}' \end{pmatrix}, \begin{pmatrix} \mathbf{b}' = \mathbf{A}'\mathbf{f}' + \mathbf{e}' \\ \mathbf{b}'_2 = \mathbf{C}'\mathbf{f}' + \mathbf{e}'_2 \end{pmatrix} \right) = 1\right].$$

Uniform: Since $\mathbf{b}' = \mathbf{u}'$ and $\mathbf{b}'_2 = \mathbf{u}'_2$, $\mathcal{A}_1$ simulates the behavior of $\mathcal{C}$ in **Game 8** and

$$\Pr\left[\mathcal{A}(\mathbf{Game}_8) = 1\right] = \Pr\left[\mathcal{A}_1\left(\begin{pmatrix}\mathbf{A}'\\\mathbf{C}'\end{pmatrix},\begin{pmatrix}\mathbf{b}'=\mathbf{u}'\\\mathbf{b}'_2=\mathbf{u}'_2\end{pmatrix}\right)=1\right].$$

Therefore,

$$\left|\Pr\left[\mathcal{A}(\mathbf{Game}_8)=1\right]-\Pr\left[\mathcal{A}(\mathbf{Game}_7)=1\right]\right| \leq \mathbf{Adv}_{\text{Norm-LWE}(m-n,m+n,\chi),t}. \tag{13}$$

Finally, in **Game 8**, the challenge ciphertext is independent of the message that the challenger $\mathcal{C}$ chooses. Hence,

$$\left|\Pr\left[\mathcal{A}(\mathbf{Game}_8)=1\right]-\frac{1}{2}\right|=0.$$

Let $m = 3n$ and sum up all the probabilities, we complete the proof.   □

4   To CCA Secure PKE

From Kiltz's generic conversion [Kil06], TBE scheme with selective-tag weak CCA security can be transformed into a PKE scheme with CCA security. Specifically, assuming that $\mathbf{TBE} = (\mathbf{T.KeyGen}, \mathbf{T.Enc}, \mathbf{T.Dec})$ is defined above and $\mathbf{SIG} = (\mathbf{SIG.Gen}, \mathbf{SIG.Sign}, \mathbf{SIG.Verify})$ is a strong one-time signature scheme, the CCA secure PKE scheme is described as follows:

$\mathbf{KeyGen}(1^n) \to (\mathsf{PKE.PK}, \mathsf{PKE.SK})$: The algorithm runs as follows:

$$(\mathsf{PK}, \mathsf{SK}) \leftarrow \mathbf{T.KeyGen}(1^n),$$
$$\mathsf{PKE.PK} \leftarrow \mathsf{PK}, \quad \mathsf{PKE.SK} \leftarrow \mathsf{SK}.$$

$\mathbf{Enc}(\mathsf{PKE.PK}, \mathbf{m}) \to \mathsf{PKE.CT}$: The algorithm runs as follows:

$$(\mathsf{VK}, \mathsf{SGK}) \leftarrow \mathbf{SIG.Gen},$$
$$\mathsf{CT} \leftarrow \mathbf{T.Enc}(\mathsf{PK}, \tau = \mathsf{VK}, \mathbf{m}),$$
$$\sigma \leftarrow \mathbf{SIG.Sign}(\mathsf{SGK}, \mathsf{CT}),$$
$$\mathsf{PKE.CT} \leftarrow (\mathsf{CT}, \sigma, \mathsf{VK}).$$

$\mathbf{Dec}(\mathsf{PKE.CT}) \to (\perp/\bar{\mathbf{m}})$: The algorithm works as follows:

Parse $\mathsf{PKE.CT}$ as $(\mathsf{CT}, \sigma, \mathsf{VK})$. Check whether $\mathbf{SIG.Verify}(\mathsf{VK}, \mathsf{CT}, \sigma) = 1$. If not, output $\perp$. Otherwise compute and output $\mathbf{m} = \mathbf{T.Dec}(\mathsf{SK}, \tau = \mathsf{VK}, \mathsf{CT})$.

5   Conclusion

In this work we use the instantiation of the LWE problem with uniform errors suggested by Micciancio and Peikert to construct a tag-based encryption scheme with selective-tag weak CCA security. It is transformed to CCA secure PKE from the generic conversion based on one-time signature. Our scheme is the first one with CCA security based on the LWE with uniform errors.

References

[ABB10] Agrawal, S., Boneh, D., Boyen, X.: Efficient lattice (H)IBE in the standard model. In: Gilbert, H. (ed.) EUROCRYPT 2010. LNCS, vol. 6110, pp. 553–572. Springer, Heidelberg (2010)

[ACPS09] Applebaum, B., Cash, D., Peikert, C., Sahai, A.: Fast cryptographic primitives and circular-secure encryption based on hard learning problems. In: Halevi, S. (ed.) CRYPTO 2009. LNCS, vol. 5677, pp. 595–618. Springer, Heidelberg (2009)

[AP12] Alperin-Sheriff, J., Peikert, C.: Circular and KDM security for identity-based encryption. In: Fischlin, M., Buchmann, J., Manulis, M. (eds.) PKC 2012. LNCS, vol. 7293, pp. 334–352. Springer, Heidelberg (2012)

[BCHK07] Boneh, D., Canetti, R., Halevi, S., Katz, J.: Chosen-ciphertext security from identity-based encryption. SIAM J. Comput. 36(5), 1301–1328 (2007)

[BGV12] Brakerski, Z., Gentry, C., Vaikuntanathan, V.: (Leveled) fully homomorphic encryption without bootstrapping. In: Goldwasser, S. (ed.) Innovations in Theoretical Computer Science, Cambridge, MA, USA, 8–10 January 2012, pp. 309–325. ACM (2012)

[BK05] Boneh, D., Katz, J.: Improved efficiency for CCA-secure cryptosystems built using identity-based encryption. In: Menezes, A. (ed.) CT-RSA 2005. LNCS, vol. 3376, pp. 87–103. Springer, Heidelberg (2005)

[BLP+13] Brakerski, Z., Langlois, A., Peikert, C., Regev, O., Stehlé, D.: Classical hardness of learning with errors. In: Boneh, D., Roughgarden, T., Feigenbaum, J. (eds.) Symposium on Theory of Computing Conference, STOC 2013, Palo Alto, CA, USA, 1–4 June 2013, pp. 575–584. ACM (2013)

[Bra12] Brakerski, Z.: Fully homomorphic encryption without modulus switching from classical GapSVP. In: Safavi-Naini, R., Canetti, R. (eds.) CRYPTO 2012. LNCS, vol. 7417, pp. 868–886. Springer, Heidelberg (2012)

[BV11] Brakerski, Z., Vaikuntanathan, V.: Efficient fully homomorphic encryption from (standard) LWE. In: Ostrovsky, R. (ed.) IEEE 52nd Annual Symposium on Foundations of Computer Science, FOCS 2011, Palm Springs, CA, USA, 22–25 October 2011, pp. 97–106. IEEE Computer Society (2011)

[CGW14] Cabarcas, D., Göpfert, F., Weiden, P.: Provably secure LWE encryption with smallish uniform noise and secret. In: Emura, K., Hanaoka, G., Zhao, Y. (eds.) ASIAPKC 2014, Proceedings of the 2nd ACM Workshop on ASIA Public-Key Cryptography, 3 June 2014, Kyoto, Japan, pp. 33–42. ACM (2014)

[CHKP10] Cash, D., Hofheinz, D., Kiltz, E., Peikert, C.: Bonsai trees, or how to delegate a lattice basis. In: Gilbert, H. (ed.) EUROCRYPT 2010. LNCS, vol. 6110, pp. 523–552. Springer, Heidelberg (2010)

[DM13] Döttling, N., Müller-Quade, J.: Lossy codes and a new variant of the learning-with-errors problem. In: Johansson, T., Nguyen, P.Q. (eds.) EUROCRYPT 2013. LNCS, vol. 7881, pp. 18–34. Springer, Heidelberg (2013)

[Gen09a] Gentry, C.: A Fully Homomorphic Encryption Scheme. Ph.D. thesis, Stanford University (2009). crypto.stanford.edu/craig

[Gen09b] Gentry, C.: Fully homomorphic encryption using ideal lattices. In: Mitzenmacher, M. (ed.) Proceedings of the 41st Annual ACM Symposium on Theory of Computing, STOC 2009, Bethesda, MD, USA, 31 May - 2 June 2009, pp. 169–178. ACM (2009)

[GPV08] Gentry, C., Peikert, C., Vaikuntanathan, V.: Trapdoors for hard lattices and new cryptographic constructions. In: Dwork, C. (ed.) Proceedings of the 40th Annual ACM Symposium on Theory of Computing, Victoria, British Columbia, Canada, 17–20 May 2008, pp. 197–206. ACM (2008)

[Kil06] Kiltz, E.: Chosen-ciphertext security from tag-based encryption. In: Halevi, S., Rabin, T. (eds.) TCC 2006. LNCS, vol. 3876, pp. 581–600. Springer, Heidelberg (2006)

[KMO10] Kiltz, E., Mohassel, P., O'Neill, A.: Adaptive trapdoor functions and chosen-ciphertext security. In: Gilbert, H. (ed.) EUROCRYPT 2010. LNCS, vol. 6110, pp. 673–692. Springer, Heidelberg (2010)

[KMP14] Kiltz, E., Masny, D., Pietrzak, K.: Simple chosen-ciphertext security from low-noise LPN. In: Krawczyk, H. (ed.) PKC 2014. LNCS, vol. 8383, pp. 1–18. Springer, Heidelberg (2014)

[KTX07] Kawachi, A., Tanaka, K., Xagawa, K.: Multi-bit cryptosystems based on lattice problems. In: Okamoto, T., Wang, X. (eds.) PKC 2007. LNCS, vol. 4450, pp. 315–329. Springer, Heidelberg (2007)

[LP11] Lindner, R., Peikert, C.: Better key sizes (and attacks) for LWE-based encryption. In: Kiayias, A. (ed.) CT-RSA 2011. LNCS, vol. 6558, pp. 319–339. Springer, Heidelberg (2011)

[MM11] Micciancio, D., Mol, P.: Pseudorandom knapsacks and the sample complexity of LWE search-to-decision reductions. In: Rogaway, P. (ed.) CRYPTO 2011. LNCS, vol. 6841, pp. 465–484. Springer, Heidelberg (2011)

[MP12] Micciancio, D., Peikert, C.: Trapdoors for lattices: simpler, tighter, faster, smaller. In: Pointcheval, D., Johansson, T. (eds.) EUROCRYPT 2012. LNCS, vol. 7237, pp. 700–718. Springer, Heidelberg (2012)

[MP13] Micciancio, D., Peikert, C.: Hardness of SIS and LWE with small parameters. In: Canetti, R., Garay, J.A. (eds.) CRYPTO 2013, Part I. LNCS, vol. 8042, pp. 21–39. Springer, Heidelberg (2013)

[MR07] Micciancio, D., Regev, O.: Worst-case to average-case reductions based on gaussian measures. SIAM J. Comput. $37(1)$, 267–302 (2007)

[OPW11] O'Neill, A., Peikert, C., Waters, B.: Bi-deniable public-key encryption. In: Rogaway, P. (ed.) CRYPTO 2011. LNCS, vol. 6841, pp. 525–542. Springer, Heidelberg (2011)

[Pei09] Peikert, C.: Public-key cryptosystems from the worst-case shortest vector problem: extended abstract. In: Mitzenmacher, M. (ed.) Proceedings of the 41st Annual ACM Symposium on Theory of Computing, STOC 2009, Bethesda, MD, USA, 31 May - 2 June 2009, pp. 333–342. ACM (2009)

[Pei10] Peikert, C.: An efficient and parallel gaussian sampler for lattices. In: Rabin, T. (ed.) CRYPTO 2010. LNCS, vol. 6223, pp. 80–97. Springer, Heidelberg (2010)

[PVW08] Peikert, C., Vaikuntanathan, V., Waters, B.: A framework for efficient and composable oblivious transfer. In: Wagner, D. (ed.) CRYPTO 2008. LNCS, vol. 5157, pp. 554–571. Springer, Heidelberg (2008)

[PW08] Peikert, C., Waters, B.: Lossy trapdoor functions and their applications. In: Dwork, C. (ed.) Proceedings of the 40th Annual ACM Symposium on Theory of Computing, Victoria, British Columbia, Canada, 17–20 May 2008, pp. 187–196. ACM (2008)

[Reg05] Regev, O.: On lattices, learning with errors, random linear codes, and cryptography. In: Gabow, H.N., Fagin, R. (eds.) Proceedings of the 37th Annual ACM Symposium on Theory of Computing, Baltimore, MD, USA, 22–24 May 2005, pp. 84–93. ACM (2005)

[RS91] Rackoff, C., Simon, D.R.: Non-interactive zero-knowledge proof of knowledge and chosen ciphertext attack. In: Feigenbaum, J. (ed.) CRYPTO 1991. LNCS, vol. 576, pp. 433–444. Springer, Heidelberg (1992)

[RS09] Rosen, A., Segev, G.: Chosen-ciphertext security via correlated products. In: Reingold, O. (ed.) TCC 2009. LNCS, vol. 5444, pp. 419–436. Springer, Heidelberg (2009)

Zero Knowledge and Secure Computations

Slow Motion Zero Knowledge Identifying with Colliding Commitments

Houda Ferradi, Rémi Géraud$^{(\boxtimes)}$, and David Naccache

École normale supérieure, Équipe de cryptographie,
45 rue d'Ulm, 75230 Paris Cedex 05, France
{houda.ferradi,remi.geraud,david.naccache}@ens.fr

Abstract. Discrete-logarithm authentication protocols are known to present two interesting features: The first is that the prover's commitment, $x = g^r$, claims most of the prover's computational effort. The second is that x does not depend on the challenge and can hence be computed in advance. Provers exploit this feature by pre-loading (or pre-computing) ready to use commitment pairs r_i, x_i. The r_i can be derived from a common seed but storing each x_i still requires 160 to 256 bits when implementing DSA or Schnorr.

This paper proposes a new concept called *slow motion zero-knowledge* (SM-ZK). SM-ZK allows the prover to slash commitment size (by a factor of 4 to 6) by combining classical zero-knowledge and a timing channel. We pay the conceptual price of requiring the ability to measure time but, in exchange, obtain communication-efficient protocols.

1 Introduction

Authentication is a cornerstone of information security, and much effort has been put in trying to design efficient authentication primitives. However, even the most succinct authentication protocols require collision-resistant commitments. As proved by Girault and Stern [13], breaking beyond the collision-resistance size barrier is impossible. This paper shows that if we add the assumption that the verifier can measure the prover's response time, then commitment collision-resistance becomes unnecessary. We call this new construction *slow-motion zero knowledge* (SM-ZK).

As we will show, the parameter determining commitment size in SM-ZK protocols is the attacker's online computational power rather than the attacker's overall computational power. As a result, SM-ZK allows a significant reduction (typically by a factor of 4 to 6) of the prover's commitment size.

The prover's on-line computational effort remains unchanged (enabling instant replies in schemes such as GPS [12]). The prover's offline work is only slightly increased. The main price is paid by the verifier who has to solve a time-puzzle per session. The time taken to solve this time-puzzle determines the commitment's shortness.

The major contribution of this work is thus a technique forcing a cheating prover to either attack the underlying zero-knowledge protocol or exhaust the

© Springer International Publishing Switzerland 2016
D. Lin et al. (Eds.): Inscrypt 2015, LNCS 9589, pp. 381–396, 2016.
DOI: 10.1007/978-3-319-38898-4_22

space of possible replies in the presence of a time-lock function that slows down his operations. When this time-lock function is properly tuned, a simple time-out on the verifier's side rules out cheating provers. It is interesting to contrast this approach to the notion of *knowledge tightness* introduced by Goldreich, Micali and Widgerson [14], and generalizations such as *precise/local ZK* introduced by Micali and Pass [18], which uses similar time-constraint arguments but to prove reduced knowledge leakage bounds.

2   Building Blocks

SM-ZK combines two existing building blocks that we now recall: three-pass zero-knowledge protocols and time-lock functions.

2.1   Three-Pass Zero-Knowledge Protocols

A Σ-protocol [6, 15, 16] is a generic 3-step interactive protocol, whereby a prover $\mathcal{P}$ communicates with a verifier $\mathcal{V}$. The goal of this interaction is for $\mathcal{P}$ to convince $\mathcal{V}$ that $\mathcal{P}$ knows some value – without revealing anything beyond this assertion. The absence of information leakage is formalized by the existence of a simulator $\mathcal{S}$, whose output is indistinguishable from the recording (trace) of the interaction between $\mathcal{P}$ and $\mathcal{V}$.

The three phases of a Σ protocol can be summarized by the following exchanges:

$$\mathcal{P} \begin{array}{c} \xrightarrow{\quad x \quad} \\ \xleftarrow{\quad c \quad} \\ \xrightarrow{\quad y \quad} \end{array} \mathcal{V}$$

Namely,

- The prover sends a *commitment* x to the verifier;
- The verifier replies with a *challenge* c;
- The prover gives a *response* y.

Upon completion, $\mathcal{V}$ may accept or reject $\mathcal{P}$, depending on whether $\mathcal{P}$'s answer is satisfactory. Such a description encompasses well-known identification protocols such as Feige-Fiat-Shamir [10] and Girault-Poupard-Stern [11].

Formally, let R be some (polynomial-time) recognizable relation, then the set $L = \{v \text{ s.t. } \exists w, (v, w) \in R\}$ defines a *language*. Proving that $v \in L$ therefore amounts to proving knowledge of a witness w such that $(v, w) \in R$. A Σ-protocol satisfies the following three properties:

- *Completeness*: given an input v and a witness w such that $(v, w) \in R$, $\mathcal{P}$ is always able to convince $\mathcal{V}$.

– *Special honest-verifier zero-knowledge*[1]: there exists a probabilistic polynomial-time simulator S which, given v and a c, outputs triples (x, c, y) that have the same distribution as in a valid conversation between $\mathcal{P}$ and $\mathcal{V}$.
– *Special soundness*: given two accepting conversations for the same input v, with different challenges but an identical commitment x, there exists a probabilistic polynomial-time extractor procedure $\mathcal{E}$ that computes a witness w such that $(v, w) \in R$.

Many generalizations of zero-knowledge protocols have been discussed in the literature. One critical question for instance is to compose such protocols in parallel [14,18], or to use weaker indistinguishably notions (e.g., computational indistinguishability).

2.2 Commitment Pre-processing

Because the commitment x does not depend on the challenge c, authors quickly noted that x can be prepared in advance. This is of little use in protocols where the creation of x is easy (e.g., Fiat-Shamir [10]). Discrete-logarithm commitment pre-processing is a well-known optimization technique (e.g., [19,22]) that exploits two properties of DLP:

1. In DLP-based protocols, a commitment is generated by computing the exponentiation $x = g^r$ in a well-chosen group. This operation claims most of the prover's efforts.
2. The commitment x being unrelated to the challenge c, can hence be computed in advance. A "pre-computed commitment" is hence defined as $\{r, x\}$ computed in advance by $\mathcal{P}$[2]. Because several pre-computed commitments usually need to be saved by $\mathcal{P}$ for later use, it is possible to derive all the r_i components by hashing a common seed.

Such pre-processing is interesting as it enables very fast interaction between prover and verifier. While the technique described in this work does not require the use of pre-processing, it is entirely compatible with such optimizations.

2.3 Time-Lock Puzzles

Time-lock puzzles [17,21] are problems designed to guarantee that they will take (approximately) τ units of time to solve. Like proof-of-work protocols [8], time-locks have found applications in settings where delaying requests is desirable, such as fighting spam or denial-of-service attacks, as well as in electronic cash [1,7,9].

Time-lock puzzles may be based on computationally demanding problems, but not all such problems make good time-locks. For instance, inverting a weak

---

[1] Note that *special* honest-verifier zero-knowledge implies honest-verifier zero-knowledge.
[2] Or for $\mathcal{P}$ by a trusted authority.

one-way function would in general not provide a good time-lock candidate [21]. The intuition is that the time it takes to solve a time-lock should not be significantly reduced by using more computers (i.e., parallel brute-force) or more expensive machines.

A time-lock puzzle is informally described as a problem such that there is a super-polynomial gap between the work required to generate the puzzle, and the parallel time required to solve it (for a polynomial number of parallel processors). The following definition formalizes this idea [5].

Definition 1 (Time-Lock Puzzle). *A time-lock puzzle is the data two PPT algorithms $T_G(1^k, t)$ (problem generator) and $T_V(1^k, a, v)$ (solution verifier) satisfying the following properties:*

- *For every PPT algorithm $B(1^k, q, h)$, for all $e \in \mathbb{N}$, there exists $m \in \mathbb{N}$ such that*

$$\sup_{t \geq k^m, |h| \leq k^e} \Pr\left[(q, a) \leftarrow T_G(1^k, t) \text{ s.t. } T_V(1^k, a, B(1^k, q, h)) = 1\right]$$

is negl(k). Intuitively, T_G generates puzzles of hardness t, and B cannot efficently solve any puzzle of hardness $t \geq k^m$ for some constant m depending on B.

- *There is some $m \in \mathcal{N}$ such that, for every $d \in \mathcal{N}$, there is a PPT algorithm $C(1^k, t)$ such that*

$$\min_{t \leq k^d} \Pr\left[(q, a) \leftarrow T_G(1^k, t), v \leftarrow C(1^k, q) \text{ s.t. } T_V(1^k, a, v) = 1 \text{ and } |v| \leq k^m\right]$$

is overwhelming in k. Intuitively, this second requirement ensures that for any polynomial hardness value, there exists an algorithm that can solve any puzzle of that hardness.

Rivest, Shamir and Wagner [21], and independently Boneh and Naor [4] proposed a time-lock puzzle construction relying on the assumption that factorization is hard. This is the construction we retain for this work, and to the best of our knowledge the only known one to achieve interesting security levels. The original Rivest-Shamir-Wagner (RSW) time-lock [21] is based on the "intrinsically sequential" problem of computing:

$$2^{2^\tau} \bmod n$$

for specified values of τ and an RSA modulus n. The parameter τ controls the puzzle's difficulty. The puzzle can be solved by performing τ successive squares modulo n.

Using the formalism above, the RSW puzzle can be described as follows:

$$T_G(1^k, t) = ((p_1 p_2, \min(t, 2^k)), (p_1, p_2, \min(t, 2^k)))$$

$$T_V(1^k, (p_1, p_2, t'), v) = \begin{cases} 1 & \text{if } (v = v_1, v_2) \text{ and } v_1 = 2^{2^{t'}} \bmod n \text{ and } v_2 = n \\ 0 & \text{otherwise} \end{cases}$$

where p_1 and p_2 are $(k/2)$-bit prime numbers. Both solving the puzzle and verifying the solution can be efficiently done if p_1 and p_2 are known.

Good time-lock problems seem to be hard to find, and in particular there exist impossibility results against unbounded adversaries [17]. Nevertheless, the RSW construction holds under a computational assumption, namely that factorisation of RSA moduli is hard.

3   Slow Motion Zero-Knowledge Protocols

3.1   Definition

We can now introduce the following notion:

Definition 2 (SM-ZK). *A* Slow Motion Zero-Knowledge *(SM-ZK) protocol* $(\sigma, \mathcal{T}, \tau, \Delta_{max})$, *where* σ *defines a* Σ *protocol,* $\mathcal{T}$ *is a time-lock puzzle,* $\tau \in \mathbb{N}$, *and* $\Delta_{max} \in \mathbb{R}$, *is defined by the three following steps of* σ:

1. *Commitment:* $\mathcal{P}$ *sends a commitment* x *to* $\mathcal{V}$
2. *Timed challenge:* $\mathcal{V}$ *sends a challenge* c *to* $\mathcal{P}$, *and starts a timer.*
3. *Response:* $\mathcal{P}$ *provides a response* y *to* $\mathcal{V}$, *which stops the timer.*

$\mathcal{V}$ *accepts iif*

– *y is accepted as a satisfactory response by σ; and*
– *x is a solution to the time-lock puzzle $\mathcal{T}$ with input (y, c) and hardness τ; and*
– *time elapsed between challenge and response, as measured by the timer, is smaller than Δ_{max}.*

3.2   Commitment Shortening

Commitments in a Σ-protocol are under the control of $\mathcal{P}$, which may be malicious. If commitments are not collision-resistant, the protocol's security is weakened. Hence commitments need to be long, and in classical Σ protocols breaking below the collision-resistance size barrier is impossible as proved by [13].

However, as we now show, commitment collision-resistance becomes unnecessary in the case of SM-ZK protocols.

4   An Example SM-ZK

While SM-ZK can be instantiated with any three-pass ZK protocol, we will illustrate the construction using the Girault-Poupard-Stern (GPS) protocol [11, 12,20], and a modification of the time-lock construction due to Rivest, Shamir and Wagner [21].

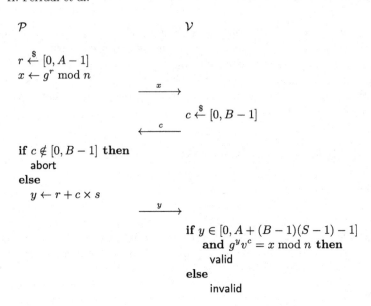

Fig. 1. Girault-Poupard-Stern identification protocol.

4.1 Girault-Poupard-Stern Protocol

GPS key generation consists in generating a composite modulus n, choosing a public generator $g \in [0, n-1]$ and integers A, B, S such that $A \gg BS$. Choice of parameters depends on the application and is discussed in [12]. Implicitly, parameters A, B, S are functions of the security parameter k.

The secret key is an integer $s \in [0, S-1]$, and the corresponding public key is $v = g^{-s} \bmod n$. Authentication is performed as in Fig. 1.

$\mathcal{P}$ can also precompute as many values $x_i \leftarrow g^{r_i}$ as suitable for the application, storing a copy of r_i for later usage. The detailed procedure by which this is done is recalled in Appendix B.

4.2 GPS-RSW SM-ZK

We can now combine the previous building-blocks to construct a pre-processing scheme that requires little commitment storage.

The starting point is a slightly modified version of the RSW time-lock function $\tau \mapsto 2^{2^\tau}$. Let μ be some deterministic function (to be defined later) and $\overline{n}$ an RSA modulus different from the n used for the GPS, we define for integers τ, ℓ:

$$f_{\tau,\ell}(x) = \left(\mu(x)^{2^\tau} \bmod \overline{n} \right) \bmod 2^\ell.$$

Here, τ controls the puzzle hardness and ℓ is a parameter controlling output size.

The function $f_{\tau,\ell}$ only differs from the RSW time-lock in two respects: We use $\mu(x)$ instead of 2; and the result is reduced modulo 2^ℓ.

The motivation behind using a function μ stems from the following observation: An adversary knowing $x_1^{2^\tau}$ and $x_2^{2^\tau}$ could multiply them to get $(x_1 x_2)^{2^\tau}$. To thwart such attacks (and similar attacks based on the malleability of RSA) we suggest to use for μ a deterministic RSA signature padding function (e.g., the Full Domain Hash [2]).

The reduction modulo 2^ℓ is of practical interest, it is meant to keep the size of answers manageable. Of course, an adversary could brute-force all values between 0 and $2^\ell - 1$ instead of trying to solve the time-lock. To avoid this situation, ℓ and τ should be chosen so that solving the time-lock is the most viable option of the two.

Under the same assumptions as RSW (hardness of factorization), and if ℓ and τ are properly tuned, $f_{\tau,\ell}$ generates a time-lock problem.

Then, we adapt a construction of M'Raïhi and Naccache [19] to GPS [11]. This is done by defining a secret J, a public hash function H, and computing the quantities:

$$x_i' = g^{H(J,i,s)} \bmod n$$

This computation can be delegated to a trusted authority. This is interesting in our case because the authority can compress these x_i' by computing $x_i = f_{\tau,\ell}(x_i')$. Note that because the authority knows the factors of $\overline{n}$, computing the x_i is fast. $\mathcal{P}$ is loaded with k pre-computed commitments $x_1, \ldots, x_k$ as shown in Fig. 2. The quantity k of pre-computed commitments depends on the precise application.

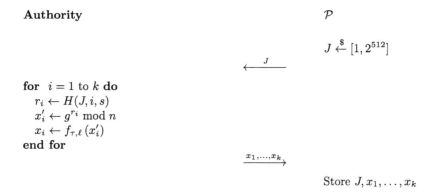

Fig. 2. Slow motion commitment pre-processing for GPS.

When $\mathcal{V}$ wishes to authenticate $\mathcal{P}$ the parties execute the protocol shown in Fig. 3.

With a proper choice of τ, ℓ we can have a reasonable verification time (assuming that $\mathcal{V}$ is more powerful than $\mathcal{P}$), extremely short commitments (e.g., 40-bit ones) and very little on-line computations required from $\mathcal{P}$.

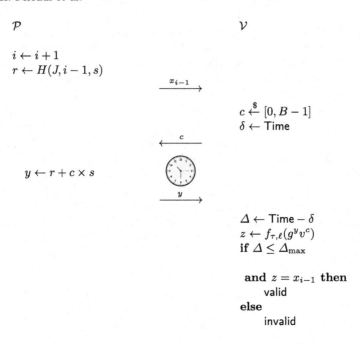

Fig. 3. Slow Motion GPS. Range tests on c and y omitted for the sake of clarity.

4.3   Choice of Parameters

What drives the choice of parameters is the ratio between:

- The time t it takes to a legitimate prover to compute y and transmits it. In GPS this is simply one multiplication of operands of sizes $\log_2 B$ and $\log_2 S$ (additions neglected), this takes time $\lambda \log(B) \log(S)$ for some constant λ (not assuming optimizations such as [3] based on the fact that operand s is constant).
- The time T it takes for the fastest adversary to evaluate once the time-lock function $f_{\tau,\ell}$. T does not really depend on ℓ, and is linear in τ. We hence let $T = \nu\tau$. Note that there is no need to take into account the size of $\bar{n}$, all we require from $\bar{n}$ is to be hard to factor. That way, the slowing effect will solely depend on τ.

In a brute-force attack, there are 2^ℓ possibilities to exhaust. The most powerful adversary may run $\kappa \leq 2^\ell$ parallel evaluations of the time-lock function, and succeed to solve the puzzle in t time units with probability

$$\epsilon = \frac{\kappa t}{2^\ell T} = \frac{\kappa \log(B) \log(S) \lambda}{\nu 2^\ell \tau}$$

A typical instance resulting in 40-bit commitments is $\{\kappa = 2^{24}, T = 1, t = 2^{-4}, \epsilon = 2^{-20}\} \Rightarrow \ell = 40$. Here we assume that the attacker has 16.7 million (2^{24}) computers capable of solving one time-lock challenge per second $(T = 1)$

posing as a prover responding in one sixteenth of a second ($t = 2^{-4}$). Assuming the least secure DSA parameters (160-bit q) this divides commitment size by 4. For 256-bit DSA the gain ratio becomes 6.4.

The time-out constant $\Delta_{\max}$ in Fig. 3 is tuned to be as small as possible, but not so short that it prevents legitimate provers from authenticating. Therefore the only constraint is that $\Delta_{\max}$ is greater or equal to the time t it takes to the *slowest legitimate prover* to respond. Henceforth we assume $\Delta_{\max} = t$.

5 Security Proof

The security of this protocol is related to that of the standard GPS protocol analysed in [12, 20]. We recall here the main results and hypotheses.

5.1 Preliminaries

The following scenario is considered. A randomized polynomial-time algorithm Setup generates the public parameters $(\mathcal{G}, g, S)$ on input the security parameter k. Then a second probabilistic algorithm GenKey generates pairs of public and private keys, sends the secret key to $\mathcal{P}$ while the related public key is made available to anybody, including of course $\mathcal{P}$ and $\mathcal{V}$. Finally, the identification procedure is a protocol between $\mathcal{P}$ and $\mathcal{V}$, at the end of which $\mathcal{V}$ accepts or not.

An adversary who doesn't corrupt public parameters and key generation has only two ways to obtain information: either passively, by eavesdropping on a regular communication, or actively, by impersonating (in a possibly non protocol-compliant way) $\mathcal{P}$ and $\mathcal{V}$.

The standard GPS protocol is proven complete, sound and zero-knowledge by reduction to the *discrete logarithm with short exponent problem* [12]:

Definition 3 (Discrete Logarithm with Short Exponent Problem). *Given a group $\mathcal{G}$, $g \in \mathcal{G}$, and integer S and a group element g^x such that $x \in [0, S-1]$, find x.*

5.2 Compressed Commitments for Time-Locked GPS

We now consider the impact of shortening the commitments to ℓ bits on security, while taking into account the time constraint under which $\mathcal{P}$ operates. The shortening of commitments will indeed weaken the protocol [13] but this is compensated by the time constraint, as explained below.

Lemma 1 (Completeness). *Execution of the protocol of Fig. 3 between a prover $\mathcal{P}$ who knows the secret key corresponding to his public key, and replies in bounded time $\Delta_{\max}$, and a verifier $\mathcal{V}$ is always successful.*

Proof. This is a direct consequence of the completeness of the standard GPS protocol [12, Theorem 1]. By assumption, $\mathcal{P}$ computes y and sends it within the time allotted for the operation. This computation is easy knowing the secret s and we have

$$g^y v^c = g^{r_i + cs} v^c = x_i' g^{cs} v^c = x_i' v^{c-c} = x_i'$$

Consequently, $f_{\tau,\ell}(g^y v^c) = f_{\tau,\ell}(x_i') = x_i$. Finally,

$$y = r + cs \leq (A-1) + (B-1)(S-1) < y_{\max}.$$

Therefore all conditions are met and the identification succeeds.   □

Lemma 2 (Zero-Knowledge). *The protocol of Fig. 3 is statistically zero-knowledge if it is run a polynomial number of times N, B is polynomial, and NSB/A is negligible.*

Proof. The proof follows [12] and can be found in Appendix A.

The last important property to prove is that if $\mathcal{V}$ accepts, then with overwhelming probability $\mathcal{P}$ must know the discrete logarithm of v in base g.

Lemma 3 (Time-Constrained Soundness). *Under the assumption that the discrete logarithm with short exponent problem is hard, and the time-lock hardness assumption, this protocol achieves time-constrained soundness.*

Proof. After a commitment x has been sent, if $\mathcal{A}$ can correctly answer with probability $> 1/B$ then he must be able to answer to two different challenges, c and c', with y and y' such that they are both accepted, i.e., $f_{\tau,\ell}(g^y v^c) = x = f_{\tau,\ell}(g^{y'} v^{c'})$. When that happens, we have

$$\mu \left(g^y v^c\right)^{2^\tau} = \mu \left(g^{y'} v^{c'}\right)^{2^\tau} \mod \bar{n} \mod 2^\ell$$

Here is the algorithm that extracts these values from the adversary $\mathcal{A}$. We write $\mathrm{Success}(\omega, c_1, \ldots, c_n)$ the result of the identification of $\mathcal{A}$ using the challenges $c_1, \ldots, c_n$, for some random tape ω.

Step 1. Pick a random tape ω and a tuple c of N integers $c_1, \ldots, c_N$ in $[0, B-1]$. If $\mathrm{Success}(\omega, c) = \mathsf{false}$, then abort.

Step 2. Probe random N-tuples c' that are different from each other and from c, until $\mathrm{Success}(\omega, c') = \mathsf{true}$. If after $B^N - 1$ probes a successful c' has not been found, abort.

Step 3. Let j be the first index such that $c_j \neq c_j'$, write y_j and y_j' the corresponding answers of $\mathcal{A}$. Output c_j, c_j', y_j, y_j'.

This algorithm succeeds with probability $\geq \epsilon - 1/B^N = \epsilon'$, and takes at most $4\Delta_{\max}$ units of time [12]. This means that there is an algorithm finding collisions in $f_{\tau,\ell}$ with probability $\geq \epsilon'$ and time $\leq 4\Delta_{\max}$.

Assuming the hardness of the discrete logarithm with short exponents problem, the adversary responds in time by solving a hard problem, where as pointed out earlier the probability of success is given by

$$\zeta = \frac{\kappa \log(B) \log(S) \lambda}{\nu 2^\ell \tau}$$

where κ is the number of concurrent evaluations of $f_{\tau,\ell}$ performed by $\mathcal{A}$. There is a value of τ such that $\zeta \ll \epsilon$. For this choice of τ, $\mathcal{A}$ is able to compute $f_{\tau,\ell}$ much faster than brute-force, which contradicts the time-lock hardness assumption. $\square$

6   Conclusion and Further Research

This paper introduced a new class of protocols, called Slow Motion Zero Knowledge (SM-ZK) showing that if we pay the conceptual price of allowing time measurements during a three-pass ZK protocol then commitments do not need to be collision-resistant.

Because of its interactive nature, SM-ZK does not yield signatures but seems to open new research directions. For instance, SM-ZK permits the following interesting construction, that we call a *fading signature*: Alice wishes to send a signed message m to Bob without allowing Bob to keep a long-term her involvement. By deriving $c \leftarrow H(x, m, \rho)$ where ρ is a random challenge chosen by Bob, Bob can convince himself[3] that m comes from Alice. This conviction is however not transferable if Alice prudently uses a short commitment as described in this paper.

A   Proof of Lemma 2

Proof. The zero-knowledge property of the standard GPS protocol is proven by constructing a polynomial-time simulation of the communication between a prover and a verifier [12, Theorem 2]. We adapt this proof to the context of the proposed protocol. The function δ is defined by $\delta(\text{true}) = 1$ and $\delta(\text{false}) = 0$, and $\wedge$ denotes the logical operator "and". For clarity, the function $f_{\tau,\ell}$ is henceforth written f.

The scenario is that of a prover $\mathcal{P}$ and a dishonest verifier $\mathcal{A}$ who can use an adaptive strategy to bias the choice of the challenges to try to obtain information about s. In this case the challenges are no longer chosen at random, and this must be taken into account in the security proof. Assume the protocol is run N times and focus on the i-th round.

$\mathcal{A}$ has already obtained a certain amount of information η from past interactions with $\mathcal{P}$. $\mathcal{P}$ sends a pre-computed commitment x_i. Then $\mathcal{A}$ chooses a commitment using all information available to her, and a random tape ω: $c_i (x_i, \eta, \omega)$.

The following is an algorithm (using its own random tape ω_M) that simulates this round:

Step 1. Choose $\overline{c_i} \xleftarrow{\$} [0, B-1]$ and $\overline{y_i} \xleftarrow{\$} [(B-1)(S-1), A-1]$ using ω_M.
Step 2. Compute $\overline{x_i} = f_{\tau,\ell}\left(g^{\overline{y_i}} v^{\overline{c_i}}\right)$.

---
[3] If y was received before $\Delta_{\max}$.

Step 3. If $c_i\left(\overline{x_i}, \eta, \omega\right) = \overline{c_i}$ then return to step 1 and try again with another pair $\left(\overline{c_i}, \overline{y_i}\right)$, else return $\left(\overline{x_i}, \overline{c_i}, \overline{y_i}\right)$.[4]

The rest of the proof shows that, provided $\varPhi = (B-1)(S-1)$ is much smaller than A, this simulation algorithm outputs triples that are indistinguishable from real ones, for any fixed random tape ω.

Formally, we want to prove that

$$\Sigma_1 = \sum_{\alpha,\beta,\gamma} \left| \Pr_{\omega_P}\left[(x,c,y) = (\alpha,\beta,\gamma)\right] - \Pr_{\omega_M}\left[(\overline{x},\overline{c},\overline{y}) = (\alpha,\beta,\gamma)\right] \right|$$

is negligible, i.e., that the two distributions cannot be distinguished by accessing a polynomial number of triples (even using an infinite computational power). Let (α,β,γ) be a fixed triple, and assuming a honest prover, we have the following probability:

$$p = \Pr_{\omega_P}\left[(x,c,y) = (\alpha,\beta,\gamma)\right]$$

$$= \Pr_{0 \le r < A}\left[\alpha = f(g^r) \wedge \beta = c(\alpha,\eta,\omega) \wedge \gamma = r + \beta s\right]$$

$$= \sum_{r=0}^{A-1} \frac{1}{A}\delta\left(\alpha = f(g^\gamma v^\beta) \wedge \beta = c(\alpha,\eta,\omega) \wedge r = \gamma - \beta s\right)$$

$$= \frac{1}{A}\delta\left(\alpha = f(g^\gamma v^\beta) \wedge \beta = c(\alpha,\eta,\omega) \wedge \gamma - \beta s \in [0, A-1]\right)$$

$$= \frac{1}{A}\delta\left(\alpha = f(g^\gamma v^\beta)\right)\delta\left(\beta = c(\alpha,\eta,\omega)\right)\delta\left(\gamma - \beta s \in [0, A-1]\right).$$

where $f = f_{\tau,\ell}$.

We now consider the probability $\overline{p} = \Pr_{\omega_M}\left[(\overline{x},\overline{c},\overline{y}) = (\alpha,\beta,\gamma)\right]$ to obtain the triple (α,β,γ) during the simulation described above. This is a conditional probability given by

$$\overline{p} = \Pr_{\substack{\overline{y}\in[\varPhi,A-1] \\ \overline{c}\in[0,B-1]}}\left[\alpha = f\left(g^{\overline{y}}v^{\overline{c}}\right) \wedge \beta = \overline{c} \wedge \gamma = \overline{y} \mid \overline{c} = c\left(f\left(g^{\overline{y}}v^{\overline{c}}\right),\eta,\omega\right)\right]$$

Using the definition of conditional probabilities, this equals

$$\overline{p} = \frac{\displaystyle\Pr_{\substack{\overline{y}\in[\varPhi,A-1] \\ \overline{c}\in[0,B-1]}}\left[\alpha = f\left(g^{\overline{y}}v^{\overline{c}}\right) \wedge \beta = \overline{c} \wedge \gamma = \overline{y}\right]}{\displaystyle\Pr_{\substack{\overline{y}\in[\varPhi,A-1] \\ \overline{c}\in[0,B-1]}}\left[\overline{c} = c\left(f\left(g^{\overline{y}}v^{\overline{c}}\right),\eta,\omega\right)\right]}$$

Let us introduce

$$Q = \sum_{\substack{\overline{y}\in[\varPhi,A-1] \\ \overline{c}\in[0,B-1]}} \delta\left(\overline{c} = c\left(f\left(g^{\overline{y}}v^{\overline{c}}\right),\eta,\omega\right)\right)$$

---

[4] The probability of success at step 3 is essentially $1/B$, and the expected number of executions of the loop is B, so that the simulation of N rounds runs in $O(NB)$: the machine runs in expected polynomial time.

then the denominator in $\bar{p}$ is simply $Q/B(A - \Phi)$. Therefore:

$$\bar{p} = \sum_{\bar{c} \in [0, B-1]} \frac{1}{B} \Pr_{\bar{y} \in [\Phi, A-1]} \left[\alpha = f\left(g^{\bar{y}} v^{\bar{c}}\right) \wedge \gamma = \bar{y} \wedge \beta = \bar{c} = c(\alpha, \eta, \omega)\right] \frac{B(A - \Phi)}{Q}$$

$$= \Pr_{\bar{y} \in [\Phi, A-1]} \left[\alpha = f\left(g^{\gamma} v^{\beta}\right) \wedge \gamma = \bar{y} \wedge \beta = c(\alpha, \eta, \omega)\right] \frac{A - \Phi}{Q}$$

$$= \sum_{\bar{y} \in [\Phi, A-1]} \frac{1}{A - \Phi} \delta\left(\alpha = f\left(g^{\gamma} v^{\beta}\right) \wedge \gamma = \bar{y} \wedge \beta = c(\alpha, \eta, \omega)\right) \frac{A - \Phi}{Q}$$

$$= \frac{1}{Q} \delta\left(\alpha = f\left(g^{\gamma} v^{\beta}\right)\right) \delta\left(\beta = c(\alpha, \eta, \omega)\right) \delta\left(\gamma \in [\Phi, A-1]\right)$$

We will now use the following combinatorial lemma:

Lemma 4. *If $h : \mathcal{G} \to [0, B-1]$ and $v \in \{g^{-s}, s \in [0, S-1]\}$ then the total number M of solutions $(c, y) \in [0, B-1] \times [\Phi, A-1]$ to the equation $c = h(g^y v^c)$ satisfies $A - 2\Phi \leq M \leq A$.*

Proof. (Proof of Lemma 4) [12, Appendix A]. Specialising Lemma 4 to the function that computes $c(f(g^{\bar{y}} v^{\bar{c}}), \eta, \omega)$ from $(\bar{c}, \bar{y})$ gives $A - 2\Phi \leq Q \leq A$. This enables us to bound Σ_1:

$$\Sigma_1 = \sum_{\alpha, \beta, \gamma} \left| \Pr_{\omega_P}\left[(x, c, y) = (\alpha, \beta, \gamma)\right] - \Pr_{\omega_M}\left[(\bar{x}, \bar{c}, \bar{y}) = (\alpha, \beta, \gamma)\right]\right|$$

$$= \sum_{\alpha, \beta, \gamma \in [\Phi, A-1]} \left| \Pr_{\omega_P}\left[(x, c, y) = (\alpha, \beta, \gamma)\right] - \Pr_{\omega_M}\left[(\bar{x}, \bar{c}, \bar{y}) = (\alpha, \beta, \gamma)\right]\right|$$

$$+ \sum_{\alpha, \beta, \gamma \notin [\Phi, A-1]} \Pr_{\omega_P}\left[(x, c, y) = (\alpha, \beta, \gamma)\right]$$

$$= \sum_{\substack{\gamma \in [\Phi, A-1] \\ \beta \in [0, B-1] \\ \alpha = f(g^{\gamma} v^{\beta})}} \left| \frac{1}{A} \delta\left(\beta = c(\alpha, \eta, \omega)\right) - \frac{1}{Q} \delta\left(\beta = c(\alpha, \eta, \omega)\right)\right|$$

$$+ \left(1 - \sum_{\alpha, \beta, \gamma \in [\Phi, A-1]} \Pr_{\omega_P}\left[(x, c, y) = (\alpha, \beta, \gamma)\right]\right)$$

$$= \left|\frac{1}{A} - \frac{1}{Q}\right| Q + 1 - \sum_{\substack{\gamma \in [\Phi, A-1] \\ \beta \in [0, B-1] \\ \alpha = f(g^{\gamma} v^{\beta})}} \frac{1}{A} \delta\left(\beta = c(\alpha, \eta, \omega)\right)$$

$$= \frac{|Q - A|}{A} + 1 - \frac{Q}{A}$$

Therefore $\Sigma_1 \leq 2|Q - A|/A \leq 4\Phi/A < 4SB/A$, which proves that the real and simulated distributions are statistically indistinguishable if SB/A is negligible. $\qquad \square$

B   GPS Commitment Pre-computation

Figure 4 described one possible way in which pre-computed commitments are generated and used for GPS. In this figure, we delegate the computation to a trusted authority. That role can be played by $\mathcal{P}$ alone, but we leverage the authority to alleviate $\mathcal{P}$'s computational burden.

To efficiently generate a sequence of commitments, the authority uses a shared secret seed J and a cryptographic hash function H. Here J is chosen by $\mathcal{P}$ but it could be chosen by the authority instead.

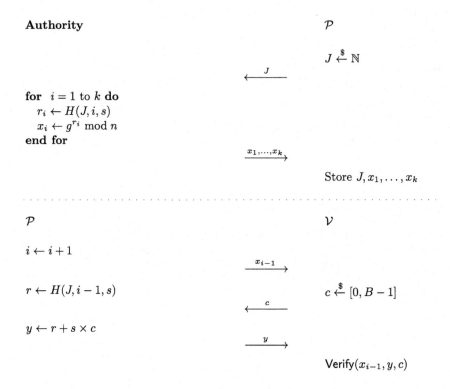

Fig. 4. Commitment pre-processing as applied to GPS. The first stage describes the preliminary interaction with a trusted authority, where pre-computed commitments are generated and stored. The second stage describes the interaction with a verifier. For the sake of clarity the range-tests on c and y were omitted. The trusted authority can be easily replaced by $\mathcal{P}$ himself.

References

1. Abadi, M., Burrows, M., Manasse, M.S., Wobber, T.: Moderately hard, memory-bound functions. ACM Trans. Internet Technol. 5(2), 299–327 (2005)
2. Bellare, M., Rogaway, P.: Random oracles are practical: a paradigm for designing efficient protocols. In: Denning, D.E., Pyle, R., Ganesan, R., Sandhu, R.S., Ashby, V. (eds.) CCS 1993, Proceedings of the 1st ACM Conference on Computer and Communications Security, Fairfax, Virginia, USA, 3–5 November, 1993, pp. 62–73. ACM (1993)
3. Bernstein, R.L.: Multiplication by integer constants. Softw. Pract. Exper. 16(7), 641–652 (1986)
4. Boneh, D., Naor, M.: Timed commitments. In: Bellare, M. (ed.) CRYPTO 2000. LNCS, vol. 1880, p. 236. Springer, Heidelberg (2000)
5. Ciobotaru, O.: On the (Non-)Equivalence of UC security notions. In: Takagi, T., Wang, G., Qin, Z., Jiang, S., Yu, Y. (eds.) ProvSec 2012. LNCS, vol. 7496, pp. 104–124. Springer, Heidelberg (2012)
6. Damgård, I.: On Σ Protocols (2010). http://www.cs.au.dk/~ivan/Sigma.pdf
7. Dwork, C., Goldberg, A.V., Naor, M.: On memory-bound functions for fighting spam. In: Boneh, D. (ed.) CRYPTO 2003. LNCS, vol. 2729, pp. 426–444. Springer, Heidelberg (2003)
8. Dwork, C., Naor, M.: Pricing via processing or combatting junk mail. In: Brickell, E.F. (ed.) CRYPTO 1992. LNCS, vol. 740, pp. 139–147. Springer, Heidelberg (1993)
9. Dwork, C., Naor, M., Wee, H.M.: Pebbling and proofs of work. In: Shoup, V. (ed.) CRYPTO 2005. LNCS, vol. 3621, pp. 37–54. Springer, Heidelberg (2005)
10. Feige, U., Fiat, A., Shamir, A.: Zero-knowledge proofs of identity. J. Cryptology 1(2), 77–94 (1988)
11. Girault, M.: An identity-based identification scheme based on discrete logarithms modulo a composite number. In: Damgård, I.B. (ed.) EUROCRYPT 1990. LNCS, vol. 473, pp. 481–486. Springer, Heidelberg (1991)
12. Girault, M., Poupard, G., Stern, J.: On the fly authentication and signature schemes based on groups of unknown order. J. Cryptology 19(4), 463–487 (2006)
13. Girault, M., Stern, J.: On the length of cryptographic hash-values used in identification schemes. In: Desmedt, Y.G. (ed.) CRYPTO 1994. LNCS, vol. 839, pp. 202–215. Springer, Heidelberg (1994)
14. Goldreich, O., Micali, S., Wigderson, A.: Proofs that yield nothing but their validity for all languages in NP have zero-knowledge proof systems. J. ACM 38(3), 691–729 (1991)
15. Goldwasser, S., Micali, S., Rackoff, C.: The knowledge complexity of interactive proof-systems (extended abstract). In: Sedgewick, R. (ed.) Proceedings of the 17th Annual ACM Symposium on Theory of Computing, 6–8 May, 1985, Providence, Rhode Island, USA, pp. 291–304. ACM (1985)
16. Hazay, C., Lindell, Y.: Efficient secure two-party protocols: techniques and constructions. Springer Science and Business Media, Heidelberg (2010)
17. Mahmoody, M., Moran, T., Vadhan, S.: Time-lock puzzles in the random oracle model. In: Rogaway, P. (ed.) CRYPTO 2011. LNCS, vol. 6841, pp. 39–50. Springer, Heidelberg (2011)
18. Micali, S., Pass, R.: Local zero knowledge. In: Kleinberg, J.M. (ed.) Proceedings of the 38th Annual ACM Symposium on Theory of Computing, Seattle, WA, USA, 21–23 May, 2006, pp. 306–315. ACM (2006)

19. M'Raïhi, D., Naccache, D.: Couponing scheme reduces computational power requirements for dss signatures. In: Proceedings of CardTech/SecurTech, pp. 99–104 (1994)
20. Poupard, G., Stern, J.: Security analysis of a practical "On the Fly" authentication and signature generation. In: Nyberg, K. (ed.) EUROCRYPT 1998. LNCS, vol. 1403, pp. 422–436. Springer, Heidelberg (1998)
21. Rivest, R., Shamir, A., Wagner, D.: Time-lock puzzles and timed-release crypto, technical report, MIT/LCS/TR-684 (1996)
22. de Rooij, P.: On schnorr's preprocessing for digital signature schemes. J. Cryptology **10**(1), 1–16 (1997)

Multi-client Outsourced Computation

Peili Li[1,2,3], Haixia Xu[1,2(✉)], and Yuanyuan Ji[1,2,3]

[1] State Key Laboratory of Information Security,
Institute of Information Engineering of Chinese Academy of Sciences,
Beijing, China
{plli,hxxu}@is.ac.cn, jiyuanyuan@iie.ac.cn
[2] Data Assurance and Communication Security Research Center
of Chinese Academy of Sciences, Beijing, China
[3] University of Chinese Academy Sciences, Beijing, China

Abstract. In this paper, we study multi-client outsourced computation where n computationally weak clients outsource their computation of a function f over their joint inputs $(x_1, \cdots, x_n)$ to remote servers. Some prior works consider outsourcing computation to an untrusted server. However these schemes either are inefficient, make the clients' status unequal or require client interactions. Based on prior works, we construct an efficient multi-client outsourced computation scheme using two servers. Our scheme avoids interactions among all the parties and it is secure against one malicious server. Furthermore it is public verifiable that any client can verify the correctness of the computation result using a public verify key.

Keywords: Multi-client · Outsourced computation · Cloud computing · Privacy · Efficient · Public verification

1 Introduction

Cloud computing has been rapidly developed recent years. The cloud not only offer data storage service but also help companies or users to accomplish their computation tasks. This trend contributes to the study of outsourced computation where a computationally weak client outsources its computation of a function f to a powerful server [6]. Outsourced computation has several real applications. For example, in cloud computing, businesses outsource their computation to a service and pay the computing time rather than purchasing and maintaining the computing devices. It helps companies save costs and can mainly focus on their key services. Outsourced computation is also useful in weak mobile devices such as smart phones, net books that would like to outsource a computation(e.g., a photo manipulation) to a remote server.

Some earlier works about outsourced computation consider single-client scenario [2,4–6,12] where the functionality only works on one client's input. While in some cases, several resource-constrained clients may want to jointly do some

© Springer International Publishing Switzerland 2016
D. Lin et al. (Eds.): Inscrypt 2015, LNCS 9589, pp. 397–409, 2016.
DOI: 10.1007/978-3-319-38898-4_23

computation. It would be desirable to study multi-client outsourced computation where n clients wish to outsource the computation of a function f over a series of joint inputs $(x_1, \cdots, x_n)$ to a remote server [3].

In outsourced computation, we can not ensure that the remote server is honest. Thus two security problems should be considered: one is keeping privacy of the outsourced data from the server, the other one is verifying the correctness of the computation result returned from the untrusted server. In addition to these two problems, in multi-client outsourced computation we should consider how to keep privacy of the clients' inputs from each other. Furthermore another problem is whether the verification of the computation can be public: any other client can verify the correctness of the computation result returned from the server using a public verify key [5]. This is important, for example, a lab assistant outsourced a computation on input x, he may need to produce a verification key that will let the patients obtain the answer from the cloud and check its correctness. In this paper, we mainly study the problems of multi-client outsourced computation and construct an efficient non-interactive multi-client outsourced computation.

1.1 Related Work

Multi-client Outsourced Computation. Choi et al. [3] first introduced the notion of multi-client verifiable computation(MVC) and constructed a MVC scheme secure against a malicious server. It is efficient in the sense of amortization. Later Gordon et al. [9] designed a MVC scheme with stronger security guarantees (satisfies a simulation-based notion of security). In their scheme the clients' cost depends on the depth of the function f being computed. Goldwasser et al. [7] showed that multi-input functional encryption can be used to design publicly verifiable multi-client outsourcing scheme(any other client can verify the correctness of computation result using a public verify key). These three works talked above all can keep privacy of the clients inputs and are secure against a malicious server. However, in both [3,9]'s schemes, only one client can do the verification and get the computation result in every execution, which makes the clients' status unequal. Although [7]'s construction is public verifiable, its privacy property relies on a completely honest client who generates the master secret key(MSK). The cost of the clients in these above three works is closely related to the function f being computed. López-Alt et al. [11] proposed a multi-key homomorphic encryption scheme which can be used in multi-client outsourced computation, but the clients need to interactively compute the decryption key in order to get the computation result.

In order to improve the clients' efficiency and avoid interactions among the clients, Peter et al. [13] designed an efficient multi-client outsourcing scheme using BCP encryption scheme and two noncolluding servers. What the clients need to do is just the encryption and decryption process, thus their cost is independent of the function f being computed. However their scheme is secure in the semi-honest model and complex interactions are needed between these two servers. Utilizing Proxy Re-Encryption scheme Wang et al. [16] constructed an efficient multi-client outsourced computation scheme that requires minimal interactions between these two servers. Similar to [13,16]'s works, [15] constructed a multi-client outsourced computation scheme using lattice-based encryption scheme and two noncolluding servers. These three works all avoid interactions among the clients and the clients' computation cost is independent of the function f being computed, but they are only secure in the semi-honest model and need server interactions. Jakobsen et al. [14] designed a framework for outsourcing multi-client computation to multiple servers. It is secure when at least one of the server is honest. Their solution is generic and can be instantiated with reactive Multiparty Computation(MPC) protocol. However,their work is only secret verifiable and the servers in [14] also need to do interactions in order to accomplish the computation.

In this work, we aim to design an efficient multi-client outsourced computation scheme that avoids interactions among the clients and the servers. And we want to make the scheme public verifiable such that all the clients can verify and obtain the computation result during one execution. Furthermore we try to make the scheme secure against a malicious server. Our basic idea comes from Choi et al.'s work [3], so we describe it in details next.

Details About Choi et al.'s Work. In their scheme, to securely compute a function f, one client(suppose client 1) generates a garbled circuit Γ of f and sends Γ to the server. Only after receiving the labels corresponding to the clients' inputs $(x_1, \cdots, x_n)$, can the server compute the garbled circuit Γ. While only client 1 who generates the garbled circuit has the input-wire labels, but it does not know the inputs of other clients and it can not send all the input-wire labels to anyone else. Thus how can the server obtain the labels corresponding to other clients' inputs? To solve this problem, They used a tool called proxy oblivious transfer protocol(proxy OT). In proxy OT, client 1 acts as a sender, client $i(i \neq 1)$ acts as a chooser and the server is a proxy. After the protocol, the proxy can only learn the labels corresponding to chooser's input and the chooser's input is kept private. This is a novel idea to extend Gennaro et al.'s one-client outsourced computation scheme [6] to multi-client outsourced computation scheme. In [3]'s construction, only the client who generates the garbled circuit can verify and get the computation result $f(x_1, \cdots, x_n)$. These n clients need to run the scheme n times in order to ensure all of them can verify and get the output. Thus all the clients need to generate the garbled circuit of function f, and the computation cost of the clients are closely related to the function f being computed.

1.2 Our Contribution

We design an efficient multi-client outsourced computation scheme using two noncolluding servers. In our scheme, the clients' inputs are keeping private from each other and the two servers. Our scheme has these following features:

- The computation cost of the clients is independent of the function f being computed. See Table 1 below.

Table 1. Comparisons with related work

Work	Client's cost
Choi et al. [6]	Only efficient in amortization
Goldwasser et al. [8]	$O(|sk_f| \cdot lk)$
Gordon et al. [7]	$O(d \cdot nlk)$
Our work	$O(lk)$

- Our scheme avoids interactions among all the parties (the clients and two servers).
- Our outsourced computation scheme is publicly verifiable that any other client can verify the correctness of the computation result.
- It is secure against one malicious server in the two-server setting(we assume that the other server is semi-honest).

In Table 1, d is the depth of the function f being computed, l is the input length and k is the security parameter. sk_f is the secret key corresponding to function f in [8]'s multi-input functional encryption scheme.

1.3   Paper Organization

The rest of this paper is organized as follows. In Sect. 2 we introduce the tool that will be used in our scheme. In Sect. 3 we talk about the model and techniques used in our scheme. In Sect. 4 we show the construction of multi-client outsourced computation scheme. Finally, in Sect. 5 we make the conclusion.

2   Preliminaries

We denote k the security parameter, $negl(\cdot)$ a negligible function and PPT the probabilistic polynomial time. Denote $Dom(F)$ the domain of the function set F and $\prod$ the protocol of multi-client outsourced computation scheme.

2.1   Garbled Circuits

Garbled circuits were first presented by Yao [17] in the context of secure two-party computation and were proven secure by Lindell and Pinkas [10]. The notion was formalized by Bellare et al. [1] and was simplified by Goldwasser et al. [8]. In this paper, we refer to [8]'s definition of the garbling scheme for simplicity.

Definition 1 (Garbling Scheme). [8] $\mathcal{C}_n$ is the set of circuit taking as input n bits. A garbling scheme for a family of circuits $\mathcal{C} = \{\mathcal{C}_n\}_{n \in N}$, is a tuple of PPT algorithms $Gb = (Gb.Garble, Gb.Enc, Gb.Eval, Gb.Dec)$ such that

1. $Gb.Garble(1^k, C)$ takes as input the security parameter k and a circuit $C \in \mathcal{C}_n$ for some n, and outputs the garbled circuit Γ and a secret key sk.

2. $Gb.Enc(sk, x)$ *takes as input the secret key* sk *and data* $x \in \{0,1\}^n$ *and outputs an encoding* c.
3. $Gb.Eval(\Gamma, c)$ *takes as input a garbled circuit* Γ, *an encoding* c *and outputs a value* y.
4. $Gb.Dec(d, Y)$ *takes as input decoding key* d *and the garbled output* Y, *maps* Y *to a final output* y.

Correctness. *For any polynomial* $n(\cdot)$, *for all sufficiently large security parameters* k, *for* $n = n(k)$, *for all circuits* $C \in \mathcal{C}_n$ *and all* $x \in \{0,1\}^n$, $Pr[(\Gamma, sk) \leftarrow Gb.Garble(1^k, C); c \leftarrow Gb.Enc(sk, x); y \leftarrow Gb.Eval(\Gamma, c) : C(x) = y] = 1 - negl(k)$.

Input and Circuit Private. *If there exist a ppt simulator* Sim_{Garble}, *such that for every ppt adversaries* A *and* D, *for all sufficiently large security parameters* k, *the following equation holds. Then the garbled scheme* Gb *for a family of circuits* $\{\mathcal{C}_n\}_{n \in N}$ *is input and circuit private.*

$$|Pr[(x, C, \alpha) \leftarrow A(1^k); (\Gamma, sk) \leftarrow Gb.Garble(1^k, C);$$
$$c \leftarrow Gb.Enc(sk, x) : D(\alpha, x, C, \Gamma, c)]$$
$$-Pr[(x, C, \alpha) \leftarrow A(1^k); (\tilde{\Gamma}, \tilde{c})$$
$$\leftarrow Sim_{Garble}(1^k, C(x), 1^{|C|}, 1^{|x|}) : D(\alpha, x, C, \tilde{\Gamma}, \tilde{c})]| = negl(k)$$

Additionally we require that the garbling scheme satisfies authenticity [3].

Authenticity. *We say that a garbling scheme provides authenticity if for any PPT adversary learning a set of labels to some input* x *is unable to produce a set of labels that corresponds to an output different from* $f(x)$.

The garbling scheme used in our protocol is a projective garbling scheme which means that the encryption algorithm treat the input in pieces instead of processing the whole input at once. In our work, each encoding only depends on one bit of the input. For example, let $x = x_1, \cdots, x_n \in \{0,1\}^n$, the encryption algorithm needs to generate encoding for each $x_i (i \in [n])$. In the rest of this paper, the word 'label' denotes the encoding in the garbling scheme.

3   Our Model and Techniques

3.1   Multi-client Outsourced Computation Scheme

Refer to the description of Choi et al.'s work [3], we give a model of multi-client outsourced computation scheme in the two-server setting. We denote the two servers S_1, S_2. Fig. 1 gives the model of the multi-client outsourced computation in two-server setting.

In Fig. 1, Z and VK_y are the messages returned by the two servers.

Security Model: In the two-server setting, we consider the security against one malicious server(S_2) and one semi-honest server(S_1). Our scheme assumes that

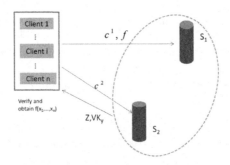

Fig. 1. Multi-client outsourced computation

the clients know which server is semi-honest in advance. Informally, we say that a multi-client outsourced computation scheme is secure if it can keep privacy the clients' inputs from each other and the two servers, and a malicious server S_2 can not forge an incorrect message Z that makes the clients accept. In this paper, we analyze its security using the real-ideal paradigm.

In the real world, all parties together with these two servers run the protocol $\prod$ to get the desired output. During the protocol, all the clients and the server S_1 follow the protocol semihonestly. An adversary $\mathcal{A}$ can corrupt server S_2 and attempts to gain more information about the clients' inputs and forge an incorrect computation result that can pass the verification.

In the ideal world, the clients send their inputs and a function f to a trusted third party(TTP). Then TTP will do the computation of function f on the clients' inputs $(x_1, \cdots, x_n)$ and obtain $f(x_1, \cdots, x_n)$. After that, TTP will send function f to the ideal world simulator $\mathcal{S}$. If $\mathcal{S}$ returns *yes*, then TTP will send $f(x_1, \cdots, x_n)$ to the clients. Otherwise, TTP will return *abort*.

We denote $\text{IDEAL}_{\mathcal{F},\mathcal{S}}(x_1, \cdots, x_n)$ the joint execution of f under the Simulator $\mathcal{S}$ in the ideal world on input $(x_1, \cdots, x_n)$. $\text{REAL}_{\prod,\mathcal{A}}(x_1, \cdots, x_n)$ the joint execution of $\prod$ under $\mathcal{A}$ in the real world on input $(x_1, \cdots, x_n)$.

Definition 2 (Multi-client Outsourced Computation). *A multi-client outsourced computation scheme in the two server setting consists the following four algorithms:*

1. *KeyGen(1^k) $\rightarrow$ (PK, SK): Takes input the security parameter k, the algorithm generates a key pair (PK, SK).*
2. *ProbGen($PK, f, x_1, \cdots, x_n$) $\rightarrow$ (c^1, c^2): Takes input the public key PK, function f being computed, and n clients' inputs ($x_1, \cdots, x_n$), the algorithm computes ciphertext c^1 and c^2, sends c^1 and f to server S_1 and c^2 to server S_2.*
3. *Compute(PK, c^1, c^2, f) $\rightarrow$ (VK_y, Z): Given the public key PK, server S_1 with input (c^1, f) and server S_2 with input c^2 corporately compute a message Z and a verify-message VK_y, and then send them to the clients.*
4. *Verify(VK_y, Z) $\rightarrow$ (y, reject): Using public verify message VK_y, the algorithm verifies the correctness of Z. If the verification passed, the corresponding result y can be recovered after the verification. Otherwise reject.*

Correctness: *A multi-client outsourced computation scheme is correct if for any function $f \in F$ and any $(SK, PK) \leftarrow KeyGen(1^k, f)$, any $(x_1, \cdots, x_n) \in Dom(F)$, if $(c^1, c^2) \leftarrow ProbGen(PK, x_1, \cdots, x_n)$ and $(VK_y, Z) \leftarrow Compute(PK, c^1, c^2, f)$, then $y = f(x_1, \cdots, x_n) \leftarrow Verify(VK_y, Z)$ holds with all but negligible probability.*

Security: *A multi-client outsourced computation scheme is secure if for any PPT adversary $\mathcal{A}$ in the real world, there exists a PPT simulator S in the ideal world such that for all $(x_1, \cdots, x_n)$ no PPT distinguisher D is able to distinguish $IDEAL_{\mathcal{F},S}(x_1, \cdots, x_n)$ from $REAL_{\prod,\mathcal{A}}(x_1, \cdots, x_n)$.*

3.2   Techniques

In this part, we talk about the techniques in our protocol. Our basic idea comes from Choi et al.'s work [3]. In Sect. 1.1 we have showed that the computation cost of the clients are closely related to the function f being computed.

To reduce the computation cost of the clients, we use a semi-honest server S_1 to generate the garbled circuit Γ of function f. One directly thinking is just let the semi-honest server S_1 to compute the garbled circuit with the labels corresponding to clients' inputs and send the computation result to the clients. While in this situation, the clients' inputs can not be kept private from S_1 for the garbled circuit is generated by S_1. Thus we need to separate the garbled circuit generation process from the computation process of the garbled circuit. Let the computation process done by another server S_2.

In our construction, let S_1 send Γ to the other server S_2. Then S_2 computes the garbled circuit Γ with the labels corresponding to $(x_1, \cdots, x_n)$. For S_1 and S_2 do not know the inputs of the clients and only S_1 know all the input-wire labels. How can S_2 get the labels corresponding to $(x_1, \cdots, x_n)$? A nature thought is to directly use [3]'s proxy OT protocol. Each client together with S_1, S_2 involved in the proxy OT protocol. While in proxy OT each client needs to run a key-exchange protocol with S_1. Every time, the clients and the server S_1 need to run the setup algorithm to generate their key pairs. To reduce the key generation process, we design a new protocol π for choosing labels.

Choose Labels. In protocol π, we use a public key encryption (PKE) scheme to encrypt the randomness elements and a one-time pad encryption scheme to encrypt the client's input. The protocol π is designed as followings:

- π.Setup(1^k): Server S_1 runs a public key encryption scheme(PKE) to generate its key pair (PK, SK).
- π.Client(x, PK): The client with its input $x(|x| = l)$. It chooses $2l$ random elements $(r_1^0, r_1^1, \cdots, r_l^0, r_l^1)$ with the same length k and a random element t ($|t| = l$). Denote $\mathbf{r} = (r_1^0, r_1^1, \cdots, r_l^0, r_l^1)$. The client computes c_1=PKE.Enc$_{PK}(\mathbf{r}, t)$ and $c_2 = (x \oplus t, r_1^{x_1}, \cdots, r_l^{x_l})$. Sends c_1 to server S_1 and c_2 to server S_2.
- π.S_1($c_1, \mathbf{k}, SK$): S_1 holds the labels $\mathbf{k} = (k_1^0, k_1^1, \cdots, k_l^0, k_l^1)$. S_1 decrypts c_1 using its secret key SK and obtains $(\mathbf{r}, t)$. Computes $w_j^{t_j} = r_j^0 \oplus k_j^0$, $w_j^{1 \oplus t_j} = r_j^1 \oplus k_j^1$ for all $j \in [l]$. Let $\mathbf{w} = (w_1^0, w_1^1, \cdots, w_l^0, w_l^1)$ and sends $\mathbf{w}$ to server S_2.

– $\pi.S_2(\mathbf{w}, c_2)$: $\mathbf{w} = (w_1^0, w_1^1, \cdots, w_l^0, w_l^1)$, parse c_2 as $(d, r_1^{x_1}, \cdots, r_l^{x_l})$. S_2 computes $k_j = w_j^{d_j} \oplus r_j^{x_j}$ for all $j \in [l]$.

The above protocol π acts as a sub-protocol of our multi-client outsourcing scheme $\prod$. In our work, let each client run the sub-protocol π with the two servers S_1 and S_2 synchronously. After executing the protocol, server S_2 can get the labels corresponding to the inputs of all clients.

Public Verification. After solving the problem of how to choose labels, server S_2 can compute the garbled circuit Γ. Another problem is how to verify the correctness of the result returned from S_2. Similar to the idea of [6], the output of the garbled circuit are the labels corresponding to the output bits. We use the garbling scheme that satisfies authenticity property. The decoding key d consists of the wire values chosen. After receiving a message Z returned from the server, the client maps the message Z to an output y using the decoding key d. If the map process passed, we say that server honestly do the computation of function f.

In this paper, We aim to design a public verify method that all the clients can verify the correctness of the computation result and obtain $f(x_1, \cdots, x_n)$. Observe that the output of the garbled circuit Γ is the label corresponding to $f(x_1, \cdots, x_n)$(For simplicity, we assume that the output of the function f is one bit). Let k_y^0, k_y^1 be the labels of the output-wire. If $f(x_1, \cdots, x_n) = 0$, Γ outputs label k_y^0, else if $f(x_1, \cdots, x_n) = 1$ outputs k_y^1. Let server S_1 compute a one-way function g on k_y^0 and k_y^1 and make $g(k_y^0), g(k_y^1)$ public. After server S_2 returns message Z to the clients, all the clients can compute $g(Z)$ and verify whether it is equal to $g(k_y^0)$ or $g(k_y^1)$. If $g(Z) = g(k_y^0)$, outputs 0, if $g(Z) = g(k_y^1)$, outputs 1. Otherwise *reject*.

4   Our Construction

Now we are ready to describe our construction of multi-client outsourced computation scheme(denote $\prod$) in the two-server setting. We use a garbling scheme and the protocol π designed in Sect. 3 as our tools. Suppose server S_1 is semi-honest. The scheme $\prod$ is designed as followings:

1. KeyGen(1^k) $\rightarrow$ (PK, SK): The semi-honest server S_1 runs π.Setup algorithm and obtains a key pair (PK, SK).
2. ProbGen($PK, f, x_1, \cdots, x_n$) $\rightarrow$ (c^1, c^2): Each client i ($i \in [n]$) runs π.Client algorithm with its input x_i to creates its ciphertext $c_i = (c_i^1, c_i^2)$, and sends c_i^1 and function f to server S_1, sends c_i^2 to server S_2. Let $c^1 = (c_1^1, \cdots, c_n^1)$, $c^2 = (c_1^2, \cdots, c_n^2)$.
3. Compute(PK, SK, c^1, c^2, f) $\rightarrow$ Z: The computation process can be divided into two parts that are done by S_1 and S_2 respectively.
   – After receiving $c^1 = (c_1^1, \cdots, c_n^1)$ and function f, S_1 first computes the garbled circuit Γ of function f. Parse (k_y^0, k_y^1) as the labels of the output-wire. S_1 computes $VK_y = (g(k_y^0), g(k_y^1))$ and makes them public(here g is a one-way function).

S_1 having SK runs $\pi.S_1$ algorithm n time to generate messages $\tau_1, \cdots, \tau_n$ respectively. τ_i is the output of $\pi.S_1$ algorithm with inputs $(c_i^1, \mathbf{k_i})$ (here $\mathbf{k_i}$ are the labels of client i's input-wires). After that S_1 sends $\tau = (\tau_1, \cdots, \tau_n)$ and Γ to S_2.

- After receiving $c^2 = (c_1^2, \cdots, c_n^2)$ from the clients and (τ, Γ) from S_1, the second server S_2 runs $\pi.S_2$ algorithm respectively to obtain the labels corresponding to each client's input. Using these labels, S_2 computes the garbled circuit Γ to obtain the output message Z and sends Z to the clients.

4. Verify(VK_y, Z) $\rightarrow$ $(y, reject)$: Using public verify message $VK_y = (g(k_y^0), g(k_y^1))$, each client computes $g(Z)$ and verifies whether it is equal to $g(k_y^0)$ or $g(k_y^1)$. If $g(Z) = g(k_y^0)$, outputs 0, if $g(Z) = g(k_y^1)$, outputs 1. Otherwise $reject$.

Efficiency Analysis. From the above construction, we can see that the clients only need to run the PKE scheme to generate their ciphertexts and then later run the one-way function to verify the correctness of the computation result. The computation cost of the clients is only depend on the their input length l and the security parameter k. The efficiency comparison with other verifiable outsourced computation schemes is given in Table 1 in Sect. 1.2.

4.1   Correctness and Security Proof

Theorem 1 (Correctness). *If the underlying garbling scheme and PKE scheme are correct, then the above scheme $\prod$ is a correct multi-client outsourced computation scheme.*

Proof: In our construction, we use the sub-protocol π to choose the labels corresponding to the clients' inputs. For the PKE and one-time pad encryption scheme in π is correct, it is easy to verify that the output of S_2 in the sub-protocol π are $(k_1^{x_1}, \cdots, k_l^{x_l})$ which are the labels corresponding to x. And we use garbling scheme to securely compute function f on the inputs $(x_1, \cdots, x_n)$. Due to the correctness of the garbling scheme, the output of our scheme $\prod$ is $f(x_1, \cdots, x_n)$. Thus our scheme is a correct multi-client outsourced computation scheme.

Theorem 2 (Security). *Suppose the garbling scheme we use is input and circuit private and provides authenticity, the PKE scheme is secure, and the one-time pad encryption scheme is secure, then our scheme $\prod$ is secure against a malicious server(S_2) and a semi-honest server(S_1).*

Proof: We prove the security using the real-ideal model described in Sect. 2 (refer to Fig. 2). $\mathcal{A}$ is an PPT adversary that corrupts server S_2.

We say that a multi-client outsourced computation scheme is secure if for any PPT adversary $\mathcal{A}$ in the real world there exists a PPT simulator $\mathcal{S}$ with black-box access to $\mathcal{A}$ such that for all $(x_1, \cdots, x_n)$ no PPT environment D is able to distinguish $\text{IDEAL}_{\mathcal{F},\mathcal{S}}(x_1, \cdots, x_n)$ from $\text{REAL}_{\prod,\mathcal{A}}(x_1, \cdots, x_n)$. The Simulator $\mathcal{S}$ in the ideal world acts as following(refer to Fig. 3 for better understanding):

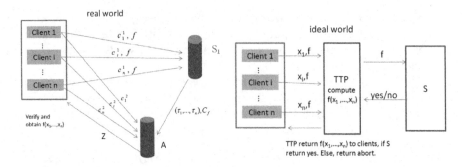

Fig. 2. Real-ideal model

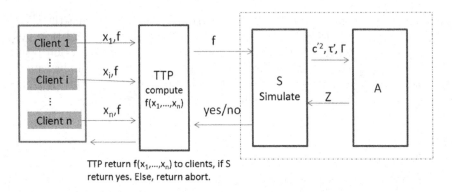

Fig. 3. Ideal world

The Simulator S

- After receiving function f from the trusted third party(TTP), the simulator S generates a garbled circuit Γ of function f. It then runs the π.client algorithm n times each with input 0^l to generates $c'^1_i(i = 1, \cdots, n)$ and $c'^2_i(i = 1, \cdots, n)$. Using the labels of the garbled circuit and the randomness chosen in running π.Client algorithm, the simulator can directly compute $\tau' = (\tau'_1, \cdots, \tau'_n)$. Let $c'^1 = (c'^1_1, \cdots, c'^1_n)$ and $c'^2 = (c'^2_1, \cdots, c'^2_n)$. Involve $\mathcal{A}$ with inputs (c'^2, τ', Γ).
- S generates the public verify message VK_y and then verifies the correctness of the output of $\mathcal{A}$ by running the verify algorithm. If the verification passed, the simulator S will return *yes* to TTP, otherwise, return *no*.
- The output of the simulator is $(c'^1, c'^2, f, \tau', \Gamma, \mathcal{A}(c'^2, \tau', \Gamma))$.

Next we show that $\text{IDEAL}_{\mathcal{F},\mathcal{S}}(x_1, \cdots, x_n)$ is indistinguishable from $\text{REAL}_{\Pi,\mathcal{A}}$ $(x_1, \cdots, x_n)$.

$\text{REAL}_{\Pi,\mathcal{A}}(x_1, \cdots, x_n)$ equals $(\bot, c^1, c^2, f, \tau, \Gamma, \mathcal{A}(c^2, \tau, \Gamma))$ if Verify algorithm return $\bot$. Otherwise $\text{REAL}_{\Pi,\mathcal{A}}(x_1, \cdots, x_n)$ equals $(\text{OUTPUT}^{\Pi}(x_1, \cdots, x_n), c^1, c^2, f, \tau, \Gamma, \mathcal{A}(c^2, \tau, \Gamma))$. Here $\text{OUTPUT}^{\Pi}(x_1, \cdots, x_n)$ denotes the output of the clients.

IDEAL$_{\mathcal{F},\mathcal{S}}(x_1,\cdots,x_n)$ equals $(\bot, S(f))$ if the simulator S returns no, otherwise IDEAL$_{\mathcal{F},\mathcal{S}}(x_1,\cdots,x_n)$ equals $(f(x_1,\cdots,x_n), S(f))$. For the simulator S has been constructed above, $S(f)$ equals $(c'^1, c'^2, f, \tau', \Gamma, \mathcal{A}(c'^2, \tau', \Gamma))$.

We define the following games:

Game 0 is the real world experiment.

Game 1 is the same as Game 0 except that the c^1 is the encryption of the randomness chosen by the simulator S.

Game 2 is the same as Game 1 expect that c^2 is the encryption of $(0^l,\cdots,0^l)$.

Game 3 is the ideal world experiment.

Lemma 1. Assume that the PKE scheme is secure, Game 0 and Game 1 are computationally indistinguishable.

Proof: The difference between Game 0 and Game 1 is that c^1 is the PKE encryption of two different messages. If there exist a PPT distinguisher D that can distinguish these two games, the there exist a PPT adversary B that can break the security of the PKE scheme.

Lemma 2. Assume that the one-time pad encryption scheme is secure and the garbled circuit is input and circuit private, Game 2 and Game 1 are computationally indistinguishable.

Proof: The only difference between Game 2 and Game 1 is how c^2 computed. In Game 1, c^2 is the encryption of $(x_1,\cdots,x_n)$, while in Game 2 c^2 is the encryption of $(0^l,\cdots,0^l)$. For the garbling scheme is input and circuit private, thus the garbled circuit do not reveal any information about the messages. For these messages $(x_1,\cdots,x_n)$ and $(0^l,\cdots,0^l)$ are encrypted by the one-time pad encryption algorithm respectively, if there exist a PPT distinguisher that can distinguish these two games, then there exist a PPT adversary B that can break the security of the one-time pad encryption scheme.

Lemma 3. Assume the garbling scheme satisfies authenticity property, then Game 3 and Game 2 are computational indistinguishable.

Proof: The garbling scheme satisfies authenticity property. If the output of the semi-honest clients in Game 2 is $\bot$, which means that the malicious adversary $\mathcal{A}$ forged an incorrect answer Z. In this case, the Simulator S in the ideal world will return no to TTP, thus the clients in the ideal world will also output $\bot$. In Game 2, if the output of the semi-honest clients return $\mathrm{OUTPUT}^\Pi(x_1,\cdots,x_n)$, it must be $f(x_1,\cdots,x_n)$ due to the authenticity of the garbling scheme. In this case, the simulator S will return yes to TTP, then the output of the clients in the ideal world is $f(x_1,\cdots,x_n)$. The distribution of Game 2 and Game 3 is indistinguishable. From the above lemmas, we get that Game 0 the real world execution is indistinguishable from Game 3 the ideal world execution.

From the above lemmas, we get that Game 0 the real world execution is indistinguishable from Game 3 the ideal world execution. $\qquad\qquad\square$

5   Conclusion

In this paper, we design an efficient multi-client outsourced computation scheme in the two-server setting model. It avoids interactions among all the parties (the clients and servers), and it is public verifiable that any client can verify the correctness of the computation result using a public verify key. The computation cost of the clients is independent from function being computed. Our scheme is proved to be correct and secure against a malicious server and a semi-honest server. In the future, it will be interesting to explore an efficient non-interactive multi-client outsourced computation scheme that is secure against two malicious servers.

Acknowledgement. This work is supported by the National Natural Science Foundation of China (No. 61379140), the National Basic Research Program of China (973 Program) (No. 2013CB338001). The authors would like to thank anonymous reviewers for their helpful comments and suggestions.

References

1. Bellare, M., Hoang, V.T., Rogaway, P.: Garbling schemes. IACR Cryptology ePrint Archive, vol. 2012, p. 265 (2012)
2. Benabbas, S., Gennaro, R., Vahlis, Y.: Verifiable delegation of computation over large datasets. In: Rogaway, P. (ed.) CRYPTO 2011. LNCS, vol. 6841, pp. 111–131. Springer, Heidelberg (2011)
3. Choi, S.G., Katz, J., Kumaresan, R., Cid, C.: Multi-client non-interactive verifiable computation. In: Sahai, A. (ed.) TCC 2013. LNCS, vol. 7785, pp. 499–518. Springer, Heidelberg (2013)
4. Chung, K.-M., Kalai, Y., Vadhan, S.: Improved delegation of computation using fully homomorphic encryption. In: Rabin, T. (ed.) CRYPTO 2010. LNCS, vol. 6223, pp. 483–501. Springer, Heidelberg (2010)
5. Dario, F., Rosario, G.: Publicly verifiable delegation of large polynomials and matrix computations, with applications. In: 2012 ACM Conference on Computer and Communication Security. ACM Press (2012). http://eprint.iacr.org/2012/281
6. Gennaro, R., Gentry, C., Parno, B.: Non-interactive verifiable computing: outsourcing computation to untrusted workers. In: Rabin, T. (ed.) CRYPTO 2010. LNCS, vol. 6223, pp. 465–482. Springer, Heidelberg (2010)
7. Goldwasser, S., Goyal, V., Jain, A., Sahai, A.: Multi-input functional encryption. IACR Cryptology ePrint Archive, vol. 2013, p. 727 (2013)
8. Goldwasser, S., Kalai, Y., Popa, R.A., Vaikuntanathan, V., Zeldovich, N.: Reusable garbled circuits and succinct functional encryption. In: Proceedings of the Forty-Fifth Annual ACM Symposium on Theory of Computing, pp. 555–564. ACM (2013)
9. Gordon, S.D., Katz, J., Liu, F.-H., Shi, E., Zhou, H.-S.: Multi-client verifiable computation with stronger security guarantees. In: Dodis, Y., Nielsen, J.B. (eds.) TCC 2015, Part II. LNCS, vol. 9015, pp. 144–168. Springer, Heidelberg (2015)
10. Lindell, Y., Benny, P.: A proof of security of Yao's protocol for two-party computation. J. Cryptol. **22**(2), 161–188 (2009)

11. López-Alt, A., Tromer, E., Vaikuntanathan, V.: On-the-fly multiparty computation on the cloud via multikey fully homomorphic encryption. In: Proceedings of the Forty-Fourth Annual ACM Symposium on Theory of Computing, pp. 1219–1234. ACM (2012)
12. Parno, B., Raykova, M., Vaikuntanathan, V.: How to delegate and verify in public: verifiable computation from attribute-based encryption. In: Cramer, R. (ed.) TCC 2012. LNCS, vol. 7194, pp. 422–439. Springer, Heidelberg (2012)
13. Peter, A., Tews, E., Katzenbeisser, S.: Efficiently outsourcing multiparty computation under multiple keys. IEEE Trans. Inf. Forensics Secur. **8**(12), 2046–2058 (2013)
14. Jakobsen, T.P., Nielsen, J.B., Orlandi, C.: A framework for outsourcing of secure computation. In: ACM Cloud Computing Security Workshop (2014)
15. Sun, Y., Wen, Q., Zhang, Y., Zhang, H., Jin, Z., Li, W.: Two-cloud-servers-assisted secure outsourcing multiparty computation. Scientific World J. **2014**, 7 (2014)
16. Wang, B., Li, M., Chow, SSM., Li, H.: Computing encrypted cloud data efficiently under multiple keys. In: IEEE Conference on Communications and Network Security (CNS), pp. 504–513. IEEE (2013)
17. Yao, A.C.: Protocols for secure computations. In: SFCS 1982 23rd Annual Symposium on Foundations of Computer Science, pp. 160–164, November 1982

Software and Mobile Security

Privacy-Enhanced Data Collection Scheme for Smart-Metering

Jan Hajny$^{(\boxtimes)}$, Petr Dzurenda, and Lukas Malina

Department of Telecommunications, Brno University of Technology, Technicka 12,
616 00 Brno, Czech Republic
{hajny,dzurenda,malina}@feec.vutbr.cz
http://crypto.utko.feec.vutbr.cz

Abstract. New types of devices, such as smart-meters, wearables and home appliances, have been connected to the Internet recently. Data they send is usually very privacy sensitive, containing personal information about, e.g., household consumption, health status or behavior profiles of family members. In this paper, we propose a cryptographic scheme for the protection of data collection systems that is secure (in the sense of data authenticity and integrity) and privacy-friendly at the same time. This functionality is achieved by designing a novel group signature that provides signature anonymity, unlinkability and untraceability while retaining features for malicious user identification. Besides the full cryptographic specification, we also provide implementation results that confirm the computational efficiency of the scheme allowing easy deployment on existing devices.

Keywords: Data collection · Sensors · Smart-metering · Privacy · Attributes · IoT · Security · Cryptography

1 Introduction

There are many applications where secure data collection from remote devices is important. The examples are temperature sensors, human activity and health sensors, machine operation status sensors, home automation equipment or implants in future. Although our scheme can be used in all these applications, we'll use the smart-metering as the example throughout the paper. The reason for our choice is that smart-metering is becoming a significant topic world-wide and a large-scale deployment of intelligent gas/water/electricity consumption meters to households is expected in a very short time. The legislation, security profiles and certification procedures are being finalized. The purpose of this paper is to show that not only the security but also privacy protection is important in data collection systems and that there are actually technical means for achieving strong privacy protection in secure data collection systems.

The scheme presented in this paper is a specific version of a group signature. It makes it possible that data originating from a smart-meter is digitally signed

© Springer International Publishing Switzerland 2016
D. Lin et al. (Eds.): Inscrypt 2015, LNCS 9589, pp. 413–429, 2016.
DOI: 10.1007/978-3-319-38898-4_24

at the device so that it cannot be altered or manipulated during the transfer. However, the signature only reveals that the data is originating from a group of authorized meters, the concrete identification number (ID) of the meter is not disclosed. To further improve privacy protection, all signatures of a single meter are mutually unlinkable. If the owner wishes so, it cannot be distinguished, whether the signed data is coming from a single smart-meter or different smart-meters. That allows energy suppliers to have a statistical overview about the consumption in a given area but prevents them from linking the consumption profile to concrete users. If a user wishes so, e.g., if smart-meters are used not only for statistical data collection but also for billing, the identity might be disclosed by the smart-meter using standard methods.

In contrast to most existing schemes, the scheme proposed here contains features for the practical revocation of invalid users and the identification of malicious users. In case the collector needs to reveal the signer's identity, he may ask a special entity that is able to de-anonymize the group signature and disclose the smart-meter's ID to the collector. However, such an action must be supported by reasonable evidence.

1.1   Model Scenario

Today's smart-meters have the ability to upload measured data to central storage. The data is used for statistical reasons, network planning, load balancing or the identification of potential problems in distribution networks. Data needs to be anonymized to preserve privacy of users. However, it is necessary to prevent malicious users from flooding the central storage with false, misleading data. The scheme presented in this paper provides mechanisms assuring that all messages delivered to the central storage are authentic (originating from real, authorized smart-meters) and integral (have not been tampered during transfer) without disclosing the identity of users. Each smart-meter has its own private key that is used to prove that it belongs to the group of authorized smart-meters. If the owner's identity needs to be disclosed (a serious problem is discovered from data measured, messages are coming from stolen/malicious meters, etc.), it is still possible, however this feature is strongly protected against misuse by cryptographic mechanisms and available only to certain authorities.

1.2   Related Work

In existing data collection schemes, privacy is usually protected by two main mechanisms, i.e., pseudonyms and group signatures.

Privacy preserving solutions based on pseudonyms have been proposed, e.g., in [14,28,29]. Some schemes need additional pseudonimization infrastructure, e.g. [29]. The work [28] uses anonymous certificates which are stored in a tamper-proof device. This approach uses a set of temporal pseudonyms and fast changing of these pseudonyms provides the privacy protection. Nevertheless, this approach is burdened by the pre-loading and storing of a large number of anonymous certificates with pseudonyms. For low-performance devices like smart-meters, the management of pseudonyms becomes a too complex operation.

The second approach is based on general group signatures. They provide user anonymity by producing message signatures on behalf of a group. Generally, group signatures guarantee the anonymity of honest users and the traceability of misbehaving users. A large number of group signatures has been proposed, e.g., in [2,3,5,10,12,20,21]. In particular, the scheme called BBS [2] serves as a fundamental building block for many security solutions (e.g., [23,31]). However most schemes lack efficient revocation mechanisms that allow the removal and de-anonymization of invalid users. Mechanisms for revocation have been proposed for schemes based on bilinear pairings [4,24]. Nevertheless, these schemes need several expensive operations, mostly bilinear pairings, modular exponentiations and multiplications and are not appropriate for applications using computationally very restricted devices such as smart-meters, wearables, wireless sensors, etc. Currently, achieving reasonable privacy of honest users and the revocability and de-anonymization of malicious users is an unresolved problem.

In addition to group signatures, the signatures based on anonymous credentials [9,17,19,26] are also an option. However, these schemes are more focused on privacy protection in access-control applications and lack features for efficient revocation and identification of malicious users too [22].

Finally, some of the schemes analyzed [1,18] rely on the hardware tamper-resistance. However, it is unrealistic to expect the presence of a tamper-proof key storage when using sensors and wearables.

1.3   Our Contribution

In this paper, we propose a novel cryptographic data collection (DC) scheme addressing the above identified weaknesses of existing systems. In particular, our scheme is designed to provide the following features.

- **Provable Security:** our DC scheme is based on provably secure cryptographic primitives, particularly the interactive proofs of knowledge about discrete logarithms [30] in the Okamoto-Uchiyama (OU) group [25]. We provide the full security analysis in Sect. 4.
- **Privacy Protection Features:** the DC scheme provides features for the protection of users' privacy and digital identity, namely the hiding of user IDs and the unlinkability and untraceability of signatures.
- **Revocation of Invalid Users:** it is possible to efficiently revoke invalid users without affecting remaining users.
- **Identification of Malicious Users:** using existing schemes, it is very difficult (if not impossible) to simultaneously provide privacy-enhancing features (signatures anonymity, untraceability and unlinkabilty) and efficient identification of malicious users. In our scheme, it is possible to obtain all these features without negative effects on communication and computation efficiency.
- **No Tamper-Proof Devices:** the scheme has cryptographic protection against collusion attacks. The user's private signing key is stored only in

user's device and cannot be used for collusion attacks even if extracted by malicious users. That makes the implementation on available hardware lacking the tamper-proof storage much easier.

- **Computational and Communication Efficiency:** our DC scheme is computationally efficient enough to run on low-resource devices like smart-meters and micro-controllers.

2 Scheme Description

In this section, we present the overview of our scheme and describe the entities and cryptographic protocols.

2.1 Notation

For various proofs of knowledge or representation, we use the efficient notation introduced by Camenisch and Stadler [7]. The protocol for proving the knowledge of discrete logarithm of c with respect to g is denoted as $PK\{\alpha : c = g^{\alpha}\}$. The proof of discrete log equivalence with respect to different generators g_1, g_2 is denoted as $PK\{\alpha : c_1 = g_1^{\alpha} \wedge c_2 = g_2^{\alpha}\}$. A signature by a traditional signature scheme (e.g., RSA) of a user U on some data is denoted as $Sig_U(data)$. The symbol ":" means "such that", "|" means "divides", "$|x|$" is the bitlength of x and "$x \in_R \{0,1\}^l$" is a randomly chosen bitstring of maximum length l.

2.2 Scheme Overview

There are three types of entities in our scheme: Meter, Collector and Revocation Authority.

- **Meter (M):** the Meter is the source of data. It acquires data from measurement, signs it using it's private key and sends it to the Collector. Meters represent users in our scheme.
- **Collector (C):** is the destination where all data is collected. It receives data from all Meters, verifies the group signatures and further processes them. The Collector is assured that data is coming from a predefined group of Meters, however the concrete IDs of the Meters remain hidden.
- **Revocation Authority (RA):** does the initial setup of the system including system parameters generation and participates on the management of meters, particularly on adding new Meters and removing invalid Meters from the system. RA participates on the de-anonymization of malicious Meters. RA is a semi-trusted authority as it can revoke Meters but is unable to de-anonymize Meters by itself only.

The entities engage in the following protocols.

- $(params, K_{RA}, K_C) \leftarrow$ Setup (k, l, m): is run by the Revocation Authority and the Collector. It inputs security parameters (k, l, m) and outputs system parameters $params$, RA's private key K_{RA} and Collector's private key K_C.

- $K_M \leftarrow$ getKey $(params, K_C, K_{RA})$: is run by the Revocation Authority, the Collector and the Meter. It inputs system parameters $params$, RA's and C's private keys K_{RA} and K_C and outputs the private key of a new Meter K_M. K_M is the Meter's private output of the protocol, no other entity learns the value of K_M although they cooperate on its generation.
- $GS_{K_M}(data) \leftarrow$ GroupSign $(params, data, K_M)$: is run by the Meter. The algorithm inputs system parameters $params$, $data$ and Meter's private key K_M and outputs the group signature $GS_{K_M}(data)$.
- $\{0, 1\} \leftarrow$ GroupVerify $(params, data, GS_{K_M}(data))$: is run by the Collector. The algorithm inputs the system parameters $params$, $data$ and Meter's group signature $GS_{K_M}(data)$ and outputs 1 if the signature is valid and the Meter has not been revoked.
- $rev \leftarrow$ Revoke $(params, GS_{K_M}(data), K_{RA})$: is run by the Collector and Revocation Authority. In case Meter needs to be removed from the system or a signature must be de-anonymized due to some misbehavior, the Collector sends the signature to Revocation Authority that revokes the Meter by putting its key-derivate rev on a public blacklist. The signatures of blacklisted Meters are always rejected by the Collectors.
- $ID_M \leftarrow$ Identify $(params, GS_{K_M}(data), K_{RA}, K_C)$: is run by the Collector and Revocation Authority. The protocol inputs the system parameters $params$, the group signature $GS_{K_M}(data)$ and private keys K_{RA}, K_C and outputs the Meter's identifier ID_M as C's private output. Only C learns the Meter's ID but RA must cooperate on that disclosure.

The overview of the architecture of the scheme is depicted in Fig. 1.

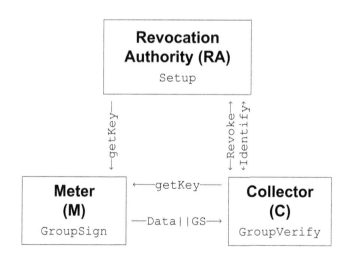

Fig. 1. Architecture of DC scheme proposed.

2.3   Description of Cryptographic Protocols

We provide the full description of all cryptographic protocols, i.e., Setup, getKey, GroupSign, GroupVerify, Revoke and Identify in this section.

Primitives Used. Two key cryptographic primitives are used in the scheme. They are the commitment schemes and the non-interactive zero-knowledge proof of knowledge protocols. Both primitives are used either in the DSA group modulo large prime p [15], denoted as $\mathbb{Z}_p^*$, or in the Okamoto-Uchiyama group modulo product n, where $n = r^2 s$, and r, s are large safe primes, denoted as $\mathbb{Z}_n^*$ [25].

The scheme presented in this paper relies on the assumption that the discrete logarithms are difficult to compute in Okamoto-Uchiyama (OU) group, similarly as in groups modulo primes or RSA composites. However, if factorization (r, s) of OU modulus $n = r^2 s$ is known, the discrete logarithms can be efficiently computed. For example, w from $c = g^w \bmod n$ can be computed as:

$$\log_g c = w = \frac{((c^{r-1} \bmod r^2) - 1)/r}{((g^{r-1} \bmod r^2) - 1)/r} \bmod r.$$

The discrete logarithm can be computed only by entities who know the factorization (r, s), as proven in [25].

Setup. $(params, K_{RA}, K_C) \leftarrow$ Setup(k, l, m) - The Setup protocol is used to generate all system parameters and private keys of C and RA. It inputs the security parameters k (length of the output of the hash function used), l (length of Meter's private key) and m (protocol error parameter).

The Collector defines a group $\mathbb{Z}_p^*$ by randomly choosing the prime modulus p and elements h_1, h_2 of large order $q : |q| = 2l$ and $q|(p - 1)$.

The Revocation Authority defines a group $\mathbb{Z}_n^*$ by randomly choosing the modulus $n : n = r^2 s$ and r, s are large safe primes such that $r = 2r' + 1$ and $s = 2s' + 1$ holds and r', s' are also primes. RA selects a random element g_1 of orders $ord(g_1) = rr'$ in $\mathbb{Z}_{r^2}^*$ and $ord(g_1) = rr's'$ in $\mathbb{Z}_n^*$. RA also randomly generates its secrets $S_1, S_2 : |S_1| = 2.5l, |S_2| = l$ and $GCD(S_1, \phi(n)) = GCD(S_2, \phi(n)) = 1$ and computes a value $g_2 = g_1^{S_2} \bmod n$.

For each group of meters, RA computes the group public key $GPK = g_1^{S_1} \bmod n$. For simplicity, we consider only one group in this paper, but many group public keys can be defined analogically.

The values $q, p, h_1, h_2, n, g_1, g_2, GPK$ are made public as system parameters $params$. The values r, s, S_1, S_2 are securely stored at RA as the K_{RA} key. C generates a keypair of any conventional signature scheme (like RSA or DSA), stores its private key as K_C and publishes its public key.

getKey. $K_M \leftarrow$ getKey$(params, K_C, K_{RA})$ - The getKey protocol is split into two parts. The first one runs between the Meter and the Collector while the second runs between the Meter and RA. The purpose of the protocol is to generate a private key of the Meter K_M in such a way that only the Meter learns the value but both C and RA contribute to its creation.

RA	Meter	Collector

$$w_1' \in_R \{0,1\}^{2l-1}, \ w_2 \in_R \{0,1\}^{l-1}$$
$$C_C = commit(w_1', w_2) = h_1^{w_1'} h_2^{w_2} \bmod p$$

$$\xrightarrow{\quad PK\{w_1', w_2 : C_C = h_1^{w_1'} h_2^{w_2}\}, Sig_M(C_C) \quad}$$

Store $(C_C, Sig_M(C_C))$

$$\xleftarrow{\quad Sig_C(C_C) \quad}$$

$$GPK' = g_1^{w_1'} g_2^{w_2} \bmod n$$

$$GPK', C_C, Sig_C(C_C),$$
$$PK\{(w_1', w_2) : C_C = h_1^{w_1'} h_2^{w_2} \wedge GPK' = g_1^{w_1'} g_2^{w_2}\}$$
$$\xleftarrow{\hspace{6cm}}$$

Random generator: $g_3 \in_R \mathbb{Z}_n^*$
$w_1'' \in_R \{0,1\}^l$
$w_3 : GPK = g_1^{w_1'} g_1^{w_1''} g_2^{w_2} g_3^{w_3} \bmod n$
Random primes: $\{w_{31}, w_{32}, \ldots, w_{3(j-1)} \in_R \{0,1\}^l\}$
$w_{3j} = w_3 (w_{31} w_{32} \ldots w_{3(j-1)})^{-1} \bmod \phi(n)$

$$\xrightarrow{\quad w_1'', (w_{31}, w_{32}, \ldots, w_{3j}), g_3 \quad}$$

Meter's private group key: $K_M = \{w_1 = w_1' + w_1'', w_2, (w_{31}, w_{32}, \ldots, w_{3j}), g_3\}$
$$GPK = g_1^{w_1} g_2^{w_2} g_3^{w_{31} \ldots w_{3j}} \bmod n$$

Fig. 2. getKey protocol in CS notation.

In the first part, the Meter randomly generates its portion of the key (w_1', w_2) and commits to these values. It digitally signs its commitment C_C using a conventional signature scheme including owner's real identity and sends it to the Collector together with the proof of construction PK. The Collector checks the proof, the signature and the identity of the Meter's owner. If the Meter is authorized to be included in the group, the Collector returns the commitment, this time signed by his K_C (again, any classical signature can be used, like RSA) together with a unique, randomly generated identifier ID_M. By this procedure, the Collector certifies that the Meter belongs to the group of meters sharing the GPK public group key. If more groups are available, the signature also contains the information about the concrete group. The Collector stores the whole protocol transcript in its database.

In the second part, the rest of the Meter's key is issued. To have a valid private key for group public key GPK, the Meter must have values (w_1, w_2, w_3, g_3) so that

$$GPK \equiv g_1^{w_1} g_2^{w_2} g_3^{w_3} \bmod n \tag{1}$$

holds. The Meter creates another commitment GPK' to his keys (w_1', w_2), this time using the Okamoto-Uchiyama group $\mathbb{Z}_n^*$. The Meter sends both commitments and Collector's signature on the commitment to RA and proves that the keys in commitments are the same. If the proof and signatures are valid, RA chooses g_3 (a random element with same properties as g_1), random w_1'' (RA's contribution to Meter's key so that $w_1 = w_1' + w_1''$) and w_3 such that the Eq. 1 holds. This can be done only by RA who knows the Okamoto-Uchiyama trapdoor secret (the factorization of n) as $w_3 = \log_{g_3}(GPK/(g^{w_1''}GPK'))$.

The w_3 is then split into j factors $\mod\phi(n)$ and sent back to the Meter as RA's portion of Meter's private key. The Revocation Authority stores the whole protocol transcript in its database.

As a result, the Meter knows a unique discrete logarithm representation $w_1, w_2, (w_{31} \ldots w_{3j})$ of public GPK. The values $w_1, w_2, (w_{31} \ldots w_{3j}), g_3$ are the Meter's private key corresponding to the group public key GPK.

The getKey protocol is fully specified and depicted in CS notation in Fig. 2.

GroupSign. $GS_{K_M}(data) \leftarrow$ GroupSign$(params, data, K_M)$ - The GroupSign algorithm inputs system parameters $params$, Meter's private key K_M and measured data $data$ and outputs a group signature $GS_{K_M}(data)$. The signature certifies that the Meter has a valid private key K_M corresponding to the group public key GPK but does not disclose any more information about the Meter, including its identifier. The protocol is the non-interactive proof of knowledge signature based on the Fiat-Shamir heuristic [13]. The Meter proves using a zero-knowledge proof of knowledge protocol that it knows the discrete logarithm representation of GPK, i.e., it knows the private key (w_1, w_2, w_3). To provide the anonymity and unlinkability of signatures, all unique values must be randomized. Therefore, the unique g_3 must be randomized to g_3' using half of the key w_3. With w_3 split into 20 values, i.e. setting the parameter j to 20, we get enough combination variants to randomize $\binom{20}{10} = 184\ 756$ signatures. Furthermore, we randomize each signature by using a unique per-signature key K_S. To provide revocation and identification features, commitments to keys (w_1, w_2, K_S) denoted as C_{12}, C_{1S} are included in the signature.

The GroupSign protocol is fully specified and depicted in CS notation in Fig. 3.

GroupVerify. $\{0,1\} \leftarrow$ GroupVerify$(params, data, GS_{K_M}(data))$ - The group signature verification is composed of the verification of the zero-knowledge proofs and the revocation check. The verification of proofs of knowledge depends on concrete implementation, the signature check used in our implementation is shown in Fig. 4. The revocation check is done by a single equation $C_{12}^{rev} \overset{?}{\not\equiv} C_{1S} \pmod{n}$ where the commitments C_{12}, C_{1S} are checked against the revocation list. If the equation holds for any of rev values present on the revocation list, the Meter has been revoked and the signature is rejected. Otherwise, the signature is accepted.

Revoke. $rev \leftarrow$ Revoke$(params, GS_{K_M}(data), K_{RA}, K_C)$: the protocol inputs the system parameters $params$, the signature $GS_{K_M}(data)$, RA's and C's private keys K_{RA}, K_C and outputs the revocation information rev. The rev value is then put on a blacklist so that all signatures can be checked against the blacklist for revoked users.

---

[1] RA's key contribution is necessary to prevent malicious users from submitting malformed keys allowing attacks on RA's secrets.

Meter   **Collector**

w_1, w_2

$W_3 = \{w_{31}, w_{32}, \ldots, w_{3j}\}$

$W_3', W_3'' : (W_3' \subset W_3) \wedge (W_3'' \subset W_3) \wedge$

$(|W_3'| = |W_3''| = j/2) \wedge (W_3' \cap W_3'' = \emptyset)$

$w_3' = \displaystyle\prod_{x \in W_3'} x$

$w_3'' = \displaystyle\prod_{y \in W_3''} y$

$g_3' = g_3^{w_3'} \bmod n$

$K_S \in_R \{0,1\}^l$

$A = GPK^{K_S} \bmod n$

$C_{12} = (g_1^{w_1} g_2^{w_2})^{K_S} \bmod n$

$C_{1S} = g_1^{K_S} \bmod n$

$C_S = g_3'^{K_S} \bmod n$

$PK = PK\{(K_S w_1, K_S w_2, K_S w_3'', K_S, w_3'') : A = g_1^{K_S w_1} g_2^{K_S w_2} g_3'^{K_S w_3''}$

$\wedge$

$A = GPK^{K_S}$

$\wedge$

$C_{12} = g_1^{K_S w_1} g_2^{K_S w_2}$

$\wedge$

$\dfrac{A}{C_{12}} = g_3'^{K_S w_3''}$

$\wedge$

$\dfrac{A}{C_{12}} = C_S^{w_3''}$

$\wedge$

$C_S = g_3'^{K_S}$

$\wedge$

$C_{1S} = g_1^{K_S}\}$

$GS_{K_M}(data) = (A || C_{12} || C_{12S} || C_S || g_3' || PK)$
$\longrightarrow$

Fig. 3. GroupSign protocol in CS notation.

In the Okamoto-Uchiyama group, which is used in this scheme, it is generally hard to compute discrete logarithms as it is in the RSA and DSA groups. But there is a trapdoor value, the factor r of modulus n, which makes the discrete logarithm computation possible. The only entity, that knows the factorization of n, thus the value of r, is RA (Revocation Authority). From $\log_{g_1} C_{12}$, the RA gets $K_S(w_1 + w_2 S2)$. From $\log_{g_1} C_{1S}$, RA gets K_S. Then, using r, RA is able to compute the revocation information $rev = (w_1 + w_2 S_2)^{-1} \bmod (rr's')$. The rev value is put on a blacklist. Then, the equation $C_{12}^{rev} \overset{?}{\equiv} C_{1S} \pmod{n}$ is evaluated during all signature verifications for all $revs$. If it holds, the Meter has been revoked and the Collector rejects the signature.

Identify. $ID_M \leftarrow \texttt{Identify}(params, GS_{K_M}(data), K_{RA}, K_C)$ - The purpose of the protocol is to de-anonymize the signature so that the concrete Meter is disclosed. This can be done only in exceptional situations, for example in cases in which malicious users need to be identified. In that case the Collector sends to the Revocation Authority the signature $GS_{K_M}(data)$. RA gets $K_S(w_1 + w_2 S2)$ and K_S in the same way as in the Revoke protocol above. Using these value, RA is able to compute $g_1^{w_1} g_2^{w_2} = g_1^{w_1 + w_2 S2}$ and find in its getKey transcript database the corresponding commitment C_C. The commitment is then sent back to Collector, who can learn the Meter's identifier ID_M and owner's identity from C_C using its getKey transcript database. Using this method, it is assured, that Collector and Revocation Authority must cooperate on the de-anonymization and only Collector learns the Meter's and owner's identity. This distribution of privileges protects users' privacy and makes the scheme less vulnerable to attacks.

3   Scheme Implementation

In Sect. 2, we used the abstract CS notation to describe the protocols. This notation allows simple description of proofs of knowledge about discrete logarithms according to their purpose. It has been shown in [8] that a large variety of proofs and their logical compositions can be constructed. However, for implementation, it is necessary to describe the protocols in details. The proof of knowledge protocols can be implemented in many ways, interactive or non-interactive. We have chosen the non-interactive implementation based on Fiat-Shamir heuristic [11] and Schnorr signatures [30]. In our construction, the challenge e is computed as a hash function of all past cryptographic values instead of receiving e from the Collector. This construction has been proven secure in the random oracle model [27]. The full specification of GroupSign and GroupVerify protocols is shown in Fig. 4.

3.1   Performance Evaluation

The scheme proposed requires no complex operations, such as bilinear pairings. Only simple modular operations are required. To compute a group signature, the Meter must compute 11 modular exponentiations and 3 modular multiplications. To verify a signature, the Collector must compute 12 modular exponentiations and 10 modular multiplications plus one exponentiation per an item on a blacklist. These operations are easily computable on contemporary micro-controllers in a very short time. We implemented the operations, i.e., modular exponentiation (Mexp), modular multiplication (Mmul), plain multiplication (Mul) and hash function (Hash), on a device with 900 MHz ARM Cortex-A7 CPU with 1 GB RAM using 6 different modular arithmetic libraries written in C. The performance results of the fastest library, GMP [16], are shown in Fig. 5. The size of the operands is shown in the graph captions and the size of modulus is 2048 b. We present the average of 100 measurements.

The results show that the most computationally demanding operation in the scheme takes around 57 ms using the fastest library (GMP [16]). We also implemented other required operations (random number generation, (modular) multiplication, hash functions, subtraction) but most of them were faster than the modular exponentiation at least by several orders of magnitude.

The total time required to compute a group signature using our experimental implementation was under 0.5 s and the time of verification on the same device was also under 0.5 s. We consider these results to be a good proof of practical implementability of our scheme on existing, off-the-shelf devices commonly used in smart-metering applications.

4   Scheme Analysis

In this section, we show that the scheme proposed in this paper is secure and privacy-friendly. We prove that group members are able to construct a valid signature on data (the scheme is *complete*), that non-members are not able to create a valid signature on data (the scheme is *sound*) and that the signature leaks no unnecessary information about signers (the scheme is *zero-knowledge*).

4.1   Security Analysis

Completeness The group members who know a valid private group key $K_M = (w_1, w_2, w_3, g_3)$ are always able to construct a signature that is accepted by the Collector's verification equations shown in Fig. 4. The completeness is proven by Eqs. 2–8. All operations are in $\mathbb{Z}_n^*$.

$$\bar{A} = A^e g_1^{z_{S1}} g_2^{z_{S2}} g_3^{\prime z_{S3}} = GPK^{eK_S} g_1^{r_{S1}} g_1^{-eK_S w_1} g_2^{r_{S2}} g_2^{-eK_S w_2} g_3^{\prime r_{S3}} g_3^{\prime -eK_S w_3''} =$$
$$= g_1^{r_{S1}} g_2^{r_{S2}} g_3^{\prime r_{S3}} = \bar{A} \tag{2}$$

$$\bar{\bar{A}} = A^e GPK^{z_S} = GPK^{eK_S} GPK^{r_S} GPK^{-eK_S} = GPK^{r_S} = \bar{\bar{A}} \tag{3}$$

$$\bar{C}_{12} = C_{12}^e g_1^{z_{S1}} g_2^{z_{S2}} = g_1^{w_1 K_S e} g_2^{w_2 K_S e} g_1^{r_{S1}} g_1^{-eK_S w_1} g_2^{r_{S2}} g_2^{-eK_S w_2} = g_1^{r_{S1}} g_2^{r_{S2}} = \bar{C}_{12} \tag{4}$$

$$\frac{\bar{A}}{\bar{C}_{12}} = \frac{A}{C_{12}}^e g_3^{\prime z_{S3}} = \left(\frac{g_1^{w_1 K_S} g_2^{w_2 K_S} g_3^{\prime w_3'' K_S}}{g_1^{w_1 K_S} g_2^{w_2 K_S}}\right)^e g_3^{\prime r_{S3}} g_3^{\prime -eK_S w_3''} = g_3^{\prime r_{S3}} =$$
$$= \frac{g_1^{r_{S1}} g_2^{r_{S2}} g_3^{\prime r_{S3}}}{g_1^{r_{S1}} g_2^{r_{S2}}} = \frac{\bar{A}}{\bar{C}_{12}} \tag{5}$$

$$\bar{div} = \frac{A}{C_{12}}^e C_S^{z_3} = \left(\frac{g_1^{w_1 K_S} g_2^{w_2 K_S} g_3^{\prime w_3'' K_S}}{g_1^{w_1 K_S} g_2^{w_2 K_S}}\right)^e C_S^{r_3} C_S^{-ew_3''} = C_S^{r_3} = \bar{div} \tag{6}$$

$$\bar{C}_S = C_S^e g_3^{\prime z_S} = g_3^{\prime eK_S} g_3^{\prime r_S} g_3^{\prime -eK_S} = g_3^{\prime r_S} = \bar{C}_S \tag{7}$$

$$\bar{C}_{1S} = C_{1S}^e (g_1)^{z_S} = g_1^{eK_S} g_1^{r_S} g_1^{-eK_S} = g_1^{r_S} = \bar{C}_{1S} \tag{8}$$

Meter   **Collector**
GroupSign   GroupVerify

w_1, w_2

$W_3 = \{w_{31}, w_{32}, \ldots, w_{3j}\}$

$W_3', W_3'' : (W_3' \subset W_3) \wedge (W_3'' \subset W_3) \wedge$

$(|W_3'| = |W_3''| = j/2) \wedge (W_3' \cap W_3'' = \emptyset)$

$w_3' = \prod\limits_{x \in W_3'} x$

$w_3'' = \prod\limits_{y \in W_3''} y$

$g_3' = g_3^{w_3'} \bmod n$

$K_S \in_R \{0,1\}^l$

$A = GPK^{K_S} \bmod n$

$C_{12} = (g_1^{w_1} g_2^{w_2})^{K_S} \bmod n$

$C_{1S} = (g_1)^{K_S} \bmod n$

$C_S = (g_3')^{K_S} \bmod n$

$r_{S1} \in_R \{0,1\}^{k+3l+m}$

$r_{S2} \in_R \{0,1\}^{k+2l+m}$

$r_{S3} \in_R \{0,1\}^{k+l+m+|w_3''|}$

$r_S \in_R \{0,1\}^{k+l+m}$

$r_3 \in_R \{0,1\}^{k+m+|w_3''|}$

$\bar{A} = g_1^{r_{S1}} g_2^{r_{S2}} g_3'^{r_{S3}} \bmod n$

$\bar{\bar{A}} = GPK^{r_S} \bmod n$

$\bar{C}_{12} = (g_1^{r_{S1}} g_2^{r_{S2}}) \bmod n$

$\bar{C}_S = (g_3')^{r_S} \bmod n$

$\bar{C}_{1S} = (g_1)^{r_S} \bmod n$

$\bar{div} = (C_S)^{r_3} \bmod n$

$e = \mathcal{H}\{data, A, \bar{A}, GPK, \bar{\bar{A}}, C_{12}, C_{1S}, C_S, \bar{C}_{12}, \bar{C}_{1S}, \bar{C}_S, \bar{div}, g_3'\}$

$z_{S1} = r_{S1} - eK_S w_1$

$z_{S2} = r_{S2} - eK_S w_2$

$z_{S3} = r_{S3} - eK_S w_3''$

$z_S = r_S - eK_S$

$z_3 = r_3 - ew_3''$

$$\xrightarrow{data || (A, C_{12}, C_{1S}, C_S, g_3', e, z_{S1}, z_{S2}, z_{S3}, z_S, z_3)}$$

Revocation check:

$$C_{12}^{rev} \stackrel{?}{\not\equiv} C_{1S} \pmod n$$

Signature check:

$$\bar{A} = A^e g_1^{z_{S1}} g_2^{z_{S2}} g_3'^{z_{S3}} \bmod n$$

$$\bar{\bar{A}} = A^e GPK^{z_S} \bmod n$$

$$\bar{C}_{12} = C_{12}^e g_1^{z_{S1}} g_2^{z_{S2}} \bmod n$$

$$\frac{\bar{A}}{\bar{C}_{12}} = \frac{A}{C_{12}}^e g_3'^{z_{S3}} \bmod n$$

$$\bar{div} = \frac{A}{C_{12}}^e C_S^{z_3} \bmod n$$

$$\bar{C}_S = C_S^e g_3'^{z_S} \bmod n$$

$$\bar{C}_{1S} = C_{1S}^e (g_1)^{z_S} \bmod n$$

$$e \stackrel{?}{=} \mathcal{H}\{data, A, \bar{A}, GPK, \bar{\bar{A}}, C_{12}, C_{1S}, C_S, \bar{C}_{12}, \bar{C}_{1S}, \bar{C}_S, \bar{div}, g_3'\}$$

Fig. 4. GroupSign and GroupVerify algorithms implementation.

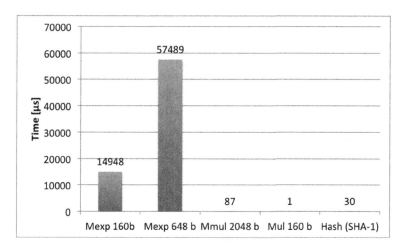

Fig. 5. Time necessary to compute modular operations on 900 MHz ARM microprocessor using GMP library.

Soundness. We implemented the group signature using the Fiat-Shamir heuristic applied to discrete logarithm proof of knowledge protocols. This is a common construction used in many other schemes [7]. Here, we follow the proof presented in [6,18].

In the detailed description of the GroupSign protocol provided in Fig. 4 it is visible that the Meter first generates randomness (values $r_{S1}, r_{S2}, r_{S3}, r_S, r_3$), then computes the commitments to these values (values $\bar{A}, \bar{\bar{A}}, \bar{C}_{12}, \bar{C}_S, \bar{C}_{1S}, d\bar{i}v$). The hash of all these values is then used as a challenge e. Since the hash is a one-way function, the Meter is unable to predict e before selecting the randomness. Thus, after committing to randomness, the Meter must be ready to response any challenge e since its value is unpredictable.

In the proof below we show that Meters that are ready to respond at least 2 challenges (e, e') know the private group key K_M. Thus, malicious users not aware of K_M are able to answer at most 1 challenge of 2^k possible challenges, thus have the probability $P = \frac{1}{2^k}$ of creating a valid signature.

Let's assume that exists a malicious Meter that is able to produce signatures for different challenges (e, e') that get accepted by the Collector, without knowing K_M. A valid signature must pass all signature checks shown in Fig. 4. Thus, both the equations must hold:

$$\bar{C}_{1S} = C_{1S}^e (g_1)^{z_S} \tag{9}$$

$$\bar{C}_{1S} = C_{1S}^{e'} (g_1)^{z'_S} \tag{10}$$

By rewriting the equations we get:

$$1 = C_{1S}^{e-e'} g_1^{z_S - z'_S} \tag{11}$$

And finally we get:

$$\log_{g_1} C_{1S} = (z_S - z_S')(e' - e)^{-1} \tag{12}$$

The Eq. 12 can be easily computed by the Meter since it knows all values e', e, z_S, z_S'. Thus, the Meter knows the $\log_{g_1} C_{1S}$. Therefore, it knows K_S.

The same method can be applied to all verification equations, proving that Meter must know $K_S, K_S w_1, K_S w_2, K_S w_3''$ to be able to construct a valid signature.

However, from these values, the discrete logarithm representation of GPK, thus K_M, can be efficiently computed. We reached the contradiction to our assumption, there are no users that can pass the verification checks without knowing K_M.

Zero-Knowledge. The protocols specified in Sect. 2 can be implemented in an interactive version with a challenge e randomly chosen by the Collector. In that case, the honest-verifier zero-knowledge (HVZK) property can be proven by constructing the standard HVZK simulator. The simulator is able to output a protocol transcript that is computationally indistinguishable from a real protocol without using any Meter's secrets. Thus, it can be proven that the protocol leaks no secret information. The simulator proceeds in a standard way, i.e. first randomly selecting the answers z, randomly generating the challenge e and computing the commitments using the Collector's verification equations.

However, we use the non-interactive version based on hash function in our implementation. The random challenge is substituted by a hash output to improve the performance of the scheme. Nevertheless, we disclose exactly the same values as in the interactive version except the hash. Therefore, assuming the hash function is secure, the protocol releases no information about the private keys.

4.2 Privacy Analysis

In the beginning of the paper we stated that the group signature is anonymous, untraceable and unlinkable to other signatures of the same Meter. We analyze these properties below.

- **Anonymity:** the only link between a particular Meter and a signature is the private key K_M. However, the key is never disclosed in any form as proven in the zero-knowledge analysis above. The g_3 value is randomized to $g_3' = g_3^{w_3'} \mod n$ for each signature.
- **Untraceability:** the Collector is unable to link the getKey protocol transcript and the GroupVerify protocol transcript. This is achieved by using different modular groups in both protocols. During the getKey protocol, the Collector learns the Meter's key commitment in $\mathbb{Z}_p^*$. During the GroupVerify protocol, the Collector learns the Meter's key commitment in the $\mathbb{Z}_n^*$ group. It is computationally unfeasible to decide whether same keys are present in

commitments using different groups unless discrete logarithm is computable. In both groups the computation of discrete logarithms is hard without the knowledge of a trapdoor.

- **Unlinkability:** no entity except Revocation Authority is able distinguish whether two signatures originate from a single Meter or two different Meters. This holds even if same data was signed. This feature was achieved by the randomization of all Meter-specific values by the session key K_S. As visible in Fig. 4, all values that would be otherwise constant are randomized by K_S. The K_S key changes for each signature. The de-randomization is possible only if discrete logarithms can be efficiently computed. The only entity that can compute discrete logarithms in $\mathbb{Z}_n^*$ is RA because it knows the factorization of n thus knows the Okamoto-Uchiyama trapdoor. This RA's ability is used to achieve the revocation and malicious user identification features.

5 Conclusion

In this paper, we introduced a novel scheme for secure and privacy-friendly data collection. The scheme assures data group authenticity and integrity, like the classical digital signatures. In addition, the privacy-enhancing features are added. The sender's identity is not disclosed to collectors and all signatures are mutually unlinkable and untraceable. In contrast to existing schemes, our scheme provides advanced features for the identification and revocation of malicious users. The attackers can be efficiently revoked from the system and their identity can be disclosed so that they are held responsible for their acting in the system. Additionally to the full cryptographic specification and security analysis, we also presented the performance analysis. The implementation results show that the scheme including revocation features is practical and easily implementable on microprocessors commonly used in existing smart-metering systems. Currently, the scheme is being piloted in smart-house applications in cooperation with industrial partners.

Acknowledgments. Research described in this paper was financed by the National Sustainability Program under grant LO1401, the Czech Science Foundation under grant no. 14-25298P "Research into cryptographic primitives for secure authentication and digital identity protection" and by Technology Agency of the Czech Republic project TA04010476 "Secure Systems for Electronic Services User Verification". For the research, infrastructure of the SIX Center was used.

References

1. Alpar, G., Hoepman, J.H., Lueks, W.: An attack against fixed value discrete logarithm representations. Cryptology ePrint Archive, Report 2013/120 (2013)
2. Boneh, D., Boyen, X., Shacham, H.: Short group signatures. In: Franklin, M. (ed.) CRYPTO 2004. LNCS, vol. 3152, pp. 41–55. Springer, Heidelberg (2004)

3. Boneh, D., Shacham, H.: Group signatures with verifier-local revocation. In: Proceedings of the 11th ACM Conference on Computer and Communications Security, CCS 2004, pp. 168–177. ACM, New York, NY, USA (2004)

4. Camenisch, J., Kohlweiss, M., Soriente, C.: An accumulator based on bilinear maps and efficient revocation for anonymous credentials. In: Proceedings of the 12th International Conference on Practice and Theory in Public Key Cryptography, PKC 2009, pp. 481–500. Irvine, Springer, Berlin, Heidelberg (2009)

5. Camenisch, J.L., Lysyanskaya, A.: A signature scheme with efficient protocols. In: Cimato, S., Galdi, C., Persiano, G. (eds.) SCN 2002. LNCS, vol. 2576, pp. 268–289. Springer, Berlin, Heidelberg (2003)

6. Camenisch, J.L., Shoup, V.: Practical verifiable encryption and decryption of discrete logarithms. In: Boneh, D. (ed.) CRYPTO 2003. LNCS, vol. 2729, pp. 126–144. Springer, Berlin, Heidelberg (2003)

7. Camenisch, J.L., Stadler, M.A.: Efficient group signature schemes for large groups. In: Kaliski Jr., B.S. (ed.) CRYPTO 1997. LNCS, vol. 1294, pp. 410–424. Springer, Berlin, Heidelberg (1997)

8. Camenisch, J., Stadler, M.: Proof systems for general statements about discrete logarithms. Technical report (1997)

9. Camenisch, J., Van Herreweghen, E.: Design and implementation of the idemix anonymous credential system. In: Proceedings of the 9th ACM Conference on Computer and Communications Security, CCS 2002, pp. 21–30. ACM, New York, NY, USA (2002)

10. Delerablée, C., Pointcheval, D.: Dynamic fully anonymous short group signatures. In: Nguyen, P.Q. (ed.) Progress in Cryptology-VIETCRYPT 2006. LNCS, vol. 4341, pp. 193–210. Springer, Heidelberg (2006)

11. Feige, U., Shamir, A.: Witness indistinguishable and witness hiding protocols. In: Proceedings of the Twenty-Second Annual ACM Symposium on Theory of Computing, STOC 1990, pp. 416–426. ACM, New York, NY, USA (1990). http://doi.acm.org/10.1145/100216.100272

12. Ferrara, A.L., Green, M., Hohenberger, S., Pedersen, M.Ø.: Practical short signature batch verification. In: Fischlin, M. (ed.) CT-RSA 2009. LNCS, vol. 5473, pp. 309–324. Springer, Heidelberg (2009)

13. Fiat, A., Shamir, A.: How to prove yourself: practical solutions to identification and signature problems. In: Odlyzko, A.M. (ed.) CRYPTO 1986. LNCS, vol. 263, pp. 186–194. Springer, Berlin, Heidelberg (1987)

14. Finster, S., Baumgart, I.: Pseudonymous smart metering without a trusted third party. In: 2013 12th IEEE International Conference on Trust, Security and Privacy in Computing and Communications (TrustCom), pp. 1723–1728, July 2013

15. Gallagher, P., Kerry, C.: FIPS PUB 186-4: Digital signature standard (DSS) (2013). http://nvlpubs.nist.gov/nistpubs/FIPS/NIST.FIPS.186-4.pdf

16. GMP: The GNU multiple precision arithmetic library (2015). https://gmplib.org

17. Hajny, J., Dzurenda, P., Malina, L.: Privacy-PAC: privacy-enhanced physical access control. In: Proceedings of the 13th Workshop on Privacy in the Electronic Society, WPES 2014, pp. 93–96. ACM, New York, NY, USA (2014). http://doi.acm.org/10.1145/2665943.2665969

18. Hajny, J., Malina, L.: Unlinkable attribute-based credentials with practical revocation on smart-cards. In: Mangard, S. (ed.) CARDIS 2012. LNCS, vol. 7771, pp. 62–76. Springer, Berlin, Heidelberg (2013)

19. Hajny, J., Malina, L., Tethal, O.: Privacy-friendly access control based on personal attributes. In: Yoshida, M., Mouri, K. (eds.) IWSEC 2014. LNCS, vol. 8639, pp. 1–16. Springer, Heidelberg (2014). http://dx.doi.org/10.1007/978-3-319-09843-2_1

20. Hwang, J.Y., Lee, S., Chung, B.H., Cho, H.S., Nyang, D.: Short group signatures with controllable linkability. In: 2011 Workshop on Lightweight Security & Privacy: Devices, Protocols and Applications (LightSec), pp. 44–52. IEEE (2011)
21. Kim, K., Yie, I., Lim, S., Nyang, D.: Batch verification and finding invalid signatures in a group signature scheme. IJ Netw. Secur. **13**(2), 61–70 (2011)
22. Lapon, J., Kohlweiss, M., De Decker, B., Naessens, V.: Analysis of revocation strategies for anonymous idemix credentials. In: De Decker, B., Lapon, J., Naessens, V., Uhl, A. (eds.) CMS 2011. LNCS, vol. 7025, pp. 3–17. Springer, Berlin, Heidelberg (2011). http://dx.doi.org/10.1007/978-3-642-24712-5_1
23. Lin, X., Sun, X., Ho, P.H., Shen, X.: GSIS: a secure and privacy preserving protocol for vehicular communications. IEEE Trans. Veh. Technol. **56**, 3442–3456 (2007)
24. Nguyen, L.: Accumulators from bilinear pairings and applications. In: Menezes, A. (ed.) CT-RSA 2005. LNCS, vol. 3376, pp. 275–292. Springer, Berlin, Heidelberg (2005)
25. Okamoto, T., Uchiyama, S.: A new public-key cryptosystem as secure as factoring. In: Nyberg, K. (ed.) EUROCRYPT 1998. LNCS, vol. 1403, pp. 308–318. Springer, Berlin, Heidelberg (1998)
26. Paquin, C.: U-prove cryptographic specification v1.1. Technical report, Microsoft Corporation (2011)
27. Pointcheval, D., Stern, J.: Security proofs for signature schemes. In: Maurer, U.M. (ed.) EUROCRYPT 1996. LNCS, vol. 1070, pp. 387–398. Springer, Berlin, Heidelberg (1996). http://dx.doi.org/10.1007/3-540-68339-9_33
28. Raya, M., Hubaux, J.P.: Securing vehicular ad hoc networks. J. Comput. Secur. **15**, 39–68 (2007)
29. Rottondi, C., Mauri, G., Verticale, G.: A protocol for metering data pseudonymization in smart grids. Trans. Emerg. Telecommun. Technol. **26**(5), 876–892 (2015). doi:10.1002/ett.2760
30. Schnorr, C.P.: Efficient signature generation by smart cards. J. Cryptol. **4**, 161–174 (1991)
31. Zhang, C., Lu, R., Lin, X., Ho, P.H., Shen, X.: An efficient identity-based batch verification scheme for vehicular sensor networks. In: INFOCOM, pp. 246–250. IEEE (2008)

A Secure Architecture for Operating System-Level Virtualization on Mobile Devices

Manuel Huber$^{(\boxtimes)}$, Julian Horsch, Michael Velten, Michael Weiss, and Sascha Wessel

Fraunhofer AISEC, Garching Near Munich, Germany
{manuel.huber,julian.horsch,michael.velten,michael.weiss,
sascha.wessel}@aisec.fraunhofer.de

Abstract. In this paper, we present a novel secure architecture for OS-level virtualization on mobile devices. OS-level virtualization allows to simultaneously operate multiple userland OS instances on one physical device. Compared to previous approaches, our main objective is the confidentiality of sensitive user data stored on the device. We isolate the OS instances by restricting them to a set of minimal, controlled functionality and allow communication between components exclusively through well-defined channels. With our secure architecture, we therefore go beyond the common deployment of Linux kernel mechanisms, such as namespaces or cgroups. We develop a specially tailored, stacked LSM concept using SELinux and a custom LSM, leverage Linux capabilities and the cgroups devices subsystem. Based on the architecture, we present secure device virtualization concepts allowing to dynamically assign device functionalities to different OS instances. Furthermore, we develop a mechanism for secure switching between the instances. We realize the architecture with a fully functional and performant implementation on the Samsung Galaxy S4 and Nexus 5 mobile devices, running Android 4.4.4 and 5.1.1, respectively. With a systematic security evaluation, we demonstrate that the secure isolation of OS instances provides confidentiality even when large parts of the system are compromised.

Keywords: Mobile device security · Security architecture · Data confidentiality · Operating System security

1 Introduction

Today's mobile devices are not only widely used, but also represent a fingerprint of their users. Essential corporate and private data is likely to be found on those devices, rising the necessity to carefully consider security aspects, such as data confidentiality protection. However, the prevalence of only few Operating System (OSs), and the pace of their development make them prone to get targeted by attackers [5,15]. The abundance of security issues makes these devices vulnerable to a large number of attacks [16,22,23,31]. Efforts were made to mitigate the susceptibility towards common attack vectors [1,3,10,14,21]. Nevertheless, an

© Springer International Publishing Switzerland 2016
D. Lin et al. (Eds.): Inscrypt 2015, LNCS 9589, pp. 430–450, 2016.
DOI: 10.1007/978-3-319-38898-4_25

adversary remains able to access all data on the phone with common privilege escalation attacks. None of the approaches features an overall secure architecture for data confidentiality, although being the utmost goal when it comes to protecting sensitive private and corporate data.

A promising way to approach data confidentiality is to provide multiple, virtualized environments on a single mobile device [6]. OS-level virtualization [19] allows to simultaneously operate multiple userland OS instances running on a single kernel instance. These virtualized instances are henceforth called *containers*. In [27,28], Wessel et al.propose such an architecture based on Android. However, their and other OS-level virtualization approaches [2,9] lack a full-fledged secure architecture for domain isolation and data confidentiality protection.

In this paper, we present a novel secure architecture for OS-level virtualization on mobile devices for Linux driven OSs. Our main objective is the confidentiality of sensitive user data at container boundaries. This means, we achieve data confidentiality when data inside a container remains confidential to other, possibly malicious containers at all time. For this purpose, we isolate containers by restricting them to a set of minimal, controlled functionality. We confine communication between components to only specific channels in our architecture. Based on this, we develop secure device virtualization concepts and a convenient mechanism for secure switching between containers. Addressing common OS-level virtualization security requirements [24], we focus on the development of an easily portable solution, suitable for real-life application. In particular, our contributions are:

- The development of a kernel-based **secure virtualization architecture** for data confidentiality, including a Secure Element (SE).
- Improved **container isolation** through confining containers to minimal, controlled functionality and to only specific communication channels. We develop a stacked Linux Security Modules (LSM) concept using Security-Enhanced Linux (SELinux) and a custom LSM. We leverage Linux capabilities and the control groups (cgroups) devices subsystem.
- A **secure device virtualization** mechanism allowing to dynamically assign hardware functionalities on a per-container basis. A classification of devices into different device categories.
- The introduction of a fast and **secure container switch mechanism** with security devices.
- A **full implementation** on the Samsung Galaxy S4 and the Nexus 5 devices.
- A **performance evaluation** for the realization on the Nexus 5 device.
- A **systematic security evaluation** of our architecture showing data confidentiality even when large parts of the system are compromised.

The paper is structured as follows. In Sect. 2, we present related work. We describe our secure architecture in Sect. 3. In Sect. 4 we present our concept for container isolation. Based on that, we elaborate the refined secure device virtualization mechanisms in Sect. 5. We develop the secure container switching mechanism in Sect. 6. In Sect. 7, we conduct a systematic security evaluation. We describe the implementation and show performance results in Sect. 8 and conclude in Sect. 9.

2   Related Work

The necessity for data confidentiality protection results from the numerous attacks on mobile device OSs, especially on the widespread Android platform [16,22,23,31]. Another threat for confidentiality are applications that leak sensitive data [13]. Approaches for security enhancements on Android are presented in [1,3,10,14,21]. These approaches focus on the middleware layer to, e.g., gain fine-grained control over the OS permission system [3,21], to restrict applications from OS resources [10,14], or to harden the OS [1]. Introducing security mechanisms on middleware level results in a highly complex system, a large Trusted Computing Base (TCB) and a very OS specific solution.

To tackle this problem, virtualization techniques create isolated environments for distinct purposes, namely *user level isolation, system-* and *OS-level virtualization*. User-level isolation [7,8,26] is an approach to create separate environments through isolating applications on the framework level. A successful attack on privileged processes results in gaining full control over the system. System virtualization [11,17,25] deals with full OS virtualization including the kernel. The approach is strongly hardware dependent, because drivers have to be reimplemented for all hardware devices. OS-level virtualization separates userland OS instances running on a single modified kernel. An attacker must compromise the kernel to break out of a container. Achieved with LXC, Jails [18], Docker [20], OpenVZ, or Linux-VServer, the technique is established on x86 and considered efficient [24,30].

Cells [2] is an OS-level virtualization approach for Android mobile devices. They introduce *device namespaces* to provide a framework for device driver virtualization on kernel-level. Device namespaces multiplex hardware driver states on a per-container basis. With the concept of active device namespaces, drivers are made aware of the current *active namespace*, i.e., the foreground container. The work puts the main focus on realizing the functionality, but lacks the consideration of security aspects. No secure architecture is provided and data confidentiality is not discussed.

Based on Cells, Condroid [9] puts the focus on efficiency. Device virtualization is mostly applied in the Android framework. More OS resources are shared among the containers, such as their read only parts and OS services. For container management, the authors port LXC and run it in a single host Android in the root namespace. This makes the solution highly specific to a certain OS version and blends domain isolation with domain interaction, resulting in a weaker security model and a larger TCB.

AirBag [29] leverages OS-level virtualization in a single phone usage model for probing and profiling of untrusted applications. The framework allows the user to install and execute new applications quarantined inside a second, untrusted container. In contrast to our approach, their objective is the preliminary analysis of Android applications before their execution in the trusted container.

The virtualization approach by Wessel et al. [27,28] forms the starting point of our work. They leverage mechanisms, like (device) namespaces and cgroups. Focusing on security aspects and integrity protection, they develop a security

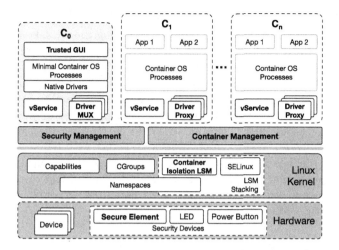

Fig. 1. The secure virtualization architecture.

infrastructure. The infrastructure provides concepts for remote management, secure communication and storage protection. In [27], the authors elaborate the integration of an SE and a device provisioning and enrollment process with a certificate infrastructure. They describe a many-to-many usage model for containers, users and devices. However, their infrastructure lacks a secure architecture for data confidentiality, secure device virtualization and container switching.

3 Architecture Overview

Figure 1 gives an overview on the components of our secure architecture. The illustration depicts different containers C_0, C_1, ..., C_n running on a single Linux kernel. We differentiate between components located in user and in kernel space. Another differentiation is between common components on a stock Linux based mobile device and between components we added. The latter ones are highlighted by bold letters. The varying background grayscale colors visualize the separation of components into different privilege levels. The dark gray colored components are in the TCB in root namespace. The mid gray C_0 is a *privileged container* in contrast to the *unprivileged containers* C_i (i.e., C_1, ..., C_n).

3.1 Hardware and Kernel Components

The hardware part consists of common hardware devices and *security devices*. We define the SE, LED and power button as security devices. These are non-virtualized hardware devices, because they serve a security critical purpose. They are not accessible to $C_{0..n}$. The LED and power button are usually available on common mobile devices. In our architecture, the power button's purpose is to securely initiate a switch between containers (see Sect. 6). The LED is a secure container indicator for the user, showing the unique color of the currently active container. $C_{0..n}$ are thus unable to disguise their identities to fake another container. We use the SE as secure storage for integrity and confidentiality protection. The SE is a passphrase-protected device, e.g., a smartcard connected via Near Field Communication (NFC). We securely virtualize the remaining devices in order to ensure a seamless user experience and the operability of the containers on the device (see Sect. 5). That includes, amongst others, graphics, input, Radio Interface Layer (RIL), sensors, or Wi-Fi device virtualization. In kernel space, we substitute the stock mobile device's kernel with our modified Linux kernel. The kernel handles multiple containers through the (device) namespaces feature. Further kernel mechanisms for container isolation and resource control we leverage in our secure architecture are capabilities, cgroups, as well as a stacked LSM with SELinux and a custom LSM.

3.2 User Space Components

Containers $C_{0..n}$, the Container Management (CM) and Security Management (SM) are located in user space. Only the SM and the CM are part of the TCB.

Security Management. The SM has the responsibility to securely and exclusively communicate with different SEs. The SM performs cryptographic operations for the CM, such as container storage key unwrapping using the SE.

Container Management. The CM configures the kernel features and acts as mediator between the containers. It has exclusive access to the LED and power button. The CM is responsible of container operations, such as to start $C_{0..n}$ or to securely switch between containers (see Sect. 6). The CM is also responsible of container storage encryption. Container storage is protected with a symmetric container key. This key is wrapped with the public key belonging to the SE's private key. When a container starts, the CM asks the SM with a provided passphrase to unwrap the container key using the SE.

Container C_0. This is a special, privileged container, comparable to dom0 in XEN [4]. C_0 is used for local container management with a Trusted GUI and for secure device virtualization (see Sect. 5). The Trusted GUI enables the user to securely enter the passphrase required for starting containers, to initiate a container switch and to make container specific and device global settings. We use the Driver MUXs as device multiplexers for user space device virtualization over container boundaries. Device drivers, often proprietary binaries, are mostly running only within a userland OS, such as Android. We therefore require C_0 to run a minimal OS for hardware device driver access.

Container C_i. These components are the isolated and unprivileged containers. The CM encapsulates $C_{0..n}$ into their specific namespaces, maintained by the kernel. During start-up of a container, the CM creates the namespaces and configures the security mechanisms. The vService in each container realizes an interface to the CM for sending commands to a container, e.g., to shutdown, suspend or resume. The Driver Proxies request device functionality from C_0's

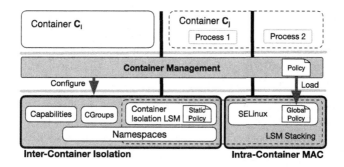

Fig. 2. Kernel mechanisms for container isolation.

Driver MUXs (see Sect. 5). This enables C_i to obtain specific device functionalities without explicit device access.

4   Container Isolation

In the following, we describe the isolation of the containers from each other and the root namespace. In order to achieve strict isolation, we restrict $C_{0..n}$ to a minimal set of functionality. We allow communication only over well-defined and protected communication channels. Figure 2 depicts a detailed view on the container isolation of C_i and C_j with the kernel mechanisms we make use of. We isolate components on intra- and inter-container basis. We support and enforce the commonly deployed LSM realization SELinux inside containers. This isolates processes inside containers to protect it from being compromised. The CM loads and enforces a global LSM policy for each container. We also require LSM mechanisms for inter-container isolation. Therefore, we use the LSM stacking mechanism[1]. This mechanism allows to register multiple LSMs in the kernel. Multiple handlers are hence called on an LSM hook to perform access control. A hook is successfully passed only if each of the handlers grants access to the kernel resource.

4.1   Communication Channels

We specify secure and exclusive communication channels between the components over well-defined interfaces. This restricts the components to interfaces exclusively used for container management and for secure device virtualization. First, we classify communication channels into three different layers of communication, as depicted in Fig. 3.

Layer 1 Communication. Layer 1 communication is on system call level, i.e., calls like open, write or ioctl, which are executed in the kernel. Any communication between components results in layer 1 communication interacting with

---
[1] http://article.gmane.org/gmane.linux.kernel.lsm/22729.

the kernel. On this layer, we prevent containers from unspecified device access with the cgroups devices subsystem based on device major minor numbers. We allow $C_{0..n}$ to directly access device drivers virtualized on kernel-level via device namespaces (see Sect. 5). To prevent components from critical system calls, we use Linux capabilities and our LSMs.

Layer 2 Communication. Layer 2 communication involves the communication between two or more processes. This layer represents all types of low-level Inter-Process Communication (IPC) over OS resources, e.g., sockets, and results in system calls. We separate this layer between containers through namespaces isolation. With our custom LSM, we selectively allow access to defined kernel resources relevant for IPC. An example is the denial of accessing certain sockets. This makes it possible to explicitly grant or refuse the establishment of communication channels.

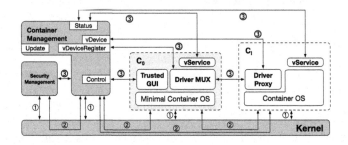

Fig. 3. Communication channels of the secure architecture.

Layer 3 Communication. This layer uses a protocol for IPC between the components. We secure the communication by message filtering and by utilizing a secure protocol. Figure 3 illustrates the following layer 3 channels we allow.

- **CM and SM:** The CM uses this channel in the root namespace to retrieve the results of the cryptographic operations that the SM executes.
- **CM and external components:** For remote device management via a backend, the CM offers a protocol on an update and remote control interface.
- **CM and vService:** To send commands to $C_{0..n}$ and to check their status, the CM communicates via the status interface with the vService inside $C_{0..n}$.
- **CM and Trusted GUI:** The CM offers a control interface for local container management. The Trusted GUI in C_0 uses this control interface.
- **CM and Driver MUX:** The Driver MUX utilizes this channel to notify the CM via the vDeviceRegister interface of the user space virtualized device functionality the multiplexer offers.
- **CM and Driver Proxy:** The Driver Proxy uses this channel to demand the CM via the vDevice interface for setting up the connection channel to the Driver MUX to obtain functionality of user space virtualized devices.

– **Driver MUX and Driver Proxy:** This channel, set up by the CM, exists for user space-based device virtualization. C_0 accesses hardware devices on layer 1 on behalf of C_i and selectively provides the functionality to C_i.

4.2   Identification and Isolation of OS Functionalities

To enforce data confidentiality across container boundaries, we prevent $C_{0..n}$ from defying namespace boundaries through other than the specified channels. In order to do so, we confine the containers to minimal OS functionalities with the kernel mechanisms (see Fig. 2). System calls represent the interface via which all components act and are thus crucial for our secure architecture. In order to achieve a global view on these resources, we investigate all system calls and their usage. Based on the whole set of system calls, we try to identify and group OS functionalities. In the following, we elaborate the protection of the functionalities using the aforementioned security mechanisms.

Mounting. We only allow containers to execute noncritical mount operations. First, we embed every container into its own mount namespace, which provides each container with isolated filesystem mount views. For managing the mount permissions of containers, we then introduce mount restrictions with our custom LSM. This prohibits mounting of non-required resources and specifies paths where a container can mount to. For example, $C_{0..n}$ are only allowed to mount sysfs to /sys and procfs to /proc. Containers are, e.g., not allowed to mount cgroups, which prevents $C_{0..n}$ from overwriting cgroups configurations. Our custom LSM performs the mount permission checks based on a static mount whitelist in our LSM policy. The list specifies the device, mount point, filesystem type and mount flags. We furthermore drop the capability CAP_SYS_ADMIN, because it comprises various critical functionality we prohibit. However, the mounting privileges are part of this capability. We therefore introduce a new capability CAP_SYS_MOUNT, which only allows a process to (un)mount and to create new mount namespaces. The new capability contains the minimal required subset of mount-related privileges former part of CAP_SYS_ADMIN.

Filesystem Access. For some of its mounted filesystems, a container should only have limited file access. To achieve this, we define protection rules with our custom LSM. We may assume fixed locations of objects in the filesystem due to the fixed mount points we defined. We utilize path-based whitelists, to specify the access permissions for filesystem locations. We define *read-write*, *read-only* and *privileged container* whitelists in our LSM policy. The LSM traverses the whitelists when the system triggers the corresponding LSM hooks, e.g., for the open or ioctl system call. An example is the access restriction to the sysfs filesystem. We allow a container to mount it in order to operate correctly, but we limit the access in this filesystem. For example, the LSM restricts an attempt from $C_{0..n}$ to set the LED color via the sysfs filesystem.

Device Access. Containers must be able to fulfill their usage purpose, which often requires virtualized device functionality from C_0, such as telephony or sensors (see Sect. 5). The goal is thus to enforce fine-granular control over device access permissions on a per-container basis. We grant or deny containers access to devices using the cgroups devices subsystem. This subsystem uses a whitelist configuration. The list specifies rules, which contain the device major minor numbers, its type, and the kind of operation allowed (e.g., `mknod`, `read`, `write`). The `/dev/random` pseudo device is an example for a device we allow a container to access. Since each container is in a different cgroup, we provide different per-container configurations. We adapt the configurations dynamically according to whether a container is in the fore- or background (see Sect. 6). With the device namespaces for kernel virtualized devices, we provide filtering mechanisms for fine-granular usage control of a device's functionality even when device node access is generally granted. Using the described mechanism allows us to permit the container to populate its own device directory. This results in less changes to containers and provides maximum compatibility. Therefore, we do not drop the capability `CAP_MKNOD` used for creating filesystem nodes. We enforce the security in using `mknod` via cgroups devices and LSM mount whitelisting.

IPC. To achieve container isolation, we generally restrict all kind of IPC between namespace boundaries. Solely for container management and secure device virtualization, we allow IPC functionality via protected and controlled communication channels (see Sect. 4.1). For inter-container isolation, IPC namespaces provide containers with dedicated resources for IPC inside containers and isolate them at container boundaries. With our custom LSM, we restrict unauthorized namespace crossing IPC. The LSM considers the PID namespaces for file-based IPC via the mounting and filesystem access restrictions. An example are checks for socket functionality with LSM hooks responsible for controlling inter-container IPC. We drop the capabilities `CAP_IPC_(OWNER, LOCK)` and `CAP_SYS_ADMIN`. These capabilities include critical IPC privileges. For instance, `CAP_IPC_LOCK` allows a process to lock memory, e.g., to prevent the OS from swapping. Such locking goes beyond the scope of a container and can lock-up the whole system if used by a malicious process.

Networking. We allow containers to individually setup the network within their boundaries. Therefore, we keep the networking capabilities `CAP_NET_*`. We embed containers into their own network namespace. Thus, the scope of the capabilities is limited to only affect the container's own subnet. However, we preserve the privilege to control the network dataflow. The CM sets up the global network configuration of the containers. The CM provides virtual network interfaces (`veth`) for each container with individual IP addresses. We conduct network package filtering and control on global level with netfilter components.

Signal Handling. With signals, a malicious container might adversely influence components of the architecture. However, a container should be capable of sending signals inside its namespace. We thus restrict containers from sending signals over namespace boundaries. We secure this functionality through PID namespaces, which ensure that signals from processes remain only visible inside a container's namespace. Therefore, we are not required to drop the capability CAP_KILL.

Resource Consumption. We provide containers access to sufficient system resources for working conveniently, but not to excessively exhaust resources. Mount namespaces provide a container with its own fixed and limited filesystem. With the cgroups CPU subsystem we determine a maximum share of the CPU resource for a process group. With the memory subsystem we ensure that a process group can only allocate a fixed maximum amount of memory. We do not need to drop the capability CAP_SYS_NICE, because even if a container changes process priorities, it cannot exceed its CPU usage limit.

Process Management. We grant a container to reduce the capabilities of its processes. We thus do not drop the capability CAP_SETPCAP. It allows processes to drop capabilities for child processes. The containers' init process has a reduced set of capabilities and is only allowed to further reduce this set. We prevent processes from process directory manipulation and accounting. For that purpose, we drop the capabilities CAP_SYS_(PACCT, CHROOT, ADMIN).

Time Management. We allow only C_0 to set the system time. Consequently, we drop the capability CAP_SYS_TIME from C_i in order to prevent them from setting the system time via system calls. However, in Android the time setting functionality works via the /dev/alarm driver. This is unfortunately not covered by CAP_SYS_TIME. We thus prohibit the access to /dev/alarm driver functionality for C_i by introducing new LSM hooks for Android alarm. With our custom LSM, we prevent time setting for C_i.

Power Management. In order to prevent containers from changing the global power state, such as from shutting down or waking the system, we drop the capabilities CAP_SYS_BOOT and CAP_WAKE_ALARM.

Kernel Module Loading. We prevent containers from loading kernel modules by dropping the capability CAP_SYS_MODULE.

Debugging. To prohibit containers from obtaining debugging control over other processes, we drop the capability CAP_SYS_PTRACE. We also enforce this with PID namespace checks in the ptrace controlling LSM hooks.

Logging. In order to prevent containers from making changes to the kernel logging functionality, we drop the capabilities CAP_AUDIT_(WRITE, CONTROL) and CAP_SYSLOG.

5   Secure Device Virtualization

Our architecture allows to securely and dynamically assign device functionalities to $C_{0..n}$ on a per-container basis. We classify each device as either a *non-virtualized, security, user space virtualized*, or *kernel virtualized* device. Depending on device type and container, we handle access to device functionality. Figure 4 depicts the virtualization and access mechanisms for the device types.

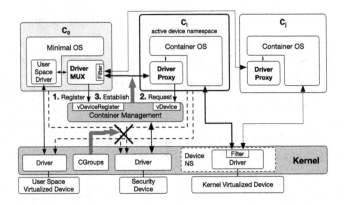

Fig. 4. Secure device virtualization mechanisms.

5.1   Non-virtualized and Security Devices

Security devices are part of the non-virtualized device category. We do not virtualize security devices, since they provide critical, security relevant functionality. In our architecture, these are the SE, LED and power button (see Sect. 3). We prohibit $C_{0..n}$ access to these devices, as depicted by the crossed dashed lines in Fig. 4. The hardware driver for accessing security devices is exclusively accessible to management components inside the TCB. We restrict access using the cgroups device access protection mechanism (see Sect. 4).

We allow access to other non-virtualized devices only to foreground containers. We enforce this device access rule during a container switch by dynamically adapting the cgroups devices whitelist (see Sect. 6). An example for such a device is the display, exclusively used by the foreground container.

5.2   Kernel Virtualized Devices

We virtualize kernel virtualized devices on kernel level using the device namespace mechanism [2]. In Fig. 4, C_i is in foreground as active device namespace, while C_j is in background. C_i is trying to access kernel virtualized devices from userland. The device driver in the kernel is addressed via the container's /dev filesystem. Examples for kernel-level virtualized devices are the alarm and input device (except for the power button), handled via the /dev filesystem. The driver decides about access to the functionality it offers based on the information about the active namespace (provided by the device namespaces). This is represented by the device namespace filter component in Fig. 4. With the cgroups devices subsystem, we have an additional driver-independent and dynamic mechanism to deny containers access to a device.

5.3   User Space Virtualized Devices

A lot of devices are accessed via proprietary user space drivers. In user space, we can re-use the existing drivers and achieve a portable solution. Thereby, we do not expand the TCB and avoid growing kernel complexity incurred by the virtualization. We place the virtualization functionality inside privileged C_0 and reuse its userland drivers. With the cgroups devices subsystem, we grant access to user space virtualized devices exclusively to C_0, highlighted by the crossed dashed lines in Fig. 4. Like C_i, C_0 is also allowed to use kernel-level virtualized devices, omitted in the illustration.

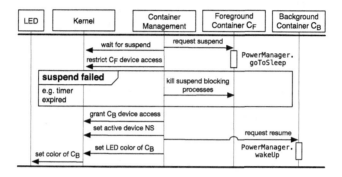

Fig. 5. The secure container switching procedure.

The Driver MUX in C_0 multiplexes the hardware device functionality for C_i. It utilizes the existing userland functionality and user space driver for hardware device access. The Driver MUX keeps track of the driver states and is aware of the different C_i. Device functionality is forwarded from C_0 over a dedicated communication channel to C_i. Userspace components in C_i are not aware of the Driver Proxy redirection. The CM sets up the channel between C_0 and C_i. The Driver MUX registers the device it virtualizes at the CM via the vDeviceRegister interface. When C_i tries to make use of a user space virtualized device's functionality, the Driver Proxy requests the CM for setting up a communication channel to the Driver MUX over the vDevice interface. Depending on whether we allow C_i access to the functionality of that device, the CM establishes the communication channel, as illustrated in Fig. 4. This can, e.g., be realized by creating a socket pair in the CM with the system call `socketpair`. The CM also informs C_0 of C_i requesting the device functionality. C_0 is thus aware of the specific container it is communicating with to securely provide C_i with different sets of functionalities for each hardware device. An example is the radio interface where C_i might be allowed to use the telephony and mobile data feature, while C_j might only be allowed to make use of the mobile data feature. The filter in the Driver MUX selectively handles device functionality access for C_i and filters non-protocol compliant data.

6   Secure Container Switch

When the user is in C_0, we use the Trusted GUI to trigger a switch to C_i. The user initiates a container switch from C_i back to C_0 by a long power button press in our concept. The CM handles the switch between containers. In the following, we describe the container switch procedure and its initiation inside C_i.

6.1   The Container Switching Procedure

Figure 5 depicts the container switch procedure. The illustration shows the switch between a foreground container C_F and a background container C_B to be put to foreground. The CM requests the suspension of C_F via the status interface to the vService. The vService triggers the suspend routine of the container OS, e.g., `PowerManager.goToSleep` on Android. The CM waits in a non-blocking mode for the OS to suspend. In the next step, the CM restricts C_F access to non-virtualized (and possibly kernel-level virtualized) devices, which are prohibited to background containers. We achieve this by dynamically reconfiguring the cgroups devices whitelist in the CM. With this mechanism, we separate device access decision making from device functionality filtering while accessing the device driver. A container could refuse or fail to suspend if certain processes do not release their resources. In that case, the CM kills those suspend blocking processes after a timeout. This forces the open devices to be closed and the container to suspension. In the next step, the CM grants C_B device access via dynamic cgroups device allocation. The following step is to switch the active device namespace to the new foreground container. The CM requests the resume of C_B via the container's vService, e.g., `PowerManager.wakeUp` on Android OS. To complete the container switch process, the CM sets the LED color according to the color of C_B via the kernel using the LED driver.

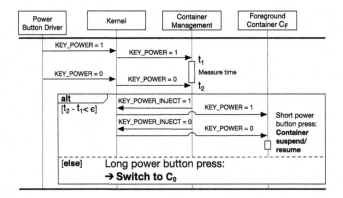

Fig. 6. The power button event capturing.

6.2   Switch Initiation in C_i

In order to securely switch to C_0 despite being in possibly malicious C_i, we use the power button. As a security device, it is exclusively accessible by the CM, meaning that power button events never arrive in $C_{0..n}$. We define the behavior, visualized in Fig. 6, as follows: Pressing the power button in C_i for more than a fixed time interval ϵ, e.g., 0.5 s, triggers a switch to C_0. Otherwise, the button triggers the suspend or resume functionality of C_F. We modify the kernel in order to forward power button events exclusively to the CM, i.e., the root namespace. The power button driver notifies the kernel of a power button pressed event (KEY_POWER, 1). The kernel forwards this event to the CM, which starts a timer at time t_1. When the power button is released, the release event (KEY_POWER, 0) arrives at the CM at time t_2. The CM then decides about the action to be carried out according to the fixed time interval ϵ. Until now, none of the events has reached any of the containers. When the user is in C_i and $t_2 - t_1 \geq \epsilon$, the CM conducts the switch to C_0. Otherwise, the CM transparently forwards the power button press and release events to C_F, resulting in either a resume or suspend. For the injection of power button events into $C_{0..n}$, we add a custom event, KEY_POWER_INJECT, to the kernel. We modify the kernel to recognize this special event type and to forward it as a common KEY_POWER event type to C_F. The power button event now appears to C_F as a common input event resulting from an input device. In case the foreground container is C_0, the CM always injects the power button events unmodified.

A malicious container spoofing C_0 could try to trick the user into believing of having switched, while stuck in the malicious one. In the worst case, the user might enter critical information inside malicious C_i. As the LED is a security device, we use it in order to securely identify C_F. The CM sets the LED color to the container's specific color.

7 Security Evaluation

We evaluate the security of our architecture regarding data confidentiality. If a container is compromised, our isolation mechanisms (see Sect. 4) ensure that the attacker with local root privileges cannot break out of the container's boundary, unable to leverage global privileges. An attacker can affect other components only through the specified communication channels. In the following, we consider the compromising of the different components and the implications on data confidentiality.

Attacker Model: In our attacker model, we consider an adversary having the capability of compromising every component outside the TCB. This includes taking full control over the privileged C_0 and the unprivileged C_i. The attacker is able to execute runtime attacks and is furthermore capable of acting as a MITM between the device and backend according to the Dolev-Yao model [12]. The attacker is also considered to be able to fully compromise a remote management backend. We also assume an attacker with physical access to the device trying to manipulate it via common physical interfaces, e.g., via USB and the touchscreen. However, we do not consider covert channels and advanced physical attacks on the device or the SE. This excludes side-channel attacks, especially cold boot, JTAG and microprobing attacks.

7.1   Compromised C_i

C_i is exposed to common attacks on the OS. In order to harden a container from compromise, we propose to limit it to trusted applications and to functionality required for its special purpose only. The processes inside C_i are isolated and protected by SELinux. Full control over the container and its data is only exposed when the attacker manages to take control over a process and to knock off SELinux. C_i cannot retrieve more device functionality via established user space virtualization channels than it is supposed to. C_0's Driver MUXs prevent this by making data routing decisions and input validation. C_i is also not capable of retrieving additional device functionality via the vDevice interface, as the CM handles setting up the connection between C_0 and C_i. The cgroups devices subsystem and device namespaces prevent C_i from prohibited device access to kernel-level virtualized devices. C_i can send fake status information or refuse commands from the CM via the status interface. However, C_i cannot deny container switching. Consequently, the overall system's behavior is not adversely affected by the compromise of C_i. Data confidentiality is retained beyond container boundaries, meaning that sensitive user data stored in other containers remains protected.

7.2   Compromised C_0

We tailored C_0 to the minimal amount of required functionality. SELinux policies further raise its security level. However, proprietary code of the container's drivers cannot be completely controlled. In case of the compromise of C_0, the attacker has access to the local control interface to the CM. The adversary can misuse device management functionality and intercept the user's passphrase entry for the SE. In this case, the attacker can start containers without user interaction when the SE is present. The attacker is also able to change settings, create and shutdown containers. C_0 has full access to user space virtualized devices and to already established Driver MUX channels. The attacker can hence drop, eavesdrop and transmit forged data from, resp., to those devices. Sensitive data transmitted over these communication channels must be encrypted to be transparent to C_0. The adversary controls the registering of device functionalities via the vDeviceRegister channel, but cannot set up new channels. The same consequences as for a compromised C_i hold regarding the status interface. Kernel-level virtualized devices and security devices cannot be impaired. The adversary also cannot take advantage of the update functionality of the CM, since the update interface is not accessible for C_0. In sum, the attacker controls many functionalities. However, the data in other containers remains confidential.

7.3   Compromised TCB

The TCB exposes full access and control over all functionalities, communication channels and data on the device. In contrast to running containers, non-running ones are still encrypted and hence remain opaque to the attacker. Only if the passphrase of the SE was intercepted and the SE is present, as well as unlocked, the adversary is able to retrieve to the containers' data. If the passphrase was not intercepted, the attacker cannot brute-force a present SE, because it locks itself after a certain amount of retries. In order to obtain control over the SE, physical access to the SE is required. The attacker has control over the backend communication channel between the device and backend. This exposes the capability to download updates and encrypted backups. However, the device cannot request container images from other devices, since the identity of the device is bound to certificates on the device itself [27].

7.4 Backend and Remote Management Link

The network communication between the device and backend is exposed to attacks according to the Dolev-Yao model. The channel is protected by TLS encryption using certificates, which prevents gaining control over this channel. In case the backend is compromised, the adversary can access the CM's control interface. The attacker is furthermore capable of carrying out denial of service attacks towards the device. The device verifies software updates through signature verification. The adversary cannot sign updates of the device's software entities, since the software-signing key and functionality are separated from the backend. Data confidentiality is hence preserved.

7.5 Physical Device Access

If the device is switched off and an attacker manages to extract all data, data confidentiality is not impaired. The storage cannot be decrypted, since the SE and its passphrase are both required for decrypting the containers. We also lock the device to prevent attackers from overwriting partitions, e.g., in firmware upgrade mode. We provide our own tailored recovery image featuring only uncritical functionality. Connecting to the USB port, an attacker has no access to the device's data and functionality. We remove functionalities, such as ADB, or mounting the device storage.

8 Implementation and Performance

We fully realized the implementation of the proposed architecture on the Samsung Galaxy S4 and Nexus 5 smartphones, where we use Android 5.1.1, resp. 4.4.4, as container OS. We verified the easy portability of the architecture by a proof-of-concept realization on the Nexus 7 tablet.

Linux Kernel. We enable the support of namespaces, capabilities and cgroups features on the device's kernel (AOSP kernel 3.4[2]). We extended the kernel with our new capability (see Sect. 4) and power key event capturing (see Sect. 6.2), as well as the LSM stacking feature (see Sect. 4). We therefore include the LSM

---

[2] https://android.googlesource.com/kernel/msm.

stacking patch, SELinux and our custom LSM implementation. We use device namespaces [2] for kernel-level virtualization, e.g., for the alarm, audio, binder and input devices.

CM. We implemented the CM as a non-blocking callback-based daemon in C using the epoll, inotify and timer kernel features. In contrast to LXC, the CM is a specifically tailored, minimalist implementation. It consists of less than 10,000 lines of code. We realized the update, control, vDevice and status interface's protocol layer with protobuf[3]. Protobuf serializes structured data transmitted over the different components and validates input. The CM processes incoming messages with callbacks. For the internal status, control and vDevice interfaces, we used UNIX domain sockets. We realized the remote control and update interfaces for the backend with TLS protected internet sockets. The CM establishes communication channels into C_i during its startup procedure. For that purpose, the CM creates a new Unix domain socket and inherits the corresponding file descriptor to the newly created root process of C_i. The root container process still under control of the CM binds the socket to a specific location in the container's filesystem. This location inside C_i can be accessed by specific processes supporting the virtualization, e.g., the vService or the Trusted GUI. The CM listens on the shared file descriptor and is hence able to accept connections from these processes over container boundaries. The CM sets up the cgroups subsystems, drops capabilities, loads the SELinux policies and revokes custom LSM privileges for C_i before delivering control to the container's `init` process. For dropping privileges, our LSM provides a special file in the `securityfs` pseudo filesystem. As soon as a process opens this file, its namespace and nested namespaces lose their LSM privileges, which is a one-way operation.

SM. We also implemented this component in C as a non-blocking callback-based daemon using the epoll, inotify and timer kernel features. The SM includes the OpenSSL library for cryptographic operations. In our implementation, we replaced the SE by a PKCS12 softtoken. With the token's private key, the CM wraps the symmetric key for container en-/decryption using `dm-crypt`. We protect the container images with a hash in a signed container configuration, including mount points, and user data images. The SM implementation comprises less than 1,500 lines of code.

C_0. This container runs a minimal Android. We kept only basic functionality and the native user space drivers and modules, such as the `rild` for accessing the radio hardware. We implemented the vService as an Android Service and the Trusted GUI as an Android system application. We realized the Driver MUXs as daemons that utilize the native drivers for user space virtualized devices, such as RIL, Wi-Fi and sensors, including GPS. The interfaces for user-space virtualization are also realized via UNIX domain sockets.

C_i. We modified the `init` process of Android to prevent `init` from firmware loading. The firmware is loaded only once into the system by C_0. We also prohibit

---

[3] https://github.com/google/protobuf.

the OS from loading the SELinux policy. We modified the Android framework to comply with a dropped resource set. For example, the stock Zygote process checks capabilities and would prevent the OS from booting.

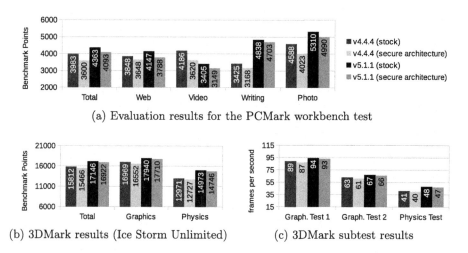

(a) Evaluation results for the PCMark workbench test

(b) 3DMark results (Ice Storm Unlimited)   (c) 3DMark subtest results

Fig. 7. Performance comparison on the Nexus 5 device between stock Android and our secure architecture using PCMark & 3DMark on Android 4.4.4 & 5.1.1 (Color figure online)

Performance Results. We ran the benchmark tools PCMark and 3DMark on the Nexus 5 with stock Android and with our secure architecture on Android 5.1.1, resp. 4.4.4. On the secure architecture, C_{0-2} were deployed and simultaneously running. We executed the benchmarks in foreground C_2. Figure 7 summarizes the performance results, which are average values over more than 30 test runs. The total number of points achieved with our secure architecture in 3DMark is close to the stock device results, because 3DMark is rather stressing the graphics hardware. 3DMark determines this figure based on the graphics and physics test results in Fig. 7b, deducted from the results in Fig. 7c. The performance impact of our architecture in PCMark is no more than 6.5 % compared to stock Android 5.1.1 (resp. 10 % with Android 4.4.4). PCMark obtains the total amount of points by aggregating over the subtest results in Fig. 7a. In general, the user experience with the secure architecture exposed no recognizable performance impact.

We measured the container switching time, C_{0-2} running, from C_0 to C_i and vice versa when C_F is not suspended. The switching procedure consumes about 330 ms to switch from C_i to C_0 and 300 ms from C_0 to C_i. High load in C_1 and C_2, such as running HD videos, caused only negligible overhead. Most time is allocated for suspending C_F. We measured the switching time in case C_F is suspended to consume only about 60 ms. Thereby, most time is spent in resuming former C_B.

9   Conclusions

We developed a secure architecture for OS-level virtualization on mobile devices. Including an SE, the main objective of the secure architecture is data confidentiality at container boundaries. To fulfill this goal, we systematically isolated the different, simultaneously running containers from each other. Therefore, we restricted the different containers to a minimal set of controlled functionality. This made it possible to confine communication of the architecture's components to only well-defined channels for container management and device virtualization. In order to realize the strict isolation, we devised a stacked LSM concept using SELinux and a specially tailored, custom LSM. We furthermore leveraged Linux capabilities and the cgroups devices subsystem. Based on that, we developed mechanisms for secure device virtualization and secure container switching sustaining a seamless user experience. Thereby, we classified devices into different categories and provided containers with distinct hardware functionalities on a per-container basis. To demonstrate the feasibility of our approach, we realized the secure architecture with a fully-functional implementation, applicable in real-life, on the Samsung Galaxy S4 and the Nexus 5 devices. The performance evaluation shows that the system performs well and that it is suitable for real-life application. In our security evaluation, we demonstrated that the architecture provides data confidentiality even when large parts of the system are compromised.

References

1. Almohri, H.M., Yao, D.D., Kafura, D.: DroidBarrier: know what is executing on your Android. In: Proceedings of the 4th ACM Conference on Data and Application Security and Privacy, CODASPY 2014, pp. 257–264. ACM (2014)
2. Andrus, J., Dall, C., Hof, A.V., Laadan, O., Nieh, J.: Cells: a virtual mobile smartphone architecture. In: Proceedings of the Twenty-Third ACM Symposium on Operating Systems Principles, SOSP 2011, pp. 173–187. ACM (2011)
3. Backes, M., Gerling, S., Hammer, C., Maffei, M., von Styp-Rekowsky, P.: AppGuard – fine-grained policy enforcement for untrusted Android applications. In: Garcia-Alfaro, J., Lioudakis, G., Cuppens-Boulahia, N., Foley, S., Fitzgerald, W.M. (eds.) DPM 2013 and SETOP 2013. LNCS, vol. 8247, pp. 213–231. Springer, Heidelberg (2014)
4. Barham, P., Dragovic, B., Fraser, K., Hand, S., Harris, T., Ho, A., Neugebauer, R., Pratt, I., Warfield, A.: Xen and the art of virtualization. In: Proceedings of the Nineteenth ACM Symposium on Operating Systems Principles, SOSP 2003, pp. 164–177. ACM (2003)
5. Becher, M., Freiling, F., Hoffmann, J., Holz, T., Uellenbeck, S., Wolf, C.: Mobile security catching up? Revealing the nuts and bolts of the security of mobile devices. In: 2011 IEEE Symposium on Security and Privacy (SP), pp. 96–111 (2011)
6. Brakensiek, J., Dröge, A., Botteck, M., Härtig, H., Lackorzynski, A.: Virtualization as an enabler for security in mobile devices. In: Proceedings of the 1st Workshop on Isolation and Integration in Embedded Systems, IIES 2008, pp. 17–22. ACM (2008)

7. Bugiel, S., Davi, L., Dmitrienko, A., Heuser, S., Sadeghi, A.R., Shastry, B.: Practical and lightweight domain isolation on Android. In: Proceedings of the 1st ACM Workshop on Security and Privacy in Smartphones and Mobile Devices, SPSM 2011, pp. 51–62. ACM (2011)

8. Bugiel, S., Heuser, S., Sadeghi, A.R.: Flexible and fine-grained mandatory access control on Android for diverse security and privacy policies. In: Proceedings of the 22nd USENIX Conference on Security, SEC 2013, pp. 131–146. USENIX Association (2013)

9. Chen, W., Xu, L., Li, G., Xiang, Y.: A lightweight virtualization solution for Android devices. IEEE Trans. Comput. **64**, 2741–2751 (2015)

10. Chin, E., Felt, A.P., Greenwood, K., Wagner, D.: Analyzing inter-application communication in Android. In: Proceedings of the 9th International Conference on Mobile Systems, Applications, and Services, MobiSys 2011, pp. 239–252. ACM (2011)

11. Dall, C., Nieh, J.: KVM/ARM: the design and implementation of the Linux ARM hypervisor. In: Proceedings of the 19th International Conference on Architectural Support for Programming Languages and Operating Systems, ASPLOS 2014, pp. 333–348. ACM (2014)

12. Dolev, D., Yao, A.C.: On the security of public key protocols. In: Proceedings of the 22nd Annual Symposium on Foundations of Computer Science, SFCS 1981, pp. 350–357. IEEE Computer Society (1981)

13. Enck, W., Gilbert, P., Chun, B.G., Cox, L.P., Jung, J., McDaniel, P., Sheth, A.N.: TaintDroid: an information-flow tracking system for realtime privacy monitoring on smartphones, pp. 1–6 (2010)

14. Enck, W., Ongtang, M., McDaniel, P.: On lightweight mobile phone application certification. In: Proceedings of the 16th ACM Conference on Computer and Communications Security, CCS 2009, pp. 235–245. ACM (2009)

15. Feizollah, A., Anuar, N.B., Salleh, R., Wahab, A.W.A.: A review on feature selection in mobile malware detection. Digit. Invest. **13**, 22–37 (2015). Elsevier Science Publishers B. V

16. Felt, A.P., Finifter, M., Chin, E., Hanna, S., Wagner, D.: A survey of mobile malware in the wild. In: Proceedings of the 1st ACM Workshop on Security and Privacy in Smartphones and Mobile Devices, SPSM 2011, pp. 3–14. ACM (2011)

17. Hwang, J.Y., bum Suh, S., Heo, S.K., Park, C.J., Ryu, J.M., Park, S.Y., Kim, C.R.: Xen on ARM: system virtualization using Xen hypervisor for ARM-based secure mobile phones. In: 5th IEEE Consumer Communications and Networking Conference, CCNC 2008, pp. 257–261 (2008)

18. Kamp, P.H., Watson, R.N.: Jails: confining the omnipotent root. In: Proceedings of the 2nd International SANE Conference, vol. 43 (2000)

19. Laadan, O., Nieh, J.: Operating system virtualization: practice and experience. In: Proceedings of the 3rd Annual Haifa Experimental Systems Conference, SYSTOR 2010, pp. 17:1–17:12. ACM (2010)

20. Merkel, D.: Docker: lightweight Linux containers for consistent development and deployment. Linux J. **2014**(239), 2 (2014)

21. Ongtang, M., McLaughlin, S., Enck, W., McDaniel, P.: Semantically rich application-centric security in Android. In: Proceedings of the 2009 Annual Computer Security Applications Conference, ACSAC, pp. 340–349. IEEE Computer Society (2009)

22. Peng, S., Yu, S., Yang, A.: Smartphone malware and its propagation modeling: a survey. IEEE Commun. Surv. Tutorials **16**, 925–941 (2014)

23. Poeplau, S., Fratantonio, Y., Bianchi, A., Kruegel, C., Vigna, G.: Execute this! Analyzing unsafe and malicious dynamic code loading in Android applications. In: Proceedings of the 20th Annual Network and Distributed System Security Symposium (NDSS) (2014)

24. Reshetova, E., Karhunen, J., Nyman, T., Asokan, N.: Security of OS-level virtualization technologies. In: Bernsmed, K., Fischer-Hübner, S. (eds.) NordSec 2014. LNCS, vol. 8788, pp. 77–93. Springer, Heidelberg (2014)

25. Rossier, D.: EmbeddedXEN: a revisited architecture of the XEN hypervisor to support ARM-based embedded virtualization. White Paper, Switzerland (2012)

26. Russello, G., Conti, M., Crispo, B., Fernandes, E.: MOSES: supporting operation modes on smartphones. In: Proceedings of the 17th ACM Symposium on Access Control Models and Technologies, SACMAT, pp. 3–12. ACM (2012)

27. Wessel, S., Huber, M., Stumpf, F., Eckert, C.: Improving mobile device security with operating system-level virtualization. Comput. Secur. (2015). http://www.sciencedirect.com/science/article/pii/S0167404815000206

28. Wessel, S., Stumpf, F., Herdt, I., Eckert, C.: Improving mobile device security with operating system-level virtualization. In: Janczewski, L.J., Wolfe, H.B., Shenoi, S. (eds.) SEC 2013. IFIP AICT, vol. 405, pp. 148–161. Springer, Heidelberg (2013)

29. Wu, C., Zhou, Y., Patel, K., Liang, Z., Jiang, X.: AirBag: boosting smartphone resistance to malware infection. In: Proceedings of the Network and Distributed System Security Symposium (2014)

30. Xavier, M.G., Neves, M.V., Rossi, F.D., Ferreto, T.C., Lange, T., De Rose, C.A.F.: Performance evaluation of container-based virtualization for high performance computing environments. In: Proceedings of the 2013 21st Euromicro International Conference on Parallel, Distributed, and Network-Based Processing, PDP 2013, pp. 233–240. IEEE Computer Society (2013)

31. Zhou, Y., Jiang, X.: Dissecting Android malware: characterization and evolution. In: Proceedings of the 2012 IEEE Symposium on Security and Privacy, SP 2012, pp. 95–109. IEEE Computer Society (2012)

Assessing the Disclosure of User Profile in Mobile-Aware Services

Daiyong Quan, Lihuan Yin$^{(\boxtimes)}$, and Yunchuan Guo

Institute of Information Engineering, Chinese Academic Sciences,
Beijing 100093, China
{quandaiyong,yinlihua,guoyunchuan}@iie.ac.cn

Abstract. Mobile-aware services can be regarded as data-sharing systems in nature. In these systems, users obtain personalized service at the cost of sharing their personal information. As a result, it will inevitably lead to the disclosure of users' profiles and raise the serious privacy concerns. To assessing the privacy risk of sharing the user profile information items, in this paper we score and measure the potential risk of users caused by sharing information for the sake of personalization services. By adopted the 3-parameter logistic model, we explore information item's sensitivity, influence and probability of proper setting as well as users' potential attitudes to measure the privacy disclosure risk. The MMLE/EM algorithm is then adopted to estimate the above parameters. Finally, experiments on synthetic and real-world data sets are conducted and the results show that the obtained scores of our approach fit well with the real-world data.

Keywords: Privacy scoring · Risk assessment · Mobile-aware sensing service · Three-parameter-logistic model

1 Introduction

With the wide application of mobile-aware services, users shared a rich of personal information to the service server for personalized service (e.g., location-based). Our smartphones, wearable devices, cars, or credit cards generate information about where we are, whom we call, or how much we spend. In addition, for scientific research and commercial purposes, some user profiles are acquired, retained and/or processed by a third party without the consent by the individual. In science, it is essential for the data to be available and shareable. Sharing data allows scientists to build on previous work, replicate results, or propose alternative hypotheses and models. However, during the process of the sharing and processing, the presence of abuse the sensitive data will lead to a privacy violation. Users' identity information, behaviors and other sensitive information will be leaked by the inference attack with the auxiliary knowledge that the adversary might have gathered. For example, the attacker can infer the patient's genetic privacy information by analysis the personalized warfarin dosing [1]. The adversary can infer the private information using social network data [2]. And an adversary can exploit an online social network with a mixture of public and private user profiles to predict the private attributes of users [3].

© Springer International Publishing Switzerland 2016
D. Lin et al. (Eds.): Inscrypt 2015, LNCS 9589, pp. 451–467, 2016.
DOI: 10.1007/978-3-319-38898-4_26

The serious privacy risk caused by the sharing the user profile's is overlooked. Users can not choose the appropriate privacy protection technology, due to the lack of accurate assessment of risk. How to measure the sharing/publishing private data leakage risks resulting from our research goals. Note that we are not concerned with database privacy, but with the privacy issues of releasing a single sensitive user profile.

In this paper, we address the privacy issues of sharing/releasing the user profile information items by scoring the potential risk of users. This score measures the users' potential privacy risk due to sharing behaviors for the sake of personalization services. Our definition of privacy score increases with the (i) sensitivity of the information items being revealed and (ii) with the impact of the revealed information items to the recipients, as well as the probability for items of profile information to be externally related. Intuitively, the more sensitive and influential of information item and the more unique (sole) the user profile in the whole data set, the greater the potential user privacy risk. We develop three-parameter-logistic model to estimate these three factors, and we show how to combine these three factors in the calculation of the privacy score. Based on this privacy score, users can select privacy enhancement technologies accordingly. Specifically, we estimate information item parameters (sensibility, impact and random setting) using EM algorithm, and use three-parameter-logistic model to calculate the probability of each item that the user is set up appropriately. The complement of the probability of each item that the user is set up appropriately is the privacy risks of each item. Finally, we put all the item privacy score together to obtain the privacy score of the user.

The contribution of this paper is threefold.

- We provided an effective methodology for computing users' privacy scores in mobile-aware settings.
- We gave EM algorithm for the computation of privacy score that will be used to guide the selection of appropriate privacy protection technologies.
- We demonstrated the effectiveness of our proposed method over synthetic data sets and real world datasets.

The rest of this paper is organized as follows. Section 2 provides the overview of the related work and the preliminaries of privacy risk assessment are in Sect. 3. We apply MMLE/EM algorithm to estimate the above parameters in Sect. 4. Section 5 presents the experimental results show that the privacy scoring fit well with real-world data. Finally, Sect. 6 provides our concluding remarks.

2 Related Work

Current the works with respect to the privacy risk assessment are achieved mainly through two ways. Firstly, privacy disclosure risk assessment is carried out through simple inquiry. For instance, whether the profile contains any sensitive information (precise location, personal physiological feature and identity) before it is released? If not, the user will have a low disclosure risk. Secondly, privacy disclosure risk assessment is conducted through technological means. Common risk assessment techniques include EBIOS (expression of needs and identification of security) [4] and PIA (privacy impact assessment) [5]. The former protects personal data as if they are valuable property and fully considers the privacy issues of the relevant data objects, while the latter reviews the potential privacy issues and risks from the perspective of all stake-holders and seeks ways to avoid and minimize the impact of privacy disclosure. To avoid contact with raw data when conducting risk assessment, privacy risk should be assessed based on privacy protection. The most commonly used method is security multiparty computation [6]. In addition, with the rapid development of social network, the leak of users' privacy is more common. In Literature [7], the sensitivity and visibility of information are assessed by the latent trait model, and privacy risk scoring mechanism can automatically give scores for corresponding operations and alert at the early stage. In Literature [15], p-link anonymous method that quantifies the similarity between user profiles is adopted, which focuses on the anonymization of information items in user profile but not on risk assessment.

Our model is inspired by the work [7], but there are some differences between them. Our model takes into account more scoring indicators. From the perspective of risk analysis elements, we extended their model. Since they focus on the sensitivity of information and ignore the attacker's inference ability, we improve our model in this respect. By setting $' = 0$, we can get their scoring model. In addition, their model is applied in social network, while our model can be used in more scenarios. Therefore, we pay more attention to the influence of information, namely, information gain. Works in [8, 9] assess the inference risk of attackers based on specific background information of social networks (links from friends and information spreading in Circle of Friends), but fail to consider the user's privacy preference, namely, the sensitivity of information. In this paper, the privacy disclosure risks in mobile-aware settings are assessed by incorporating the sensitivity of information item, impact of the information item and the random setting of the information of a user profile.

3 Preliminaries and Notions

In this section, we give the preliminaries for privacy risk assessment. And then, we briefly introduce several notions involved in privacy scoring.

3.1 User Profiles

When users share their personal data with the mobile aware service system, attackers (other common users in the service system or non-trusted third parties) can collect user information to construct user profile set. User profile is a vector of the user information items. We assume that user i has a profile consisting of m profile items (also called information items). We use $up_i = \{it_1, it_2, \cdots, it_m\}$ denote this vector. For each profile item, such as it_i, user set a privacy level that determines their willingness to disclose information associated with this item. At present, the domain of the privacy level in this paper is $\{0, 1\}$. It will be extended $\{0, 1, \cdots, l\}$ in the near future. Let $UP = \{up_1, up_2, \cdots, up_n\}$ be n user profiles. The $n \times m$ matrix U stores the privacy levels of all n users for all m profile items. $U(i, j) = 1$ means the ith user is willing to disclose the contents of *jth* information item. While $U(i, j) = 0$ means the user will not to public the sensitive information associated with this item. For example, we assume user i and let $j = \{$mobile-phone number$\}$ be a single profile item. Naturally, setting $U(i, j) = 1$ is a more risky behavior than setting $U(i, j) = 0$. Making i's mobile phone publicly available increases i's privacy risk.

In general, when $U(i,j) < U(i,\hat{j})$, it reveals the user i is unwilling to disclose contents of information item j, with respect to another information item $\hat{j}$. Similarly, when $U(i,j) < U(i',j)$, it means the user i to be more conservative about the privacy settings of information item j than the other user i'. Then we introduce several related vector.

Row vector U_i represents the privacy level settings on all information items of the *ith* user. It indicates how much the user cares about the sensitive information items. Column vector U_j represents the privacy level on the *jth* information item by all users. It embodies the sensitivity of the information item itself. For instance, a 2×3 privacy setting matrix is denoted as $U = \begin{bmatrix} 0 & 1 & 1 \\ 0 & 0 & 1 \end{bmatrix}$. The first row vector $\begin{bmatrix} 0 & 1 & 1 \end{bmatrix}$ corresponds to the user who does not really care about his/her privacy and thus most of the items have a privacy weight of 1. The second row vector $\begin{bmatrix} 0 & 0 & 1 \end{bmatrix}$ corresponds to the user who values his/her sensitive information items and thus most of the items have a privacy weight of 0. As mentioned above, the column vector of the user data embodies sensitivity of the information item itself. The first column vector $\begin{bmatrix} 0 & 0 \end{bmatrix}^T$ represents that the first item of information is highly sensitive and all the users will not sharing this item. The second column vector $\begin{bmatrix} 1 & 0 \end{bmatrix}^T$ represents that the second item of information is moderately sensitive. Some users make it public while the rest make it unavailable. The third column vector $\begin{bmatrix} 1 & 1 \end{bmatrix}^T$ represents that the third item of information is not sensitive. All users will share it without privacy risk.

3.2   Notions of Privacy Scoring

Privacy attitude quantifies how concerned the user is about his or her privacy. By setting the information item public or not, the user reveal their privacy attitude implicitly. For instance, those users who hardly release any sensitive items of information have a serious attitude towards the privacy. Conversely, those users who arbitrarily share the location on social networks in real time hold in contempt of the privacy risk.

Sensitivity of a profile item quantifies the privacy level. The bigger value of the privacy level, the more sensitive the item is. Intuitively, the higher the sensitivity of an item, the less number of people are willing to disclose it.

The user profile in mobile-aware system has a multi-attribute characteristic. Users leave their nickname, IM contact number, gender, friends and other items of information on Weibo account. They also share the items: real name, IM contact number, graduate institutions and so on in Renren[1]. For instance in Zhenai.com[2], some users make their email, marital status, education level, income, hobbies and expertise, sexual orientation and other information items available. For each item of information, the user can set the privacy settings. The privacy is mainly addressed by restricting, through an access-control mechanism. Which information item is fully open, which information is totally hidden, and which information items are open to the friends while some ones close to the stranger. The following examples illustrate the sensitivity of a profile item.

Example 1. Assume user i and two profile items $j = \{$mobile-phone number$\}$ and $j' = \{$age$\}$. $U(i,j) = 1$ is a much more risky setting than $U(i,j') = 1$. The mobile-phone number is more sensitive than the age. The sensitivity of an item depends on the item itself. In the above example, phone number is more sensitive than the age. Moreover, it depends on the user's privacy preference. Such as, in dating platform sexual orientation item is more sensitive than the phone number item.

Impact of the information item quantifies the information receiver's information gain. In the mobile-aware era, the public of certain information items of user profile will incur the dramatic changes in cognitive for the receivers. The disclosure of the information items shocks the public. In this paper, we adopt the slope of the curve that characterized the probability of setting correctly as the impact of the information items.

Random setting of the information item quantifies the random guess. We often consider users' settings for different profile items as random process. There are still a small chance that the user with no privacy protection ability setting the items correctly. In such case, the user makes a guess to configure his profile items.

4   Assessment Method of Privacy Disclosure Risks

This section, we elaborate our assessment method of privacy disclosure risks. We present the mathematic model and how to estimate the parameters of a specific information item given the user profile matrix.

4.1   Active Privacy Assessment Framework

In this subsection, we present a general framework, as depicted in Fig. 1. In this paper, we focus on the privacy risk assessment.

User profiles are the input and the risk scores of users are the output. The framework is composed of two core components: privacy risk assessment component and Privacy Enhancement Technologies component. The former is our topic in this paper, and the latter is our follow-up study. We deem PETs as important, albeit orthogonal, to our work.

Risk assessment task should have two functions: (1) the capability to actively determine the level of privacy risks prior to data sharing (2) the ability to guide users to

---

[1] a leading real-name social networking internet platform in China.

[2] a social network for dating.

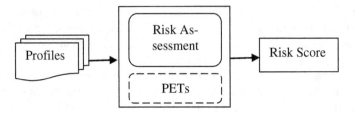

Fig. 1. Privacy risk assessment and management framework

select proper privacy protection technologies once data sharing has been decided. It should be noted that efforts should be made to protect relevant data during the risk assessment.

4.2 Mathematical Model

We assume users' settings for different profile items as random variables described by a probability distribution. In this case, the matrix U (CF. Sect. 3.1) observed is just one sample that follows probability distribution. For binary matrix whose value range is limited to $\{0, 1\}$, p_{ij} is used to represent the probability of choosing $U(i,j) = 1$ by the user i, namely, $p_{ij} = Pr\{U(i,j) = 1\}$. The privacy disclosure risk indicator is embodied by privacy risk score. The greater the value is, more likely the user's privacy will be disclosed. Our basic idea is that privacy disclosure risk scoring obeys the following principles:

More sensitive the information item is, the greater the value will be; more influential the released information item is, the greater the value will be; more likely the user profile is correlated with external data, the greater the value will be. Thus, privacy disclosure risk score is a monotonic function of three parameters: information sensitivity, influence, correlation possibility.

Inspired by the psychological test process, we take the user's setting of information item in related profile as a test. The 3-parameter logistic model is adopted to calculate the probability that the information item meets privacy requirements after the setting, namely, $p_{ij} = \Pr(U_{ij} = 1 | \theta_i, \alpha_j, \beta_j, \eta_j)$.

$$p_{ij} = \eta_j + \frac{1 - \eta_j}{1 + e^{-\alpha_j(\theta_i - \beta_j)}} \tag{1}$$

The main parameter symbols and their descriptions are shown in Table 1.

Table 1. Parameter description

Symbol	Descriptions
α	The influence of the released information item
β	The sensitivity of information item
θ	The user's awareness of privacy protection
	The risk that user profile is correlated with external data
ρ_{ij}	The probability that the user i's information item j is properly set

p_{ij} is a function of three parameters $(\alpha_j, \beta_j, \eta_j)$, which also involves the user i's privacy attitude θ_i. For a given user profile UP_i with parameters $(\alpha_j, \beta_j, \eta_j)$, the plot of the above equation in Eq. (1) is called the characteristic curve of the information item. The individual risk score of the information item j for the user i is quantified and expressed as:

$$fen(i,j) = 1 - p_{ij} \qquad (2)$$

It shows that the privacy disclosure risk faced by the user i in terms of the information item j equals to the probability of improper setting of this information item j. When we put all the items' score together, we obtain the privacy score of a user.

$$fen(i) = \sum_{j=1}^{m} fen(i,j) \qquad (3)$$

To calculate the privacy score, we need to estimate the parameters $(\alpha_j, \beta_j, \eta_j)$ of all information items $j \in \{1, \cdots, m\}$ and the privacy attitude θ_i of all the users $i \in \{1, \cdots, n\}$. Based on Eq. (1), the probability p_{ij} is computed, namely, $p_{ij} = \Pr[U(i,j) = 1]$. If the values of the parameters are known, the computing p_{ij}, for every i and j, is trivial. Thereby, we can obtain the user privacy score according to Eq. (3). In this following subsection, we will present in details how to estimate the parameters of a specific information item j. that is, how to estimate the specific $\theta, \alpha, \beta, \eta$.

4.3   Estimation of Parameters

For different user i and different information item j, the distribution of $u(i,j)$ also varies, so the Method of Moments does not work. Therefore, maximum likelihood estimation (MLE) method is often adopted to estimate parameters due to its inherent advantages [12].

We have n user profiles with privacy settings, whose generation process is independently and identically distributed. Thus, the likelihood function is:

$$L(\theta; \alpha, \beta | U) = \prod_{i=1}^{n} \prod_{i=1}^{m} P_{ij}^{u_{ij}} (1 - P_{ij})^{1 - u_{ij}} \qquad (4)$$

As we know, $p_{ij}(1 - P_{ij}) > 0$, taking its log will be helpful for further calculation. The log-likelihood function is expressed as:

$$\ell(\theta; \alpha, \beta | U) = \log L(\theta; \alpha, \beta, \eta | U)$$
$$= \sum_{i=1}^{n} \sum_{j=1}^{m} (u_{ij} \log P_{ij} + (1 - u_{ij}) \log(1 - P_{ij})). \qquad (5)$$

By calculating the conditional extremum of Eq. (5), the maximum likelihood estimation value of the parameter is obtained. We estimate parameters in two cases: (1) the parameters of the information item are known, and the user's privacy attitude needs to be estimated. That is, $(\vec{\alpha}, \vec{\beta}, \vec{\eta})$ are given and $\vec{\theta}$ is to be calculated. (2) The user's privacy attitude $\vec{\theta}$ and the parameters $(\vec{\alpha}, \vec{\beta}, \vec{\eta})$ all remain unknown. We need to estimate these parameters and the user's privacy attitude in accordance with the privacy setting matrixes of n users. In the latter case, EM algorithm is adopted to achieve marginal maximum likelihood estimation (MMLE/EM). Due to space limitations, we elaborate the latter case.

The first case: the parameters of the information item are known, and the user's privacy attitude needs to be estimated. For each information item j, logarithmic likelihood function (5) has an unknown parameter θ, and θ_i's first-order derivative is 0, expressed as:

$$f = \frac{\partial l}{\partial \theta_i} = 0 \ i = 1,\dots,n \tag{6}$$

Where the parameter expression with respect to the privacy attitude θ is a nonlinear equation and the iterative method must be used to solve it, such as Newton-Raphson (N_R) [10] or dichotomy method. In this paper, we use Newton-Raphson method.

The second case: Our basic ideas are as follows: First of all, under the condition that the initial value of the user's privacy attitude is given, the parameter values of the information item are estimated; based on these values, the user's privacy attitude value is estimated; further, the privacy attitude value is used for upgrading the parameters of the information item. Repeat the iterations until the convergence is achieved. The main feature of EM algorithm lies in that each step of iteration consists of two steps: firstly, get the Expectation Step (abbreviated as E Step); secondly, get the Maximization Step (abbreviated as M Step). To be specific, a priori of the user's privacy attitude is firstly given and then integration (marginalized means of privacy attitude) is conducted.

Let $\zeta = (\alpha,\beta,\eta)$ and $p\left(u_i|\vec{\zeta}\right) - \int L\left(u_i|\theta_i\vec{\zeta}\right) g\left(\theta_i|\vec{\zeta}\right) d\theta_i$. According to Bayes Theorem, the conditional distribution of privacy attitude θ for u_i (namely, θ's posterior distribution) is $h\left(\theta|u_i, \vec{\zeta}\right) = \frac{L\left(u_i|\theta_i,\vec{\zeta}g(\theta_i)\right)}{P\left(u_i|\vec{\zeta}\right)} = \frac{L\left(u_i|\theta_i,\vec{\zeta}g(\theta_i)\right)}{P\left(u_i|\vec{\zeta}\right)g(\theta_i)d\theta_i}$. Then the marginal likelihood function based on the user's privacy setting U is $M = \prod_{i=1}^{n} P\left(u_i|\vec{\zeta}\right)$. The equivalent form of the logarithmic marginal likelihood function is:

$$\ln M = \sum_{i=1}^{n} \ln P\left(u_i|\vec{\zeta}\right) \tag{7}$$

It can be seen from the above equation that M's first-order derivatives all have integral symbols. Gauss-Hermite quadrature can be used to convert integration into summation.

E Step: Calculating $E_\theta\left[\ln L\left(U,\theta|\zeta\right)|\ U,\zeta^{(p)}\right]$, and we denote it as $\Omega\left(\zeta|\zeta^{(p)}\right)$.

$$\Omega\left(\zeta|\zeta^{(p)}\right) = \sum_{i=1}^{n}\left\{E_\theta\left[\ln L(u_i|\theta_i,\zeta)|u_i,\zeta^{(p)}\right] + E_\theta\left[\ln g(\theta_i)|u_i,\zeta^{(p)}\right]\right\} \qquad (8)$$

M Step: $\Omega\left(\zeta^{(p+1)}|\zeta^{(p)}\right) \geq \Omega\left(\zeta|\zeta^{(p)}\right)$. To obtain the $\zeta^{(p+1)}$, we must solve the nonlinear equation where $f_k = 0, k = 1,2,3$, $f_1 = \frac{\partial \ln M}{\partial \alpha_j}$, $f_2 = \frac{\partial \ln M}{\partial \beta_j}$, $f_3 = \frac{\partial \ln M}{\partial \eta_j}$, $j = 1,\ldots,m$.

As $f_k = 0$, $k = 1,2,3$ are integrals, we will adopt the form of numerical integral and set $g(\theta)$ as a standard normal distribution. By using Gauss-Hermite quadrature to convert integration into summation, the following equation can be obtained:

$$f_j \approx (1 - \eta_j)\sum_{k=1}^{q}(x_k - \beta_j)\left(u_{ij} - \hat{P}_{kj}\right)\hat{W}_{kj}\hat{h}\left(x_k|u_{ij},\zeta^{(p)}\right) \qquad (9)$$

where $x_1, x_2\ldots,x_q$ are integral nodes and $A(x_k)$ is the quadrature coefficient. $P_{ij} = \eta_j + \frac{1-\eta_j}{1+\exp^{(-\alpha_j(X_i-\beta_j))}}$, $P_{ij}^* = \frac{P_{ij}-\eta_j}{1-\eta_j}, Q_{ij}^* = 1 - P_{ij}^*$, $w_{ij} = \frac{P_{ij}^* Q_{ij}^*}{P_{ij}Q_{ij}}$, $\hat{h}(x_k|u_{ij},\zeta^{(p)}) = \frac{L_i(x_k)A(x_k)}{\sum_{k=1}^{q}L_i(x_k)A(x_k)}$ Let $r_{kj}^{(p)} = \sum_{i=1}^{n}u_{ij}\hat{h}(x_k|u_{ij},\zeta^{(p)})$ and $f_k^{(p)} = \sum_{i=1}^{n}\hat{h}(x_k|u_{ij},\zeta^{(p)})$, the above numerical integrations of f_j can be:

$$f_1 \approx (1 - \eta_j)\sum_{k=1}^{q}(x_k-\beta_j)\hat{w}_{kj}\left(r_{kj}^{(p)} - \hat{P}_{kj}f_k^{(p)}\right) \qquad (10)$$

$$f_2 \approx (1 - \eta_j)\alpha_j\sum_{k=1}^{q}\hat{w}_{kj}\left(r_{kj}^{(p)} - \hat{P}_{kj}f_k^{(p)}\right) \qquad (11)$$

$$f_3 \approx -\frac{1}{1-\eta_j}\sum_{k=1}^{q}\frac{r_{kj}^{(p)} - P_{kj}f_k^{(p)}}{\hat{P}_{kj}} \qquad (12)$$

$f_k^{(p)}$ means the expectation value of the user with privacy setting vector u_i and privacy protection capability x_k among all n users based on the given $\zeta^{(p)}$. $r_{kj}^{(p)}$ represents the expectation number of users with privacy setting vector u_i and privacy protection capability x_k who properly set the *jth* information item based on the given $\zeta^{(p)}$.

To solve the equations, $F(\zeta) = \{f_1(\zeta) = 0, f_2(\zeta) = 0, f_3(\zeta) = 0\}$, we must compute the second-order partial derivatives $\frac{\partial \Omega(\zeta|\zeta^{(p)})}{\partial t_i \partial t_j} = g_{ij}, t_i, t_j \in \{\alpha_j, \beta_j, \eta_j\}$. Obviously, second-order derivatives can be further derived by first-order derivative: $g_{ij} = \frac{\partial f_i}{\partial t_j}, i, j = 1, 2, 3$. We assume that the parameter value of the information item ζ has worked out the estimated value at the previous iteration. Assume that the newly estimated value is always more close to the true value than the previous estimated value, and we take the newly obtained parameter estimated value as the true value of ζ. Then, $f_k^{(p)}$ and $r_{kj}^{(p)}$ will be known in the next iteration. According to the meanings of $f_k^{(p)}$ and $r_{kj}^{(p)}$, we have the equation $E\left(r_{kj}^{(p)} - f_k^{(p)} \hat{P}_{kj}\right) = 0$. The second-order derivative of $\ln M$ is more simper, which greatly reduce the solving process.

$$E_{g11} \approx \frac{\partial f1}{\partial \alpha j} = \frac{\frac{-1}{(1-\eta_j)^2} \sum_{k-1}^{q} (xk - \beta j)^2 f_k^{(p)} \left(\hat{P}_{kj} - \eta_j\right)^2 \hat{Q}_{kj}}{\hat{P}_{kj}} \tag{13}$$

Likewise, the values of $E_{g22}, E_{g23}, E_{g31}, E_{g32}, E_{g33}$ can also be estimated.

N-R method can be used for iteration to linearize the nonlinear equation step by step, and one set of linear equations will be solved in the each iteration. The iteration equation is $\zeta_{k+1} = \zeta_k - J^{-1}F(\zeta_k), k = 0, 1, 2, \cdots$, which can be transformed into:

$$J(\zeta_k)(\zeta_{k+1} - \zeta_k) + F(\zeta_k) = 0 \tag{14}$$

Where is the Jacobian matrix that consists of second-order derivatives.

$$J = \begin{pmatrix} \frac{\partial^2 \ln L}{\partial \alpha_j^2} & \frac{\partial^2 \ln L}{\partial \alpha_j \partial \beta_j} & \frac{\partial^2 \ln L}{\partial \alpha_j \partial \eta_j} \\ \frac{\partial^2 \ln L}{\partial \beta_j \partial \alpha_j} & \frac{\partial^2 \ln L}{\partial \beta_j^2} & \frac{\partial^2 \ln L}{\partial \beta_j \partial \eta_j} \\ \frac{\partial^2 \ln L}{\partial \eta_j \partial \alpha_j^2} & \frac{\partial^2 \ln L}{\partial \eta_j \partial \beta_j} & \frac{\partial^2 \ln L}{\partial \eta_j^2} \end{pmatrix} = \begin{pmatrix} \frac{\partial f_1}{\partial \alpha_j} & \frac{\partial f_1}{\partial \beta_j} & \frac{\partial f_1}{\partial \eta_j} \\ \frac{\partial f_2}{\partial \alpha_j} & \frac{\partial f_2}{\partial \beta_j} & \frac{\partial f_2}{\partial \eta_j} \\ \frac{\partial f_3}{\partial \alpha_j} & \frac{\partial f_3}{\partial \beta_j} & \frac{\partial f_3}{\partial \eta_j} \end{pmatrix}$$

Let $\Delta \zeta = \zeta_{k+1} - \zeta_k$, and the iteration Eq. (14) can be transformed into:

$$\begin{cases} J(\zeta_k)(\Delta \zeta_k) = -F(\zeta_k) \\ \zeta_{k+1} = \zeta_k + \Delta \zeta \end{cases} \tag{15}$$

For the each iteration, we just need to solve the linear equation: $\Delta \zeta = -J^{-1}(\zeta_k)F(\zeta_k)$. The termination condition for the iteration is $\| \Delta \zeta \| < \varepsilon$, or we can also set the maximum number of iterations as the termination condition.

4.4   Calculation and Aggregation of Individual Risk Score

The privacy disclosure risk faced by the user i in terms of the information item j equals to the probability of improper setting of this information item j. If the user's privacy attitude and the parameters of the information item are known, the probability of proper setting of information item can be easily calculated in accordance with Eq. (1). Likewise, it is also easy for us to calculate the individual risk score based on Eq. (2). After knowing the risk score of each information item, we can get the privacy score for each user through aggregation of the risk scores of all information items.

5 Experiment and Analysis

In the section, we made some experiments on synthetic data sets and real data sets to verify the effectiveness of our proposed method of scoring. Our experiment environment is as follows. The operating system is Windows 7, and synthetic data is generated by WinGen3 [13], EM algorithm is implemented with JAVA language and the matplotlib of Python plotting package is used.

5.1 Data Set

Now we provide the brief description of the synthetic and real-world data sets used in our experiments. A 256×16 binary matrix for UP (CF. Sect. 3.1) is generated by the Win-Gen tool, among which the rows correspond to the user and the columns correspond to the information items. For every information item j (the total number of j is 16), we have generated an influence α from (0, 2) and an information sensitivity β from (6, 8) randomly. Afterwards, we have generated the privacy attitude of user uniformly from (0, 14). For the information item j of the user i, we have generated the probability p_{ij} with $\eta_j + \frac{1-\eta_j}{1+e^{-\alpha_j(\theta_i - \beta_j)}}$.

The real-world data set Data comes from user profile on social network platforms. We have gathered the privacy settings of user files on Sina Weibo and Tencent Wechat in order to overcome the insufficient information items. 300 Followers of the Weibo account of our research team member have been surveyed to collect their privacy item settings of their own Weibo accounts and their settings of privacy option files of their own Wechat accounts. There are 16 information items in the questionnaire, of which the information items are: "Show my contact information (telephone, mobile phone)? Show my e-mail? Show your income? Friend confirmation? Allow QQ friends recommendation? Find me by QQ number? Public Moments? Allow 10 pictures of my Moments for others? Show my location? Show my input state? Show my QQ Music playlist? Show my QQ game? Show my course dynamics?".

5.2 Analysis of Experimental Results from Synthetic Data Set

The horizontal axis in Fig. 2 means the user's awareness of privacy protection, namely, privacy attitude θ: The greater the value θ is, the more the users care about their privacy and more reluctant the user will be to release the profile items. Thus, $p(\theta)$ value will be greater and the privacy disclosure risk faced by the user is smaller. The smaller the value θ is, the careless user will attach to privacy protection and the smaller $p(\theta)$ value will be. In this case, the probability of setting proper privacy protection is small but will not possibly be 0. A user who does not care about privacy protection will also possibly set proper privacy protection by coincidence. The vertical axis refers to the possibility that the user sets proper information items (expressed in the form of probability).

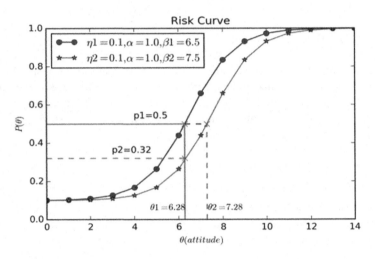

Fig. 2. Probability of different sensitivities (Color figure online)

Under the circumstance that the influence α of information item and the random guess setting η are given, Fig. 2 also compares the sensitivity β of different information items. As mentioned above, β represents the sensitivity of information item, which is an inherent attribute of information item. For users with the same awareness of privacy protection (the same privacy attitude), if the inherent sensitivity of the information item is different (for instance, $\beta_2 > \beta_1$), the information item with greater sensitivity is less likely to be made public, namely, $p_2(= 0.32) < p_1(0.5)$. The red $\times$ in the figure suggests the probability $p_1(= 0.5)$ of proper setting under the same privacy setting; as the sensitivity $\beta_2(= 7.5) > \beta_2(= 6.5)$, the privacy attitude $\theta_2(= 7.28) > \theta_1(= 6.28)$.

Figure 3 illustrates the different influence α of information items with the random guess probability η that the users set proper information items with the help of common sense and the sensitivity β of information item are given. As mentioned in Sect. 3.2, α means the slope of the curve that characterized the probability of setting correctly as the impact of the information items.

When the probability is 0.55, more steeper the curve is (in Fig. 3, the slope of green solid line is greater than that of blue solid line), the greater the impact of the released information item on the information recipient is, namely, $\alpha_2(= 2.0) > \alpha_1(= 1.0)$. When the user's awareness $\theta(= 7.5)$ of privacy protection for information item is higher than the information sensitivity $\beta(= 7.5)$, cautious users are more likely to properly deal with information item with great influence (the right part of the equilibrium point in Fig. 3 where the green line is above the blue line); When the user's awareness $\theta(= 7.5)$ of privacy protection for information item is lower than the

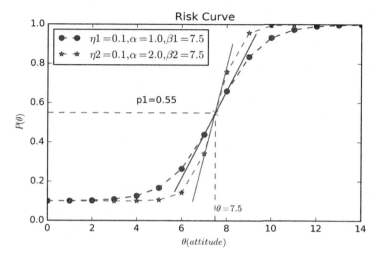

Fig. 3. Probability of different influences (Color figure online)

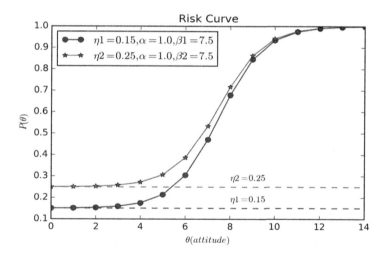

Fig. 4. Probability of different guess (Color figure online)

information sensitivity $\beta (= 7.5)$, careless users are less likely to properly deal with information item with great influence (the left part of the equilibrium point in Fig. 3 where the green line is below the blue line).

Figure 4 demonstrates the probability varies with the random guess probability η that the users set proper information items under the circumstance that the influence α of information item and the information sensitivity β are known. It can be seen from Fig. 4 that the curve's intercept on the vertical axis is greater than 0, suggesting that any user may properly set the information items regardless of his/her awareness of privacy protection, even if this kind of probability is minimal. The greater the value η is, easier the information item is protected ($\eta_2(0.25) > \eta_1(0.15)$). It can be seen from the line close to the origin of coordinates in Fig. 4 that, even if the user has not developed any awareness of privacy protection, the information item will still be disclosed due to its inherent attributes (impact and sensitivity), and the so-called "perfect privacy" actually does not exist. It can be seen from the graph in the upper right corner in Fig. 4 that, when the user has a strong awareness of privacy protection, the possibility that the information items are randomly and properly set nearly has no influence on the probability $p(\theta)$ (the blue and green lines overlap).

5.3   Experimental Analysis on Real-World Data Set

256 valid users are selected from the survey results (excluding the users whose privacy setting of information item is all 0 or all 1). Based on EM algorithm, the curves for 16 information items are obtained.

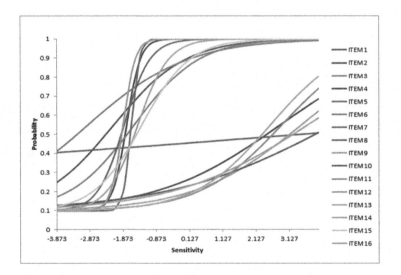

Fig. 5. Characteristic curves of information items (Color figure online)

It can be seen from Fig. 5 that more sensitive the information item is, more closer the curve will get to the lower right corner. With reference to Fig. 6, we can find that the information items 4, 8, 9, 10, 11, 13 and 16 all have sensitivity greater than the threshold value 2, which are the ones the users do not want to release in our experiments. It should be noted that the information item 8 has the highest sensitivity and that its corresponding characteristic curve in Fig. 5 is not perfect.

Figure 6 illustrates the sensitivity of 256 users. The peak points of the curve correspond to the information item number is 4, 8, 9, 10, 11, 13 and 16. The red dotted line is that we specify sensitive threshold. It is worth mentioning that the information item 8 has the highest sensitivity.

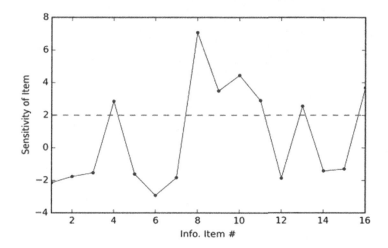

Fig. 6. Sensitivity of information item (Color figure online)

Figure 7 shows the 256 users' privacy score. Most users have a score ranging from 5 to 7. The user No. 240 has the lowest score (2.27 points) and the user No. 79 has the highest score (11.88 points). By checking their privacy settings, we found that they both set the information items with high sensitivity as to be open to the public. We check their privacy settings and find that they all set the information items with high sensitivity as to be open (namely, 1).

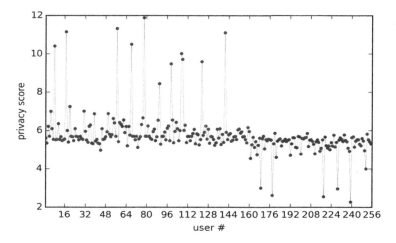

Fig. 7. The users' privacy risk scores

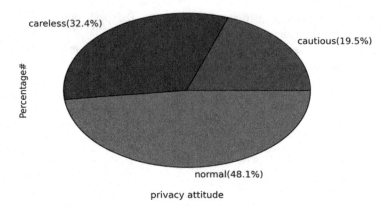

Fig. 8. The privacy attitude of users

Figure 8 demonstrates the statistics of the involved users' privacy attitude. The 256 users' attitudes distributed within $(-1,1)$ are classified in accordance with the threshold values 0.4 and -0.15. Thus, we get 50 cautious users and 83 careless users, which tally with the statistical data in the random street interviews of a talk show "Who Dominates My Privacy?" [16].

6 Conclusions

Our goal is to enhance the user's awareness of privacy protection in mobile-aware settings. By scoring the privacy disclosure risk of user profile, the users are able to select appropriate privacy protection techniques and tools. By adopting the 3-parameter logistic model, we explore information item's sensitivity, influence and probability of random setting as well as the user's potential attitude (how much the user cares about the sensitive information items) to measure the user's privacy disclosure risk. MMLE/EM methods are adopted to estimate the above parameters. Finally, our ideas are verified on synthetic and real-world data sets.

According to the experiments on synthetic data set, the user's privacy attitude obeys normal distribution; but in terms of real-world data set, the followers we have selected do not obey this distribution, which is probably due to the fact that they pay more attention to their own privacy. Nevertheless, our experimental results show that the model parameters estimated based on these assumptions fit well with real-world data.

Acknowledgements. This work was supported by the National High Technology Research and Development Program of China (2013AA014002) and "Strategic Priority Research Program" of the Chinese Academy of Sciences (XDA06030200).

References

1. Fredrikson, M., Lantz, E., Jha, S., Madison, W., Lin, S., Clinic, M., Page, D., Ristenpart, T.: Privacy in pharmacogenetics: an end-to-end case study of personalized warfarin dosing (2014)
2. Lindamood, J., Heatherly, R., Kantarcioglu, M., et al.: Inferring private information using social network data. In: Proceedings of the 18th International Conference on World Wide Web, pp. 1145–1146. ACM (2009)
3. Zheleva, E., Getoor, L.: To join or not to join: the illusion of privacy in social networks with mixed public and private user profiles. In: Proceedings of the 18th International Conference on World Wide Web, pp. 531–540. ACM (2009)
4. Methodology for Privacy Risk Management: How to Implement the Data Protection Act [R/OL], 09 May 2012. http://www.piawatch.eu/node/1539
5. Office of the Privacy Commissioner. Privacy impact assessment guide. Australian Government[R/OL], 16 July 2008. http://www.privacy.org.nz/news-and-publications/guidance-notes/privacy-impact-assesssment-handbook
6. Clifton, C., Kantarcioglu, M., Vaidya, J., et al.: Tools for privacy preserving distributed data mining. ACM SIGKDD Explor. Newsl. 4(2), 28–34 (2002)
7. Liu, K., Terzi, E.: A framework for computing the privacy scores of users in online social networks. ACM Trans. Knowl. Discovery Data (TKDD) 5(1), 6 (2010)
8. Mislove, A., Viswanath, B., Gummadi, K.P., et al.: You are who you know: inferring user profiles in online social networks. In: Proceedings of the Third ACM International Conference on Web Search and Data Mining, pp. 251–260. ACM (2010)
9. Ryu, E., Rong, Y., Li, J., et al.: Curso: protect yourself from curse of attribute inference: a social network privacy-analyzer. In: Proceedings of the ACM SIGMOD Workshop on Databases and Social Networks, pp. 13–18. ACM (2013)
10. Ypma, T.J.: Historical development of the Newton-Raphson method. SIAM Rev. 37(4), 531–551 (1995)
11. Su, H.: Parallel trust-region logistic regression over large scale mobile data. J. Comput. Res. Dev. 414–419 (2010) (in Chinese)
12. Fisher, R.A.: On the mathematical foundations of theoretical statistics. Philos. Trans. Roy. Soc. Lond. Series A Containing Pap. Math. Phys. Charact. 222, 309–368 (1922)
13. http://www.umass.edu/remp/software/simcata/wingen/downloadsF.html
14. Woodruff, D.J., Hanson, B.A.: Estimation of item response models using the EM algorithm for finite mixtures (1996)
15. Zhu, Y., Xiong, L., Verdery, C.: Anonymizing user profiles for personalized web search. In: Proceedings of 19th International Conference on World Wide Web (WWW), pp. 1125–1126 (2010)
16. http://weibo.com/p/1001603866724062655904

Interactive Function Identification Decreasing the Effort of Reverse Engineering

Fatih Kilic[1,2]($\boxtimes$), Hannes Laner[2], and Claudia Eckert[1,2]

[1] Technische Universität München, Munich, Germany
{kilic,eckert}@sec.in.tum.de,
{fatih.kilic,claudia.eckert}@aisec.fraunhofer.de
[2] Fraunhofer AISEC, Garching near Munich, Munich, Germany
hannes.laner@aisec.fraunhofer.de

Abstract. Today's software is growing in size and complexity. Consequently analysing closed-source binaries becomes time-consuming and labour-intensive. In the common use case, the analyst is only interested in specific functions of the given application. Identifying the relevant functions is difficult since no related meta information is given. In this paper we present a framework which speeds up the reverse-engineering process using interactive function identification. We use the benefits of Dynamic Binary Instrumentation as base to collect the executed function calls. We support the analyst in filtering the relevant functions for specific functionality. Our approach is divided into three process steps. Real-time data gathering, user defined information processing/filtering and graphical representation. We show a significant speed up in the reverse engineering process using our framework. We reduce the number of executed functions to be viewed by the analyst more than 90 % and due to visual components we help the analyst pre-selecting the functions on an abstract level.

Keywords: Reverse engineering · Information visualisation · Security · IP protection

1 Introduction

Reverse engineering is a challenging task, requiring time and experience. Analysing given binary executables can be used to find code blocks with certain properties, possible bottlenecks and vulnerable spots of a system or to identify components and their relationships. In order to accomplish this, reverse engineers are forced to look into the assembly of the binaries, since most of the software in use is closed-source.

The main goal of reverse engineering is to determine the functionality of a given binary or to locate a specific functionality inside the executable. This allows modifications or to bypass certain functionality. Usually the gained information about the software behaviour and structure resides in the reverse engineers mind. Therefore, mastering reverse engineering takes a lot of time and is

© Springer International Publishing Switzerland 2016
D. Lin et al. (Eds.): Inscrypt 2015, LNCS 9589, pp. 468–487, 2016.
DOI: 10.1007/978-3-319-38898-4_27

labour-intensive even for an experienced reverse engineer, making reverse engineering a costly activity.

There exist tools for static as well as for dynamic analysis of executables to assist the reverse engineer. Static tools can be disassemblers, string searching tools, signature comparing utilities and others. IDA Pro [1] is an example for a well-known static analysis tool. It is a disassembler, which includes function name resolving of known API-calls and allows the to view the assembly in a flow chart, representing the branches of the code.

Dynamic analysis is mainly performed using debuggers, allowing the reverse engineer to inspect the executed code at runtime, set breakpoints and look at the content of registers at a given execution. Furthermore binaries can be analysed during execution with the help of binary instrumentation, allowing the analyst to dynamically insert additional code into the execution flow of the application. Using dynamic analysis it is possible to extract variables, memory accesses, function calls and more during execution. Also with the already existing tools the reverse engineering process takes a considerable amount of time. Identifying certain functionality requires experience and patience.

We developed a framework to achieve progress in this problem. We use Dynamic Binary Instrumentation (DBI) as base to collect all executed functions, identified by their Virtual Function Address (VFA), inside the target application. The gathered information is saved to a database and is processed by using set operations on the data and represented using our framework. Using graphical visualisation techniques we display the data in a manner to the reverse engineer so that he can deduce the applications behaviour and structure, thus increasing the efficiency of his reverse engineering process.

In summary, our contributions are the following. Our framework allows to

- efficiently log function calls of a target application.
- provide labelling of program states.
- process the gathered data using common set operations.
- reduce the amount of functions the analyst has to check.
- visualise the processed information with highlighting.

The rest of the paper is structured as follows. Section 2 describes previous work done towards our topic. In Sect. 3, we present the design of our framework. Section 4 shows the possibilities of data gathering and presents our solution. Then we discuss the information processing performed by our framework in Sect. 5. The information visualisation is presented in Sect. 6. We describe the implementation of our framework in Sect. 7 and evaluate our framework with a set of experiments in Sect. 8, showing how we identify the location of functionality in selected applications. We conclude in Sect. 9.

2   Related Work

Software visualisation displays software structure, behaviour or evolution using information visualisation. An overview over the existing software visualisation tools is given by Diehl [15]. For example, Rigi [20] displays program structure and interaction. Rigi allows to analyse and document large software systems, when there the source code is available. The information about the system's evolution is visualised as directed graph to represent software modules. SeeSoft [17] can be used to visually represent the evolution of software source code. Eick et al. focused heavily on the software engineering part such as version control and static structure analysis, but also use profilying as a dynamic analysis to round up Seesoft. Reniers et al. [24] present their tool for software maintenance. The toolset is designed to keep track of software structure, metrics and code duplicates. These informations are represented visually. Trinius et al. [25] use visual analysis to quickly identify malware samples and classify the samples according to their behaviour as illustrated by the tool with treemaps and thread graphs. To do so they use a sandbox report and visualise it to the analyst. Using this approach it simplifies the analyst's work of classifying new malware samples into the families of already known malware.

Reverse engineering a given binary is time consuming. Different researchers try to develop tools to make the reverse engineering process more efficient and to speed it up. Quist et al. [23] presented a method using dynamic analysis to visually represent the execution flow of a program (focusing on malware), making the process of reverse engineering easier. They use the Ether hypervisor framework to monitor the execution of the target application and display the data to the reverse engineer in a processed manner highlighting the often executed portions. A visual reverse engineering system was already presented by Conti et al. [12] indicating that visual utilities speed up the work of analysts noticeably. Conti et al. present a way to analyse binary files, allowing the reverse engineer to gain insight into unfamiliar formats and structures.

Determining relevant functions and parts is the main task of the reverse engineer. Our tool is stepping here, providing the reverse engineer additional analysis functionality to visualise and identify interesting functions. For our framework, we use visualisation to display gathered reverse engineering data to speed up the reverse engineers labour.

Previous work on function identification focused on identifying cryptographic algorithms using dynamic analysis or static analysis of the binary. Wang et al. [26] try to identify cryptographic functionality using DBI. The assumption about the programs behaviour that the message is processed after the decryption causes problems when it comes to identifying block ciphers, because they only get processed at the end of the message. Caballero et al. [11] extend the method introduced by Wang et al. and scan for repeatedly called functions. This approach is able to identify more algorithmic procedures, but also leads to false positives in loop-intensive applications. The proposed method of Caballero et al. was further developed by Gröbert et al. [18], who introduced a divide-and-conquer algorithm, analysing parts of the target application's source and merge them back together later on.

DBI is already widely used for performance analysis of all kinds like callgrind [14] for call graphs and cache performance analysis and Dr. Memory [8] to find memory leaks. In general there are different approaches to do DBI. One way is probe-based like Dyninst [10]. In this approach, so called trampolines are added in the executable and when they are executed they jump to the instrumentation instructions. This makes the code not transparent because the original instructions are overwritten with trampolines and are no more in the memory. A program can notice this and so prohibit the reverse engineering. But as an advantage trampolines can be executed fast without much overhead. The more flexible approach for analysis is the jit-based approach which is used by frameworks like DynamoRIO, Pintool and Valgrind [13,14,16]. It means that just before a block of instructions of the original application is executed, it is analysed and new instructions are dynamically inserted. This means that the main work of the frameworks is done during runtime of the analysed application.

We provide reverse engineers a tool, which allows to identify functions of their interest and does not rely on algorithms but on program state classifications to filter the specific VFAs with the interaction of the analyst.

3   Application Design

Our Framework is divided into three components as illustrated in Fig. 1. The *Extractor*-Module of our framework creates a real-time call trace of the target application. The interactive usage of the *Processor*-Module allows the to filter the functions executed in a specific state of the program. The *Visualiser*-Module is responsible for visualising the information returned by the *Processor*-Module.

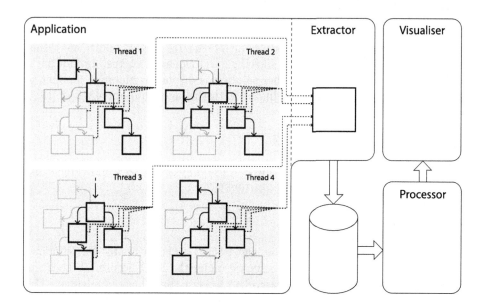

Fig. 1. Framework design

During the execution of the target application our *Extractor*-Module collects data on the loaded modules and the executed functions. By doing so, the *Extractor*-Module creates a call trace, which we want to enrich by a state property. The application will be in different states based on the actions the is triggering. A state in our case could be for example opening a file, establishing a connection or just idling. The analyst is able to set human understandable labels to mark the states that are interesting. This can be also a set of different states for example key press and execute at once by pressing the *Enter* key on the keyboard. By doing so the analyst marks executed functions of the current performed actions of the applications with a label. This label can also be interpreted as a state of the program, in which a set of functions are called to perform a certain action and lead into another state. We allow the analyst to set his labels with the help of our *Extractor*-Module, which adds the label information to every called function. This is performed using DBI. This labelling process is central to our approach, since we want to visualise functions executed in different program states later.

As further step after collecting data of calls made during execution with the respected label, we allow the to set different filters to the collected data. The may specify a concatenation of these filters using common set operators, allowing him to quickly identify functions of program states, find stubs used in various states or display call traces of the threads. This functionality is implemented in our *Processor*-Module.

The *Visualiser*-Module is using the visual perception of human beings to display the results of the *Processor*-Module. It uses different optical elements like colors, shapes and sizes for fast visual recognition of important data. As one solution we use boxes for different functions, displaying the amount of calls for the function performed in the selected state filter. It adds colouring to the functions, representing *Labels*, *Modules* or *ThreadIDs* to give additional visual information to the reverse engineer allowing him to identify interesting code parts inside the executable faster. Another solution is displaying the high amount of results as coloured node graph optimized with the algorithm ForceAtlas2 [19].

4   Information Gathering

In this section we first will discuss different frameworks for DBI and explain the decision for a framework we use in our tool set.

4.1   DynamoRio

The framework supports x86 (32-bit and 64-bit) and is available for Windows and Linux. It is developed as free software under the BSD license. To keep the application code transparent DynamoRIO follows three basic guidelines. The first says to keep as much as possible unchanged from the original application. So if you, for example, count the mov instructions in the program executed in the virtual environment it is very likely that the number is very near to the mov instructions execute in the original application. If it is necessary to change something, the application should not notice the change (second guideline). Finally the third guideline states that DynamoRIO does not make assumptions about the architecture and the operation system besides the minimum needed.

To generate basic blocks, DynamoRIO copies small parts of the executable into the code cache and applies small modifications. This is called copy-and-annotate. To optimize the execution, often sequentially executed basic blocks are combined and put into the trace cache which is separated by the normal basic block code cache. There the trace can be executed as one unit without executing management overhead, which decreases execution time. Also an indirect branch lookup is inlined and so it is possible to execute more basic blocks with indirect branches without doing a context switch back to the Manager. Another optimization DynamoRIO applies is the delay of interrupts. This is important because in contrast to Valgrind, for DynamoRIO it is not easy to determine the current machine context at every point. So if possible the interrupt is delayed to a point where the state of the application is accessible to DynamoRIO. For example if a timer signal is received and the execution is currently in the middle of a basic block in the code cache the application gets the timer interrupt later.

When it comes to the development of a tool for DynamoRIO (called clients), DynamoRIO provides the possibility to change the instructions of a clean C call before inserting it. Of course then the tool developer has to care himself about transparency and can deliberately destroy transparency. To avoid problems of shared libraries, DynamoRIO loads the library used in the instrumentation code separately.

4.2 Pin

Pin is a proprietary framework from Intel which can be used free of charge for non-commercial use. It supports x86 (32-bit and 64-bit), Itanium and ARM architecture. It focuses on an easy to use high level C/C++ API [13]. So Pin follows a call-based model which means you do not insert single instructions, only calls to C/C++ functions. Internally Pin works often like DynamoRIO and tries to improve certain points. In difference to DynamoRIO and Valgrind Pin automatically inlines code for performance optimization (mainly execution time) and takes care about register saving. Also it can be dynamically attached or detached to a program execution.

For optimization reasons Pin uses a Just-in-time-Compiler which directly compiles form the ISA code to the same ISA (for example x86 code to x86 code). During compilation, registers can be re-allocated. For example, if the instrumentation code, which should be included, needs registers also needed by the application, it can be avoided to save and restore registers by re-allocation of registers. This re-allocation must now be handled when going from one basic block (or trace) to another. Pin tries do only the minimal reconciliation needed. To achieve this for every trace entry we remember the register bindings and consider them if we compile a new trace which targets this trace. No reconciliation is needed if we compile a trace which is a target of a single other trace. In this case we just use the binding of the traces targeting the trace currently compiled. Because of this technique, the executed instructions differ much more from the original instructions, compared to DynamoRIO.

4.3   Valgrind

Valgrind is an Open Source framework under the GPL licence and is available for many different ISAs like x86, ARM, PPC, MIPS. To combine the different ISAs, Valgrind uses an intermediate representation (IR). So first the code is translated to the IR and there the instrumentation code can be easily added platform independently. To execute a basic block in the IR the block again has to be translated back to the original ISA. This procedure is split into eight phases and is called disassemble-and-resynthesise.

Valgrind takes more basic blocks together to a superblock, which only has one entry but can have more exits, to reduce the management overhead. So Valgrind is one of the most flexible DBI frameworks which of course also brings some drawbacks demonstrated in the next section [14,22].

4.4   Instrumentation Tool Selection

For our framework we decided to use Intel's Pin [2] for three reasons. Pin can be attached and detached from the target application during execution, which is quite useful to allow the analyst to instrument only the part of the target application he chooses to.

Compared to DynamoRIO, Pin is more stable. Memory intensive 32-bit applications can crash because DynamoRIO has a memory overhead which then could exceed the addressable memory [9,16]. Also in difference to DynamoRIO, Pin chains basic blocks incrementally, which means that at the end of a basic block with an indirect jump, Pin adds new targets dynamically to the chain. Compared to DynamoRIO, which collects one trace and saves it, this approach is more flexible.

Valgrind is a much more comprehensive framework than Pin. Due to its great potential Valgrind looses out when it comes to performance. Regarding performance of an application with instrumentation, Pin is doing really well, compared to Valgrind and DynamoRIO, as stated in the Pin white paper [21].

5   Information Processing

In this section we present the *Processor*-Module of our framework. The gathered data contains the information about every executed function with a specific *Label*, *ThreadID* and *Module* matched to the respective VFA. The *Module* and *ThreadID* for each function is extracted from the application and is saved by the *Extractor*-Module. The *Label* information is set by the analyst using the *Extractor*-Module. A function with VFA x is part of the set label:y if and only if a call to the VFA occurred within the time the specified y as the current program state. Since the labelling is a time dependent property, the values in a set of a specific label are not limited to the specific functions the analyst is looking for. The target application is always executing some functions like updating the Graphical User Interface (GUI) of the application or something alike. Therefore the VFAs for a specific label may also have occurred in other defined labels. This has to be considered during processing.

Trying to identify a specific functionality is quite challenging. During the execution of the process we define our labels of the program states. Figure 2 shows the execution of an application. The x-Axis shows the time in seconds, the y-Axis states the amount of unique function calls within a second span of time. The time where label *connect* was specified is marked by falling hachures (\\), the label *userinput* with rising hachures (//) and label *init* was specified without hachures. The colours indicate the called functions belonging to a label or to an intersection of labels. We are interested in extracting unique calls of functions belonging only to the label *connect* coloured in red. The red colour displays the amount of functions uniquely called during *connect*. As we can see in Fig. 2 during label *connect* there are also function calls, which are also executed in states with other labels. The goal of the *Processor*-Module is to process these interesting VFAs from the gathered data.

The *Processor*-Module provides the possibility to filter the collected functions by *ThreadID*, *Module* and the used *Label*, gathered during the execution. The main goal of this filtering functionality is to determine specific VFA that

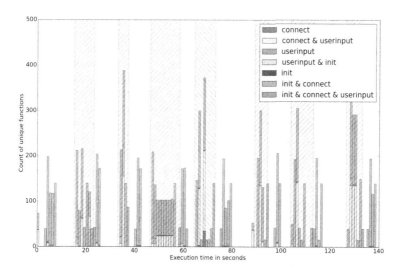

Amount of unique VFAs per second. Colours show the belonging of the function to labels or label intersections. // hachures show the time period where label *userinput* was set by the analyst, \\ the label *connect*, *init* was set otherwise

Fig. 2. Executed functions with label

occurs within specified circumstances (states). To do so we allow the analyst to chain together the filtering options for *Label*, *Module* and *ThreadID* with common operations on sets. These operations are union, intersection and set difference ($\cup$, $\cap$, $\backslash$).

To extract those functions from the gathered data we will use the set operations of our *Processor*-Module. As shown in Fig. 3 after the extracting process we end up with 440 (=39+1+204+196) functions (red label set) called during the label *connect*. To reduce the amount of VFAs to the interesting application parts, the reverse engineer has the possibility to exclude the other two labels using our *Processor*-Module with its set difference operator, ending up with 39 VFAs uniquely called by the functionality he is interested in. Figure 3 shows the number of functions belonging to one or more labels. There are 39 functions, which exclusively belong to the label *connect*, while there are 196 functions called during *connect* as well as *userinput*. 204 functions have been called during all three labels. We can filter for the functions exclusively used in *connect* by using label:connect\label:userinput\label:init. In this example we used three labels. It is possible to use more labels, leading to a more detailed states, which allows the analyst identify further functionality of the target application.

After applying the analyst's filter the *Processor*-Module calculates the call count for each function accordingly from the gathered call trace. The number of calls can help the analyst to understand the internal structure of the target application more quickly. We calculate the number of calls for the VFAs to give the analyst lead towards the function he is looking for. Since the analyst knows

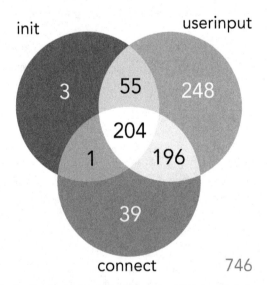

The numbers indicate the exclusive belonging of a VFA to a label or an intersection of labels.

Fig. 3. Venn diagram of executed functions

how often he triggered his action during the program execution, the call count allows him to conclude if the VFA is relevant for him. E.g., if he triggered the connect functionality during the gathering process once he is most likely looking for a function also called once.

6   Graphical Representation

After the analyst specified the filter he wants to apply to the collected function call data in the *Processor*-Module it is presented to him by our *Visualiser*-Module. The main goal of the *Visualiser*-Module is to show the big data collection from the *Extractor*-Module in a manner that the analyst can quickly derive information about the program structure and behaviour by visualising the applied filter.

Since a program under analysis will execute a few hundred calls in a short period of time, common reverse engineers have to rely on their experience where to look for certain functionality. Using our framework after an experimental execution with label setting can speed up this process by reducing the virtual addresses the reverse engineer has to check for a functionality.

We implemented a box view and a graph view to visualise the filter of the analyst. The result set of the VFAs of the applied filter is presented to the analyst by displaying boxes for every VFA as shown in Fig. 4 by default. The VFA itself is not the first thing we want the analyst to notice. We want him to quickly identify functions with a low call count and highlight this property with text size.

Fig. 4. Graphical representation

Displaying the number of calls of a specific VFA gives the analyst a general idea about the structure of the program. VFAs with a smaller count call, especially when filtering for labels, tend to be the function the analyst is looking for. As a second focus after the call count of the individual VFAs we allow the analyst to bind function properties (label, module, thread) to colour the background of the box or the border. This gives the analyst multiple ways of analysing the gathered data and allows him to notice VFA properties visually. We use only two colour informations at the same time to not overload the user with visual effects. Further we have three properties we allow to be visualised with colours.

The analyst removes one by setting his filter. Due to this two colour informations are enough. We provide a colour set as default, but the analyst may change the colours individually. The method of visualisation of the properties is stated in Table 1.

E.g., after applying the filter for *connect* exclusive functions label:connect\ label:userinput\label:init we end up with 39 boxes in our *Visualiser*-Module. Since we used a very heavy label restricting filter we bind the module property of the VFA to the background colour and the thread information to the border colour. This way we can quickly determine the main module responsible for the label we filtered for. We end up with 37 boxes of the module *putty.exe*, which represent the main VFAs for the connect functionality inside the PuTTY application.

Using this individual colouring option allows the analyst to familiarize with the VFAs and their properties on a visual basis, without having to remember properties in numerical form.

If the analyst is interested in a specific thread behaviour he may set the filter to that thread, bind the background color of the boxes to the label property and the border colour to the module. The eye now can catch labels in use for the thread easily, thus recognise the functionalities executed by the various threads.

Analysing modules for their purpose can also be achieved with the help of our *Visualiser*-Module. We just have to set the filter to a specific module. We bind the thread to the background, to identify the threads the module is used in quickly. Further the border colour may be set according to the label, giving us an idea in which program states the module is in use.

On selection of a function box the analyst is provided with additional properties regarding the selected VFA. These properties include lists of labels and threads the VFA occurred, the module and the overall call count of the function (see Table 1).

The purpose of the graph view is to give a quick overview over the target application. It can also help the analyst when the result set of his filter is too large to narrow down his filter to the interesting functionality. In the node graph view the function VFAs are represented by nodes of the tree, the size of the nodes

Table 1. Function details and visual representation

Function property	Value	Visualisation
Function ID	385	–
Function address	0x01351780	Text
Function name	ssh2_setup_pty	–
Module:	putty.exe	Colour
Thread ID(s):	4	Colour
Label:	connect	Colour
Count:	2	Text and Size

visually represents the call count of the function if the analyst chooses to do so, allowing him to find VFAs with a lower call count more quickly. Further we allow the analyst to colour the nodes according to label, module or thread. We only display the call count inside the nodes and provide the additional properties (Table 1) of each function only on selection of the analyst.

Figure 5 shows a representation of all the called VFAs (represented as nodes) related to their calling thread by edges. The rectangle nodes represent the different thread Id's. This allows the analyst to gain insight into VFAs used by one specific thread. For example the functions west of thread 2 were only called by thread 2. Thread 5 has its specific functions displayed north of the node and thread 6 to its east. On the other hand the analyst can quickly determine VFAs, which are used by more threads. In Fig. 5 these would be the nodes between the threads 0, 5, 7, 6 and 2. Thread 0 and 7 only use functions shared with other threads. Such shared VFAs are most likely low-level functions, such as *strlen* or similar. Low-level and shared functions generally are called more frequently than specific functions, therefore we want to visualise the number of calls as a central focus of attention. The call count of VFAs is represented by the number in the nodes, as well as by the node size. The lower the call count the bigger is the node. This helps the analyst determine interesting functions as described in Sect. 5 in this graph more easily. Especially the VFAs with an amount of one or two calls are very promising to represent a certain functionality. From the graph the analyst can see how many different VFAs are used by the threads. Thread 4 calls the most amount of functions. The color of the node represents the module, the function is belonging to. For example, the analyst can derive that thread 3 only uses one module, the dark green one. The orange module is only used by thread 4. Overall Fig. 5 shows the analyst relations between threads, the functions and their modules, but also provides visual highlighting of VFAs with a low amount of calls.

7   Implementation

In this section we describe how we implemented our framework on a machine with an Intel Core i7-4600U CPU 2.10 GHz CPU and 8 GB RAM. We used Windows 7 Professional with Service Pack 1 as Operating System (OS).

7.1   Dynamic Binary Instrumentation

In order to keep track of the called functions as well as logging additional information about the functions and modules we use Intel's Pin Framework (Pin 2.14 kit 71293 E) [2], because of its overall performance and stability described in Sect. 4.4

The *Extractor*-Module is implemented as a Pintool, which has the purpose of saving the execution flow of the target application but also allow us to label certain states of execution.

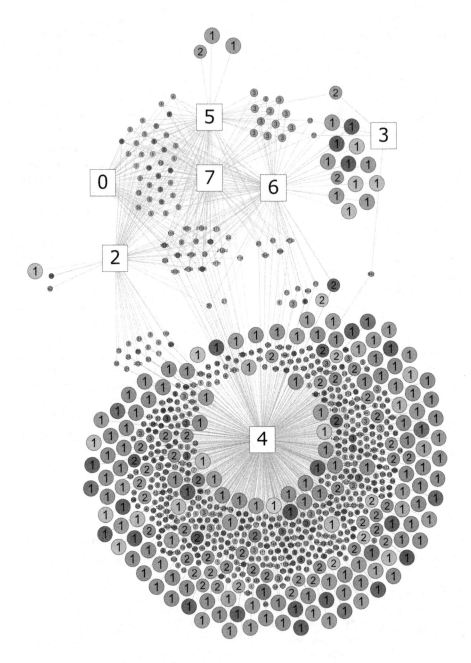

Nodes: VFAs; Rectangle nodes: ThreadIDs; Colours: Module; Number: Call count
Shows belonging of VFA to a module and threads.

Fig. 5. Graphical representation with nodes stating the call count.

The labelling process needs us to allow to interact with the Pintool while it is attached to the target process. To enable a simple communication with the Pintool we print the address of an allocated variable of the Pintool to a file before adding any kind of instrumentation to the target application. This address is then used by our labelling functionality. When the analyst sets a new label we simply open a handle to the target application process and use *WriteProcessMemory* to the passed address. From this point on the *Extractor*-Module will use the new label, with whom is written to the database. The overhead by establishing an IPC we would have even bigger. Thus we write the value into the memory ourselves.

After passing the address of the label variable we add two Instrumentations to the target application: *IMG_AddInstrumentFunction* is called whenever an image is loaded by the target process. Therefore this instrumentation allows us to gather data about the loaded modules and save them to our database. *RTN_AddInstrumentFunction* allows us to insert a function at routine granularity. This way we can enumerate the existing procedures, save the interesting data about these functions and add a further instrumentation using *RTN_InsertCall* to log every call of the function.

7.2 Buffering Data

Keeping track of all the functions called within the target application is a challenging task. Since we also want to instrument very fast and thread intensive applications, the amount of calls in the target application quickly may exceed the amount of calls we can save to our hard disk. If we would pause the process on every function call to write the desired information to our hard disk, we would end up with a very slow reacting binary. The amount of function calls may exceed the number of functions we can save to hard disk in a given time period. We want to buffer the gathered data in memory and write them to hard disk to improve performance. Since we gather data from different threads we have to make sure we are thread safe. The Boost Library [4] includes a single-producer-single-consumer lock-free queue. This means one thread is allowed to push to the queue and one to pop from it without worries about race conditions. We use a *boost::lockfree::spsc_queue* and keep the data in memory. We save objects representing function calls in our queue from multiple threads so we have to use a lock to take fall into the single-producer requirements for the queue. Furthermore we have to pause the target process if our queue is full. Otherwise we would miss some function calls in our execution flow. Writing to the database will be done by an additional thread in the target process. This allows us to push multiple function calls onto memory during program execution without great performance issues. The additional thread will pop multiple strings of gathered data about the function calls from the queue, concatenate them and then store them to the database on hard disk. This way, we lower the performed hard disk accesses. We plotted the time related to the objects of called functions kept in memory for the worst-case-scenario under full CPU usage. As we can see in Fig. 6 a usage of more than two hundred objects will not gather any benefit to the performance, since the time used per object settles down at about 0.02 ms afterwards.

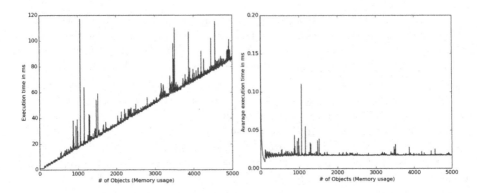

Fig. 6. Time/memory plot for the worst-case-scenario

If the target application exits we pause the process, write the remaining data from the queue to our database and then quit the process.

7.3 Gathered Data

We want to save information about the modules loaded by the target process, like the name and path to the module for recognition purposes and the module base to calculate relative VFAs of functions. Further we are interested in properties of the called functions such as their name, the module they belong to, the VFA and the amount of calls of the function. During execution of the application we monitor the timestamp of a call, the VFA of the currently called function, the ID of the issuing thread and the label specified by the analyst.

All this data is saved to a single SQLite database (SQLite Version 3.8.10.1) [3]. We decided to use SQLite, because it is a lightweight solution. Furthermore SQL allows us to perform queries to evaluate the gathered data very efficiently. The scheme of this database is illustrated in Fig. 7.

When a module is loaded by the process we save it in the table *modules*, storing its name, the path to the module and the module base address. Functions present in the executable are saved with eventual given name and VFA and a

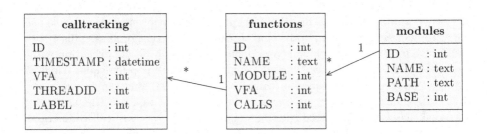

Fig. 7. UML-Diagram of database

reference to a module they belong to. The call count is calculated after the extracting process, in the *Processor*-Module. The table *calltracking* is the main table where we store our call trace. For every function call we insert a new ID, with a timestamp, save the VFA of the called function and add the thread who called the function and reference the label as an integer variable.

7.4 Information Processing

As a next step the analyst can use the *Processor*-Module, which can be used to perform common set operations to the gathered data. Since the data already resides in an SQLite database we use the SQL implementation of the set operations to reach our goal.

Union of two sets of filters is performed by concatenating two select statements, according to the desired filter together, together with the UNION operator. Intersections of sets are performed using the INTERSECT operator and the set difference is calculated by using EXCEPT.

After the gathering process we can make use of the *calltracking*-table of the database to apply filters to the label and the different threads. Since this table quickly has a few hundred thousands of entries we make filtering more effective by creating a temporary copy of the *calltracking*-table, restricting it to unique entries of *VFA*, *ThreadID* and *Label*. Our main output is the table *functions* where the VFA and the count resides. These are the main informations we want the analyst to catch immediately. We also apply module filters to the *functions*-table, since a join of the large *calltracking*-table with the functions table takes an unnecessary amount of time.

We also provide parentheses to be inserted into the filter to gain even more possibilities and lower possible mistakes in longer statements. E.g. the analyst can specify the following operation chaining together multiple set operations as he chooses.

$$((\text{label:key\_Enter} \setminus \text{label:key\_A}) \cap \text{threadid:7})) \cap \text{module:gpg.exe}$$

To provide the use of parentheses in filtering statements we use temporary tables. These are created for any statement between an opening and a closing bracket, parsing the input to a statement without any brackets, removing the innermost once at a time with regular expression search. This is also necessary because SQLite does not support the use of brackets with multiple select statements. Therefore standard SQLite statements could not reach the details we would like. The temporary tables are dropped before a new filter is applied.

7.5 Information Visualisation

After applying the desired filter to the function calls, the analyst quickly wants to determine functions by their VFAs with special attention to functions with a low call count. A low amount of calls is indicating a higher probability of being the function, the analyst is looking for. This is due to the analyst looking for a functionality, which is most likely implemented in a wrapper function. These are called less often than the according helper functions.

The *Visualiser*-Module of our application shows VFAs in boxes. These are implemented as custom widgets in Qt (Qt Version 5.4.1 32-bit), allowing us to bind the border colour or the background colour to the analyst specified preferences. On selection of a box we display the additional information about the VFA in a small pop-up window.

After the database selection we make sure to generate an appropriate amount of colours for the function properties and only show them if the analyst chooses to do so, as we do not want to overload the interface with information the analyst is not interested in.

The *Visualiser*-Module uses the box visualisation as the primary way of representing the functions according to the analysts filter settings. The node graph view as described in Sect. 6 still has to be implemented in future work.

8   Experiments

In this section we present some experiments performed with our framework. We show how we identified specific functions in open source software, explain the labels we set during the execution and provide the results of the performed tests on the framework's performance.

We used as a ground truth open source applications to validate the detected functions in the source file. We selected three applications using security related functions like cryptographic functions or hash functions. We have chosen the well known applications Putty [7], GPG [6] and OpenSSL [5]. To verify our results easier we compiled the test application ourselves with debug information in order to get function names for the found function VFAs.

We present the results of function identification in Putty, GPG and OpenSSL in Table 2. We set labels during execution for Putty trying to identify the functions relevant for establishing a SSH connection. We processed the data with our framework showing only functions that occurred during state *connect* and excluded all the other labels. In GPG we looked for functions responsible for creating an RSA key. OpenSSL is analysed for the relevant functions for the creation of the PEM file of a newly created key.

Step-by-step we are now going through the example looking for PuTTY's SSH connection functionality.

We started in label *init* and changed it to *analystinput* after a few seconds of not interacting with the application. In *analystinput* we typed the analyst name, hit the ENTER key and typed the password for the connection we are going to establish. We used the *init* label between the actions. We change the label to *connect* and hit the ENTER key straight afterwards. After the connection is completed we change back to *init* and perform some more actions within the console like listing the files of the current directory and closing the connection using the *analystinput* label. To identify the functions related to the connect functionality we used the following filter:

$$(\text{label:connect} \setminus \text{label:init} \setminus \text{label:analystinput}) \cap \text{module:putty.exe}$$

Table 2. Experiments

APP-Name	# executed	# filtered	# interesting functions	Relevance	Reduction
Putty	746	37	18 ssh functions	48.6 %	95.04 %
GPG	335	23	8 cipher functions	34.8 %	93.13 %
			6 hash functions	26.1 %	
OpenSSL	735	29	24 cipher/hash functions	82.8 %	96.05 %

The other applications were analysed in a similar manner in order to identify a main functionality.

In Table 2 the columns state the application name, the number of executed function and the number of filtered functions, which we found using an appropriate filter on the application. Further we number the interesting functions representing the functionality we try to identify.

The column *relevance* of Table 2 indicates how much of the filtered functions by our process also are interesting functions a reverse engineer might be looking for. The column *reduction* shows the percentage of the executed functions the analyst does not have to look into when using our framework filtering. With the help of our tool and using good labelling the analyst can reduce his work by more than 90 %, assuming he would check every executed function otherwise.

9   Conclusion

We proposed our framework for interactive function identification decreasing the effort of reverse engineering. We have shown how we speed up the reverse engineering process by using three steps in the analysing process. We used the Pintool to get the data of all executed functions during the test. We used a time memory trade-off to speed up the logging functionally for real-time performance. We processed the data by supporting common set operations. A filter can be applied to labels, threads and module of a function and concatenated together using common set operations to restrict the desired functionality of the binary even further using our *Processor*-Module. We implemented a graphical representation for the output of the huge amount of data to identify the results fast using visual components reducing the functions We used an iterative and interactive approach to give the analyst the possibility to set human readable labels using our *Extractor*-Module. We reduced the number of functions to be checked by the analyst by more than 90 %.

References

1. Hex-Rays IDA. https://hex-rays.com/products/ida/index.shtml. Accessed 10 Aug 2015
2. Intel's pin framework. https://software.intel.com/en-us/articles/pin-a-dynamic-binary-instrumentation-tool. Accessed 10 Aug 2015

3. Sqlite. http://sqlite.org. Accessed 10 Aug 2015
4. Boost   spsc_queue.   http://www.boost.org/doc/libs/1_55_0/doc/html/boost/ lockfree/spsc_queue.html. Accessed 10 Aug 2015
5. Openssl. https://www.openssl.org/. Accessed 10 Aug 2015
6. Gpg. https://www.gnupg.org. Accessed 10 Aug 2015
7. Putty. http://www.chiark.greenend.org.uk/~sgtatham/putty/. Accessed 10 Aug 2015
8. Bruening, D., Zhao, Q.: Practical memory checking with Dr. Memory. In: Proceedings of the 9th Annual IEEE/ACM International Symposium on Code Generation and Optimization, CGO 2011, pp. 213–223. IEEE Computer Society, Washington, DC, USA (2011)
9. Bruening, D., Zhao, Q., Amarasinghe, S.: Transparent dynamic instrumentation. In: Proceedings of the 8th ACM SIGPLAN/SIGOPS Conference on Virtual Execution Environments, VEE 2012, pp. 133–144. ACM, New York (2012)
10. Buck, B., Hollingsworth, J.K.: An API for runtime code patching. Int. J. High Perform. Comput. Appl. **14**(4), 317–329 (2000)
11. Caballero, J., Poosankam, P., Kreibich, C., Dispatcher, S.D.: Enabling active botnet infiltration using automatic protocol reverse-engineering. In: Proceedings of the 16th ACM Conference on Computer and Communications Security, pp. 621–634 (2009)
12. Conti, G., Dean, E., Sinda, M., Sangster, B.: Visual reverse engineering of binary and data files. In: Goodall, J.R., Conti, G., Ma, K.-L. (eds.) VizSec 2008. LNCS, vol. 5210, pp. 1–17. Springer, Heidelberg (2008)
13. Corporation, I.: Pin tools. https://software.intel.com/sites/landingpage/pintool/ docs/62732/Pin/html/. Accessed 10 Aug 2015
14. Developers, V.: Valgrind user manual. http://valgrind.org/docs/manual/manual. html. Accessed 10 Aug 2015
15. Diehl, S.: Software Visualization: Visualizing the Structure, Behaviour, and Evolution of Software. Springer Science and Business Media, Heidelberg (2007)
16. DynamoRIO. Dynamorio API. http://dynamorio.org/docs/. Accessed 10 Aug 2015
17. Eick, S.G., Steffen, J.L., Sumner Jr., E.E.: Seesoft-a tool for visualizing line oriented software statistics. IEEE Trans. Softw. Eng. **18**(11), 957–968 (1992)
18. Gröbert, F., Willems, C., Holz, T.: Automated identification of cryptographic primitives in binary programs. In: Sommer, R., Balzarotti, D., Maier, G. (eds.) RAID 2011. LNCS, vol. 6961, pp. 41–60. Springer, Heidelberg (2011)
19. Jacomy, M., Heymann, S., Venturini, T., Bastian, M.: ForceAtlas2, a continuous graph layout algorithm for handy network visualization. Medialab Center Res. **560** (2011)
20. Kienle, H.M., Müller, H.A.: Rigian environment for software reverse engineering, exploration, visualization, and redocumentation. Sci. Comput. Program. **75**(4), 247–263 (2010)
21. Luk, C.-K., Cohn, R., Muth, R., Patil, H., Klauser, A., Lowney, G., Wallace, S., Reddi, V.J., Hazelwood, K.: Pin: building customized program analysis tools with dynamic instrumentation. In: Proceedings of the ACM SIGPLAN Conference on Programming Language Design and Implementation, PLDI 2005, pp. 190–200. ACM, New York (2005)
22. Nethercote, N., Seward, J.: Valgrind: a framework for heavyweight dynamic binary instrumentation (2007)
23. Quist, D., Liebrock, L.M., et al.: Visualizing compiled executables for malware analysis. In: 6th International Workshop on Visualization for Cyber Security, VizSec 2009, pp. 27–32. IEEE (2009)

24. Reniers, D., Voinea, L., Ersoy, O., Telea, A.: The solid* toolset for software visual analytics of program structure and metrics comprehension: from research prototype to product. Sci. Comput. Program. **79**, 224–240 (2014)
25. Trinius, P., Holz, T., Göbel, J., Freiling, F.C.: Visual analysis of malware behavior using treemaps and thread graphs. In: 6th International Workshop on Visualization for Cyber Security, VizSec 2009, pp. 33–38. IEEE (2009)
26. Wang, R., Wang, X., Zhang, K., Li, Z.: Towards automatic reverse engineering of software security configurations. In: Proceedings of the 15th ACM Conference on Computer and Communications Security, CCS 2008, pp. 245–256. ACM, New York (2008)

Author Index

Printed in the United States
By Bookmasters